Legume Crop Genomics

Editors

Richard F. Wilson
USDA-ARS-NPS
Beltsville, Maryland

H. Thomas Stalker
North Carolina State University
Raleigh, North Carolina

E. Charles Brummer
Iowa State University
Ames, Iowa

AOCS PRESS
Champaign, Illinois

AOCS Mission Statement
To be the global forum for professionals interested in lipids and related materials through the exchange of ideas, information science, and technology.

The paper used in this book is acid-free and falls within the guidelines established to ensure permanence and durability.

Library of Congress Cataloging-in-Publication Data
Legume crop genomics / editors, Richard F. Wilson, H. Thomas Stalker, E. Charles Brummer

p. cm.
ISBN 1-893997-48-0 (alk. paper)
1. Legumes—Genome mapping. 2. Legumes—Genetics. I. Wilson, Richard F., 1947- II. Brummer, E. C. (E. Charles) III. Stalker, H. T. (Harold Thomas), 1950-

SB317.L43L435 2004
633.3'04233—dc22

2004003151

Printed in the United States of America
08 07 06 05 04 5 4 3 2 1

Preface

As we take initial steps into the 21st century, we find ourselves standing at the threshold of the greatest period of scientific discovery yet known to *Homo sapiens* L. As typically happens at the dawn of historically great eras in the advancement of knowledge, this realization comes to most of us with deliberate speed until our attention is captivated by an event of immense significance, such as the "small step for man, giant leap for mankind" on the lunar surface. In a similar vein, the Human Genome Project represents the seminal event that shines light upon the threshold at which we now stand, and signifies man's quest to understand the mechanisms that enable "life" as we know it. The increasingly familiar term for such scientific scrutiny is *genomics*.

The advent of the genomics era did not spring upon us overnight, like some magic beanstalk. Rather, genomics is a dynamic, technology-driven process that may have originated with man's initial efforts to accelerate nature's selection for enhanced competence among biological organisms. Nevertheless, the success of the Human Genome Project has catapulted the evolution of genomics into a more sophisticated science that may now be synonymous with efforts to obtain the primary DNA-oligonucleotide sequence of entire genomes, within nearly every organism on Earth. Yet it is important to understand that generating DNA-sequence information from a multitude of chromosomal fragments is only the beginning of the journey to gain insight to the structure, organization, and function of genes in a given species. Related-species specific strategies must be devised to achieve useful assembly and interpretation of the derived genomic information. In many ways these necessary strategies are akin to tactics for piecing together an intricate jigsaw puzzle. However, as is evident throughout the plant kingdom, the complexity of genomic analysis of legume crops is often elevated by the presence of multiple sets of genomes, where each genomic complement represents a slightly different puzzle. Thus, assembly of legume crop genomes of varying size and complexity requires a high level of well-defined research strategies to ensure proper DNA-sequence analyses.

Rising scientific enthusiasm for genomic investigations among living organisms also creates an extraordinarily competitive environment for appropriations that would finance these rather expensive ventures. Hence, the most successful genomic strategies not only develop and deploy innovative fundamental technologies that help move genomic science forward, but also build research coalitions among scientists to improve the efficiency of genomic investigations among related species. For example, the U.S. Legume Crops Genomics Initiative (USLCGI) was organized under the auspices of the American Soybean Association, United Soybean Board, National Peanut Foundation, USA Dry Pea and Lentil Council, the National Dry Bean Council, and the Alfalfa Council to facilitate communication and cooperation among scientists with an interest in genomic research on soybeans, peanuts, pea and lentils, common bean, alfalfa, and model-legume crops. USLCGI was founded on the premise that the development of an integrated legume genomics research system would enhance the ability to leverage information across legume crops and model species. In keeping with that mission, demonstrated cooperative interaction of this nature not only ensures timely research progress in all legume crops associated with USLCGI, but also enhances the competitive position of USLCGI within the framework of the National Plant Genome Initiative, which is coordinated by the Interagency Working Group on Plant Genomes, Committee of Science, National Science & Technology Council.

With that background, one purpose of this volume is to document the initial research strategies, the development of genomic tools and resources, and the legume-community consensus on the research objectives that will guide the genomic characterization of major legume crops. However, another goal is to present this technical theme in a manner that should help many readers answer the question, "What is genomics?" In that regard, Chapter 2, by Weissinger and Allina, provides a strong foundation for developing an understanding of the scope and of the jargon of genomics. Chapter 3, by Doyle, shows how the many different legume species are related to one another; and Chapters 9 and 10, by Young and by Schlueter *et al.*, delve into the question of how the genomes of major legume crops have evolved. These chapters establish a broad context for the following discussions of how DNA sequences are used to identify genes, to determine gene function, and to reconstruct major portions of gene-rich regions in legume chromosomes. Then, perhaps most important of all, the Chapters 17–19 give insight into some of the practical applications of legume crop genomics. These examples include ways to (i) expedite genetic enhancement of seed-constituent quality in legume crops, (ii) control major pests that limit legume crop productivity, such as soybean cyst nematode, and (iii) abate allergenic response among individuals who may be sensitized to specific proteins in legume crop species. These final chapters should help readers formulate an opinion on the question, "Why is genomic research needed?" In considering that query, we hope this volume has shown that genomics represents a quantum leap in technology; and that the application of this technology in legume-crop enhancement will ensure that U.S. agriculture remains competitive in domestic and global markets for legumes and legume crop products.

Richard F. Wilson
H. Thomas Stalker
E. Charles Brummer

Contents

Chapter 1

Development and Status of the U.S. Legume Crops Genomics Initiative

H. Roger Boerma[a] and Marc Curtis[b]

[a] Center for Soybean Improvement, University of Georgia, Athens, GA 30602-6810;

[b] American Soybean Association, 12125 Woodcrest Executive Dr., Suite 100, St. Louis, MO 63141-5009.

Introduction

Legumes, together with cereals, have been fundamental to the development of modern agriculture. Since the dawn of civilization, many legume species have been instrumental in supplying human food (e.g., soybean, *Glycine max* (L.) Merr.; common bean, *Phaseolus vulgaris* L.; pea, *Pisum sativum* L.; peanut, *Arachis hypogaea* L.; lentil, *Lens culinaris* Medik; and chickpea, *Cicer arietinum* L.), edible oils (peanut, soybean), and animal fodder and forage (alfalfa, *Medicago sativa* L.; and clovers, *Trifolium* spp.) (1). Legumes are second only to grasses in importance for human and animal dietary needs. Worldwide, grain and forage legumes are grown on 12–15% of the arable land (about 180 million ha) (2). They provide 33% of the nutritional nitrogen requirements of humans (3). In 2002, U.S. soybean production was 79 million metric tons and alfalfa production was 74 million metric tons (4, 5). The direct value of U.S. production of soybean and alfalfa in 2002 was $20 billion.

Agriculturally important legumes provide added value through their symbiosis with bacteria that annually fix 40 to 60 million metric tons of atmospheric nitrogen (6). This unique ability of legumes reduces the dependence of farmers on expensive chemical fertilizer, reduces our dependence on petroleum products, and improves soil and water quality. One of the driving forces behind sustainable agriculture and protection of the environment is effective management of nitrogen in farming systems. Intensive farming practiced in developed countries is predicated on using large amounts of nitrogen fertilizer. This practice has led to significant deterioration of water, soil, and air quality. As the world's population approaches 10 billion within the next half century, nitrogen needs for increased crop production will exacerbate current environmental problems. Increased cultivation and productivity of legumes would ameliorate environmental degradation, reduce the depletion of nonrenewable resources, and provide adequate nitrogen for the population (1). Legumes clearly play a significant role in protecting human health, increasing farm profitability, and mitigating environmental problems.

The legume species, pea, was the experimental organism for Mendel's pioneering genetics research. Since the rediscovery of Mendel's findings, crop legumes have been the focus of intensive genetic studies to improve yield, quality, resistance to biotic and abiotic stresses, and to extend the geographic range of adaptation. As a result, certain legume crops have well-studied genetic systems characterized by classical biochemical and physical markers, cytogenetic analysis, chemically induced mutations, and DNA marker–based genome linkage maps. Yet in many of these well-defined systems, comprehensive genetic analysis is limited due to the large size of the genomes of legume crops. Furthermore, few of the basic tools required for modern genome analysis, including polymerase chain reaction (PCR) based DNA markers, expressed sequence tag (EST) databases, or bacterial artificial chromosome (BAC) libraries, have been developed for most legume species. No crop legume has an integrated genetic, physical, and transcript map (7). Furthermore, although efficient transformation has recently been achieved in several of the crop legumes, some of the important crop species remain recalcitrant to transformation (8).

The genomes of most crop legumes are large and relatively complex. For example, soybean is an ancient polyploid and alfalfa an autotetraploid with genome sizes of approximately 1,200 megabases (Mb) and 1,600 Mb, respectively (9). The genome size of cultivated peanut is 2,800 Mb and pea is 4,000 Mb. These larger sizes significantly complicate the development of ordered physical maps of the genome, as well as the identification and location of important genes. Legume genomes contain considerable regions of repetitive DNA, making their complete sequencing financially tenuous at the current cost per finished base. Syntenic relationships within botanical families, however, make it possible to use plant species with much smaller genomes to facilitate understanding of those with large genomes. For example, the recent complete sequencing of the smaller genomes of *Arabidopsis thaliana* (L.) Heyn. (128 Mb) and rice, *Oryza sativa* L. (425 Mb), provide the platform for genome analysis of more complex species such as canola (*Brassica napus* L.), broccoli (*Brassica oleracea* L. var. *italica*), corn (*Zea mays* L.), and wheat (*Triticum aestivum* L. Thell.). Information from the *Arabidopsis* and rice genomes is rapidly being translated across the complex species to enhance disease and pest resistance, yield, and compositional quality of the seed.

To expedite and simplify genome analysis of crop legumes, it has been proposed that parallel analysis of a legume with a smaller genome be considered (10). Recently, studies sponsored by the NSF Plant Genome Program have shown that the barrel medic, *Medicago truncatula* Gaertner, is an ideal candidate for parallel analysis with crop legumes. Barrel medic is a diploid; has a relatively small genome (about 450 Mb); has a rapid generation time; is self-compatible; and appears to have synteny with alfalfa and also to a somewhat lesser degree with pea and soybean (11–13). Comparative analysis with other legume crops would provide additional advantages by complementing genetic knowledge available in the different legume species. For example, soybean is a major crop with significant prior study of genetics and crop and seed physiology. Common bean benefits from relatively well-developed genetic studies and ample polymorphism within the cultigen. Peanut possesses a unique reproductive physiology, which can contribute to a greater understanding of plant reproductive biology.

Development of the Initiative

The U.S. Legume Crops Genomics Initiative (USLCGI) was organized to facilitate communication and cooperation among growers and scientists with an interest in genomics research on U.S. legume crops. This initiative was founded on the premise that the development of an integrated legume genomics research plan would enhance scientists' ability to leverage information across legume crops and model species. In addition, the initiative had as its goal the enhancement of federal funding for genomics research for all U.S. legume crops. During the late 1990s independent groups of legume researchers developed crop-specific genomic plans (14). From a review of these plans, it became clear that achieving the proposed research was an enormous and expensive challenge given the large number of important U.S. legume crops. In addition, few of these legume crops possessed the necessary genetic and genomics resources to effectively utilize existing genomics data from model organisms and the rapidly evolving approaches to whole genome evaluations.

Given the importance of legume crops, their limited genomic and genetic resources, and the inadequate level of research funding, it was decided to convene a workshop of U.S. legume genomic scientists to review the current status of legume genomics research and determine the highest priority research for the major U.S. legume crops (15). In July 2001, 26 legume scientists with knowledge of structural and functional genomics, DNA markers, transformation, bioinformatics, and legume crop improvement were invited to participate in a workshop held at Hunt Valley, Maryland. These scientists came from public and private organizations in 20 states and represented many of the leading geneticists in the United States working with legumes. During this 2-day workshop, scientists reached consensus on six critical areas of genomic research in which progress was needed across all legume species: (i) genome sequencing of strategic legume species, (ii) physical map development and refinement, (iii) functional analysis, (iv) development of DNA markers for comparative mapping and breeding, (v) characterization and utilization of legume biodiversity, and (vi) development of legume data resources (see Table 1.1). At the same time that the critical areas were defined, it was recognized that the level of resource development for the various crop species was significantly different, with no single species being advanced in all areas. Because these six areas encompass a wide research spectrum, it was determined that improved coordination among scientists was critical to achieve significant progress across crop legume species.

Following the Hunt Valley priority-setting workshop, leaders of six U.S. commodity organizations (American Soybean Association, United Soybean Board, National Peanut Foundation, USA Dry Pea and Lentil Council, the National Dry

TABLE 1.1
Priority Genomics Research Goals Established at the 2001 Hunt Valley Workshop

- Genome Sequencing of Strategic Legume Species
    - Sequence gene-rich regions of soybean, common bean, and peanut
    - Produce whole-genome sequencing of barrel medic (*Medicago truncatula*)

- Physical Map Development and Refinement
    - Develop physical maps in peanut and common bean
    - Refine and/or complete physical maps in barrel medic and soybean
    - Develop transcript maps of known genes (ESTs)
    - Integrate maps among taxonomically key species

- Functional Analysis: Transcriptional and Genetic
    - High-throughput stable and transient gene transfer systems
    - Assignment of gene function
    - Expression analysis
    - Proteomics, metabolomics, and metabolic reconstruction
    - Gene knockout systems

- Development of DNA Markers for Comparative Mapping and Breeding
    - Universal set of PCR-based legume Sequence Tag Sites
    - Establishment of local and global similarities of different gene arrangements
    - Quantitative trait loci (QTL) discovery
    - Determination of levels of genome conservation among the legume crops

- Characterization and Utilization of Legume Biodiversity
    - Broadening and refining legume phylogeny
    - Establishing levels of biodiversity within the different legume crops
    - Comparing closely related species for gene discovery
    - Crop–microbe co-evolution
    - Multigene family evolution
    - Domestication of legume crops
    - Preserving and utilizing germplasm

- Development of a Legume Data Resource
    - Development of a legume-wide database (molecular, genetic, expression, diversity, and breeding) publicly available through the Web
    - Leveraging data from each legume species for greater efficiency

Bean Council, and the Alfalfa Council) met to review the newly established research priorities. The grower leaders agreed to accept the scientists' consensus on research priorities and to request formal approval from their grower organizations to work together as a unified group in support of the priority genomics research on legume crops as part of a U.S. Legume Crops Genomics Initiative (15). During the fall of 2002, the outcome of the Hunt Valley Workshop was presented to the various legume crops grower organizations. Each organization voted to accept the genomic research priorities and to participate as a formal member of the planned USLCGI.

Activities of the USLCGI

At the March 2002 USLCGI meeting, the grower leaders and species coordinators (researchers selected at the Hunt Valley workshop to represent each major legume crop) agreed on the formal structure of the USLCGI Steering Committee. The committee consisted of two grower members and one research member for each of the following legume commodities: (1) alfalfa and clovers, (2) dry bean, (3) dry pea and lentil, (4) peanut, and (5) soybean. In addition, one research scientist was selected to represent model legumes. Each commodity polled its researchers to nominate their research representative for the USLCGI Steering Committee. The grower members for each crop were selected by their respective grower organization(s), and the nominated research representatives were approved by their respective grower organization. The current membership of the USLCGI Steering Committee is shown in Table 1.2.

TABLE 1.2
U.S. Legume Crops Genomics Steering Committee Membership and Area of Representation

- Alfalfa and clovers
    - Mr. Mark McCaslin, Forage Genetics, Prior Lake, MN
    - Dr. Charlie Brummer,[a,b] Iowa State University, Ames, IA
    - Open position
- Dry pea and lentil
    - Mr. George Anderson, USA Dry Pea & Lentil Council, Moscow, ID
    - Mr. Todd Scholz, USA Dry Pea & Lentil Council, Moscow, ID
    - Dr. Norman Weeden,[b] Montana State University, Bozeman, MT
- Dry bean
    - Mr. Tom Grebb, Central Bean Co., Inc., Quincy, WA
    - Mr. Steve Antonius, ADM Edible Bean Research, Caldwell, ID
    - Dr. Paul Gepts,[b] University of California, Davis, CA
- Model legumes
    - Dr. Nevin D. Young,[b] University of Minnesota, St. Paul, MN
- Peanut
    - Mr. Joe Boddiford, Georgia Peanut Commission, Tifton, GA
    - Mr. Howard Valentine, American Peaunt Council, Big Canoe, GA
    - Dr. Tom Stalker,[b] North Carolina State University, Raleigh, NC
- Soybean
    - Mr. Jim Stallstrom, United Soybean Board, Winthrop, MN
    - Mr. Marc Curtis,[c] American Soybean Association, Leland, MS
    - Dr. Randy Shoemaker,[b] USDA-ARS, Ames, IA

[a] Vice-chair of the USLCGI Steering Committee.
[b] Research member of the USLCGI Steering Committee.
[c] Chair of the USLCGI Steering Committee.

While agricultural research has been exemplary in linking basic research with the needs of agriculturalists, the USLCGI is innovative in bringing grower leadership and genetic and genomic researchers of U.S. legume crops together to develop the research agenda cooperatively. Historically, the grower leadership of each crop has acted independently of other commodities to move its own genomics research agenda forward. Given the apparent similarity of the genomes of legumes, however, there are opportunities to focus a research agenda that would not only benefit individual legume crops, but would also effectively move forward the science of this important plant family and result in the enhanced sustainability and profitability of U.S. agriculture.

The USLCGI Steering Committee has continued to meet regularly to formulate bylaws, elect its leadership, develop additional plans, and communicate the importance of the USLCGI to Federal funding organizations (Table 1.3). For example, meetings with Under Secretary Joe Jen and Deputy Under Secretary Rodney Brown of the U.S. Department of Agriculture and representatives from the National Science Foundation have served to communicate the purpose of the USLCGI and the high-

TABLE 1.3
Major Activities Related to the Development and Organization of the U.S. Legume Crops Genomics Initiative (USLCGI)

Date	Activity
July 2001	Legume Genomics Workshop and associated grower leader meeting, Hunt Valley, MD
December 2001	Approval of Hunt Valley research priorities by individual grower organizations
January 2002	Presentation of Hunt Valley research priorities in the Legumes Workshop at the Plant & Animal Genome X, San Diego, CA
March 2002	USLCGI organizational meeting of grower leaders and species coordinators, Baltimore, MD
April 2002	Election of USLCGI Steering Committee by grower organizations and researchers
June 2002	USLCGI Steering Committee meeting to initiate development of bylaws and meeting with USDA administrators, Washington, DC
October 2002	USLCGI Steering Committee meeting for development of bylaws, election of chair, vice-chair, and executive committee, and for meeting with NSF administrators, Washington, DC
November 2002	Initial issue of *USLCGI Update* (Diane Bellis, editor), a newsletter to keep USLCGI Steering Committee members, researchers, and grower leaders informed on important legume genomics issues and opportunities
January 2003	USLCGI Steering Committee meeting to plan the NSF Research Coordination Network Grant and development of a Legume Genomics Monograph, San Diego, CA
June 2003	Submitted NSF Research Coordination Network Grant (PIs: T. Stalker, C. Brummer, P. Gepts, R. Shoemaker, N. Weeden, and N. Young)
November 2003	USLCGI Steering Committee meeting to develop legislative strategy and meet with NIH, NSF, and USDA staff, Washington, DC

priority genomics research needs for legume crops. In addition, the Steering Committee has served as the conduit through which research priorities are related to commodity-funding organizations at both the state and national levels. In fact, one can see the effect of the USLCGI in revised genomics plans of each individual legume crop (16).

Because the research members of the USLCGI were nominated by their peers and approved by their respective grower organizations, they have credibility with their fellow researchers and a good working relationship with the grower members on the committee. By partnering researchers with grower leaders, the USLCGI has created broad-based support for legume crops genomics research in many states. Continued communication among commodity leaders of the major U.S. legume crops and the scientific community will be critical for applying basic scientific information to solve real-world problems in legume crop improvement.

The USLCGI has already shown that a diverse group of commodity organizations can work together to achieve a common goal. The steering committee has established its research goals and organizational structure, elected its leadership, developed previously nonexistent lines of communication, and worked together to submit a networking grant to the National Science Foundation (Table 1.3). The final impact of the USLCGI on the coordination and funding of legume crops research has not yet been determined, but the initiative is positioned to benefit society by more efficiently meeting the nutritional needs of humans while protecting the environment and increasing the sustainability and profitability of U.S. agriculture.

References

1. Graham, P.H., and C.P. Vance, Legumes: Importance and Constraints to Greater Use, *Plant Physiol. 131*:872–877 (2003).
2. Food and Agricultural Organization of the United Nations, FAOSTAT Agricultural Data, http://apps.fao.org/page/collections?subset=agriculture (2003).
3. Vance, C.P., P.H. Graham, and D.L. Allan, Biological Nitrogen Fixation: Phosphorus a Critical Future Need, in *Nitrogen Fixation: From Molecules to Crop Productivity*, edited by F.O. Pedrosa, M. Hungria, M.G. Yates, and W.E. Newton, Kluwer Academic Publishers, Dordrecht, The Netherlands, 2000, pp. 506–514.
4. Golbitz, P., *2003 Soya & Oilseed Bluebook*, Soyatech, Inc., Bar Harbor, Maine, 2003, pp. 321–364.
5. United States Department of Agriculture, National Agricultural Statistics Service, Crop production 2002 summary, http://usda.mannlib.cornell.edu/reports/nassr/field/pcp-bban/cropan03.txt (2003).
6. Smil, V., Nitrogen in Crop Production, *Global Biogeochem Cycles 13*:647–662 (1999).
7. VandenBosch, K.A., and G. Stacey, Summaries of Legume Genomics Projects from around the Globe: Community Resources for Crops and Models, *Plant Physiol. 131*:1–26 (2003).
8. Somers, D.A., D.A. Samac, and P.M. Olhoft, Recent Advances in Legume Transformation, *Plant Physiol. 131*:892–899 (2003).
9. Arumuganathan, K., and E.D. Earle, Nuclear DNA Content of Some Important Plant Species, *Plant Mol. Biol. Rep. 9*:211–215 (1991).

10. National Research Council, The National Plant Genomics Initiative: Objectives for 2003–2008, www.nap.edu/books/0309085217/html/ (2002).
11. Endre, G., A. Kereszt, Z. Kevei, S. Mihacea, P. Kaló, and G.B. Kiss, A Receptor Kinase Gene Regulating Symbiotic Nodule Development, *Nature 417*:962–966 (2002).
12. Gualtieri, G., O. Kulikova, E. Limpens, D.J. Kim, D.R. Cook, T. Bisseling, and R. Geurts, Microsynteny between Pea and *Medicago truncatula* in the SYM2 Region, *Plant Mol. Biol. 50*:225–235 (2002).
13. Yan, H.H., J. Mudge, D.J. Kim, D. Larsen, R.C. Shoemaker, D.R. Cook, and N.D. Young, Estimates of Conserved Microsynteny among the Genomes of *Glycine max*, *Medicago truncatula* and *Arabidopsis thaliana*, *Theor. Appl. Genet. 106*:1256–1265 (2003).
14. Boerma, H.R., D. Buxton, M. Kelly, and K. Van Amburg, Soybean Genomics White Paper January 2000, http://129.186.26.94/Genomics/Soybean_Genomics.html (2000).
15. Boerma, H.R., J. St. John, and J. Yezak Molen, U.S. Legume Crops Genomics Workshop White Paper, www.legumes.org/ (2001).
16. Bellis, D., H.R. Boerma, E. Ready, and R. Wilson, Strategic Plan for Soybean Genomics 2003 to 2007, http://129.186.26.94/external_links.html (2003).

Chapter 2

An Introduction To Plant Genomics

Arthur K. Weissinger and Sandra M. Allina

Department of Crop Science, North Carolina State University, Raleigh, NC 27695-7620

What Is Plant Genomics?

Genomic science is a wholly new approach by which we are seeking to understand the blindingly complex structures and interactive processes that generate a biological phenotype. Although it is not uncommon to encounter definitions of genomics that are little more than lists of the analytical tools used in molecular biology research, this new science is far more than such a definition might imply. These technologies are, of course, essential tools with which to pursue important biological questions, but focus on the technology alone has tended to obscure the fundamental significance of this new discipline and the basic paradigm shift that it implies.

The goal of plant genomic science is to develop an integrated body of knowledge about how the plant genome is organized, how transcription of the genome is regulated, and how the resulting gene products interact to create the plant structure, its products, and its activities. Plant genomic science seeks to understand changes in whole-plant phenotypes in terms of identifiable changes occurring at the level of the gene and its transcription products.

Such an ambitious undertaking is made possible by the convergence of molecular biology and the new field of computational science, bioinformatics. Amazingly rapid progress has been made in the development of powerful new molecular procedures, such as molecular mapping, large-scale DNA sequencing, and microarray analysis of mRNAs, that enable scientists to acquire very large amounts of data regarding the structure of the genome and the regulation of its transcription. However, the central feature that differentiates genomic science from earlier attempts to understand the complex relationship between genomic anatomy and metabolic function is its use of computers and complex statistical techniques to record and correlate data derived from multiple diverse sources. Computer analysis of molecular data makes possible a detailed understanding of how the genome is organized. Further, it allows us to understand how the expression of the genome is modulated in various tissues under different environmental and physiological conditions. It also helps us to make sense of the genetic variation that is observed among organisms from different taxonomic groups. Web-based, broadly accessible databases, combined with computer applications for data management and manipulation, are as integral to genomic science as are the molecular technologies that are more commonly associated with it.

Two additional new fields of study are closely related to the study of genomics, although they are outside the scope of this volume. These are proteomics and metabolomics. The total protein complement of a plant is known as the proteome and thus, proteomics is the characterization and quantification of all the proteins in a specific biological pathway, organelle, cell, tissue, organ, or organism. The goal of proteomics is to develop a global understanding of all proteins and their interactions that are involved in the structures and mechanisms of a cell, tissue, or organ. Metabolomics seeks to characterize the myriad of small metabolites that plants produce (the metabolome) and their complex interactions. Its goal is to find measurable physiological states associated with the biological response of a plant to specific conditions in order to elucidate the metabolic products and pathways involved in the response.

Ultimately, the combined results of this research enable us to understand how the information contained in the genetic code operates, through complex regulated systems of transcription, translation, and metabolic function, to produce the phenotype. With this knowledge, we can greatly improve the effectiveness of our efforts to address long-standing problems of crop improvement. We can also use this new knowledge to improve the quality of crop commodities and broaden the applications for which they are used. This offers the possibility of expanding the role of agriculture far beyond the food, feed, and fiber commodities it has traditionally provided, to the production of higher-value products such as pharmaceutical proteins and secondary metabolites.

The Science of Plant Genomics

Genomic science parallels more traditional forms of biological science in that it seeks to understand both the structure and function of the genome. The study of the structure and organization of the genome is referred to as structural genomics, whereas the study of how the genome is expressed is described as functional genomics. The organization of the genome is studied using methods capable of determining its structure with increasing resolution. These methods are capable of producing extremely dense maps of molecular markers and of determining the complete nucleotide sequences of the entire genome. The expression of genes also can be studied by using relatively simple procedures to observe the transcription of a single gene in a tissue taken at a single moment. By using far more sophisticated technology, the dynamic, coordinated transcription of a whole galaxy of genes can be monitored in tissues at different moments in development or under differing environmental or physiological conditions. Ultimately, the mechanisms by which the information carried in the genome is translated into functioning living systems can be studied using an array of different protocols and instruments to observe how proteins and other biological components interact to carry out complex biological functions.

Molecular Markers

Molecular markers are identifiable polymorphic DNA sequences that vary among individuals and/or divergent taxa. They are widely used to study the organization of plant genomes and for the construction of genetic maps. The presence or absence of particular markers and the variation in their chromosomal locations are used to determine the phylogenetic relationships among plant taxa.

Molecular markers are finding broad application in plant breeding, where they are used to "tag" genes that affect phenotypic traits that are difficult to observe directly or ones that are strongly affected by environment. They are also used to identify genes affecting quantitative traits. Finally, molecular markers can facilitate the identification and cloning of genes.

Molecular Marker Technologies

Numerous methods have been developed to exploit DNA polymorphisms for use as practical genetic markers. For example, restriction fragment length polymorphisms (RFLPs) (1) are visualized by treating genomic DNA with restriction endonucleases to cleave the DNA into fragments. Different fragment arrays are produced when DNA from different individuals is treated with the same enzyme. After separation of these fragments by gel electrophoresis, the DNA is transferred to a membrane (blot) and hybridized with a short DNA sequence (probe) that is radio-labeled or labeled with chemiluminescent compounds. The probe binds (hybridizes) only homologous DNA fragments on the blot. When the blot is exposed to photographic film, hybridized DNA is visualized as a band on the film. Because each band represents a small segment of the genomic DNA, it serves as a marker for the chromosomal segment from which it was derived.

Numerous other procedures have been developed to visualize DNA polymorphisms, and some of these are discussed in detail in following chapters. All of these rely on the fact that there are small but detectable differences between the DNA sequences of any two individuals and that these differences tend to increase as the genetic distance between these individuals increases. Some techniques, such as RAPD (randomly amplified polymorphic DNA) and AFLP (amplified fragment length polymorphisms) incorporate polymerase chain reactions (PCR) to enhance sensitivity (2–5).

Microsatellite markers (also known as simple sequence repeats [SSR]) are based on the presence of small tandemly repeated sequences (typically 1 to 5 nucleotides) that occur throughout plant genomes. The number of repeated units in a given segment of DNA often varies between individuals. However, the sequences flanking these microsatellites tend to be highly conserved, and this permits differences in their lengths to be detected by PCR using primers complementary to the flanking regions (6,7).

Most variation among plants occurs as single nucleotide polymorphisms (SNPs), differences in single DNA bases between otherwise homologous sequences in different individuals (8). Several kinds of SNPs have been described. Perhaps the easiest of these to understand are single nucleotide differences occurring between alleles at corresponding loci in two individuals. If these alleles are available as cloned DNA, they can be sequenced in their entirety and their sequences compared to identify one or more nucleotide differences between them. In theory, it would then be possible to select a plant line carrying one or the other of these alleles by isolating that particular locus from each line in question, sequencing the alleles, and comparing the sequences to determine which allele each plant contains. A somewhat simplified version of this can be accomplished by isolating only a part of the corresponding locus from each plant, for example, by PCR amplification of an internal part of the locus. These PCR products could then be sequenced and compared, as they would be for the entire locus.

Because sequencing many genes is too expensive and laborious to be practical, another means of comparison has been developed based on the fact that for a PCR amplification to work properly, the 3' terminal nucleotide of the primer used in the amplification must match exactly to the corresponding nucleotide in the template sequence. It therefore becomes possible to differentiate between two alleles containing a single nucleotide difference by designing a primer that is complementary to a segment of the template but whose 3' terminal nucleotide is complementary with only one of the two possible nucleotides identified at the polymorphic site. Thus, if PCR amplification of one of these alleles is attempted with the primer, amplification occurs. When this same primer is used to PCR the other allele, though, no amplification takes place. This difference can then be detected by simply running the PCR products on an electrophoretic gel to score the plants for presence or absence of the anticipated amplification product, thus identifying which allele of the pair each plant carries.

In practice, the approach discussed in the previous paragraph has been used in a few cases almost exactly as described (9), but very few alleles are actually cloned and thus available for such comparisons. If this were the only method in which SNPs could be used, they would be of very limited utility. Rather, it is possible to do much the same sort of comparison with undefined, but largely homologous, sequences derived from two different individuals by preparing and sequencing cDNAs from different individuals. Large numbers of randomly chosen cDNAs, referred to as expressed sequence tags (EST) can be derived from two or more individuals and then sequenced using high throughput automated sequencers (10). All of the sequences can then be compared by appropriate software to identify pairs of cDNAs with complete sequence homology except for single nucleotide differences between them. Because sequence data is then available for each EST, it is possible to make primers as described previously with which to differentiate between two individuals at the virtual loci represented by the ESTs. Further, analysis of the very large numbers of amplification products can be accomplished using microarray technology instead of electrophoretic gels. This approach permits high-speed multiplexed analysis of many samples and also allows the entire process, including data acquisition and management, to be automated.

As with any other form of molecular marker, it is possible to develop statistical correlations between specific SNPs and genes governing phenotypic traits. These markers then can be used to guide selection for associated genes even when the genes themselves are unknown, or in the case of Quantitative Trait Loci (QTL), if the contribution of any single gene to the phenotype is difficult to measure.

Cleaved amplified polymorphic sequences (CAPS), like several other marker techniques, exploit differences in restriction sites that occur within otherwise homologous DNA sequences among two or more individuals (9,11). In this case, however, known or deduced differences in sequences are used to make primer sets that amplify stretches of sequence surrounding a restriction recognition sequence in which a mutation exists in one individual but not in the other. The amplification products are then digested with the same enzyme, and one is cleaved, whereas the other is not. Again, this technology applies equally well to known, cloned alleles at a single locus and to homologous ESTs derived from two different individuals.

Observation of whole-plant traits is often difficult because of environmental factors, ploidy, dominance relationships among alleles, and epigenetic effects. Molecular markers are unaffected by environmental factors, epigenetic effects, or dominance effects, but their observation can be complicated by differences in ploidy. Further, development of correlations between markers and phenotypic characteristics can be complicated by all of these factors.

It is important to note that it is sometimes impossible to find polymorphic markers for a trait of interest using a specific marker technology, and this may dictate the use of another system that offers better resolution. For example, it may be impossible to find polymorphic RFLP markers correlated with a specific trait, but it may be possible to find useful markers with another marker system, such as AFLP.

Use of Molecular Markers to Construct Genetic Maps

Genetic maps are charts that elucidate the spatial arrangement of genetic elements among and within plant chromosomes. Mapping studies determine and record the linkage relationships among genetic elements and are used both in the study of fundamental biological phenomena and in applied plant breeding programs. For example, a genetic map used in a fundamental study might reveal that certain genes are almost always clustered together in tight linkage groups, even in different species that are distantly related. Such an observation is often consistent with the conclusion that the proteins encoded by the linked genes are functionally interrelated and that this genetic linkage is adaptive. Similarly, a genetic map might be used in a breeding program to facilitate the replacement of a deleterious allele at a genetic locus when that locus is linked to another locus at which a desirable allele is already present. By knowing the linkage distance between the two loci, it is possible to estimate more accurately how many progeny in a segregating population must be examined in order to find individuals in which a recombination event has resulted in the replacement of the deleterious allele.

Traditional genetic maps are derived through statistical analysis of recombination data obtained from segregating progenies. However, a major limitation of traditional mapping procedures is their reliance on observation of whole-plant traits, some of which are strongly affected by environmental factors, and some of which are conditioned by complex genetic interactions. Maps based on DNA molecular markers circumvent many of these difficulties because they are based on the segregation of identifiable polymorphic DNA sequences, rather than on traits that are conditioned by proteins or secondary gene products (12).

Statistical techniques similar to those used for constructing genetic maps from phenotypic data are used to derive maps based on DNA markers. Such maps show the linkage relationships among markers, without reference to any phenotypic traits. It is also possible to develop statistical correlations between molecular markers and observable phenotypic characteristics to produce a map in which both DNA polymorphisms and genes encoding observable traits are organized according to their linkage relationships.

Use of Molecular Markers in Gene Cloning

Molecular markers can be used to facilitate the process of gene cloning. Although it is unlikely that a molecular marker is derived from the DNA in a gene that directly governs a specific trait, markers are often found that are correlated with the trait. This implies that it is located at a site that is physically close to the gene. Ideally, at least two linked markers can be identified that flank the gene encoding the trait. In this case, it is sometimes possible to identify one or a small number of BAC clones (see "Genetic Libraries," later in this chapter) that share sufficient homology with these marker sequences to hybridize with them. This suggests that the gene of interest resides on a chromosomal segment that is near enough to the clone to permit it to be reached by chromosome walking (13).

Marker-Assisted Breeding

Molecular marker technologies are now widely used to improve the efficiency and precision of breeding programs (some examples of which are discussed in subsequent chapters). Molecular markers are particularly useful when it is necessary to select for traits with low heritability. If a molecular marker is tightly linked with an allele at a locus affecting a phenotypic trait of interest, that marker can serve as a proxy for the allele. The marker can be used to monitor the inheritance of the gene(s) affecting the trait, even if it is difficult to score the trait directly. This approach is very useful in breeding for characteristics such as fungal resistance, where direct observation of the trait is greatly complicated by environmental factors that affect development of disease symptoms (14–16).

Molecular markers may be useful for breeding traits governed by numerous genes, each of which makes a small contribution to development of the characteristic. It is notoriously difficult to breed for these quantitative traits because the QTL that govern them are often distributed throughout the genome, and it is difficult to identify individuals in which favorable alleles are present simultaneously at all loci that govern the trait.

Unfortunately, for some traits (such as grain yield), virtually every gene in the organism affects the trait either directly or indirectly. For other traits, a very large number of genes is involved. In both of these cases, the majority of the genes involved are either unknown, or the specific character trait that each affects is difficult to measure. It is in these situations that molecular markers, combined with bioinformatic tools, truly excel.

Statistical correlations can be developed between defined phenotypes and sets of molecular markers. Although no single marker may be correlated absolutely with the phenotype, very strong correlations between several markers and the phenotype indicate that segments of chromosomes harboring the markers are associated with the phenotype. This implies that selection for the presence of these markers selects for increased frequency of linked QTLs affecting the trait. This technique is now being exploited by numerous breeding programs (17–19).

To facilitate the process of identifying markers associated with genes affecting a particular trait, a procedure known as bulk segregate analysis can be used. In this procedure, bulk DNA samples are prepared from each of two plant populations, one of which expresses the trait of interest, the other of which does not. Ideally, both populations are very closely related to one another; in some cases, the populations may represent segregating progenies from crosses of parents with and without the specific trait. Each of these bulk samples contains a random sample of all of the DNA from the plants in the population it represents. Marker techniques such as RAPD or RFLP can then be used to compare the samples. Any polymorphisms observed between the two samples are likely candidates to represent genes involved in the trait of interest, but verification can be accomplished by observing banding patterns of individuals within the bulks with known phenotypes. These DNAs can then be used either to monitor populations over time to aid in selection, or they can be used to facilitate cloning and characterization of genes governing the trait (20).

DNA Sequencing

DNA sequence analysis permits the determination of genome anatomy with the greatest degree of resolution possible because it identifies all of the bases in the sequenced segment of DNA. There are numerous approaches to the determination of DNA sequences, each with a somewhat different goal, and each requiring a slightly different methodology.

The ultimate goal of many plant genomic programs is to acquire the complete sequence of the plant genome. This goal has been accomplished for the model plant, *Arabidopsis* (21); there also are several draft sequences available for rice (22–24). It has been possible to sequence these genomes because of their relatively small size and low genomic complexity. The genome of *Arabidopsis* contains only 125 megabase pairs (Mb), and that of rice contains about 425 Mb. These genomes also contain low levels of repetitive DNA. Conversely, most legumes have relatively large and complex genomes. The pea genome consists of approximately 4,000 Mb, and the genomes of peanut and soybean contain 2,800 Mb and 1,200 Mb, respectively (25). They also contain high levels of repetitive DNA (26,27). It is unlikely that, at this time, the entire genomes of leguminous plants will be sequenced. Instead, the sequencing effort will concentrate on gene-rich areas.

Other programs carry out DNA sequencing in support of molecular marker development, to identify genes and elucidate their structures, to compare the structures of genes with analogous function taken from different plants, to analyze the structure of isolated genes in order to understand how they might need to be altered to make them suitable for use in plant transformation, and to accomplish a host of other applications. Each of these goals dictates a slightly different approach to choosing the particular DNAs that are to be sequenced, the vectors into which they are cloned, and the extent to which the whole sequence must be analyzed.

Genetic Libraries

A genetic library contains a collection of DNA molecules that can represent either a specially chosen subset of the genome of an organism or virtually all of its DNA. Typically, these sequences are integrated into either prokaryotic or eukaryotic vector systems that permit the library to be maintained and reproduced. The choice of the type of library is determined by the applications for which it is intended.

Genomic libraries can theoretically contain the entire genome of an organism. The most inclusive libraries are constructed from genomic DNA that either has been cleaved at random intervals with a restriction endonuclease or has been sheared mechanically and sub-cloned into an appropriate library vector (shotgun libraries). Genomic libraries are typically cloned into yeast artificial chromosome (YAC) (28–30) or bacterial artificial chromosome (BAC) vectors. Libraries prepared using BACs (31–33) are probably the most useful of these because they tend to be more stable, and they also typically contain DNA derived from only a single chromosomal site in the plant from which the library is prepared. These vectors are capable of carrying clones up to several hundred kilobase pairs (kb) in length and so can accommodate enough DNA to encompass numerous genes.

Genomic libraries are typically used to achieve research objectives such as the cloning of entire genes, including their introns and associated regulatory elements. They are especially useful for the recovery and characterization of promoters and other elements needed to achieve appropriate expression of transgenes. Because the DNA segments are usually quite large, they can also be used in certain forms of gene cloning in which chromosome walking is used to locate genes associated with identifiable molecular markers.

Cot-based cloning and sequencing offers a way in which to increase the relative frequency of actively transcribed genes present in a genomic DNA library (34,35). This method employs a technique developed 35 years ago to study the complexity of genomes (36). Briefly, sheared DNA is heat denatured and then is cooled to allow it to reassociate. The rate at which reassociation occurs is dependent upon the frequency at which a given sequence occurs in the genomic DNA. DNAs that reassociate at different rates can then be separated from one another by chromatography. Recently, Peterson *et al.* (34,35) took advantage of this phenomenon to separate the genomic DNA into three discrete fractions: highly repetitive, moderately repetitive, and single/low copy fractions. Because cot-based cloning greatly increases the efficiency with which functional genes can be cloned and sequenced, it makes the sequencing of the genes in large, highly complex plant genomes much more feasible than it would be otherwise.

cDNA libraries contain a copy of only the genomic DNA that is transcribed. That is because cDNA libraries are prepared from cDNA that is prepared by reverse transcription of mRNA isolated either from the whole plant or from a specific tissue or organ at a specific developmental stage, or after a certain type of treatment. In theory, no two cDNA libraries will contain exactly the same DNA sequences, whereas two genomic libraries could contain the same DNA sequences.

cDNA libraries are typically prepared in library vectors based on bacterial plasmids, phagemids, or cosmids that hold much smaller amounts of DNA than BACs or YACs (37,38). This is possible because cDNA clones are typically short (ca. 1 kb or less) because they are equivalent in size to the mRNA from which they are prepared, which in turn is equivalent to, or smaller than, the transcribed portion of the gene from which the RNA is transcribed.

The clones derived from cDNA libraries are often transformed into a host plant in which the product of the mRNA (most often a protein) can be studied or used to create a novel phenotype in the host. cDNA libraries, like genomic libraries, can also be sequenced; however, they provide the sequence only of the genes that are transcribed in a specific tissue under defined conditions.

Expressed sequence tag (EST) libraries are a special type of cDNA library in which at least a partial sequence data has been obtained from each of the clones (39,40). These sequences are then entered into a searchable database, in which it is possible to identify ESTs with homology to known genes or DNA motifs. ESTs identified in this manner can then be obtained for use as probes with which to identify, recover, and ultimately clone the gene or genes of interest. These ordered (identifiable) cDNA clones can also be spotted on substrates for use in microarray studies of gene expression (see "Microarray Analysis" later in this chapter).

Sequencing Methods

Virtually all DNA sequence analysis carried out at present employs some version of the sequence analysis chemistry developed by Sanger (41). In this procedure, replicas of the DNA to be sequenced are produced enzymatically from a single-stranded template representing one of the two strands of the target sequence. The reaction mixture contains DNA primers, which are short, single-stranded DNA molecules that are complementary to a single-stranded target region in either the target DNA or the library vector adjacent to the insert DNA. It also contains a mixture of all four deoxyribonucleotide triphosphates (dNTPs) needed for synthesis of a DNA strand homologous with the target sequence and a small amount of each of the four dideoxynucleotide triphosphates (ddNTPs), molecules that are virtually identical to dNTPs except that they have an H rather than an OH group at their 3' termini. A different fluorescent label is incorporated into each of the four ddNTPs so that the reaction products from each reaction mixture are fluorescently labeled.

Synthesis of a DNA strand complementary to the target sequence is initiated at the 3' terminus of the primer, and dNTPs are incorporated sequentially to form a new complementary DNA strand. Occasionally, a ddNTP is incorporated into the growing molecule in place of the corresponding dNTP, resulting in chain termination because the ddNTP lacks the 3' terminal OH required for chain elongation. The result is a collection of reaction products of various lengths, with each size class containing a homogeneous mixture of products resulting from the termination of chain elongation occurring at the same nucleotide position in the target DNA. When these are separated by size, and the results obtained from each of the four reaction mixtures are combined, it is possible to deduce the identity of the base at each position along the chain based on the fluorescence of the terminal ddNTP on each molecule. The length of each size class is proportional to the distance of the termination from the 5' terminus of the growing molecule, so that the shorter the length of the newly synthesized strand, the shorter the distance between the 3' terminus (the site of chain termination) and the 5' terminus of the molecule.

Robotic devices can be used both to carry out all of the preparatory chemistry and to load electrophoretic instruments (42). Originally, reaction products were separated on polyacrylamide slab gels (41), but these gels ran relatively slowly because of their inability to dissipate heat, which limited run current. In more modern instruments, much greater efficiency and much faster run times are achieved using capillary array electrophoresis (43,44). This method employs an array of capillary tubes in which electrophoresis is carried out to separate reaction products.

Much of the chemistry and handling of the samples is fully automated. By combining modern sequencing instruments and software, it is possible to obtain large amounts of highly accurate sequence data very quickly. These innovations make it possible to obtain sequence data on large numbers of DNA molecules quickly and to automatically compare the sequence data with publicly accessible databases to identify similarities between the newly acquired data and existing published sequences. Numerous organizations, both public and private, offer custom DNA sequencing services. Usually, sequence data are recorded directly to a computer file and may include statistical estimates of the accuracy of the reported data.

Transcriptomics: Studying Expression of the Genome

The transcriptome is the aggregate of all mRNA transcripts present in a cell or tissue at a specific point in time, and the study of these RNAs is sometimes referred to as transcriptomics. The transcriptome represents that portion of the genome that is transcribed in that cell or tissue under the specific conditions present at the time the mRNA is observed. It is intuitively obvious that the transcriptome is different in different cells and tissues in an organism at any given moment and also that it varies over time within a single tissue over the course of its development. This variation exists within the plant, even though it is presumed that all somatic cells are genetically identical, but they express different gene products at various times, depending on the required function of a given tissue and the environmental conditions to which it is exposed.

An important goal of genomic science is to understand how gene expression is modulated within the plant to allow it to react appropriately to various environmental stimuli. Further, the plant needs appropriate amounts of products at the proper time in the course of its life cycle. Understanding this process is of paramount importance if we are to be able to manipulate gene expression in order to improve the agronomic performance and increase the value of crop plants.

The primary approach to understanding gene expression is to study how the transcription of different genes varies among tissues, and within a single tissue under defined environmental and/or physiological conditions. The amount of any specific mRNA in a total RNA sample can be detected by using any of several different methods. These include northern blot analysis, reverse transcription–polymerase chain reaction (RT-PCR), nuclease protection assays (NPA), and microarray analysis. Using these methods, it is possible to detect specific RNAs and to estimate their levels of expression relative to one another. From these data it is possible to deduce the contribution of each of the genes represented in the assay to the phenotype of the test tissue. Implicitly, this provides knowledge of which genes' functions might be altered to modify the function of the tissue or plant to better serve the needs of agriculture.

Northern Blots

Northern blots permit the detection of mRNA transcripts from known genetic sequences (45). Essentially, total cellular RNA is separated by electrophoresis, transferred to a membrane, and hybridized to a radiolabeled probe of a known DNA sequence. If mRNA transcribed from that DNA sequence is in the RNA pool, it will hybridize to the probe and then can be visualized through autoradiography.

The use of northern blots to observe and quantify transcripts is simple, inexpensive, reliable, and easily applicable to any species. However, there are limitations to the usefulness of northern blot analysis. It is typically used to observe the expression pattern of a single gene at a time, but it has proved cumbersome to study the expression patterns of multiple genes, even though they may be coordinately transcribed. Also, RNA blot analysis is not well suited to quantitative studies of transcription.

RT-PCR

RT-PCR is typically reserved for situations in which the RNA species of interest is produced in amounts too small to be detected easily by standard RNA hybridization methods or when only very small amounts of tissue are available for test (46–48). In this procedure, total RNA is reverse transcribed to produce cDNA, resulting in the production of many different cDNAs, which could theoretically represent every mRNA in the original sample.

Following reverse transcription, the cDNAs are subjected to PCR using a pair of primers that are homologous to known sequences in the gene (and transcript) of interest. Typically only a single primer pair is employed. It is also important to note that the PCR conditions for amplification of the target cDNA must be carefully optimized to ensure maximum sensitivity with minimum background. Following PCR, the amplification products are separated according to size by agarose gel electrophoresis, stained with ethidium bromide, and photographed with transmitted UV light.

RT-PCR is limited by the need to know the precise sequence of the transcript(s) of interest. Additionally, the technique is best suited for detection of a single RNA species in each PCR reaction, although it is possible to subject the cDNA product mixture to PCR with multiple primer pairs in different PCR reactions.

Competitive, quantitative RT-PCR can be employed to quantify specific RNA transcripts and detect variation in their expression levels under different experimental conditions, which among other things, opens up the possibility of studying the biological relevance of small expression differences of specific RNAs (47,48). To perform competitive PCR, known quantities of PCR competitors, for example the Competicon™ (Ambion), are added to PCR reactions containing the reverse-transcribed target RNA (cDNA). These competitors should amplify in a like manner to the target cDNA because they include the same primer binding sites and produce a very similar size fragment. Thus, a competition between the target template and the competitor template is produced. A comparison of the ratio of target and competitor PCR products can be used to determine the initial ratio of the templates, and because the amount of competitor was known, the quantity of the target RNA in the initial sample can be calculated.

Nuclease Protection Assays

The nuclease protection assays (NPA) is an extremely sensitive method for the detection, quantification, and mapping of specific RNAs (49). NPA is carried out by hybridizing one or more radiolabeled, single-stranded antisense RNA probes (produced by *in vitro* transcription) to an RNA sample in a small volume of liquid. Although hybridization takes place in a manner analogous to that in northern blots, hybridization is much more efficient in a small volume of liquid than it is when the target RNA is bound to a membrane. The solution can contain up to about 100µg of total or poly(A)-RNA. After hybridization, any remaining probe and unbound sample RNA are removed with a mixture of nucleases that digest unbound probe and target. The nucleases are then inactivated, and the remaining hybridized RNAs (the protected RNA species) are precipitated.

It is also possible to use single-stranded DNA probes in place of the antisense RNA probes, but if this is done, the digestion step must include S1 nuclease. The products remaining after nuclease digestion (the protected RNAs) are separated according size by polyacrylamide gel electrophoresis. The gel is then placed directly in contact with photographic film to acquire an image of the bound RNAs. It is also possible to use probes labeled with nonradioactive label. If such probes are used, the bound probe is transferred from the gel to a membrane after electrophoresis, and the probe is visualized by carrying out a secondary reaction on the membrane. RNA levels can be quantified by comparing the intensity of probe bands with that of either an internal control or a known amount of a specific sense-strand RNA.

NPA is a powerful tool for studying specific RNA transcripts rapidly and inexpensively. Unlike northern blots, this method permits the simultaneous detection of multiple (up to 12) RNA species by introducing multiple probes into the original binding reaction. Quantification is the same as for a single RNA species but requires the use of specific internal or external controls for each species to be measured. The procedure is also significantly more sensitive than northern blotting and is less expensive and time consuming than RT-PCR.

Microarray Analysis

Microarray analysis allows the transcripts of many genes to be detected and quantified simultaneously (50). It is especially useful for observing the relative level of specific transcripts present in a tissue when the plant is exposed to differing environmental or physiological conditions. Importantly, this methodology permits the simultaneous observation of the transcription of numerous sequences. When microarrays are constructed using an ordered collection of cDNAs, they can also be an important tool for the identification and, ultimately, the isolation of genetic sequences. The modulation of the transcription of these genes is associated with defined responses, developmental stages, certain phenotypic traits, etc.

Single-stranded DNAs, typically derived either from cDNA libraries or through direct oligonucleotide synthesis, are immobilized in an array of very small spots on the surface of a glass or plastic slide. This operation is normally carried out with the aid of a robotic device capable of spotting the DNAs onto the slide with great precision. The number of DNAs applied to the slide varies according to the particular system employed, but current technology permits the spotting of about 10,000 DNAs per slide. If the oligonucleotides are synthesized in place on the surface of the slide, it is possible to create as many as 200,000 discrete spots on the slide. Using one state-of-the-art device, it is possible to print 50 slides, each with 10,000 spots, within 6 hours (51).

RNA to be tested against the microarray is extracted from the test sample, and cDNAs are prepared from it by reverse transcription. A fluorescent label is incorporated into the DNA probes during synthesis of the cDNA. This process theoretically produces cDNAs representing each mRNA in direct proportion to the number of copies of the RNA transcript present in the test sample.

These fluorescently labeled cDNAs are then applied to the bound DNA array. If an individual cDNA in the sample contains a significant amount of sequence homology with a single-stranded DNA on the array, it hybridizes with the corresponding homologous sequence. Hybridization frequency, and thus the amount of fluorescent dye bound to a spot on the array, occurs in direct proportion to the number of cDNA molecules with homology to bound DNA, and to the extent of homology between test cDNA and bound sequences. The array is then scanned, resulting in an image of the hybridized array. During the scan, the level of fluorescence given off by each spot in the array can be quantified, providing a measurement of the relative abundance of the RNAs in the original sample that is bound to the array.

Microarray analysis permits the simultaneous observation of the mRNAs transcribed from many genes. Because of this, it is well suited for studies in which the objective is to observe and quantify numerous mRNAs simultaneously. For example, it is possible to observe how the transcription of many genes changes between two tissue sources, each subjected to different environmental stimuli prior to harvest. Such an observation could reveal that certain genes are transcribed more or less in a coordinate manner in response to a change of environment, i.e., their rates of transcription change at the same time in response to the same stimulus, although the direction and magnitude of such change may be dissimilar (52).

When microarrays are constructed using an ordered cDNA array, they can also be an important tool for the isolation of genetic sequences, the modulation of whose transcription is associated with defined responses, developmental stages, certain phenotypic traits, etc. cDNAs that are identified as being homologous with highly represented transcripts isolated from a tissue under specific conditions may encode proteins that are abundant in that tissue. The cDNA can also be used to prepare probes that, in turn, can be used to identify full-length coding sequences present in cDNA or genomic DNA libraries.

Bioinformatics: Interpreting the Findings of Genomic Science

Bioinformatics is a new branch of computational and information science with the goal of organizing and making comprehensible the vast amount of information derived through analysis of the genome, transcriptome, proteome, and metabolome. Because this body of information is so vast and complex, it could easily fall prey to the same fate that has always limited the progress of knowledge: The data could continue to accumulate at great cost and effort, yet not be applied either to further our understanding of fundamental biological processes or to improve our crops. The activity of compiling the information in such a way that a model of a functioning organism is obtained will constitute a tremendous ordering force. The definition of molecular blueprints that describe cellular events will structure the experimental findings and foster quantitative predictive ability.

Simply put, bioinformatics is the fusion of statistical and computer science with molecular biology in order to store, organize, analyze, and integrate the many kinds of data produced through genomic science. Importantly, the advent of bioinformatics marks a paradigm shift in biological science, away from its earlier emphasis on the elucidation of individual metabolic pathways and the description of isolated genetic systems, to the development of a coordinated understanding, and ultimately prediction, of biological functions. The ultimate goal is to enable the construction of models in which all of these data can be synthesized into predictive models that will allow us to understand the multitude of systems and processes that translate the information contained in the genome into the dynamic whole-plant phenotype.

The ability to process vast amounts of genomic data will make it possible to compare whole genomes to look for variation among different species. This makes it possible to study the evolution and phylogenetic relationships among crop species with great precision.

Genomic/Bioinformatic Sites on the World Wide Web

Three major sequence repositories have been established that act as archives for every published DNA sequence, and all are accessible through the World Wide Web. They are the DNA Databank of Japan (DDBJ), maintained by the Japanese National Institute of Genetics; the database of the European Molecular Biology Laboratory (EMBL), maintained by the European Bioinformatics Institute; and GenBank, which is maintained by the National Center for Biotechnology Information (NCBI), in the United States. The information contained in these databases is updated and exchanged regularly. The sequence data contained in these databases have been increasing geometrically for more than 20 years. For example, the GenBank DNA sequence database contained approximately 6.8×10^5 bases in 1982, approximately 8.3×10^7 by 1992, and had grown to about 2.7×10^{10} by 2002. In October 2003, GenBank contained approximately 3.6×10^{10} bases (53).

In order to make this mass of data more manageable, species-specific databases have been developed for several important crops. For example, SoyBase (http://soybase.agron.iastate.edu/) is a genomic database maintained by USDA-ARS and Iowa State University. It attempts to collect and present all available data about *Glycine* spp. It contains a massive amount of information, including molecular maps, identified quantitative trait loci, characterized molecular probes, proteomic and metabolomic data, information about germplasm resources, and a comprehensive list of citations relevant to the genus. A few of the many other plant-specific databases on the Web include the Legume Information System (www.comparative-legumes.org/); the Plant Genome Data & Information Center (www.nal.usda.gov/pgdic/); GrainGenes (http://wheat.pw.usda.gov/index.shtml); and The *Arabidopsis* Information Resource (www.*arabidopsis*.org/).

Transformation: Direct Modification of the Plant Genome

Transformation is the deliberate modification of the plant phenotype by the introduction of foreign genetic material. All transformation systems that are commonly used to alter important crop species share certain key elements. They require that the genetics of the intended alteration to the plant be understood well enough to know what genetic elements must be introduced to effect the desired change. This also implies that these genetic elements are either available or can be obtained. They must then be assembled into transformation vectors—specialized plasmids that carry the gene of interest, a promoter capable of driving expression in the tissue(s) of interest, a terminal poly-adenylation sequence, and other regulatory elements—in a form suitable for introduction into the plant (54). Often, none of these elements are derived from plants, and it is frequently necessary to alter certain aspects of them, such as codon usage in the gene of interest, to allow them to function properly in plants. These synthetic genes are normally carried within a circular plasmid that also contains bacterial sequences required for maintenance of the plasmid in bacteria.

Most transformation systems require the use of cultured tissues as recipient material into which the vector is introduced. Importantly, the cultured tissues must be capable of regeneration to form whole plants, and the resulting plants must be fertile if the transgene is to be of use in a breeding program. Frequently, tissue culture is the limiting step, and it can be the major determinant to efficiency of a transformation system.

All systems of transformation require methods by which the transforming DNA can be introduced into cells of the cultured tissue. Most commonly, this is carried out by genetically altered *Agrobacterium tumefasciens*, which transfers part of its DNA into plant cells as a normal part of its life cycle. Efficient *Agrobacterium*–based transformation systems have been developed for many dicotyledonous, and some monocotyledonouos crop species (55–59).

Alternatively, naked DNA, usually carried as part of a plasmid that is designed for high-copy number replication in bacteria, is introduced directly into plant cells by mechanical means. This is most commonly done by microprojectile bombardment or electroporation of protoplasts, which are cells from which the walls have been removed by treatment with a mixture of enzymes (60–64).

Following integration of the transforming sequences into the genome, transgenic tissues must be identified either through the use of selective agents or though their expression in the tissues of observable markers. Protocols must be available for the regeneration of whole plants from transgenic tissues. Finally, biochemical and/or molecular methods must be available with which to test the transgenic plant for proper expression of the newly introduced transgene.

Genomic science is providing valuable tools and genetic materials that permit the direct manipulation of genes and regulatory elements in order to alter existing plant traits and to produce new traits not normally found in crop species. Transformation can serve as a tool for practical plant improvement by introducing genes into the plant to alter existing traits or to produce wholly new characteristics not normally found in the crop. Ultimately, transgenics could be developed in which major metabolic pathways are altered through the introduction of multiple genetic elements encoding products affecting numerous processes in metabolic pathways.

Transformation is also used as a tool for fundamental research by testing the effects produced when new or modified genetic elements are introduced into the plant genome. In this regard, high-throughput plant transformation is an essential part of genomic science, as are efficient DNA sequencing, microarray analysis, or any of the other methods more traditionally thought of as the tools of genomics.

Theoretically, transformation circumvents difficulties associated with other methods of introducing foreign genes, such as wide or bridging crosses, because only defined genetic elements are transferred without associated genetic material that could have adverse effects on the performance of adapted varieties. It theoretically permits the virtually instantaneous alteration of a genetic trait. Introgressive hybridization, by comparison, can require numerous generations of selective breeding to incorporate a new trait as well as to eliminate deleterious genes that may be tightly linked with desirable ones.

Currently, transformation is limited in most instances to the introduction of only one or a small number of genes, usually affecting only a single metabolic step. Two of the great hopes for genomic science are that it will provide both an enhanced understanding of the mechanisms involved in complex metabolic systems and that it will facilitate the isolation and characterization of the genetic elements needed to manipulate these systems through transformation.

Genomics and Crop Improvement

The results of genomic science allow us to understand variation within evolving populations at the molecular level and are invaluable for understanding and manipulating complex genetic traits. Genomic science is working to help us understand how important characteristics of crop plants are inherited and how a multitude of gene products interact in an integrated fashion to produce the plant phenotype. Ultimately, this will facilitate both conventional breeding and transformation efforts to produce crop varieties with enhanced agronomic performance and commodity value. It will lead to a more complete understanding of the genetic and metabolic mechanisms responsible for important traits and will permit the breeder to manipulate them with greater certainty. This will allow breeding programs to achieve more rapid rates of gain and will result in the development of better adapted, higher yielding crop varieties. Importantly, the results of this science will facilitate the development of value-added products that could expand and stabilize agricultural markets, which ultimately are the economic foundation of society.

All of genomic science will be of little importance to the future of plant improvement unless collaboration is fostered between genomic scientists and applied plant breeders who are able to translate fundamental observations into practical plant improvement protocols. Few people are trained to carry out both fundamental research in molecular biology and applied plant breeding. Clearly, active collaboration between genomic scientists and breeders is necessary to realize the potential this new science has to revolutionize crop improvement.

References

1. Beckman, J.S., and M. Soller, Restriction Fragment Length Polymorphisms and Genetic Improvement of Agricultural Species, *Euphytica 35*:11–124 (1986).
2. Brummer, E.C., J.H. Bouton, and G. Kochert, Analysis of Annual *Medicago* Species Using RAPD Markers, *Genome 38*:362–367 (1995).

3. Vos, P., R. Hogers, M. Bleeker, M. Reijans, T. van de Lee, M. Hornes, A. Frijters, J. Pot, J. Peleman, M. Kuiper, and M. Zabeau, AFLP: A New Technique for DNA Fingerprinting, *Nucleic Acids Res. 23*:4407–4414 (1995).
4. Barcaccia, G., E. Albertini, S. Tavoletti, M. Falcinelli, and F. Veronesi, AFLP Fingerprinting in *Medicago* spp.: Its Development and Application in Linkage Mapping, *Plant Breed. 118*:335–340 (1999).
5. Jenczweski, E., J.M. Prosperi, and J. Ronfort, Differentiation Between Natural and Cultivated Populations of *Medicago sativa* (Leguminosae) from Spain: Analysis with Random Amplified Polymorphic DNA (RAPD) Markers and Comparison to Allozymes, *Mol. Ecol. 8*:1317–1330 (1999).
6. Diwan, N., A.A. Bhagwat, G.B. Bauchan, and P.B. Cregan, Simple Sequence Repeat DNA Markers in Alfalfa and Perennial and Annual *Medicago* Species, *Genome 40*:887–895 (1997).
7. Rongwen, J., M.S. Akkaya, A.A. Bhagwat, U. Lavi, and P.B. Cregan, The Use of Microsatellite DNA Markers for Soybean Genotype Identification, *Theor. Appl. Genet. 90*:43–48 (1995).
8. Kanazin, V., H. Talbert, D. See, P. DeCamp, E. Nevo, and T. Blake, Discovery and Assay of Single-nucleotide Polymorphisms in Barley (*Hordeum vulgare*), *Plant Mol. Biol. 48*:529–537 (2002).
9. Paris, M., M.G.K. Jones, and J.K. Eglinton, Genotyping Single Nucleotide Polymorphisms for Selection of Barley β-amylase Alleles, *Plant Mol. Biol. Rep. 20*:149–159 (2002).
10. Schmid, K.J., T.R. Sörensen, R. Stracke, O. Törjék, T. Altmann, T. Mitchell-Olds, and B. Weisshaar, Large-scale Identification and Analysis of Genome-wide Single-nucleotide Polymophisms for Mapping in *Arabidopsis thaliana*, *Genome Res. 13*:1250–1257 (2003).
11. Bäumler, S., F.G. Felsenstein, and G. Schwarz, CAPS and DHPLC Analysis of a Single Nucleotide Polymorphism in the Cytochrome *b* Gene Conferring Resistance to Strobilurins in Field Isolates of *Blumeria graminis* f. sp. *Hordei*, *J. Phytopathol. 151*:149–152 (2003).
12. Cregan, P.B., T. Jarvik, A.L. Bush, R.C. Shoemaker, K.G. Lark, A.L. Kahler, N. Kaya, T.T. VanToai, D.G. Lohnes, J. Chung, and J.E. Specht, An Integrated Genetic Linkage Map of the Soybean Genome, *Crop Sci. 39*:1464–1490 (1999).
13. Spiegelman, J.I., M.N. Mindrinos, C. Fankhauser, D. Richards, J. Lutes, J. Chory, and P.J. Oefner, Cloning of the *Arabidopsis* RSF1 Gene by Using a Mapping Strategy Based on High-density DNA Arrays and Denaturing High-performance Liquid Chromatography, *Plant Cell 12*:2485–2498 (2000).
14. Dekkers, J.C., and F. Hospital, The Use of Molecular Genetics in the Improvement of Agricultural Populations, *Nat. Rev. Genet. 3*:22–32 (2002).
15. Horejsi, T., J.E. Staub, and C. Thomas, Linkage of Random Amplified Polymorphic DNA Markers to Downy Mildew Resistance in Cucumber (*Cucumis sativus* L.), *Euphytica 115*:105–113 (2000).
16. Kota, R., M. Akbari, and B. Raymond Lyon, Identification of Molecular Markers Linked with Enhanced Tolerance to Verticillium Wilt in Cotton (*Gossypium hirsutum* L.). *Plant and Animal Genome VI Conference*, www.intl-pag.org/6/abstracts/158.html (1998).
17. Fulton, T.M., T. Beck-Bunn, D. Emmatty, Y. Eshad, J. Lopez, V. Petiard, J. Uhig, D. Zamir, and S.D. Tanksley, QTL Analysis of an Advanced Backcross of *Lycopersicon pe-*

ruvianum to the Cultivated Tomato and Comparisons with QTLs Found in Other Wild Species, *Theor. Appl. Genet. 95*:881–894 (1997).
18. Romagosa, I., F. Han, S.E. Ullrich, P.M. Hayes, and D.M. Wesenberg, Verification of Yield QTL Through Realized Molecular Marker-assisted Selection Responses in a Barley Cross, *Mol. Breed. 5*:143–152 (1999).
19. Foolad, M.R., L.P. Zhang, A.A. Khan, D. Niño-Liu, and G.Y. Lin, Identification of QTLs for Early Blight (*Alternaria solani*) Resistance in Tomato Using Backcross Populations of a *Lycopersicon esculentum* × *L. hirsutum* cross, *Theor. Appl. Genet. 104*:945–958 (2002).
20. McClean, P., *Specialized Mapping Tools*, www.cc.ndsu.nodak.edu/instruct/mcclean/plsc731/mapping/mapping4.htm (1998).
21. The *Arabidopsis* Genome Initiative, Analysis of the Genome Sequence of the Flowering Plant *Arabidopsis thaliana*, *Nature 408*:796–815 (2000).
22. Goff, S.A., D. Ricke, T.H. Lan, G. Presting, R. Wang, M. Dunn, J. Glazebrook, A. Sessions, P. Oeller, H. Varma, D. Hadley, D. Hutchison, C. Martin, F. Katagiri, B. M. Lange, T. Moughamer, Y. Xia, P. Budworth, J. Zhong, T. Miguel, U. Paszkowski, S. Zhang, M. Colbert, W.L. Sun, L. Chen, B. Cooper, S. Park, T.C. Wood, L. Mao, P. Quail, R. Wing, R. Dean, Y. Yu, A. Zharkikh, R. Shen, S. Sahasrabudhe, A. Thomas, R. Cannings, A. Gutin, D. Pruss, J. Reid, S. Tavtigian, J. Mitchell, G. Eldredge, T. Scholl, R.M. Miller, S. Bhatnagar, N. Adey, T. Rubano, N. Tusneem, R. Robinson, J. Feldhaus, T. Macalma, A. Oliphant, and S. Briggs, A Draft Sequence of the Rice Genome (*Oryza sativa* L. ssp. *japonica*). *Science 296*:92–100 (2002).
23. Yu, J., S. Hu, J. Wang, G.K. Wong, S. Li, B. Liu, Y. Deng, L. Dai, Y. Zhou, X. Zhang, M. Cao, J. Liu, J. Sun, J. Tang, Y. Chen, X. Huang, W. Lin, C. Ye, W. Tong, L. Cong, J. Geng, Y. Han, L. Li, W. Li, G. Hu, X. Huang, W. Li, J. Li, Z. Liu, L. Li, J. Liu, Q. Qi, J. Liu, L. Li, T. Li, X. Wang, H. Lu, T. Wu, M. Zhu, P. Ni, H. Han, W. Dong, X. Ren, X. Feng, P. Cui, X. Li, H. Wang, X. Xu, W. Zhai, Z. Xu, J. Zhang, S. He, J. Zhang, J. Xu, K. Zhang, X. Zheng, J. Dong, W. Zeng, L. Tao, J. Ye, J. Tan, X. Ren, X. Chen, J. He, D. Liu, W. Tian, C. Tian, H. Xia, Q. Bao, G. Li, H. Gao, T. Cao, J. Wang, W. Zhao, P. Li, W. Chen, X. Wang, Y. Zhang, J. Hu, J. Wang, S. Liu, J. Yang, G. Zhang, Y. Xiong, Z. Li, L. Mao, C. Zhou, Z. Zhu, R. Chen, B. Hao, W. Zheng, S. Chen, W. Guo, G. Li, S. Liu, M. Tao, J. Wang, L. Zhu, L. Yuan, and H. Yang, A Draft Sequence of the Rice Genome (*Oryza sativa* L. ssp. *indica*), *Science 296*:79–92 (2002).
24. Barry, G.F., The Use of the Monsanto Draft Rice Genome Sequence in Research, *Plant Physiol. 125*:1164–1165 (2001).
25. Bennett, M.D., and I.J. Leitch, Nuclear DNA Amounts in Angiosperms—583 New Estimates. *Ann. Bot. 80*:169–196 (1997).
26. Goldberg, R.B., DNA Sequence Organization in the Soybean Plant, *Biochem. Genet. 16*:45–68 (1978).
27. Gurley, W.B., A.G. Hepburn, and J.L. Key, Sequence Organization of the Soybean Genome, *Biochemica Biophysica Acta 561*:167–183 (1979).
28. Grill, E., and C. Somerville, Construction and Characterization of a Yeast Artificial Chromosome Library of *Arabidopsis* Which is Suitable for Chromosome Walking, *Mol. Gen. Genet. 226*:484–490 (1991).
29. Ward, E.R., and G.C. Jen, Isolation of Single-copy-sequence Clones from a Yeast Artificial Chromosome Library of Randomly-sheared *Arabidopsis thaliana* DNA, *Plant Mol. Biol. 14*:561–568 (1990).

30. Dunford, R., L. Vilageliu, and G. Moore, Stabilization of a Yeast Artificial Chromosome Containing Plant DNA Using a Recombination-deficient Host, *Plant Mol. Biol.* *21*:1187–1189 (1993).
31. Shizuya, H., B. Birren, U.-J. Kim, V. Mancino, T. Slepak, Y. Tachiiri, and M. Simon, Cloning and Stable Maintenance of 300-kilobase-pair Fragments of Human DNA in *Escherichia coli* Using an F-factor-based Vector. *Proc. Natl. Acad. Sci. USA* *89*:8794–8797 (1992).
32. Wang, G.-L., T. E. Holsten, W.-Y. Song, H.-P. Wang, and P.C. Ronald, Construction of a Rice Bacterial Artificial Chromosome Library and Identification of Clones Linked to the Xa-21 Disease Resistance Locus, *Plant J. 7*:525–533 (1995).
33. Cregan, P.B., J. Mudge, E.W. Fickus, L.F. Marek, D. Danesh, R. Denny, R.C. Shoemaker, B.F. Matthews, T. Jarvik, and N.D. Young, Targeted Isolation of Simple Sequence Repeat Markers Through the Use of Bacterial Artificial Chromosomes, *Theor. Appl. Genet.* *98*:919–928 (1999).
34. Peterson, D.G., S.R. Wessler, and A.H. Paterson, Efficient Capture of Unique Sequences from Eukaryotic Genomes, *Trends Genet. 18*:547–550 (2002).
35. Peterson, D.G., S.R. Schulze, E.B. Sciara, S.A. Lee, J.E. Bowers, A. Nagel, N. Jiang, D.C. Tibbitts, S.R. Wessler, and A.H. Paterson, Integration of Cot Analysis, DNA Cloning, and High-Throughput Sequencing Facilities Genome Characterization and Gene Discovery, *Genome Res. 12*:795–807 (2002).
36. Britten, R.J., and D.E. Kohne, Repeated Sequences in DNA, *Science 161*:529–540 (1968).
37. Schuler, G.D., Sequence Mapping by Electronic PCR. *Genome Res. 7*:541–550 (1997).
38. Heidecker, G., and J. Messing, Sequence Analysis of Zein cDNAs Obtained by an Efficient mRNA Cloning Method, *Nucleic Acids Res. 11*:4891–4906 (1983).
39. Delseny, M., M. Raynal, M. Laudie, F. Varoquaux, P. Comella, H.J. Wu, R. Cooke, and F. Grellet, Sequencing and Mapping the *Arabidopsis* Genome: A Weed Model for Real Crops, *Symp. Soc. Exp. Biol. 50*:5–9 (1996).
40. Van der Hoeven, R., C. Ronning, J. Giovannoni, G. Martin, and S. Tanksley, Deductions About the Number, Organization, and Evolution of Genes in the Tomato Genome Based on Analysis of a Large Expressed Sequence Tag Collection and Selective Genomic Sequencing, *Plant Cell 14*:1441–1456 (2002).
41. Sanger, F., S. Nicklen, and A.R. Coulson, DNA Sequencing with Chain-terminating Inhibitors, *Proc. Natl. Acad. Sci. USA 74*:5463–5467 (1977).
42. Swerdlow, H., B.J. Jones, and C.T. Wittwer, Fully Automated DNA Reaction and Analysis in a Fluidic Capillary Instrument, *Anal. Chem. 69*:848–855 (1997).
43. Huang, X.C., M.A. Quesada, and R.A. Mathies, DNA Sequencing Using Capillary Array Electrophoresis. *Anal. Chem. 64*:2149–2154 (1992).
44. Marziali, A., and M. Akeson, New DNA Sequencing Methods. *Ann. Rev. Biomed. Eng.* *3*:195–223 (2001).
45. Alwine, J.C., D.J. Kemp, and G.R. Stark, Method for Detection of Specific mRNAs in Agarose Gels by Transfer to Diazobenzyloxymethyl-paper and Hybridization with DNA Probes, *Proc. Natl. Acad. Sci. USA 74*:5350–5354 (1977).
46. Lynas C., S.D. Cook, K.A. Laycock, J.W. Bradfield, and N.J. Maitland, Detection of Latent Virus mRNA in Tissues Using the Polymerase Chain Reaction, *J. Pathol.* *157*:285–289 (1989).
47. Ohan, N.W., and J.J. Heikkila, Reverse Transcription-polymerase Chain Reaction: An Overview of the Technique and its Applications, *Biotechnol. Adv. 11*:13–29 (1993).
48. Ambion, *RT-PCR: The Basics*, www.ambion.com/techlib/basics/rtpcr/index.html (2003).

49. Ambion, *The Basics: What is a Nuclease Protection Assay?*, www.ambion.com/techlib/basics/npa/ (2003).
50. Thibaud-Nissen, F., R.T. Shealy, A. Khanna, and L.O. Vodkin, Clustering of Microarray Data Reveals Transcript Patterns Associated with Somatic Embryogenesis in Soybean, *Plant Physiol. 132*:118–136 (2003).
51. TeleChem International, Inc., http://arrayit.com/ (2003).
52. Maguire, T.L., S. Gimmond, A. Forrest, I. Iturbe-Ormaetxe, K. Meksem, and P. Gresshoff, Tissue-specific Gene Expression in Soybean (*Glycine max*) Detected by cDNA Microarray Analysis, *J. Plant Physiol. 159*:1361–1374 (2002).
53. GenBank, Regular Release, NIAS DNA Bank, www.dna.affrc.go.jp/htdocs/growth/gb_tbl.html (2003).
54. Bevan, M., Binary Agrobacterium Vectors for Plant Transformation, *Nucleic Acids Res. 26*:8711–21 (1984).
55. Barton, K.A., A.N. Binns, A.J. Matzke, and M.D. Chilton, Regeneration of Intact Tobacco Plants Containing Full Length Copies of Genetically Engineered T-DNA to R1 Progeny, *Cell 32*:1033–1043 (1983).
56. Fraley, R.T., S.G Rogers, R.B. Horsch, P.R. Sanders, J.S. Flick, S.P. Adams, M.L. Bittner, L.A. Brand, C.L. Fink, J.S. Fry, G.R. Galluppi, S.B Goldberg, N.L. Hoffman, and S.C. Woo, Expression of Bacterial Genes in Plant Cells, *Proc. Natl. Acad. Sci. USA 83*:4803–4807 (1983).
57. Tyagi, A.K., and A. Mohanty, Rice Transformation for Crop Improvement and Functional Genomics, *Plant Sci. 158*:1–18 (2000).
58. Vain, P., J. Buyser, V. Bui Trang, R. Haicour, and Y. Henry, Foreign Gene Delivery into Monocotyledonous Species, *Biotechnol. Adv. 13*:653–671(1995).
59. Sharma, K.K., and V.V. Anjaiah, An Efficient Method for the Production of Transgenic Plants of Peanut (*Arachis hypogaea* L.) Through *Agrobacterium tumefaciens*-mediated Genetic Transformation, *Plant Sci. 159*:7–19 (2000).
60. Tomes, D.T., A.K., Weissinger, M. Ross, B.J., Higgins, S. Schaaf, J. Malone-Schoneberg, M. Staebell, P. Flynn, J. Anderson, and J. Howard, Transgenic Tobacco Plants and Their Progeny Derived by Microprojectile Bombardment of Tobacco Leaves, *Plant Mol. Biol. 14*:261–268 (1990).
61. Klein, T.M., M. Fromm, A. Weissinger, D. Tomes, S. Schaaf, M. Sletten, and J.C. Stafford, Transfer of Foreign Genes into Intact Maize Cells with High-velocity Microprojectiles, *Proc. Natl. Acad. Sci. USA 85*:4305–4309 (1988).
62. Ozias-Akins, P., J.A. Schnall, W.F. Anderson, C. Singsit, T.E. Clemente, M.J. Adang, and A.K. Weissinger, Regeneration of Transgenic Peanut Plants from Stable Transformed Embryogenic Callus, *Plant Sci. 93*:185–194 (1993).
63. Davey, M.R., E.L. Rech, and B.J. Mulligan, Direct DNA Transfer to Plant Cells. *Plant Mol. Biol. 13*:273–285 (1989).
64. Somers, D.A., D.A. Samac and P.M. Olhoft, Recent Advances in Legume Transformation, *Plant Physiol. 131*:892–899 (2003).

Chapter 3

Legume Phylogeny: Context for the Family, the Major Groups, and the Relationships of Crop and Model Species

Jeff J. Doyle

Department of Plant Biology, Cornell University, Ithaca, NY 14853

Introduction

Leguminosae (or Fabaceae) is the third largest family of flowering plants, after Orchidaceae (the orchid family) and Asteraceae (Compositae; the sunflower family), with over 600 genera and 20,000 species. It is an extraordinarily diverse family of worldwide distribution, with representatives occurring from alpine and arctic regions to the equatorial tropics. Legumes include herbaceous plants, such as the temperate to subtropical crop species pea, vetch, soybean, and common bean, and also large, woody lianas, such as *Wisteria* and 100-meter tall trees of tropical forests.

Members of the Leguminosae are characterized by the familiar fruit, technically termed a legume, that gives the family its name. The variability in legume fruits is a good example of the diversity of the family. As Cronquist (1) puts it, the fruit of Leguminosae is "commonly dry and dehiscent along both sutures (i.e., a typical legume) but sometimes indehiscent (and then sometimes winged) or breaking transversely into 1-seeded joints." Thus, there are many members of the family whose fruits would not be recognized readily as legumes by most people. What may be more surprising is that there are many plants that most people would never think of as being related to peas or soybeans but whose unmistakable legume fruits reveal them to be members of Leguminosae. In part this is because of the diversity in habit (herb, shrub, vine, tree), but also because the flowers of legumes are highly variable. Mention a legume flower and most people will visualize a butterfly-like (papilionoid) flower, with its keel, wings, and standard, whether large and showy like that of a pea or much smaller like that of a soybean. Indeed this type of flower is found in the majority of species in the family, but these species are all part of a single lineage among the many that possess the legume fruit. Other members of the family have variously modified flowers, including flowers that look much less like that of a pea than like that of an old-fashioned rose, radially symmetrical and with an open aspect and five similar looking petals. The mimosoid lineage of legumes has flowers that are often individually very small, aggregated, and with numerous colored stamens that form the attraction for insect visitors. But both of these groups, with their nonpapilionoid flowers, produce legume fruits and are members of the single evolutionary lineage called Leguminosae.

The purpose of this chapter is to describe what is known about the phylogenetic relationships of the legume family in order to provide a context for understanding its incredible diversity. The literature on legume phylogenetics is rich, with an increasing number of molecular systematics studies at all levels, but it is particularly aimed at elucidating relationships among and within major divisions of the family, such as subfamilies and tribes (groups of genera). The seminal series *Advances in Legume Systematics* has featured several volumes on legume phylogeny at these levels (2–5). As is true of the field of systematics as a whole, the pace of phylogenetic discovery in Leguminosae has been accelerated by the increasing availability of molecular data, and many of the studies described in these volumes involve gene phylogenies. Again as elsewhere in plant systematics, the trend has been from phylogenies based on single genes, usually the chloroplast gene *rbcL* (6,7) to studies using more than one sequence, often involving both nuclear and chloroplast markers (8).

Over the last few years, the tremendous progress in plant genomics has broadened from a focus on *Arabidopsis* and *Oryza* to new models such as *Medicago truncatula* Gaertner and has led to an increased appreciation that the next generation of research will involve moving outward from the models to the rest of plant diversity. The Leguminosae family contains traditional genetic models such as pea, genomics models such as *M. truncatula* and *Lotus japonicus* (Regel) K. Larsen, and potential crop models such as *Glycine.* It also contains many other species of major or minor economic importance that are the focus of this and other volumes (9–11). How long a jump is it from legumes to *Arabidopsis*? How much of legume diversity is covered either by current models or by crop species that could become models? How are the many different domesticated legume species related to one another? These questions can be answered using the results of phylogenetic studies.

Phylogenetic Analysis of Sequence Data: What It Can and Cannot Do

Before talking about phylogenetic results, it is important to explain some of the methodology, assumptions, and terminology of systematic biology (12). The ultimate object of a molecular phylogenetic analysis is to identify hierarchical patterns among the organismal taxa (species, genera, families, etc.) of interest by sequencing one or more genes from plants sampled to represent the relevant taxa. The choice of gene generally is based not on function but on qualities such as structure (introns present vs. absent) and rate of change among the taxa: slowly evolving genes without introns are useful for questions involving ancient divergences, whereas more rapidly evolving introns or spacers are useful for questions involving closely related taxa such as species within a genus. The gene can be from the nucleus, chloroplast, or mitochondrion. Sequences are analyzed by any of several phylogenetic approaches—maximum parsimony, distance, maximum likelihood, or Bayesian methods—to produce one or more phylogenetic trees. Finding the unknown "true" tree can be a tricky business, in part because such phenomena as unequal evolutionary rates or biases in base composition among sequences can mislead phylogeny reconstruction algorithms. Moreover, an important assumption of such analyses is that there is a hierarchical pattern among the sequences in the first place; extensive recombination, for example, violates this assumption.

When, as is usually the case, a phylogenetic hypothesis is produced for the sequences, this gene tree is usually taken to have the same shape (topology) as the tree relating the plant species, genera, or families from which the sequences were obtained: The gene tree is identical to the so-called species tree or taxon tree. In other words, the implicit assumption is that the genes are perfect stand-ins for the genera or species. This assumption is violated by various phenomena, including introgressive hybridization, the unwitting use of a mixture of orthologous and paralogous sequences or stochastic processes of drift and fixation of polymorphisms during speciation. Under such circumstances, gene trees will not track the actual history of the taxa from which sequences were obtained.

In phylogenetic parlance, a tree is a hierarchical structure composed of taxa (here defined as the units of analysis, i.e., DNA sequences in a typical molecular systematic study) connected to nodes that represent ancestors. An ancestor plus all of its descendants forms what is called a monophyletic group; such a group is often said to be natural; a natural group forms a clade in a phylogenetic tree. The genus *Pisum*, which includes *P. sativum* L., the garden pea, is a monophyletic group of species all derived from a common ancestor. In contrast, an ancestor plus some but not all of its descendants is said to be paraphyletic. Paraphyletic groups are a problem because they are "unnatural," making it impossible to talk about phylogenetic relationships in a meaningful way. A legume example is the genus *Vicia*, which includes *V. faba* L., the broad bean or fava bean. The genus *Pisum* shares a more recent common ancestor with some *Vicia* species than those *Vicia* species do with other *Vicia* species (13). Therefore, the ancestor of all species classified in *Vicia* is also the ancestor of *Pisum* species; the exclusion of *Pisum* from *Vicia* makes *Vicia* paraphyletic (hence, unnatural).

The goal of phylogeny reconstruction is to identify monophyletic groups. Changes to the taxonomic nomenclature often follow in the wake of new understanding of phylogenetic relationships, but they necessarily lag behind and are often unwelcome when they are made, particularly when they involve familiar groups. A classic example is the tomato, which is known to the vast majority of researchers as *Lycopersicon esculentum* Mill. The genus *Lycopersicon* was found to be nested in the very large genus *Solanum*, thus making the tomato more closely related to the potato, *Solanum tuberosum* L., than the potato is to other species of the paraphyletic genus *Solanum*. To remedy this situation, *Lycopersicon* was submerged into *Solanum*, and the correct name for the tomato is now *Solanum lycopersicon* L. Given progress in legume phylogeny, such nomenclatural changes are likely to occur in Leguminosae in the future.

The Context: Where Legumes Fit in the Flowering Plants

In a landmark paper in 1993, Chase, Soltis, Olmstead, and colleagues published a phylogenetic analysis of 500 sequences of the chloroplast gene *rbcL*, representing much of the diversity of flowering plants (14). Since then, the *rbcL* dataset has continued to expand and has been augmented by sequences from additional genes; this, with analytical advances, has produced a relatively stable phylogenetic hypothesis for the angiosperms (15; Fig. 3.1).

Prior to these molecular phylogenetic studies, there had been little agreement as to the identity of the closest relative of the Leguminosae based on morphological, anatomical, and chemical data (16). Even so, the finding by Chase *et al.* (14) that the Polygalaceae (milkwort family) was sister to the legumes came as a surprise to most systematists because this family of 10 genera, with about 1,000 species, was not among the leading candidates. Later studies (15) have added to the support for an order Fabales (Fig. 3.1) comprising Leguminosae, Polygalaceae, the small mainly tropical family Surianaceae, and the genus *Quillaja* (Chilean soap tree), which was formerly classified in the Rosaceae (rose family).

Fabales is part of a "nitrogen-fixing clade" that includes the eight other flowering plant families known to form nitrogen-fixing symbioses with phylogenetically diverse soil bacteria (rhizobia). Most families in the nitrogen-fixing clade are not N-fixers, and even in some that do include genera that form nitrogen-fixing nodules, such genera are in the minority (17). This distribution, together with the morphological and anatomical evidence for the diversity of nodules, has suggested that nodulation has arisen independently several times in the N-fixing clade (17). Soltis *et al.* (18) suggested that the ancestor of the N-fixing clade evolved the predisposition to nodulate, which was exploited independently by diverse members of the clade, including legumes.

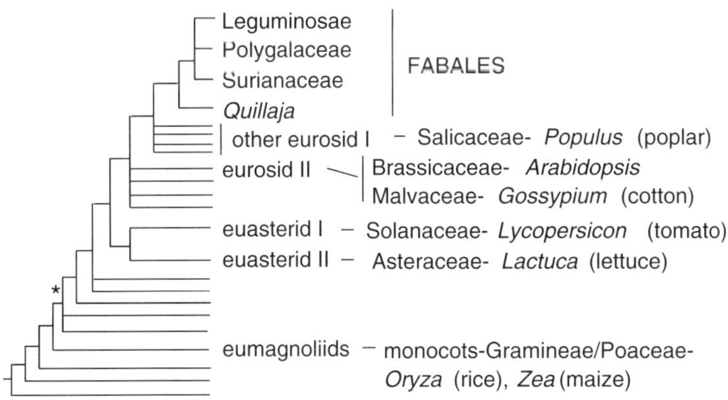

Figure 3.1. Phylogeny of flowering plants (angiosperms) showing the position of Leguminosae as part of the order Fabales within the rosid I clade of eudicots. Some important model groups are shown; the asterisk marks the ancestor of all eudicots, including legumes. This tree is taken from Doyle and Luckow (37; used by permission of the American Society of Plant Biologists) and is based on trees from Soltis *et al.* (15).

The N-fixing clade is part of a rosid I group that includes yet additional families. The entire rosid I clade, in turn, is sister to a rosid II group, among whose members are Brassicaceae (mustard family), to which *Arabidopsis* belongs, and Salicaceae (willow or poplar family), to which *Populus* belongs. More distantly related are the asterid groups, among which is the Solanaceae, which includes the tomato (*S. lycopersicon*). All of these groups, together with a number of other families, comprise the eudicots. The monocots, including, notably, the grasses (Poaceae or Gramineae: wheat, maize, etc.) are more distantly related. Yet more distantly related are diverse angiosperm families such as water lilies (Nymphaeaceae) and the controversial *Amborella*, which is generally thought to be sister to all other extant flowering plants (19).

Leguminosae or Fabaceae? A Monophyletic Group by Any Name

The monophyly of the Leguminosae was not questioned by recent taxonomists, who were swayed primarily by the distinctiveness of the legume fruit. Early broad-scale phylogenetic analyses of angiosperms (14) supported the monophyly of the family but sampled only a handful of legume species. A series of phylogenetic studies using the chloroplast *rbcL* gene and aimed at resolving relationships within the family (6,7,20–23) included a study that also analyzed a broad sample of nonlegume families and genera and showed that the Leguminosae was indeed monophyletic (6).

There has long been confusion about the name of the family. In the Code of Botanical Nomenclature, taxa of any rank above genus are to be named for a genus that is a member of the group in question. Thus, Solanaceae is named for *Solanum* and Brassicaceae is named for *Brassica*. As already mentioned, "Leguminosae" comes from the fruit type that typifies the family; there is no genus by this name in the family. But Leguminosae is one of a handful of large and well-known families for which the illegal name has been officially conserved; other examples are Compositae and Gramineae (grasses). In each case, however, a standard name also exists for each family: Asteraceae for Compositae (named for the genus *Aster*), and Poaceae for Gramineae (*Poa*). The other name for Leguminosae is Fabaceae, named for *Faba,* an old name for the genus *Vicia*. Lewis and Schrire (24) have concluded that the best name for the family is, in fact, Leguminosae, but both names are common in the scientific literature and are acceptable. A practical consequence of all of this is that some databases use only one name or the other, so if you can't find information about legumes under one name, try the other.

Major Evolutionary Lineages in Leguminosae: Two Natural Groups and One Paraphyletic Assemblage

The other major nomenclatural issue has been whether to recognize some of the distinct groups within Leguminosae as families. Three major groups of legumes have been recognized and have been accorded status as either subfamilies within a single family (Leguminosae or Fabaceae) or as three separate families. These are the mimosoid legumes, recognized either as the subfamily Mimosoideae or the family

Mimosaceae; the papilionoid legumes, classified either as subfamily Papilionoideae (or Faboideae) or as its own family, Fabaceae (or Papilionaceae); and the caesalpinioid legumes, subfamily Caesalpinoideae or family Caesalpiniaceae.

Paraphyly is an issue at this most basic level of the family. Both the mimosoid and papilionoid groups have long been considered natural, with few if any adjustments. In contrast, the caesalpinioids have been considered to comprise a number of unrelated lineages. Phylogenetic work has confirmed these suspicions. These hypotheses have been tested in a series of molecular phylogenetic analyses that have mainly relied on data from the chloroplast genome (6,21,23,25,26).

Subfamily Caesalpinioideae is a diverse, mostly tropical assemblage of 162 genera and approximately 3,000 species. Phylogenetic analyses have shown that it consists of several distinct groups, some of which correspond to sets of genera (tribes) recognized in earlier taxonomic treatments (27). One of these groups, the tribe Cercideae, appears to have diverged from the common ancestor of the remainder of the family (Fig. 3.2). The largest genus of Cercideae is the pantropical *Bauhinia*, with around 250 species, among them showy ornamentals with orchid-like flowers. *Bauhinia* also includes the marama bean, *B. esculenta* Burchell (= *Tylosema esculentum* (Burchell) A. Schreiber), an African species that produces a large, edible tuber and seeds described as having nutritive value equal to peanut or soybean (9). Cercideae also includes the small genus *Cercis* (redbud, Judas tree) with species in temperate eastern and western North America, Europe, and Asia.

Estimates for the age of the origin and early evolution of the legumes vary, but a rich Eocene macrofossil (leaves, fruits, flowers) record shows that the major lineages of the family already existed by around 50 million years ago (mya), suggesting a still older divergence of the earliest lineages such as Cercideae (28,29). How much earlier is unknown. The earliest definitive legume pollen dates from around 75 mya (29), but although the macrofossil record from the late Cretaceous (ca. 100 mya) has yielded a wealth of other angiosperm families (30–32), legumes have yet to be found from this period.

Moving along the family tree toward the familiar model legumes, the next lineages to diverge (Fig. 3.2) also contain taxa classified as Caesalpinioideae, notably the large tropical tribes Detarieae and Amherstieae, which together compose a monophyletic group (26,27). Among members of this group are large rainforest trees and several genera that produce high quality wood (e.g., *Intsia*). *Copaifera* and *Hymenaea* are important sources of copal resins used in varnishes and lacquers. Tamarind (*Tamarindus indica* L.) is grown, primarily in India, for the sweet pulp of its pods and also provides gum and wood of good quality.

The next major feature of the legume phylogenetic tree is a dichotomy between the papilionoid lineage and a large lineage dominated by caesalpinioids that also contains the Mimosoideae (Fig. 3.2). These two large groups not only include the bulk of genera and species of the family, but also all of the taxa known to nodulate (33). The Mimosoideae, if not a monophyletic group as currently circumscribed, almost certainly is monophyletic if its closest caesalpinioid relatives are included

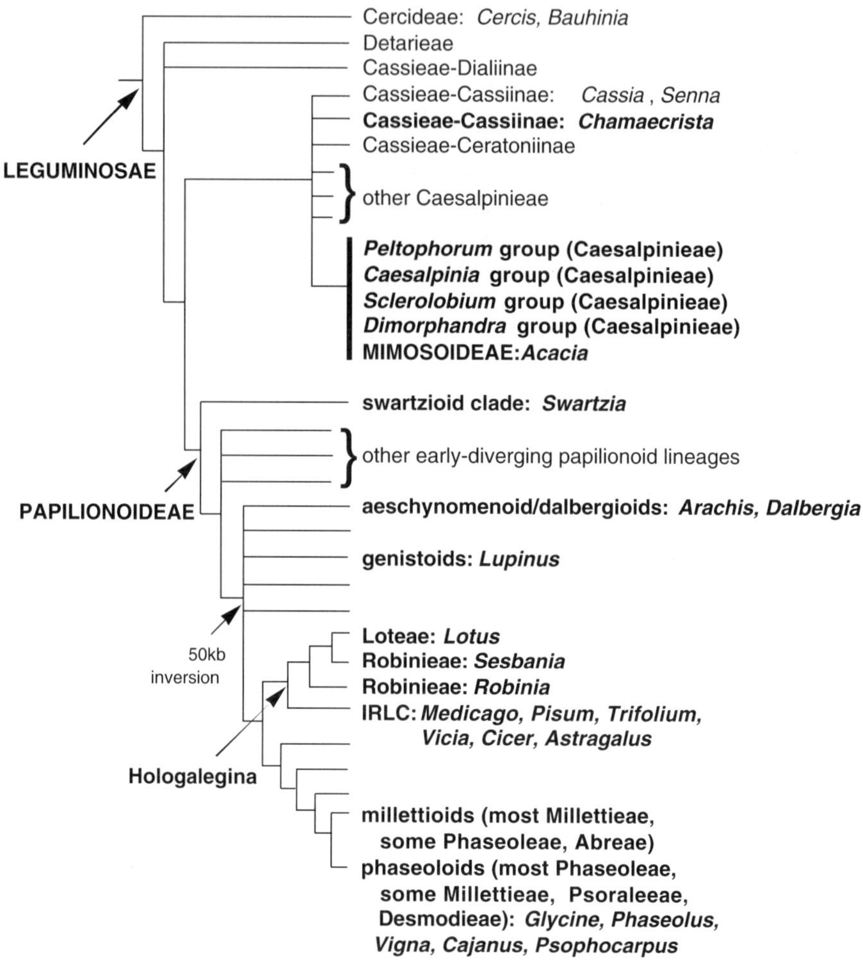

Figure 3.2. Overview of phylogenetic relationships in the Leguminosae (37, used by permission of the American Society of Plant Biologists). Leguminosae not indicated as being either Mimosoideae and Papilionoideae are classified as Caesalpinieae, which is a paraphyletic group. Only taxa mentioned in the text are listed by name; bold type indicates taxa that include at least some species that nodulate.

(26,34,35). Many mimosoids have small flowers aggregated in heads, with small petals and showy, colored stamens. Pollen in many groups is shed in aggregates rather than as individual grains. Ninety-five percent or more mimosoids nodulate, as do their close caesalpinioid relatives (33). Nodulation occurs in a handful of other caesalpinioids besides those in the large clade in which mimosoids are embedded. It is thus possible that the nitrogen-fixing symbiosis evolved multiple times in this clade separately from another origin in the papilionoids (33,36,37).

The Mimosoideae consists of approximately 77 genera and 3,000 species and includes such groups as acacias and mesquite (*Prosopis*) but relatively few cultivated taxa. The large genus *Acacia* includes species used for timber, forage, perfume (*A. farnesiana* (L.) Willd.), tannins for the leather industry (*A. mearnsii* DeWild.), and gum arabic (*A. senegal* (L.) Willd.). *Leucaena* has considerable potential as a fast-growing forage tree with nutritive quality comparable to that of alfalfa. *Desmanthus virgatus* (L.) Willd. also has possibilities as a forage, and several other genera include promising fast-growing tree species such as *Albizia*, *Enterolobium*, *Samanea*, and *Mimosa* (9). Relationships in Mimosoideae have recently been studied using molecular data (34).

Nearly all of the Caesalpinioideae not included in Cercideae or the Detarieae–Amherstieae clade are part of the clade that includes Mimosoideae. Economically important or potentially useful taxa in this group include carob (*Ceratonia siliqua* L.), grown for its sugar-rich pods, and the shrub *Cordeuxia edulis*, called *ye-eb* in its native Somalia, where its seeds are a source of protein, starch, fat, and sugar (9). Honeylocust (*Gleditsia triacanthos* L.) is a temperate North American tree species with potential as a forage crop. *Haematoxylum campechianum* L., logwood, is the source of a reddish-purple dye; various species of *Caesalpinia* also produce dyes that were much sought after in the European colonial period, one of which gave its name to Brazil (9).

The sister group to this large caesalpinioid/mimosoid clade is the Papilionoideae, which phylogenetic analyses support as a monophyletic group. Thus, of the three subfamilies, two, Mimosoideae and Papilionoideae, are monophyletic or nearly so. In contrast, as has been suspected for many years (3), Caesalpinioideae is paraphyletic. Therefore, just as all vertebrates (including humans) could be called fishes (because all the earliest-diverging lineages of vertebrates are organisms that we call fishes), all legumes could be called caesalpinioids because both mimosoids and papilionoids are nested within a sea of caesalpinioids. The key point for legume biologists is that the bulk of phylogenetic diversity within Leguminosae is to be found among genera classified as Caesalpinioideae. These are the extant members of lineages that diverged from one another and from the lineage that gave rise to the papilionoid legumes many millions of years ago. Compared with these plants, *Medicago*, *Lotus*, and *Glycine* are very close relatives; how close is discussed later in this chapter.

It is useful to illustrate one of the implications of the structure of legume phylogenetic relationships. With the current availability of large EST collections and the future promise of genomic sequences from one or more legumes, there is considerable interest in comparing genomic repertoires across diverse taxa. For example, a set of genes has been identified that are shared between *M. truncatula* and other well-studied legumes (*Glycine max* (L.) Merr. or *Lotus japonicus*) but that do not occur in the complete sequence of *Arabidopsis thaliana* (G. May, Noble Foundation, Ardmore, OK, personal communication). Such genes are of potential interest and may be involved in functions unique to legumes. But as Fig. 3.1 illustrates, it is a long way from *Medicago* and *Glycine* to *Arabidopsis*. Thus, some of these genes could be specific to legumes and could have homologues in the genomes of *Cercis* and all of the lineages between Cercideae and *Glycine*. But others may occur only in papilionoids or only in the subset of Papilionoideae that includes *Medicago* and *Glycine*. Yet others may occur throughout the N-fixing clade or in all rosid I families. Only considerable sampling, guided by our knowledge of phylogenetic relationships, will reveal whether some of these genes are legume specific.

The Familiar Legumes: Subfamily Papilionoideae

The Papilionoideae is by far the largest subfamily, with 476 genera and about 14,000 species (38). It is, as already mentioned, a monophyletic group, and it is estimated that all papilionoids shared a common ancestor around 45 to 50 mya (penalized likelihood estimate of Wojciechowski (35)). Phylogenetic studies of the subfamily have confirmed the existence of several major groups, but some groups of genera do not fall into these major lineages, and in general the early divergences in the subfamily are weakly supported (35). This is unfortunate because although 90% or more of the subfamily's members nodulate, the majority of those that do not nodulate appear to belong to these early-diverging lineages (6,33; Fig. 3.2); resolving their relationships would shed light on the issue of how often nodulation has evolved or been lost during the evolution of the subfamily. For example, what is possibly the earliest-diverging lineage in the subfamily is the swartzioid clade (Fig. 3.2), so named because its members have been classified in Swartzieae, a tribe that has been moved back and forth between Caesalpinioideae and Papilionoideae over the last century and a half of taxonomic work. This group is of interest because at least some of its members are known to nodulate (33). If the basal placement of the swartzioid clade is confirmed, it would suggest either that the swartzioids evolved nodulation separately from other papilionoids or that the ancestral papilionoid was able to nodulate and there were subsequently several independent losses of nodulation among descendants of this ancestor.

Other independent early-diverging lineages include some genera traditionally classified as Dalbergieae, but primarily include various genera of Sophoreae, independently or in groups with other taxa such as Dipterygeae and nonswartzioid Swartzieae. The tonka bean (*Dipteryx odorata* Willd.) in the Dipterygeae was for centuries important as the source of a vanilla scent used in foods and tobacco. Among other members of these basal genera is *Styphnolobium japonicum* (L.) Schott (= *Sophora japonica* L.), the commonly planted pagoda tree. An ancestor near the base of the papilionoids experienced a 50 kb inversion in its chloroplast genome (39), a character of most papilionoids, including all four of the major lineages in the subfamily (Fig. 3.2). Each of these four lineages includes some model legumes or economically important species. They are (i) the aeschynomenoid-dalbergioid group, (ii) the genistoid group, (iii) the Hologalegina, and (iv) the phaseoloid-millettioid group. The relationships among these four lineages remain unresolved, except that the last two are sister taxa. The names of these groups are derived from tribes recognized in taxonomic treatments of the subfamily (38,40).

Within these lineages, some tribes retain their identities by comprising more or less monophyletic groups (e.g., the bulk of Desmodieae forms a monophyletic group within the phaseoloid/millettioid lineage), whereas others are scattered among several lineages (e.g., Millettieae). Some tribes, such as Sophoreae, were already appreciated as being para- or polyphyletic in earlier treatments of the family (41,42) and have been confirmed as such. Others, such as Millettieae, are not nearly as unnatural as might have been expected from earlier treatments (41). The following is a description of each group, emphasizing relationships surrounding model or economically important species.

The Peanut and Its Allies: The Aeschynomenoid-Dalbergiod Group

This is the smallest of the major lineages, with 44 genera and around 1,100 species (43). It appears to be sister to the tribe Amorpheae, a North American group of eight genera and around 240 species (35). One interesting feature of the aeschynomenoid-dalbergioid group is that its members share "aeschynomenoid" nodules, which are morphologically distinctive within the family (33,43). The clade is thought to have shared a common ancestor around 34 mya (35). The aeschynomenoid-dalbergioid group primarily comprises members of the two tribes that give it its name (Aeschynomeneae and Dalbergieae). However, the group also includes two genera (*Brya* and *Cranocarpus*) traditionally classified as Desmodieae, a tribe that belongs to the phaseoloid-millettioid lineage. Several genera typically classified in Dalbergieae are not part of the aeschynomenoid-dalbergioid lineage but instead fall into the unresolved basal portion of the papilionoid phylogenetic tree (6,35). Among Dalbergieae that do fall into the aeschynomenoid-dalbergioid group are the various species of *Dalbergia* that are the source of rosewood and species of *Pterocarpus* that provide such luxury woods as narra and muninga (9).

The other large group of genera in this lineage traditionally has been considered the tribe Aeschynomeneae. However, Aeschynomeneae and Dalberieae do not form monophyletic groups within the aeschynomenoid-dalbergioid lineage based on phylogenetic analyses of molecular and morphological data, and Lavin *et al.* (43) instead recognize three major clades that they consider cryptic in the sense that they are not defined by traditionally recognized morphological characters. Of these three clades, one is named for a third tribe, the Adesmieae, which comprises a single genus of over 200 species centered in temperate South America, particularly Argentina. A second clade, the *Dalbergia* clade, includes that genus as well as several others from both Dalbergieae and Aeschynomeneae. Lastly, the *Pterocarpus* clade includes Dalbergieae, the two misplaced Desmodieae, and several Aeschynomeneae.

Among the members of Aeschynomeneae that comprise the *Pterocarpus* group is *Arachis*, the genus that includes the peanut (*Arachis hypogaea* L.). The sister genus to *Arachis* is *Stylosanthes*, which includes the forage crop Townsville stylo (*S. humilis*, H.B. and K.), and these two genera, together with other Aeschynomeneae (*Chapmannia, Geoffroea*), form a strongly supported clade (43). The origin of the peanut, a tetraploid, has been debated for many years, but recent work seems to have resolved this issue by using phylogenies of two nuclear genes encoding different fatty acid desaturases to identify its A and B genomes (44). These are shown to derive from two species, *A. duranensis* Krapov. and W.C. Gregory and *A. ipaensis* Krapov. and W.C. Gregory. Jung *et al.* (44) concluded that these two diploid species, and not the tetraploid *A. monticola* Krapov. and Rigoni, are the progenitors of *A. hypogaea.*

The Lupine and Its Allies: The Genistoids

A genistoid alliance was recognized by Polhill (3,41), and molecular phylogenies have confirmed the existence of a lineage roughly corresponding to this group (6,21,23,25,35,45). The core genistoid group recognized in these studies is composed of the tribes Genisteae, Thermopsideae, Euchresteae, Crotalarieae, Liparieae, and Podalyrieae. Of these, the first three are primarily northern temperate and the last three are predominantly southern African. Genisteae, Thermopsideae, and Crotalarieae are monophyletic, and the single genus of Euchresteae is related to some Sophoreae. Podalyrieae and Liparieae form a clade together with some other genera previously classified as Sophoreae. The genistoid alliance of Polhill (3,41) included the exclusively or primarily Australian tribes Bossiaeae and Mirbelieae, but recent molecular work has shown that these are not part of the core genistoid clade (45). Another tribe, the Brongniartieae, appears to be related to the core genistoids (35).

These core genistoid clades comprise 53 genera and around 1,500 species, concentrated in several large genera such as *Crotalaria* (ca. 600 species, including *C. juncea* L., sunnhemp or Indian hemp), *Aspalathus* (250 species), and *Lupinus* (250 species). The core genistoids are estimated to have diverged from a common ancestor around 29 mya (35). They are characterized by the presence of quinolizidine alkaloids, a trait its members share with some members of Sophoreae that appear to be the closest relatives of the genistoid lineage (7,35). The genistoids are particularly rich in chemical variation, and the distributions of alkaloids, cyanogenic glycosides, and phenolic compounds provide characters for hypothesizing relationships at all taxonomic levels in the group (46). Some of these compounds are related to the economic utility of several species. This is true in a negative way for some lupines, for example, where extensive leaching is necessary to remove alkaloids before cooking. On the other hand, phenolic compounds are the primary components of the increasingly popular rooibos tea made from leaves of the African *Aspalathus linearis* (N.L. Burm.) R. Dahlgr. in the Crotalarieae (47).

Genisteae appears to be a monophyletic tribe, and relationships within it have recently received attention (48). It includes among its members the brooms (e.g., *Ulex*) and the lupine (*Lupinus*), a large genus with several cultivated species both in the Old World (e.g., lupini bean = *L. albus* L.) and the New World (*L. mutabilis* Sweet, the tarwi or chocho that was a major crop of indigenous people of the Andes). Relationships in this large genus are complex and require further work, despite molecular studies such as those of Käss and Wink (25).

Medicago, Lotus, and Much More: The Hologalegina

This is the largest group of papilionoids—indeed of legumes—in terms of numbers of species, with nearly 4,000 species in 75 genera. This distinction is mainly due to the presence of a single genus, *Astragalus* (milk-vetch), which has over 2,000 species, 1/10 of all of legume species diversity. The Hologalegina appear to have diverged from a common ancestor around 37 to 38 mya (35); therefore, this is the best current estimate for the divergence of *Lotus* and *Medicago*.

Hypotheses of relationships in this group have gone through several changes over the last quarter century. In his 1981 treatment of Papilionoideae, Polhill included a large grouping termed the "temperate herbaceous group" or "epulvinate series." The former name referred to the distribution and habit of the genera and tribes included in this group, whereas the latter name referred to the absence of the pulvinus, the leaf-base swelling so characteristic of other legumes. This group included the large galegoid group of tribes (Galegeae, Vicieae, Trifolieae, etc.) as well as *Lotus* and its allies (Loteae and Coronillieae). In 1990, Lavin *et al.* (49) surveyed the legumes for a major feature of the chloroplast genome, the large inverted repeat (IR), which was known to have been lost in some species of the galegoid group but retained in soybean (50). It was subsequently found that the entire galegoid group had lost the IR, but that the chloroplast genome of *Lotus* and its allies possessed the IR (49). This led to a revision of the concept of a comprehensive temperate herbaceous group.

More recently, however, molecular phylogenetic analyses of chloroplast genes have shown that the *Lotus* group and the galegoids (IRLC in Fig. 3.2) are indeed sister taxa, although they diverged from one another around 38 mya (35). This has led to the recognition of this entire group as a single taxon called Hologalegina. Whether this is where the story will end remains to be seen. Data from several single-copy nuclear genes have been equivocal in their support for this grouping (H-K. Choi, D.R. Cook, and J.J. Doyle, unpublished), and some (albeit with limited sampling) have instead grouped Loteae with the phaseoloid group (H-K. Choi and D.R. Cook, University of California–Davis, personal communication). Such a placement is of interest, given the fact that the nodule type prevalent in *Lotus* and its allies is the determinate desmodioid type found in phaseoloids such as soybean, rather than the indeterminate caesalpinioid or astragaloid types typical of galegoid genera (33,36). However, unlike nodules of *Glycine* and its allies but like nodules of other Hologalegina, *Lotus* nodules transport amides rather than ureides (51).

The Loteae, with around 10 genera and 300 species, is of interest because it is to this tribe that *Lotus japonicus*, which is one of the major legume models (Fig. 3.3), belongs. Phylogenetic relationships within Loteae do not follow traditional generic boundaries. Instead the tribe is a classic case of a single large genus—in this case *Lotus*, with over 100 species—that is paraphyletic, with some of its numerous separate lineages traditionally recognized as separate genera, such as *Hammatolobium*

(52). Among these is the genus *Coronilla*, which includes *C. varia* L. (crown vetch), a species planted widely as a soil bank stabilizer in the United States. In earlier classifications (3), *Coronilla* was, in fact, placed in a separate tribe (Coronilleae) from *Lotus*. The model legume, *L. japonicus*, is part of a species complex that includes *L. corniculatus* L. (birdfoot trefoil), a common roadside weed in eastern North America that is originally of European origin. The complex, like the genus as a whole, includes

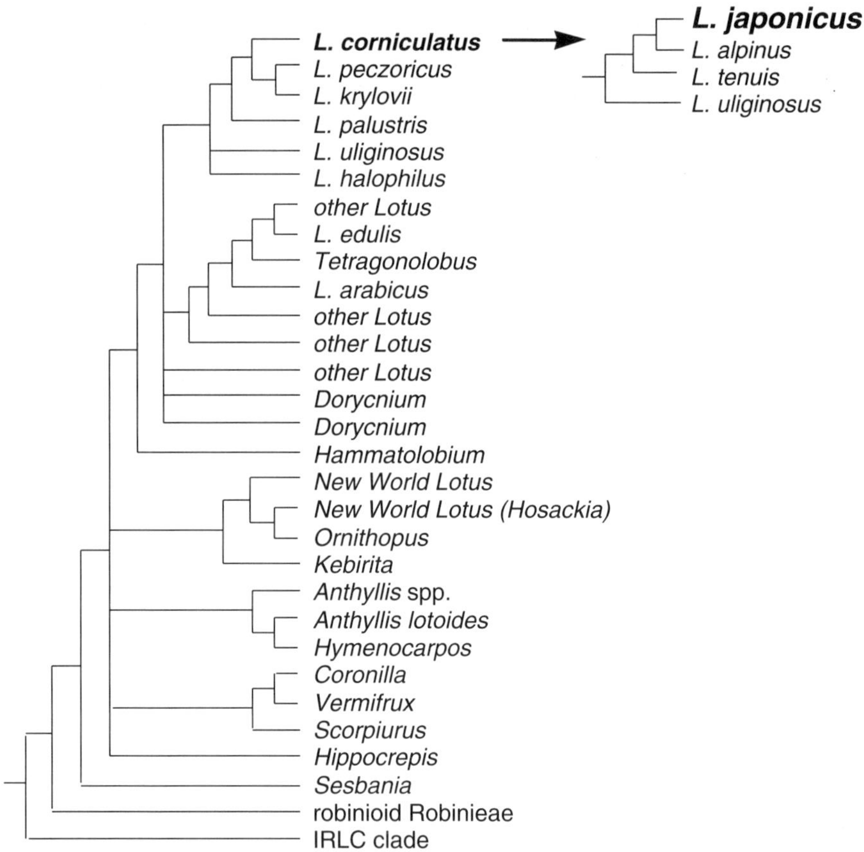

Figure 3.3. Relationships in the Loteae-Robinieae clade of Hologalegina, emphasizing the phylogenetic neighborhood of *Lotus japonicus*. The topology of the more inclusive tree is primarily from the ribosomal ITS study of Allan *et al.* (52); the base of the tree is from Wojciechowski (35). Taxa designated as "*L.*" are *Lotus* species; *Lotus* species are Old World with the exception of two related groups from North and South America. *Lotus* is clearly paraphyletic relative to several segregate genera. *Lotus japonicus* is a diploid member of the *L. corniculatus* complex; the smaller tree indicates its relationships to other diploid species in this complex inferred from chloroplast DNA data (55).

both diploids and polyploids and has been the focus of several biosystematic studies (53,54). Within the *L. corniculatus* complex, chloroplast DNA evidence indicates that *L. japonicus* is most closely related to *L. alpinus* Schleicher (55). The *L. corniculatus* complex is part of a weakly supported clade that includes such Old World *Lotus* species as *L. palustris* Willd. and *L. krylovii* Schischk. and Sergievsk. (52). The seeds and young pods of one species of *Lotus*, *L. tetragonolobus* L. (winged pea = *Tetragonolobus purpureus* (D.C.) Schleicher) are eaten in southern Europe.

The sister groups to Loteae in chloroplast phylogenies (35,56) are *Sesbania* and the robinioid group of the tribe Robinieae. This is of note because, as a primarily woody group, Robinieae was not part of the Temperate Herbaceous Group of tribes (3). *Sesbania*, which includes the stem-nodulating species *S. sesban* (L.) Merr. and *S. grandiflora* (L.) Poir (used in agroforestry in Asia), has traditionally been considered part of Robinieae (3,41). However, molecular phylogenetic studies suggest that *Sesbania* is more closely related to Loteae than to core Robinieae (35; Fig. 3.3). Like Loteae, *Sesbania* and Robinieae retain both copies of the chloroplast IR. Members of the core robinioid group of Robinieae include the black locust (*Robinia pseudoacacia* L.), a familiar temperate-zone tree in North America; relationships have been studied using nuclear and chloroplast sequence data by Lavin *et al.* (8), who use the phylogeny to address some long-standing biogeographic questions.

The bulk of Hologalegina is found in the galegoid group of tribes: Vicieae, Trifolieae, Galegeae, Hedysareae, Cicereae, and Carmichaelieae (Fig. 3.4). This is the group of tribes that, along with a few genera classified in the tribe Millettieae (including the familiar *Wisteria*), are united by the loss of one copy of the chloroplast IR (49,57–59), and thus are often referred to as the IR-loss clade (IRLC). The IRLC group is thought to have shared a common ancestor between 25 and 29 mya (35). Of the galegoid tribes, Galegeae (22 genera, ca. 2,500 species), Hedysareae (seven genera, ca. 250 species), and the small New Zealand tribe Carmichaelieae (five genera, ca. 45 species) contain few familiar or economically important genera. Galegeae does include the huge genus *Astragalus* (tragacanth is a gum produced by some species) and also *Glycyrrhiza glabra* L., the source of licorice; Hedysareae includes the forage crop sainfoin (*Onobrychis viciaefolia* Scop.). Cicereae contains the single genus *Cicer*, among whose ca. 40 species is chickpea or garbanzo (*C. arietinum* L.).

Of greatest economic and scientific interest are the Vicieae and Trifolieae, which together comprise 11 genera and nearly 800 species. Genera of Vicieae are *Pisum*, one of whose two species is *P. sativum* (pea), *Vicia* (140 species of vetches, including *V. faba*, the broad bean or fava bean), *Lens* (five species including *L. culinaris* Medik., the lentil), the single species of *Vavilovia*, and *Lathyrus* (about 150 species, including *L. odoratus*—sweet pea—and *L. sativus*—grass pea). Trifolieae comprises seven genera, including *Trigonella* (which includes *T. foenum-graecum* L., whose seeds are the source of the spice fenugreek), the large genus *Trifolium* (clovers; ca. 250 species), *Melilotus* (sweet clovers), and *Medicago*, among whose 60 or so species are *M. sativa* L. (alfalfa) and *M. truncatula*. Recent phylogenetic studies using chloroplast genes nest a monophyletic Vicieae within Trifolieae, with

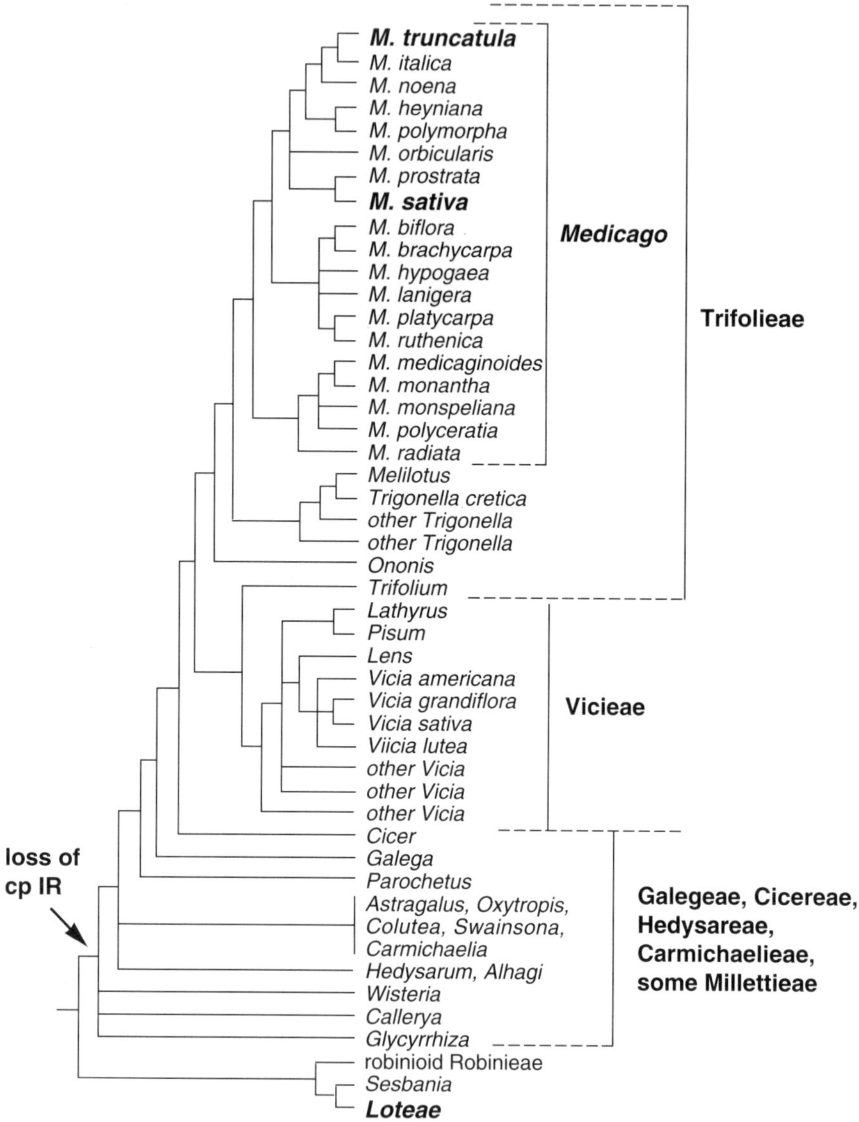

Figure 3.4. Relationships in the inverted repeat loss clade (IRLC) of Hologalegina, emphasizing the phylogenetic neighborhood of *Medicago truncatula*. Tribal names are given; Trifolieae is paraphyletic (due to the placement of the large genus *Trifolium* as sister to Vicieae). The topology of Vicieae and Trifolieae clade is from analysis of the chloroplast gene *matK* by Steele and Wojciechowski (13); all *Medicago* species used in that analysis are shown. Relationships among genera from tribes other than Vicieae and Trifolieae are not shown in detail and are taken from Wojciechowski (13). The loss of the large chloroplast inverted repeat in an ancestor of this clade is indicated.

Trifolium sister to Vicieae (13; Fig. 3.4). This position of Vicieae makes Trifolieae paraphyletic. Within Vicieae, the genus *Vicia* also appears to be paraphyletic, with *Pisum*, *Lens*, and *Lathyrus* nested among its species (13).

Relationships within several IRLC genera have been studied using molecular data. One of the first molecular systematic studies in plants was a chloroplast DNA RFLP study of *Pisum* (60), and similar studies were conducted in *Lathyrus* (61). Astragalus has been the subject of several studies, including that of Wojciechowski *et al.* (62), who used both nuclear and chloroplast sequences to show that it is largely monophyletic and to identify clades within this huge genus.

Generic boundaries involving *Medicago* have been modified in recent years. Bena (63) transferred several species of *Trigonella* to *Medicago* based on phylogenetic analysis of nuclear ribosomal spacer regions (external transcribed spacer [ETS] and internal transcribed spacer [ITS]) and in a series of studies investigated relationships in *Medicago* (64–66). In these analyses, *M. truncatula* was strongly supported as part of a clade with *M. littoralis* Rhode ex Loisel and *M. tornata* (L.) Mill. Differences in sampling make it difficult to compare the results of the Bena studies with the ribosomal ITS study of Downie *et al.* (67) or the chloroplast study of Steele and Wojciechowski (13), neither of which sampled *M. littoralis* or *M. tornata*. Those studies both placed *M. italica* E.H.L. Krause, a species not sampled by Bena (63), as sister to *M. truncatula* (Fig. 3.4). Cultivated alfalfa is part of a complex in which various taxa are recognized as either species or varieties, and where hybridization, polyploidy (auto-, allo-, or both), and a long history of movement of plants out of their natural ranges by humans has led to a confusing situation much in need of phylogenetic elucidation (68,69). Relationships among the principal taxa in the complex based on a few molecular markers are thus far equivocal (Doyle and Brummer, unpublished data; K. Steele, Arizona State University East, Mesa, AZ, personal communication). *Medicago sativa* and *M. truncatula* are relatively closely related within *Medicago*, a finding that is consistent with the colinearity of their linkage maps (70).

Soybean, Common Bean, and Many Minor Domesticates: The Phaseoloid-Millettioid Group

The last major lineage of papilionoids (Fig. 3.2) is the largest in terms of the number of its genera (around 160) and, next to Hologalegina, is largest in species as well (over 2,300). The phaseoloid-millettioid clade shared a common ancestor approximately 41 mya (35). Half of the genera in this group belong to the tribe Phaseoleae in Polhill's (40) treatment, making it the largest tribe in that classification. Phaseoleae is one of the largest in number of species, with over 1,500. Among Phaseoleae are such familiar and economically or scientifically important genera as *Glycine* (*G. max* = soybean), *Phaseolus* (most importantly *P. vulgaris* L., the common bean; but also *P. lunatus* L., lima bean; *P. acutifolius* A. Gray, tepary bean; *P. coccineus* A. Gray, scarlet runner bean), *Vigna* (*V. radiata* (L.) Wilczek, mungbean; *V. unguiculata* (L.) Walp., cowpea; *V. mungo* (L.) Hepper, black gram; *V. aconitifolia* (Jacq.) Marechal, moth bean; *V. umbellata* (Prain) Maréchal, rice bean; *V. vexillata* (L.) A. Rich, wild mung; *V. angularis* (Willd.) Ohwi and Ohashi, adzuki bean; *V. subterranea* (L.) Verdc., Bambara groundnut or groundbean), *Cajanus* (*C. cajan* (L.) Millsp., pigeonpea), and other minor cultivated genera such as *Lablab* (*L. purpureus* L., lablab or hyacinth bean), *Sphenostylis* (*S. stenocarpa* (Hochst. ex. A. Rich.) Harms., African yam bean), *Macroptilium* (*M. atropurpureum* (DC) Urb., siratro), *Macrotyloma* (*M. uniflorum* (Lam.) Verdc., horse gram; *M. geocarpum* (Harms) R. Marechal and J.C. Baudet, Kersting's groundnut), *Psophocarpus* (*P. tetragonolobus* (L.) DC., winged bean), *Canavalia* (*C. ensiformis* (Jacq.) DC, jackbean; *C. gladiata,* swordbean), *Centrosema pubescens* Benth. (centro), *Apios* (*A. americana,* groundnut or Indian potato), *Flemingia* (*F. vestita* Benth. ex Baker, a minor African tuber crop), *Pueraria* (various tuber-bearing species and kudzu [*P. monticola,* formerly known as *P. lobata* (Willd.) Ohwi]), and *Pachyrhizus* (*P. erosus* (L.) Urban and *P. tuberosus* (Lan.) Spreng, yam bean or jicama). Phaseoleae include such ornamentals as the many species of *Erythrina*, collectively called coral-tree.

The second tribe that gives its name to this lineage is Millettieae, formerly Tephrosieae (71,72). Millettieae, with around 40 genera and 850 species, is a pantropical woody tribe, in contrast to the largely herbaceous Phaseoleae. Among its members are trees and shrubs that provide fish poisons and the insecticide rotenone (e.g., *Lonchocarpus*, *Derris*, and *Tephrosia*). As noted earlier, *Wisteria* and a few other genera formerly included in Millettieae lack the chloroplast inverted repeat and have been removed from the tribe; they are properly placed with the IRLC clade of Hologalegina (6,35).

Millettieae was viewed as comprising basal elements of several different evolutionary lineages, particularly Phaseoleae (3). Phaseoleae, however, was considered to be a more natural assemblage (42). Molecular phylogenetic studies have shown that both Phaseoleae and Millettieae are unnatural (6,58). These analyses agree on the existence of two major clades, one a core Millettieae clade with only a few Phaseoleae near its base, and the other comprising mostly Phaseoleae but with a few Millettieae at its base. The millettioid clade includes the monogeneric Abreae, among whose 15 or so species is *Abrus precatorius* L. (rosary pea or precatory pea), whose highly poisonous seeds are used as beads, sometimes with deadly consequences. Phaseoleae in the millettioid clade are primarily genera classified as subtribe Diocleinae, which includes *Canavalia* (sword or jack bean). Phylogenetic relationships among Millettieae have been studied using both nuclear (58,73) and chloroplast (74) markers, and it is these studies that have largely been responsible for our understanding of the disparate relationships of genera previously classified in this tribe.

The phaseoloid clade includes Phaseoleae and also two other tribes, Desmodieae and Psoraleeae. Relationships within this clade based on *rbcL* are relatively weakly resolved in many cases, but it appears that both Desmodieae and Psoraleeae comprise monophyletic groups (after the transfer of anomalous

Desmodieae to the aeschynomenoid-dalbergioid group) nested among genera classified as Phaseoleae. Psoraleeae consists of fewer than ten genera and around 150 species, including some tuber-bearing species in Australia and North America (9). The much larger Desmodieae, with around 30 genera and over 500 species, includes the large genus *Desmodium* (ca. 300 species) with its hitchhiking segmented fruits (loments) covered with Velcro-like hairs, some of whose species have promise as forage crops (9). A more established forage genus is *Lespedeza* (40 species); relationships in *Lespedeza* and allies have been studied by Nemoto *et al.* (75).

As noted previously, there are numerous genera of economic significance in the phaseoloid-millettioid lineage, ranging from important crop plants to minor or regional domesticates. Among the most important worldwide is the soybean. The genus *Glycine* is fundamentally polyploid with $2n = 40$ in *G. max* and other diploid members of the genus. A paleopolyploid origin was hypothesized on the basis of this number, which is approximately double that of most Phaseoleae, which are generally $2n = 20$ or 22 (76), and a duplicated genomic structure has been recognized for some time (77,78). Recent studies analyzing distances among duplicated sequences suggest that this event took place around 15 mya (79), but whether the event was fundamentally autopolyploid or allopolyploid remains to be confirmed. This is because only one nuclear gene phylogeny currently is available for *Glycine* and its generic allies, the gene encoding the chloroplast-expressed isozyme of glutamine synthetase (ncpGS) (80). It is not clear whether this gene was duplicated as a consequence of the polyploid event or whether it was a later, single-gene duplication (80). What is needed are phylogenies of several nuclear genes, each sampled from *Glycine* and related genera, and including both *Glycine* copies resulting from the polyploid event, in order to have a chance of distinguishing between auto- and allopolyploidy (80).

Relationships around *Glycine* (Fig. 3.5) remain rather poorly understood. Phylogenies of chloroplast noncoding regions (81) and ncpGS (80) both agree in placing the pantropical genus, *Teramnus*, as the closest generic relative of *Glycine*, and in identifying *Amphicarpaea* (hog-peanut, a small genus with species in eastern North America and eastern Asia) as another rather close relative. The Lee and Hymowitz (81) study concentrated on genera classified by Lackey (42) as Phaseoleae subtribe Glycininae, but molecular phylogenetic studies have shown this group to be highly unnatural (6,82,83). Kajita *et al.* (6) and Doyle *et al.* (80) identified a number of other genera not classified in that subtribe as being more closely related to *Glycine* than are some members of Glycininae. These include two genera transferred to Glycininae in Polhill's (1994) classification from Phaseoleae subtribe Diocleinae, *Pachyrhizus* (jicama) and *Calopogonium* (calpo). Other groups more closely related to *Glycine* than both tribal classifications of Papilionoideae and subtribal classifications of Phaseoleae would suggest are Psoraleeae—(78), *Bituminaria* and *Otholobium* in Fig. 3.5—and the core monophyletic group of genera classified as Phaseoleae subtribe Phaseolinae by Lackey (42); this is the group that includes *Phaseolus* and *Vigna*.

The soybean and its progenitor, the wild and weedy *G. soja*, are considered to be conspecific, with *G. max* simply representing a cultigen modified by human selection so as to differ from its wild progenitor (84). Both taxa are annuals native to northeastern

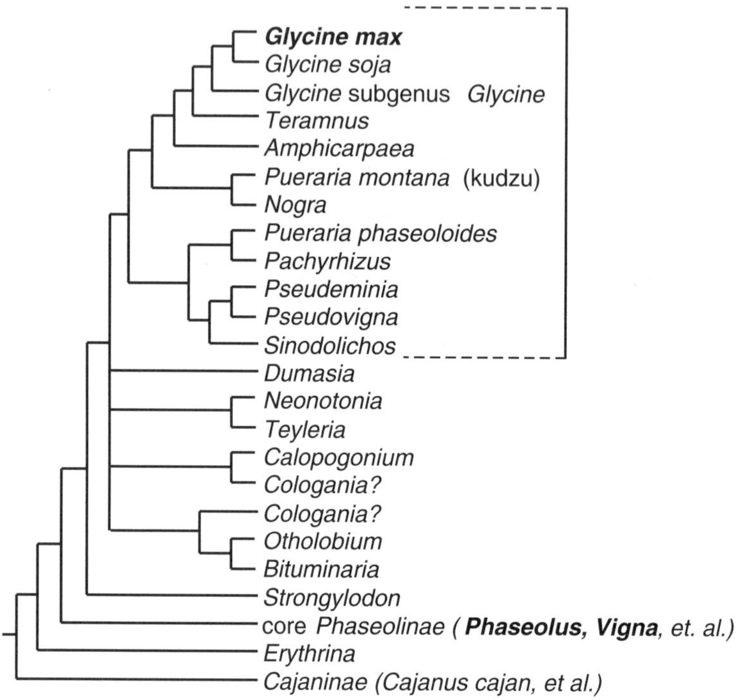

Figure 3.5. The phylogenetic neighborhood of the soybean (*Glycine max*). The tree is a depiction of relationships inferred from the chloroplast genes *rps16* (81) and *rbcL* (6) and the nuclear gene encoding the chloroplast-expressed isozyme of glutamine synthetase (ncpGS (80)). Most of the detail is from *rps16* and ncpGS; *rbcL* provides the overall context within the phaseoloid-millettioid clade. Sampling was not identical in the *rps16* and ncpGS studies, and the topologies favored by each differ in several points. *Dumasia* was strongly supported as part of the clade marked "A" in the ncpGS tree but was weakly supported as sister to a *Calopogonium-Cologania* clade for *rps16*. The *Calopogonium-Cologania* grouping was strongly supported by *rps16,* whereas *Calopogonium* grouped with *Neonotonia-Teyleria* and *Cologania* with *Otholobium-Bituminaria* (not sampled for *rps16*) for ncpGS. The basal portions of this tree are supported only by the *rbcL* study. Soybean (*G. max*) is sister to the wild Asian annual species, *G. soja,* with the annuals sister to the perennial, mostly Australian species of subgenus *Glycine*. *Pueraria* is polyphyletic according to both ncpGS and *rps16*, which sampled numerous other species of that genus (81).

Asia, and both are classified as subgenus *Soja*. The secondary germplasm pool of soybean is a group of over 20 diploid ($2n$ = 38, 40) perennial species in subgenus *Glycine*. These species have been the subject of numerous phylogenetic studies (85,86), as has a complex of recently evolved allopolyploids (some no more than 30,000 years old) formed from them (87).

Phaseolus and *Vigna* are closely related and are part of the core Phaseolinae group linked by the presence of a 78-kb inversion in their chloroplast genomes. This mutation is not found in a few genera classified as Phaseolinae by Lackey (42), notably *Psophocarpus*, the genus that includes winged bean (*P. tetragonolobus*). Chloroplast phylogenies (6) suggest that *Psophocarpus* may not even be as closely related to this core group as was previously thought (82,83).

The *Vigna-Phaseolus* complex has traditionally been a difficult group taxonomically, and both molecular and other data suggest that *Vigna*, found in Africa, Asia, and the New World, is paraphyletic relative to both the exclusively New World *Phaseolus* and to other genera such as *Ramirezella* (88). Relationships in this entire complex have been the subject of several molecular systematic studies that have revealed major groupings within *Vigna* and elucidated their relationships to *Phaseolus* and other genera (89–92). Goel *et al.* (92) have identified the likely progenitors of *V. mungo, V. radiata, V. umbellata,* and *V. unguiculata* from among wild *Vigna* species.

There have been several molecular systematic studies on relationships among *Phaseolus* species (88,89,93). Delgado *et al.* (88) concluded on the basis of extensive sampling of ribosomal ITS sequences that *Phaseolus* is monophyletic and identified numerous monophyletic groups within it. They further concluded that the five most commonly cultivated species of *Phaseolus* all are related to the *P. vulgaris* and *P. lunatus* groups, and that the gene pools traditionally recognized for these two species could not be detected with ITS data, despite the existence of considerable sequence variation for this region.

Indigofereae is sister to the remainder of the phaseoloid-millettioid lineage (6,35). The tribe includes the large genus *Indigofera* (over 700 species), which has been the subject of recent molecular and morphological phylogenetic study (94). Several species of *Indigofera* provide indigo dye; interestingly, this dye is found in other genera outside of *Indigofera*, such as *Lonchocarpus* in the millettioid group (9). A second member of *Indigofereae*, *Cyamopsis tetragonoloba* (L.) Taubert, is the source of guar gum, used in many processed foods.

Model Species and the Diversity of the Leguminosae

With this summary of relationships in the Leguminosae, it is possible to ask how well existing legume models capture the diversity of the family. There are various criteria that can be used in answering this question. The most obvious is simply to compare the number of genera or species that are descended from the common ancestor of the model species with the total number of genera or species in the family (Fig. 3.2). By this criterion, the common ancestor of the three existing principal model legumes, *M. truncatula, L. japonicus,* and *G. max* is the ancestor of around 7,000 of the 14,000 species in subfamily Papilionoideae and around 250 of its 476 genera, so this group speaks for around 50% of papilionoid diversity (Fig. 3.2). If one considers the entire family, then the numbers are lower: around 35% of species diversity and 33% of generic diversity.

Adding more models could increase coverage, of course, but in terms of this numerical criterion not all additions would help equally. *Phaseolus*, for example, although very useful for learning more about the evolutionary processes in the neighborhood of *Glycine*, would not increase coverage of the entire subfamily or family. Similarly, *Pisum* adds nothing to the overall coverage given its relatively close relationship to *Medicago*. This highlights the necessity of including a phylogenetic criterion in evaluating the utility of model species. To sample more fully the diversity of the Papilionoideae, models outside the Hologalegina and phaseoloid-millettioid groups are required. Adding *Lupinus*, a genus that has been studied relatively well for nodulation (95), would increase the coverage to include the genistoid clade. Similarly, peanut (*Arachis hypogaea*) or a diploid relative would complement existing models by adding a member of the aeschynomenoid-dalbergioid group. Thus, with two additional models, the vast bulk of the Papilionoideae could be represented.

To truly cover the entire diversity of the Leguminosae, however, lineages that diverged early from the ancestors of the Papilionoideae should be sampled. The Cercideae, being sister to the remainder of the entire family, would be an obvious choice and is represented by the temperate zone genus *Cercis*, which is highly accessible to many legume researchers in North America, Europe, and Asia.

The Value of Sampling Legume Diversity: Genome Size as an Example

Doyle and Luckow (37) used genome size as an example of how additional sampling within the family would be useful and how Cercideae in particular would be of interest to study. They pointed out that small-genome legumes occur in several lineages outside the Hologalegina, where both *Lotus* and *Medicago* belong. The Plant C-value database at the Royal Botanic Gardens, Kew, Plant DNA (www.rbgkew.org.uk/cval/homepage.html) has over 500 records for legume species, but the vast majority are from Papilionoideae, and the distribution is very uneven even for that subfamily. Genome sizes range from 368 to over 26,000 Mb/1C, with 23 values at or below the sizes given for *M. truncatula* and *L. japonicus* (466 Mb). Many of these are from species or genera related to *Lotus* or *Medicago*, but there are also several from members of the phaseoloid-millettioid clade, including species of *Phaseolus* and *Vigna*, and the smallest value in the family is from a genus closely related to these taxa, *Lablab niger* Medik., a relative of *L. purpureus* L. (hyacinth bean).

The majority of C-values reported for the papilionoids are from these two groups. There are no C-values reported from early-diverging papilionoid taxa, which would be of particular interest in inferring the condition of the earliest ancestors of the subfamily. There are also relatively few records for the two other major papilionoid lineages. The genistoid group is thought to be fundamentally polyploid (76), so it is not surprising that none of the values for its species in the Kew database are among the smallest in the family. However, it is interesting that the smallest genome reported for the genistoid lineage is relatively small (539 Mb in *Cytisus nigricans*

L.), suggesting that the original diploid genome size of the group could be quite small.

Outside the Papilionoideae there are few records, but even so there are members of Mimosoideae on the list of genomes as small as or smaller than *L. japonicus* and *M. truncatula*. *Prosopis* (mesquite) species range from 392 to 490 Mb, and here again the case can be made that because these species are $2n = 28$, they are probably polyploid, making the original genome sizes potentially under 200 Mb, which is getting close to the size of *Arabidopsis*.

This brings us back to Cercideae, at the base of the entire family (Fig. 3.2). C-values are available only from a few species of *Bauhinia*. Those currently in the database are nearly 600 Mb, but recent sampling (L. Hanson and I. Leitch, Royal Botanic Gardens, Kew, UK, personal communication) has identified species with C-values around 400 Mb. *Bauhinia* is thought to be polyploid, at $2n = 28$ (76). *Cercis*, in contrast, is $2n = 14$ and is thus considered fundamentally diploid, so it is possible that it could have a genome as small as 200 Mb.

Regardless of how small the original genome size of Cercideae was, the most parsimonious argument is that the genome size was relatively small in the ancestor of the entire family, and that small genome sizes have been retained in taxa scattered throughout the Leguminosae. There have been large increases in genome size in species that have chromosome numbers typical of diploids, for example in the Vicieae. Based on chromosome numbers, some lineages are polyploid, including many mimosoids and groups of papilionoids such as the genistoid group. Genomic studies are suggesting that even some legumes with low chromosome numbers and small genomes may be fundamentally polyploid, as is suspected in the case of *Arabidopsis* (96). For example, Schlueter *et al.* (79) have presented evidence for additional polyploid events in the genomes of *Medicago* and *Glycine*; one event, dated to between 40 and 50 mya, may have occurred in the common ancestor of these taxa (B. Pfeil, J. Doyle, J. Schlueter, and R. Shoemaker, unpublished data; also see Chapter 10 of this volume). If so, the entire clade comprising Hologalegina plus the phaseoloid-millettioid group—over a third of the Leguminosae—is fundamentally polyploid.

Another Argument for More Legume Models: Nodulation

Nodulation is arguably the most significant biological phenomenon in the Leguminosae. It is nodulation that sets legumes apart from other flowering plant families that include model taxa—grasses, Solanaceae, and Brassicaceae, for example. As noted earlier, nodulation is not unique to the legumes, nor is it universal in the family. This is one of the features that makes it especially interesting, particularly from an evolutionary perspective. The earliest diverging lineages of the family, such as Cercideae, do not nodulate, and neither do the families that are the closest relatives of the legumes, such as Polygalaceae. Nodules clearly evolved independently in legumes, but the question is, which legumes? Was there a single origin, or was the nodule invented multiple times in different groups? The question is complicated by at least two problems: uncertainties in the phylogenetic tree and the possibility that many independent losses of this energy-expensive habit could have occurred. The former problem is solvable by obtaining more data and sampling more widely in the family. In the foreseeable future, we will have a very good idea of the phylogenetic tree for the Leguminosae; this is a certainty. The second problem is more difficult because it has no simple solution. Even with a fully resolved phylogeny, the possibility will remain that a hypothesis requiring two or three gains and a small number of losses is less realistic than a hypothesis that requires only a single origin followed by independent losses in many different lineages. Simply counting steps (gains plus losses), however, and accepting the simplest (most parsimonious) answer will not necessarily produce the correct solution.

In considerations of the evolution of nodulation outside the Leguminosae, it has been easier to accept a multiple-origins scenario for two reasons (36): First, the distribution of nodulation is much more patchy within some families than it is in the legumes. Secondly, consideration of anatomical and developmental data in a phylogenetic framework is consistent with the hypothesis of multiple origins. In the legumes, we know a great deal about nodules in a handful of papilionoid taxa, primarily in the phaseoloid-millettioid group (*Glycine* and *Phaseolus*) and Hologalegina (*Lotus* and *Medicago*). These desmodioid and caesalpinioid nodules, though very different in their development and chemistry, are almost certainly homologous and share a common origin with the aeschynomenoid nodules of the aeschynomenoid-dalbergioid lineage and with the caesalpinioid and *Lupinus* nodule types found in the genistoid lineage (33). Much less is known, in general, about these nodules, particularly the aeschynomenoid type, and it would be helpful for comparative studies of nodulation to include representatives from these groups.

Little is known about nodules outside the papilionoids, and this is unfortunate. On the basis of gross morphology and determinate development, the type of nodule found in Mimosoideae and in nodulating caesalpinioids such as *Chamaecrista* is considered to be caesalpinioid and is thus shared with many papilionoids, including *Medicago*, other Hologalegina, many genistoids, and scattered taxa in other lineages of the Papilionoideae (33). But are these nodules all homologous? In other words, are they all derived from a single common ancestor from which nodules evolved? This brings us back again to the question of origins. But it would certainly be worth knowing much more about these nodules, particularly at the molecular level. What are the gene repertoires of nodules in mimosoids and caesalpinioids, and how do they compare with those of papilionoid taxa? Of the nodulation genes identified by Fedorova *et al.* (97) from nodule EST collections of *M. truncatula*, how many are found in these other legume nodules? How many of the genes that are found in legume nodules but that cannot be identified in the genome of *Arabidopsis* are expressed in nodules of nonpapilionoid legumes or in nonlegume nodulators? Can phylogenetic relationships of nodule-expressed genes belonging to gene families be used to infer origins of nodules (36,98)?

This brings us back to the idea that legume models outside of those already in existence would be very useful for understanding what it means to be a legume. McKey (99) suggested that what makes legumes unique is not so much nodulation as their invention of a nitrogen-demanding lifestyle, which provides significant ecological advantages that humans have harnessed for a variety of purposes. Most obviously among these is agriculture, in the form of both food and forage crops. But humans have found nitrogen-containing compounds from legumes useful for other purposes, such as stunning fish. Learning more about the biochemistry of diverse legumes could be of great value in understanding what it is about the family that makes it so highly successful and so useful to humans.

Acknowledgments

I am grateful for the many people with whom I have interacted in the legume systematics community over the years, especially my students, postdocs, and collaborators. In particular I would like to thank Jane Doyle and Tony (A.H.D.) Brown. Roger Polhill was for many years the guiding force in legume systematics and was the impetus for my involvement in higher-level legume systematics. My legume research has for many years been funded by the NSF Systematic Biology Program, currently in award DEB-0089483; I also acknowledge NSF DBI Plant Genome award 0321664.

References

1. Cronquist, A., *An Integrated System of Classification of Flowering Plants,* Columbia University Press, New York, 1981.
2. Crisp, M.D., and J.J. Doyle, *Advances in Legume Systematics, Part 7: Phylogeny,* Royal Botanic Gardens, Kew, United Kingdom, 1995.
3. Polhill, R.M., and P.H. Raven, *Advances in Legume Systematics, Part 1,* Royal Botanic Gardens, Kew, United Kingdom, 1981.
4. Herendeen, P., and A. Bruneau, *Advances in Legume Systematics, Part 9,* Royal Botanic Gardens, Kew, United Kingdom, 2000.
5. Klitgaard, B.B., and A. Bruneau, *Advances in Legume Systematics, Part 10: Higher Level Systematics,* Royal Botanic Gardens, Kew, United Kingdom, 2003.
6. Kajita, T., H. Ohashi, Y. Tateishi, C.D. Bailey, and J.J. Doyle, *rbcL* and Legume Phylogeny, with Particular Reference to Phaseoleae, Millettieae, and allies, *Syst. Bot. 26*:515–536 (2001).
7. Wink, M., and G.I.A. Mohamed, Evolution of Chemical Defense Traits in the Leguminosae: Mapping of Distribution Patterns of Secondary Metabolites on a Molecular Phylogeny Inferred from Nucleotide Sequences of the *rbcL* Gene, *Biochem. Syst. Ecol. 31*:897–917 (2003).
8. Lavin, M., M.F. Wojciechowski, P. Gasson, C. Hughes, and E. Wheeler, Phylogeny of Robinioid Legumes (Fabaceae) Revisited: *Coursetia* and *Gliricidia* Recircumscribed, and a Biogeographical Appraisal of the Caribbean Endemics, *Syst. Bot. 28*:387–400 (2003).
9. National Academy of Sciences, *Tropical Legumes: Resources for the Future,* National Academy of Sciences, Washington DC, 1979.
10. Smartt, J., *Grain Legumes, Evolution and Resources,* Cambridge University Press, New York, 1990.

11. Pickersgill, B., and J.M. Lock, *Advances in Legume Systematics, Part 8; Legumes of Economic Importance*, Royal Botanic Gardens, Kew, United Kingdom, 1996.
12. Doyle, J.J., and B.S. Gaut, Evolution of Genes and Taxa: A Primer, *Plant Mol. Biol. 42*:1–23 (2000).
13. Steele, K.P., and M.F. Wojciechowski, Phylogenetic Analyses of Tribes Trifolieae and Vicieae Based on Sequences of the Plastid Gene matK (Papilionoideae: Leguminosae), in *Advances in Legume Systematics, Part 10: Higher Level Systematics,* edited by B.B. Klitgaard, and A. Bruneau, Royal Botanic Gardens, Kew, United Kingdom, 2003, pp. 355–370.
14. Chase M.W., D.E. Soltis, R.G. Olmsead, D. Morgan, D.H. Les, M.R. Duvall, R. Price, H.G. Hills, Y. Qiu, K.A. Kron, J.H. Rettig, E. Conti, J.D. Palmer, M.T. Clegg, J.R. Manhart, K.J. Sytsma, H.J. Michaels, W.J. Kress, M.J. Donoghue, W.D. Clark, M. Hedren, B.S. Gaut, R.K. Jansen, K-J. Kim, C.F. Wimpee, J.F. Smith, G.R. Furnier, S.H. Straus, Q. Xiang, G.M. Plunkett, P.S. Soltis, L.E. Eguiarte, G.H. Learn, S.C.H. Barrett, S. Graham, and V.A. Albert, Phylogenetics of Seed Plants: An Analysis of Nucleotide Sequences from the Plastid Gene *rbcL*, *Ann. Mo. Bot. Gard. 80*:528–580 (1993).
15. Soltis, D.E., P.S. Soltis, M.W. Chase, M.E. Mort, D.C. Albach, M. Zanis, V. Savolainen, W.H. Hahn, S.B. Hoot, M.F. Fay, M. Axtell, S.M. Swensen, L.M. Prince, W.J. Kress, K.C. Nixon, and J.A. Farris, Angiosperm Phylogeny Inferred from 18S rDNA, *rbcL,* and *atpB* Sequences, *Bot. J. Linn. Soc. 133*:381–461 (2000).
16. Dickison W.C., Relationships of the Leguminosae, in *Advances in Legume Systematics, Part 1,* edited by R.M. Polhill and P.H. Raven, Royal Botanic Gardens, Kew, United Kingdom, 1981, pp. 35–54.
17. Swensen, S.M., The Evolution of Actinorhizal Symbioses: Evidence for Multiple Origins of the Symbiotic Association, *Am. J. Bot. 83*:1503–1512 (1996).
18. Soltis, D.E., P.S. Soltis, D.R. Morgan, S.M. Swensen, B.C. Mullin, J.M. Dowd, and P.G. Martin, Chloroplast Gene Sequence Data Suggest a Single Origin of the Predisposition for Symbiotic Nitrogen Fixation in Angiosperms, *Proc. Natl. Acad. Sci. USA 92*:2647–2651 (1995).
19. Zanis, M.J., D.E. Soltis, P.S. Soltis, S. Matheews, and M.J. Donoghue, The Root of the Angiosperms Revisited, *Proc. Natl. Acad. Sci. USA 99*:6848–6853 (2002).
20. Käss E., and M. Wink, Molecular Evolution of the Leguminosae: Phylogeny of the Three Subfamilies Based on *rbcL*-Sequences, *Biochem. Syst. Ecol. 24*:365–378 (1996).
21. Käss E., and M. Wink, Molecular Phylogeny and Phylogeography of *Lupinus* (Leguminosae) Inferred from Nucleotide Sequences of the *rbcL* Gene and ITS 1 + 2 Regions of rDNA, *Plant. Syst. Evol. 208*:139–167 (1997).
22. Doyle, J.J., J.A. Chappill, C.D. Bailey, and T. Kajita, Towards a Comprehensive Phylogeny of Legumes: Evidence from *rbcL* Sequences and Non-molecular Data, in *Advances in Legume Systematics, Part 9,* edited by P. Herendeen and A. Bruneau, Royal Botanic Gardens, Kew, United Kingdom, 2000, pp. 1–20.
23. Doyle, J.J., J.L. Doyle, J.A. Ballenger, E.E. Dickson, T. Kajita, and H. Ohashi, A Phylogeny of the Chloroplast Gene *rbcL* in the Leguminosae: Taxonomic Correlations and Insights into the Evolution of Nodulation, *Am. J. Bot. 84*:541–554 (1997).
24. Lewis, G.P., and B.D. Schrire, Leguminosae or Fabaceae?, in *Advances in Legume Systematics, Part 10: Higher Level Systematics,* edited by B.B. Klitgaard, and A. Bruneau, Royal Botanic Gardens, Kew, United Kingdom, 2003, pp. 1–3.

25. Käss, E., and M. Wink, Phylogenetic Relationships in the Papilionoideae (Family Leguminosae) Based on Nucleotide Sequences of cpDNA (*rbcL*) and ncDNA (ITS 1 and 2), *Mol. Phyl. Evol. 8*:65–88 (1997).
26. Bruneau, A., F. Forest, P.S. Herendeen, B.B. Klitgaard, and G.P. Lewis, Phylogenetic Relationships in the Caesalpinioideae (Leguminosae) as Inferred from Chloroplast *trnL* Intron Sequences. *Syst. Bot. 26*:487–514 (2001).
27. Herendeen, P.S., A. Bruneau, and G.P. Lewis, Phylogenetic Relationships in Caesalpinioid Legumes: A Preliminary Analysis Based on Morphological and Molecular Data, in *Advances in Legume Systematics, Part 10: Higher Level Systematics,* edited by B.B. Klitgaard, and A. Bruneau, Royal Botanic Gardens, Kew, United Kingdom, 2003, pp. 37–62.
28. Taylor, D.W., Paleobiogeographic Relationships of Angiosperms from the Cretaceous and Early Tertiary of the North American Area, *Bot. Rev. 56*:279–415 (1990).
29. Herendeen, P.S., W.L. Crepet, and D.L. Dilcher, The Fossil History of the Leguminosae: Phylogenetic and Biogeographic Implications, in *Advances in Legume Systematics, Part 4, The Fossil Record,* edited by P.S. Herendeen and D.L. Dilcher, Royal Botanic Gardens, Kew, United Kingdom, 1992, pp. 303–316.
30. Schonenberger, J., E.-M. Friis, K.L. Matthews, and P.K. Endress, Cunoniaceae in the Cretaceous of Europe: Evidence from Fossil Flowers, *Ann. Bot. 88*:423–437 (2001).
31. Takahashi, M., P.S. Herendeen, and P.R. Crane, Lauraceous Fossil Flower from the Kamikitaba Locality (Lower Coniacian; Upper Cretaceous) in Northeastern Japan, *J. Plant Res. 114*:429–434 (2001).
32. Gandolfo, M.A., K.C. Nixon, and W.L. Crepet, Triuridaceae Fossil Flowers from the Upper Cretaceous of New Jersey, *Am. J. Bot. 89*:1940–1957 (2002).
33. Sprent, J.I., *Nodulation in Legumes,* Royal Botanic Gardens, Kew, United Kingdom, 2001.
34. Luckow, M., J.T. Miller, D.J. Murphy, and T. Livshultz, A Phylogenetic Analysis of the Mimosoideae (Leguminosae) Based on Chloroplast DNA Sequence Data, in *Advances in Legume Systematics, Part 10: Higher Level Systematics,* edited by B.B. Klitgaard, and A. Bruneau, Royal Botanic Gardens, Kew, United Kingdom, 2003, pp. 197–220.
35. Wojciechowski, M.F., Reconstructing the Phylogeny of Legumes (Leguminosae): An Early 21st Century Perspective, in *Advances in Legume Systematics, Part 10: Higher Level Systematics,* edited by B.B. Klitgaard, and A. Bruneau, Royal Botanic Gardens, Kew, United Kingdom, 2003, pp. 5–35.
36. Doyle J.J., Phylogenetic Perspectives on Nodulation: An Evolving View of Plants and Symbiotic Bacteria, *Trends Plant Sci. 3*:473–478 (1998).
37. Doyle, J.J., and M.A. Luckow, The Rest of the Iceberg: Legume Diversity and Evolution in a Phylogenetic Context, *Plant Physiol. 131*:900–910 (2003).
38. Lewis, G.P., B.D. Schrire, B.A. Mackinder, and J.M. Lock, *Legumes of the World,* Royal Botanic Gardens, Kew, United Kingdom, 2004.
39. Doyle, J.J., J.L. Doyle, J.A. Ballenger, and J.D. Palmer, The Distribution and Phylogenetic Significance of a 50kb Chloroplast DNA Inversion in the Flowering Plant Family Leguminosae, *Mol. Phyl. Evol. 5*:429–438 (1996).
40. Polhill, R.M., Classification of the Leguminosae, in *Phytochemical Dictionary of the Leguminosae, Volume 1: Plants and their Constituents,* edited by F.A. Bisby, J. Buckingham, and J.B. Harborne, Chapman and Hall, London, 1994, pp. xxxv–lvii.

41. Polhill, R.M., Papilionoideae, in *Advances in Legume Systematics, Part 1,* edited by R.M. Polhill and P.H. Raven, Royal Botanic Gardens, Kew, United Kingdom, 1981, pp. 191–208.
42. Lackey, J.A., Phaseoleae, in *Advances in Legume Systematics, Part 1,* edited by R.M. Polhill and P.H. Raven, Royal Botanic Gardens, Kew, United Kingdom, 1981, pp. 301–327.
43. Lavin, M., R.T. Pennington, B.B. Klitgaard, J.I. Sprent, H.C. de Lima, and P.E. Gasson, The Dalbergioid Legumes (Fabaceae): Delimitation of a Pantropical Monophyletic Clade, *Am. J. Bot. 88*:503–533 (2001).
44. Jung, S., P.L. Tate, R. Horn, G. Kochert, K. Moore, and G.A. Abbott, The Phylogenetic Relationship of Possible Progenitors of the Cultivated Peanut, *J. Hered. 94*:334–340 (2003).
45. Crisp, M.D., S. Gilmore, and B.-E. van Wyk, Molecular Phylogeny of the Genistoid Tribes of Papilionoid Legumes, in *Advances in Legume Systematics, Part 9,* edited by P. Herendeen and A. Bruneau, Royal Botanic Gardens, Kew, United Kingdom, 2000, pp. 249–276.
46. van Wyk, B.-E., The Value of Chemosystematics in Clarifying Relationships in the Genistoid Tribes of Papilionoid Legumes, *Biochem. Syst. Ecol. 31*:875–884 (2003).
47. van Heerden, F.R., B.-E. van Wyk, A.M. Viljoen, and P. Steenkamp, Phenolic Variation in Wild Populations of *Aspalathus linearis* (Rooibos Tea), *Biochem. Syst. Ecol. 31*:885–895 (2003).
48. Ainouche, A., R.J. Bayer, P. Cubas, and M.-T. Misse, Phylogenetic Relationships and Evolution within Tribe Genisteae (Papilionoideae) with Special Reference to Genera Ulex and Lupinus, in *Advances in Legume Systematics, Part 10: Higher Level Systematics,* edited by B.B. Klitgaard, and A. Bruneau, Royal Botanic Gardens, Kew, United Kingdom, 2003, pp. 239–252.
49. Lavin, M., J.J. Doyle, and J.D. Palmer, Evolutionary Significance of the Loss of the Chloroplast DNA Inverted Repeat in the Leguminosae Subfamily Papilionoideae, *Evolution 44*:390–402 (1990).
50. Palmer, J.D., B. Osorio, J. Aldrich, and W.F. Thompson, Chloroplast DNA Evolution Among Legumes: Loss of a Large Inverted Repeat Occurred Prior to Other Sequence Rearrangements, *Curr. Genet. 11*:275–286 (1987).
51. Takane, K., S. Tajima, and H. Kouchi, Structural and Expression Analysis of Uricase mRNA from *Lotus japonicus, Mol. Plant Microbe Interact. 13*:1156–1160 (2000).
52. Allan G.J., E.A. Zimmer, W.L. Wagner, and D.D. Sokoloff, Origin and Classification of Tribe Loteae: Insight Based on Phylogenetic Analysis of Internal Transcribed Pacer Regions of nrDNA, in *Advances in Legume Systematics, Part 10: Higher Level Systematics,* edited by B.B. Klitgaard, and A. Bruneau, Royal Botanic Gardens, Kew, United Kingdom, 2003, pp. 371–393.
53. Grant, W.F., and E. Small, The Origin of the *Lotus corniculatus* (Fabaceae) Complex: A Synthesis of Diverse Evidence, *Can. J. Bot.* 74:975–989 (1996).
54. Gauthier, P., R. Lumaret, and A. Bedecarrats, Genetic Variation and Gene Flow in Alpine Diploid and Tetraploid Populations of *Lotus* (*L. alpinus* (D.C.) Schleicher/*L. corniculatus* L.). I. Insights from Morphological and Allozyme Markers, *Heredity 80*:683–693 (1998).
55. Gauthier, P., R. Lumaret, and A. Bedecarrats, Chloroplast DNA Variation in the Genus *Lotus* (Fabaceae) and Further Evidence Regarding the Maternal Parentage of *Lotus corniculatus* L., *Theor. Appl. Genet. 95*:629–636 (1997).

56. Wojciechowski, M.F., M.J. Sanderson, K.P. Steele, and A. Liston, Molecular Phylogeny of the "Temperate Herbaceous Tribes" of Papilionoid Legumes: A Supertree Approach, in *Advances in Legume Systematics, Part 9,* edited by P. Herendeen and A. Bruneau, Royal Botanic Gardens, Kew, United Kingdom, 2000, pp. 277–298.
57. Liston, A., Use of the Polymerase Chain Reaction to Survey for the Loss of the Inverted Repeat in the Legume Chloroplast Genome, in *Advances in Legume Systematics, Part 7: Phylogeny,* edited by M.D. Crisp and J.J. Doyle, Royal Botanic Gardens, Kew, United Kingdom, 1995, pp. 31–40.
58. Hu, J.-M., M. Lavin, M.F. Wojciechowski, and M.J. Sanderson, Phylogenetic Analysis of Nuclear Ribosomal ITS/5.8 S Sequences in the Tribe Millettieae (Fabaceae): *Poecilanthe-Cyclolobium,* the Core Millettieae, and the *Callerya* Group, *Syst. Bot. 27*:722–733 (2002).
59. Hu, J.-M., and S.P. Chang, Two New Members of the *Callerya* group (Fabaceae) Based on Phylogenetic Analysis of *rbcL* Sequences: *Endosamara racemosa* (Roxb.) Geesink and *Callerya vasta* (Kosterm.) Schot., *Taiwania 48*:118–128 (2003).
60. Palmer, J.D., and D. Zamir, Chloroplast DNA Evolution and Phylogenetic Relationships in *Lycopersicon*, *Proc. Nat. Acad. Sci. USA, 79*:5006–5010 (1982).
61. Asmussen, C.B., and A. Liston, Chloroplast DNA Characters, Phylogeny, and Classification of *Lathyrus* (Fabaceae), *Amer. J. Bot. 85*:387–401 (1998).
62. Wojciechowski, M.F., M.J. Sanderson, J.-M. Hu, Evidence on the Monophyly of *Astragalus* (Fabaceae) and its Major Subgroups Based on Nuclear Ribosomal DNA ITS and Chloroplast DNA *trnL* Intron Data, *Syst. Bot. 24*:409–437 (1999).
63. Bena, G., Molecular Phylogeny Supports the Morphologically Based Taxonomic Transfer of the "Medicagoid" Trigonella Species to the Genus *Medicago* L., *Plant Syst. Evol. 229*:217–236 (2001).
64. Bena, G., M.F. Jubier, I. Olivieri, and B. Lejeune, Ribosomal External and Internal Transcribed Spacers: Combined Use in the Phylogenetic Analysis of *Medicago* (Leguminosae), *J. Mol. Evol. 46*:299–306 (1998).
65. Bena, G., B. Lejeune, J.M. Prosperi, and I. Olivieri, Molecular Phylogenetic Approach for Studying Life-History Evolution: The Ambiguous Example of the Genus *Medicago* L., *Proc. R. Soc. Lond., B, Biol. Sci. 265*:1141–1151 (1998).
66. Bena, G., J.M. Prosperi, B. Lejeune, and I. Olivieri, Evolution of Annual Species of the Genus Medicago: A Molecular Phylogenetic Approach, *Mol. Phyl. Evol. 9*:552–559 (1998).
67. Downie, S.R., D.S. Katz-Downie, E.J. Rogers, H.L. Zujewski, and E. Small, Multiple Independent Losses of the Plastid *rpoC1* Intron in *Medicago* (Fabaceae) as Inferred from Phylogenetic Analyses of Nuclear Ribosomal DNA Internal Transcribed Spacer Sequences, *Can. J. Bot. 76*:791–803 (1998).
68. Lesins, K., and I. Lesins, *Genus* Medicago *(Leguminosae): A Taxogenetic Study,* Kluwer, Dordrecht, The Netherlands, 1979.
69. Small, E., and M. Jomphe, A Synopsis of the Genus Medicago (Leguminosae), *Can. J. Bot. 67*:3260–3294 (1989).
70. Cook, D.R., D.J. Kim, N.D. Young, G. Kiss, T.H.N. Ellis, T. Bisseling, and T. Huguet, 2002. Annual Report for NSF Plant Genome Project DBI-0110206: Towards the Complete Gene Inventory and Function of the *Medicago truncatula* genome. Available online at www.medicago.org/documents/NSF2002_report/NSF_2002_report.pdf (Verified June 18, 2003).

71. Geesink, R., Tephrosieae, in *Advances in Legume Systematics, Part 1,* edited by R.M. Polhill and P.H. Raven, Royal Botanic Gardens, Kew, United Kingdom, 1981, pp. 245–260
72. Geesink, R., Scala Millettiearum, *Leiden Botanical Series 8*:1–131 (1984).
73. Lavin, M., E. Eshbaugh, J.-M. Hu, S. Mathews, and R.A. Sharrock, Monophyletic Subgroups of the Tribe Millettieae (Leguminosae) as Revealed by Phytochrome Nucleotide Sequence Data, *Am. J. Bot. 85*:412–433 (1998).
74. Hu, J.-M., M. Lavin, M.F. Wojciechowski, and M.J. Sanderson, Phylogenetic Systematics of the Tribe Millettieae (Leguminosae) Based on Chloroplast *trnK/matK* Sequences and its Implications for Evolutionary Patterns in Papilionoideae, *Am. J. Bot. 87*:418–430 (2000).
75. Nemoto T., and H. Ohashi, Diversity and Evolution of Anatomical Structure of Loments in Tribe Desmodieae (Papilionoideae), in *Advances in Legume Systematics, Part 10: Higher Level Systematics,* edited by B.B. Klitgaard, and A. Bruneau, Royal Botanic Gardens, Kew, United Kingdom, 2003, pp. 395–412.
76. Goldblatt P., Cytology and Phylogeny of Leguminosae, in *Advances in Legume Systematics, Part 1* edited by R.M. Polhill, and P.H. Raven, Royal Botanic Gardens, Kew, United Kingdom, 1981, pp. 427–463.
77. Zhu, T., J.M. Schupp, A. Oliphant, and P. Keim, Hypomethylated Sequences: Characterization of the Duplicate Soybean Genome, *Mol. Gen. Genet. 244*:638–645 (1994).
78. Shoemaker, R.C., K. Polzin, J. Labate, J. Specht, E.C. Brummer, T. Olson, N. Young, V. Concibido, J. Wilcox, J.P. Tamulonis, G. Kochert, and H.R. Boerma, Genome Duplication in Soybean (*Glycine* subgenus *Soja*), *Genetics 144*:329–338 (1996)
79. Schlueter, J.A., P. Dixon, C. Granger, and R.C. Shoemaker, Mining the EST Databases to Determine Evolutionary Events in the Legumes and Grasses, *23rd Stadler Genetics Symposium Proceedings* (in press). See also Chapter 10, this volume.
80. Doyle, J.J., J.L. Doyle, and C. Harbison, Chloroplast-expressed Gutamine Synthetase in *Glycine* and Related Leguminosae: Phylogeny, Gene Duplication, and Ancient Polyploidy, *Syst. Bot. 28*:567–577 (2003).
81. Lee, J., and T. Hymowitz, A Molecular Phylogenetic Study of the Subtribe Glycininae (Leguminosae) Derived from the Chloroplast DNA *rps16* Intron Sequences, *Am. J. Bot. 88*:2064–2073 (2001)
82. Doyle, J.J., and J.L. Doyle, Chloroplast DNA Phylogeny of the Papilionoid Legume Tribe Phaseoleae, *Syst. Bot. 18*:309–327 (1993).
83. Bruneau A., J.L. Doyle, and J.J. Doyle, Phylogenetic Evidence in Phaseoleae: Evidence from Chloroplast Restriction Site Characters, in *Advances in Legume Systematics, Part 7: Phylogeny,* edited by M.D. Crisp and J.J. Doyle, Royal Botanic Gardens, Kew, United Kingdom, 1995, pp. 309–330.
84. Hymowitz, T., and R.J. Singh, Speciation and Cytogenetics, in *Soybeans: Improvement, Production, and Uses, 3rd ed.,* edited by J.E. Specht and H.R. Boerma, *Agronomy Monograph No. 16,* American Society of Agronomy, Madison, WI, 2004.
85. Doyle J.J., J.L. Doyle, and A.H.D. Brown, A Chloroplast DNA Phylogeny of the Wild Perennial Relatives of Soybean (*Glycine* subgenus *Glycine*): Congruence with Morphological and Crossing Groups, *Evolution 44*:371–389 (1990).
86. Kollipara, K.P., R.J. Singh, and T. Hymowitz, Phylogenetic and Genomic Relationships in the Genus *Glycine* Willd. Based on Sequences from the ITS Region of Nuclear rDNA, *Genome 40*:57–68 (1997).

87. Doyle, J.J., J.L. Doyle, J.T. Rauscher, and A.H.D. Brown, Evolution of the Perennial Soybean Polyploid Complex (*Glycine* subgenus *Glycine*): A Study of Contrasts, *Biol. J. Linn. Soc. Lond.* (in press).
88. Delgado-Salinas A., T. Turley, A. Richman, and M. Lavin, Phylogenetic Analysis of the Cultivated and Wild species of *Phaseolus* (Fabaceae), *Syst. Bot. 24*:438–460 (1999).
89. Delgado-Salinas, A., A. Bruneau, and J.J. Doyle, Chloroplast DNA Phylogenetic Studies in New World Phaseolinae (Leguminosae: Papilionoideae: Phaseoleae), *Syst. Bot. 18*:6–17 (1993).
90. Vaillancourt, R.E., N.F. Weeden, A. Bruneau, and J.J. Doyle, Chloroplast DNA Phylogeny of Old World Vigna (Leguminosae), *Syst. Bot. 18*:642–651 (1993).
91. Doi, K., A. Kaga, N. Tomooka, and D.A. Vaughan, Molecular Phylogeny of Genus *Vigna* subgenus *Ceratotropis* Based on rDNA ITS and *atpB-rbcL* Intergenic Spacer of cpDNA Sequences, *Genetica 114*:129–145 (2002).
92. Goel, S, S.N. Raina, and Y. Ogihara, Molecular Evolution and Phylogenetic Implications of Internal Transcribed Spacer Sequences of Nuclear Ribosomal DNA in the *Phaseolus-Vigna* Complex, *Mol. Phyl. Evol. 22*:1–19 (2002).
93. Llaca, V., A. Delgado-Salinas, and P. Gepts, Chloroplast DNA as an Evolutionary Marker in the *Phaseolus vulgaris* Complex, *Theor. Appl. Genet. 88*:646–652 (1994).
94. Schrire, B.D., M. Lavin, N.P. Barker, H. Cortes-Burns, I. von Senger, and J.-H. Kim, Towards a Phylogeny of *Indigofera* (Leguminosae-Papilionoideae): Identification of Major Clades and Relative Ages, in *Advances in Legume Systematics, Part 10: Higher Level Systematics,* edited by B.B. Klitgaard and A. Bruneau, Royal Botanic Gardens, Kew, United Kingdom, 2003, pp. 269–302.
95. Swiderski, M.R, Z. Zaborowska, and A.B. Legocki, Identification of New Nodulin cDNAs from Yellow Lupine by Differential Display, *Plant Sci. 151*:75–83 (2000).
96. Vision, T.J., D.G. Brown, and S.D. Tanksley, The Origins of Genomic Duplications in *Arabidopsis, Science 290*:2114–2117 (2000).
97. Fedorova, M., J. van de Mortel, P.A. Matsumoto, J. Cho, C.D. Town, K.A. VandenBosch, J.S. Gantt, and C.P. Vance, Genome-wide Identification of Nodule-specific Transcripts in the Model Legume *Medicago truncatula, Plant Physiol. 130*:519–537 (2002).
98. Doyle J.J., Phylogeny of the Legume Family: An Approach to Understanding the Origins of Nodulation, *Ann. Rev. Ecol. Syst. 25*:25–349 (1994).
99. McKey, D., Legumes and Nitrogen: The Evolutionary Ecology of a Nitrogen-demanding Lifestyle, in *Advances in Legume Systematics, Part 5: The Nitrogen Factor* edited by J.I. Sprent, and D. McKey, Royal Botanic Gardens, Kew, United Kingdom, 1994, pp 211–228.

Chapter 4

Genomics and Genetic Diversity in Common Bean

Phillip McClean,[a] James Kami,[b] and Paul Gepts[b]

[a]Department of Plant Sciences, North Dakota State University, Fargo, ND 58105; and

[b]Department of Agronomy and Range Science, University of California, Davis, CA 95616-8515

The Agronomic and Experimental Importance of *Phaseolus Vulgaris*

Common bean (*Phaseolus vulgaris* L.; $2n = 2x = 22$) is the most important edible food legume. It represents 50% of the grain legumes consumed worldwide. In some countries, such as Brazil and Mexico, it is the primary source of protein in the human diet. As such, common bean is a very important nonprocessed food crop in third world countries and contributes significantly to the world's protein diet. The importance of this crop as an international protein source is reflected by the fact that the dry bean export market alone has a value of $1.8 billion to the U.S. economy (1). In addition, the cash value of the crop at the U.S. farm gate is $1 billion. Lately, domestic bean consumption has increased because of the rising importance of ethnic foods, high levels of certain minerals and vitamins, and the perceived health benefits related to the blood-cholesterol–lowering effects of beans (2,3).

Many genomic features contribute to the attractiveness of common bean as an experimental crop species. The genome size, estimated to be about 450 to 650 million base pairs (Mb)/haploid, is comparable to rice (4), which generally is considered to have the smallest genome among major crop species. Nearly all loci are single copy (5–7), and the traditionally large families, such as resistance gene analogs (8) and protein kinases (9), are of moderate size.

From a population genetics perspective, the major subdivisions of wild common bean progenitors are known, and the domesticated gene pools have been defined. Based on phaseolin seed storage protein variation (10,11), marker diversity (12,13), and morphology (14), two major gene pools of wild common bean were identified. The Middle American gene pool extends from Mexico through Central America and into Colombia and Venezuela, whereas the Andean gene pool is found in southern Peru, Chile, Bolivia, and Argentina. The two domesticated gene pools appear to converge in Colombia (10). Based on a novel phaseolin type, a third, possibly ancestral gene pool based in southern Ecuador and northern Peru was described (15,16).

Two major domestication events, one in Mesoamerica (possibly west-central Mexico) and the other in the southern Andes, appear to have resulted in the Middle American and Andean gene pools that mirror the geographic distribution of the wild progenitors. Following domestication, gene pool divergence led to the development of three races within each of the two major domesticated gene pools (17). The Middle American gene pool, consisting of races Durango, Jalisco, and Mesoamerica, is represented by the medium and small seeded pinto, pink, black, white, and some snap beans. The Andean gene pool, consisting of races Nueva Granada, Peru, and Chile, is represented by the large-seeded kidney, cranberry, and many snap beans.

Phaseolus research benefits from the existence of two extensive germplasm collections: one at the USDA Plant Introduction Station in Pullman, WA (about 13,000 accessions), and the other at the International Center of Tropical Agriculture (CIAT) in Colombia (about 25,000 accessions). Each collection is freely distributed. Both collections include a broad sample of wild and domesticated *P. vulgaris,* and core collections representing the genetic diversity of the species are available (18,19). Such collections represent a rich source of genetic variability to study species-wide diversity and to apply new genetic marker tools to uncover new sources of disease resistance.

Phaseolus Phylogeny

Recent plant phylogenies are primarily derived using sequence data from genes representing each of the three plant genomes. As with all nitrogen-fixing species, the data have placed *Phaseolus* as a member of the Fabaceae (also known as Leguminosae) family. The family is one of three (along with Polygalaceae and Surianaceae) that comprise the order Fabales. In turn, the Fabales are one of the four orders of the Eurosid I clade. The Fabaceae is subdivided into three subfamilies, and *Phaseolus* is a member of the Papilionoideae subfamily. This is by far the largest subfamily, consisting of 476 genera (20), and all of the main economic legume crops are found among the nearly 14,000 species of this subfamily. It is estimated that this subfamily appeared about 50 million years ago (mya) (21), and it resolves as monophyletic with weak support (22,23).

Four main groups are found within the Papilionoideae subfamily. Two of these, the Hologalegina and Phaseoloid/Millettoid groups, appear to be sister taxa that appeared 35 to 40 mya. Tribes that contain important economic species are assigned to each group. The primary tribe within the Hologalegina group, the IRLC tribe, contains the *Pisum, Cicer, Trifolium,* and *Medicago* genera. Estimates place the appearance of this tribe at about 25 mya. The Phaseoloid/Millettoid group contains the Phaseoleae and Millettieae tribes. The Phaseoleae tribe is economically important because it contains the *Phaseolus, Vigna,* and *Glycine* genera. Further analysis places *Phaseolus* and *Vigna* in the Phaseolinae subtribe, whereas *Glycine* is a member of the Glycininae subtribe.

The most thorough phylogeny of the genus *Phaseolus* was developed by Delgado-Salinas *et al.* (24) using 5.8S rDNA, the corresponding internal transcribed spacer regions one and two, and morphological data. The two species *Macroptilium atropurpureum* (DC) Urb. and *M. erythroloma* appear to be the most related species to the genus *Phaseolus*. Further, all of the 55 species in the analysis formed a monophyletic clade. Within this clade, *P. microcarpus* forms the earliest branch. A total of nine species groups were defined. Among the cultivated species, *P. lunatus* L. was in a group with other South American and oceanic island *Phaseolus* species. The other four cultivated species, *P. vulgaris* L., *P. coccineus* L., *P. polyanthus* (L.) B.S.P., and *P. acutifolius* A. Gray were members of the same group. This research is significant because it provides an experimental basis upon which future hypotheses regarding *Phaseolus* phylogeny can be drawn when other genes are considered.

DNA Sequencing

To compare sequence data available for *Phaseolus* with that available for other members of the Fabaceae family, we performed a series of searches of the Genbank database (summarized in Table 4.1). Currently, nearly 700,000 Fabaceae nucleotide sequences are found in GenBank. Because of the significant recent funding, 87% of the sequences are expressed sequence tag (EST) products. Of total Fabaceae sequences, 76% are *Glycine max* (L.) Merr. and *Medicago truncatula* Gaertner EST sequences. A limited number of EST sequences are available for *Phaseolus.* Twenty thousand one hundred twenty ESTs represent globular stage gene expression in *P. coccineus* (25). The next largest class of sequences in the database represents sequences obtained from genomic surveys. The vast majority of these are end sequences of bacterial artificial chromosome (BAC) and transformation competent bacterial artificial chromosome (TAC) clones. For this class of sequences, 60% represent TAC end sequence data from *Lotus corniculatus* L. (26), whereas 26,000 BAC sequences are available for *G. max* (27). For *Phaseolus,*

TABLE 4.1
Summary of Fabaceae Nucleotide Sequence Found in GenBank (13 Dec., 2003) (Nucleotide Queries Submitted to Entrez at NCBI, www.ncbi.nlm.nih.gov)

	Total[a]	EST[b]	Genomic survey[c]	Gene CDS[d]	Gene complete CDS[e]
Fabaceae	696,895	601,418	77,152	5,592	1,858
Arachis hypogaea	1,563	1,346	0	67	24
Cicer arietinum	513	25	0	217	0
Glycine max	374,174	344,542	26,208	1,411	642
Lotus corniculatus	83,450	36,311	46,569	162	45
Lupinus spp.	2,897	2,492	0	159	81
Medicago sativa	1,548	879	0	329	106
Medicago truncatula	192,804	187,763	3,885	241	74
Phaseolus spp.	22,028	20,807	162	443	119
Phaseolus vulgaris	1,592	575	162	375	96
Pisum sativum	4,623	3,037	154	747	305
Vigna spp.	888	207	89	224	115
Vicia spp.	588	1	65	164	35

[a]query: species [orgn].
[b]query: species [orgn] est cdna.
[c]query: species [orgn] genomic survey.
[d]query: species [orgn] cds NOT est.
[e]query: species [orgn] complete NOT est.

all of the genomic survey sequence data represent the work of Murray *et al.* (28). They sequenced and annotated both ends of the Bng RFLP clones originally used to develop the Florida linkage map (5).

EST data provide a very useful glimpse of the expressed portion of genes in the genome. The data provide a catalog of potential genes to be found once a genome is sequenced. Because EST sequences are named by standard annotation procedures, such as BLAST analysis, they do not represent experimental data. In contrast, GenBank has a large collection of Fabaceae coding sequences (CDSs) that were characterized using genetic or biochemical procedures to confirm the nature of the gene product. If the sequence extends from the start to a stop codon, it is labeled a complete CDS. About one-third of the Fabaceae CDS are complete CDS. Of all the Fabaceae species, *G. max* and *Pisum sativum* L. contain the most completely described sequences.

With 119 genes, *Phaseolus* contains more experimentally defined complete genes than the recently studied model legumes *M. truncatula* (74 genes) and *L. corniculatus* (45 genes). This is certainly indicative of the long history of *Phaseolus* as an experimental organism. For example, the *P. vulgaris* phaseolin storage protein gene was the first plant gene shown to contain an intron (29). The extensive analysis of this gene family has provided important information regarding gene expression in plants (30). Early studies on such *Phaseolus* genes as phenylalanine ammonia lyase (31), chitinase (32), chalcone synthase (33), and chalcone isomerase (34) provided important insight into the physiological response to disease attack in plants.

A number of experiments have positioned genes on the *P. vulgaris* linkage map. Murray *et al.* (28) obtained end sequences of the Bng RFLP clones. Subsequent annotation using the sequences as query for Blast searches of GenBank (Blast E-value $< 1 \times 10^{-5}$) identified 87 genes. Because the Bng clones have previously been mapped (5), the genetic location of these genes is known. Yu *et al.* (35–37) and Blair *et al.* (38) used deposited GenBank sequence data to discover simple repeats that were used to develop microsatellite makers. The mapping of more than 50 of the microsatellite markers has genetically defined the position of these genes. In addition, Rivkin *et al.* (8) defined the genetic location of nine disease-resistance–related sequences, whereas Ferrier-Cana *et al.* (39) placed eight similar genes at a single gene cluster. Collectively, the genetic location of nearly 150 genes is now available in *Phaseolus*.

For a species without a major current genomic program, other approaches are necessary for gene discovery. One of the most useful is homology-based PCR cloning procedures, which can provide insights into the evolutionary relationship of *Phaseolus* genes to other plant genes. Rivkin *et al.* (8) used this procedure to clone nine *Phaseolus* resistance gene analogs (RGA) homologous to previously cloned resistance genes. Each RGA exhibited a higher degree of homology with a sequence from soybean than another *Phaseolus* RGA (8), demonstrating that ancestor legumes contained lineage specific sequences that were transmitted. Because many other Fabaceae RGA sequences recently have been reported, we reanalyzed the data by including those additional sequences. The *Phaseolus* RGA sequences are found in five highly supported different lineages (Fig. 4.1), and each lineage consists of multiple legume species. This confirms previous observations regarding the evolution of this important gene family in legumes.

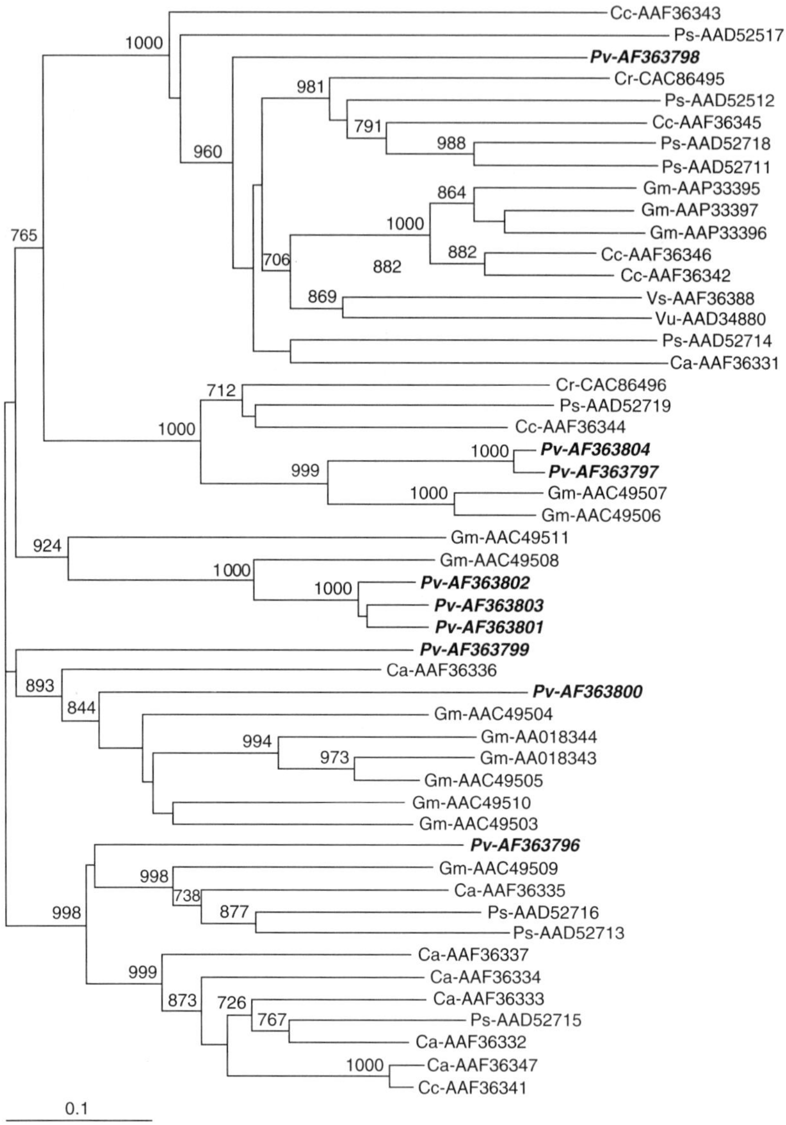

Figure 4.1. Phylogenetic analysis of resistance gene analog sequences in the Fabaceae. RGA amino acid sequences for Fabaceae species that ran from the P-loop to the kinase 2a and 3 domains were selected from GenBank. These were aligned, and neighbor-joining analysis was performed. Those nodes supported with bootstrap values great than 700/1000 are noted. Each RGA is designated by its species abbreviation followed by the GenBank accession number. The abbreviations are: Ca (*Cicer arietinum*), Cc (*Cajanus cajan*), Gm (*Glycine max*), Ps (*Pisum sativum*), Pv (*Phaseolus vulgaris*), Vs (*Vigna subterranea*), and Vu (*Vigna unguiculata*). The *Phaseolus* sequences are in bold italics.

A similar approach was used to study various kinases in *Phaseolus*. By carefully designing primers to the catalyitic domain, Vallad *et al.* (9) cloned 25 sequences related to the *Pto* resistance gene from tomato, the only resistance gene known to exclusively contain kinase functionality. This approach successfully cloned *Pto*-like kinase sequences that were distinct from the many kinase known to exist in eukaryotic genomes. Phylogenetic analysis also determined that the *Pto*-like proteins form a distinct class of kinases. In addition, this research revealed a highly conserved kinase subdomain unique only to plant species.

These PCR-based cloning experiments focused on the relationship of *Phaseolus* to other legumes. Recently, McClean *et al.* (40) used intron sequences to study intraspecies and intergenic relationships with *Phaseolus*. Unique insertion/deletion events and nucleotide polymorphisms defined 20 *P. vulgaris* haplotypes, and small red Middle American beans formed a distinct group not seen when other sequences are used to study relationships within the species. Collectively, these experiments describe the overall utility of PCR-based gene cloning to extend the number of genes cloned in *Phaseolus*.

Genetic Markers

Allozymes, seed proteins, restriction fragment length polymorphisms (RFLPs), random amplified polymorphic DNA (RAPDs), amplified fragment length polymorphic DNA (AFLPs), microsatellites, and inter-simple sequence repeats (ISSRs) have been used to locate genes in common bean. The major applications have been the elucidation of geographic patterns of genetic diversity, molecular linkage mapping, marker-assisted selection, the development of contigs, and positional cloning of genes or gene clusters.

Allozymes were first described in common bean by Kami *et al.* (16,41–42). Further analyses were conducted by Koenig and Gepts (13), Singh *et al.* (43), and Debouck *et al.* (15). The major contribution of allozymes was to clarify the geographic boundaries of major gene pools in wild common bean. They helped define that wild beans of Colombia actually belong to the Mesoamerican gene pool, whereas the wild beans of Ecuador and northern Peru belong to an intermediate gene pool distinct from the Mesoamerican and Andean gene pools (15).

Allozymes also have been useful to identify eco-morpho-geographical races in common bean, particularly in the Mesoamerican gene pool. When specific allozymes were used as prior classification criteria for canonical analyses of phenotypic data, distinct groups were revealed that had characteristic morphological and agronomic traits and were distributed in different ecogeographical areas (44). These groups have been formalized as races, with three races each in both the Mesoamerican and Andean gene pools (17).

Isozymes have also been used to characterize genetic diversity in other *Phaseolus* species, notably runner bean (*P. coccineus*) (45–47), tepary bean (*P. acutifolius*) (48,49), and lima bean (*P. lunatus*) (50–54). In spite of their usefulness in identifying overall patterns of genetic diversity, isozymes have largely been superseded by markers that are more numerous, polymorphic, and better distributed in the genome.

Seed protein analyses have considered primarily the two largest seed protein fractions: phaseolin and the APA proteins (arcelin, phytohaemagglutinin, and α-amylase inhibitor). Both are coded by a single, although complex, locus, consisting of multiple genes in tandem (6,55–58). Phaseolin electrophoretic diversity has been particularly useful to demonstrate the existence of multiple domestications in common and lima beans (10,59) and of a single domestication in tepary bean (60). The value of these markers is that different electrophoretic types can reflect multiple changes at the molecular level. Therefore, independent mutations leading to the same electrophoretic type are unlikely, and each type may have a single origin (61).

Both the phaseolin and APA loci code for important agronomic traits. Phaseolin is the major seed storage protein in common bean. The phaseolin locus is not only a structural locus for phaseolin (62,63) but also a quantitative trait locus (QTL) for phaseolin levels in the seed and for seed weight (64). Based on Sax's description of bean seed color phenotypes, it is likely that the seed weight QTLs actually correspond to the phaseolin locus, given its linkage with the *P* locus, which controls flavonoid pigments in the bean plant (64,65). The APA locus contains sequences that provide resistance against seed weevils (56,66,67). Both the protein and the corresponding DNA sequence have been used as markers for direct selection in breeding either for increased phaseolin or to transfer bruchid resistance from wild to domesticated beans (62,63,68). In spite of their advantages as a biochemical marker, seed proteins suffer the same disadvantage as allozymes in that their genome coverage is limited.

Molecular markers, based on direct or indirect DNA sequence analyses, have increasingly become the marker of choice in beans. In common bean, RFLPs have been used to confirm patterns of genetic diversity identified previously with biochemical markers. They confirmed the multiple domestication scenario for this species (12). RFLPs, however, were principally used as framework markers to develop molecular linkage maps in common bean. The maps of Vallejos *et al.* (5), Gepts *et al.* (69), Nodari *et al.* (70), and Adam-Blondon *et al.* (71) were based primarily on RFLPs, although they also contained isozyme and seed protein markers. Recently, these maps have been correlated into a consensus map based on the recombinant inbred population BAT93 × Jalo EEP558 (Fig. 4.2) (6,72). Five hundred RFLP markers have been mapped in common bean. RFLP maps were instrumental in gaining insight into the inheritance of resistance to common bacterial blight (70,73,74). In addition, common bean was the first dicot in which the inheritance of the domestication syndrome was analyzed by QTL analysis (65). Recently, 150 RFLP markers have been transformed into sequence-tagged sites or SSRs (28).

RAPDs have been used extensively to develop molecular linkage maps but also to characterize genetic diversity. Although this type of marker suffers from lack of reproducibility among laboratories, it can be used within one laboratory with more consistent results. Numerous maps based primarily on RAPDs have been developed, primarily to map genes controlling disease resistances and to tag specific disease resistance genes (75,76). Some of the RAPD markers tagging disease resistance genes have been transformed in sequence-tagged sites (35,77–81).

RAPD markers have also been used to investigate genetic diversity and relationships in the common bean gene pool, not only in the centers of origin (81–84) but also outside the centers (83,85). One of the more salient findings is the identifi-

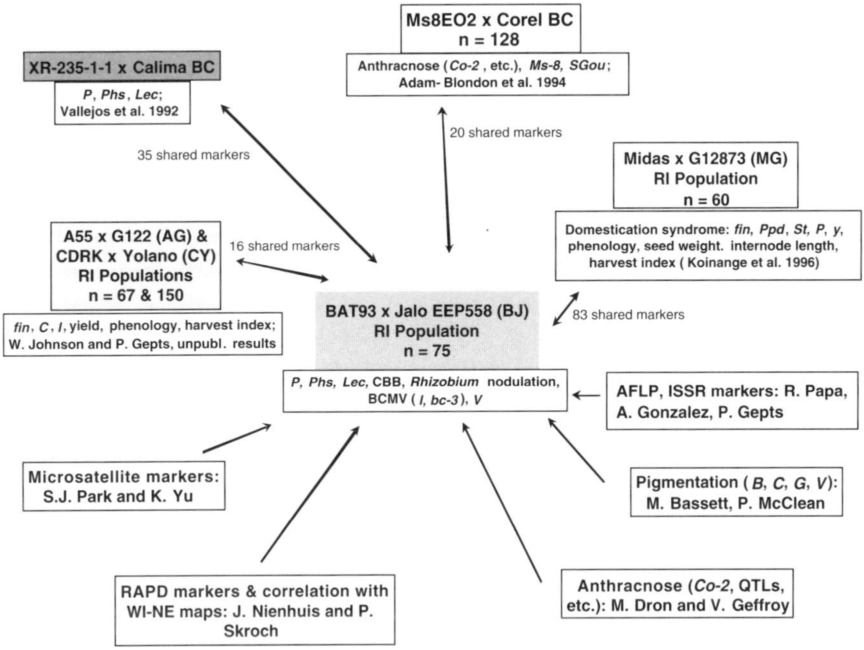

Figure 4.2. The central role of the BAT93 × Jalo EEP558 recombinant inbred population in the integration of molecular linkage maps of common bean (146).

cation of a fourth Middle American race, Guatemala (82). Freyre *et al.* (86) used RAPDs to determine the genetic relationships among wild common bean spanning the distribution range from northern Mexico to northwestern Argentina. They showed that although the Mesoamerican gene pool is geographically structured with Mexican, Central American, and Colombian components, the southern Andean gene pool was apparently unstructured. The intermediate gene pool (Ecuador and northern Peru) was distinct from the Mesoamerican and southern Andean gene pools.

One of the solutions to the lack or reproducibility of RAPDs has been the use of other PCR-based markers with longer primers. AFLPs have proven very useful to characterize wild common bean and lima bean germplasm (19,87) in order to infer the predominant direction of gene flow between domesticated and wild beans as well as to compare the spatial differentiation of domesticated and wild beans (88), and to document the high level of polymorphism and the ecogeographical differentiation among farmer-grown varieties in central Italy (89). In addition, they have been useful to distinguish closely related common bean genotypes for DNA fingerprinting (R. Papa and P. Gepts, unpublished data). Because AFLPs quickly generate a high number of markers, they have been used to develop low-density linkage maps, as well (90,91).

Microsatellite or SSR markers have been developed in recent years from published sequences (37,38,92) and from microsatellite-enriched libraries (38,93,94). These molecular markers (about 150) have been used principally to increase the density of existing maps, especially the core linkage map established in the BAT93 × Jalo EEP558 (36,38). Additional microsatellite markers are needed, however, to increase their density on the linkage map. In addition, targeted identification of microsatellites tagging genes of agronomic importance should be conducted, and the codominance of these markers should be utilized to further characterize genetic diversity in common bean. One the questions raised is the extent of homozygosity in a predominantly selfing species such as common bean in contrast with a predominantly outcrossing species such as runner bean, *P. coccineus*.

About 15 molecular linkage maps have been established in different mapping populations in common bean (reviewed in 75,76). Genes, whether major genes or QTLs, for a wide variety of traits have been located or tagged on these maps. Traits include, but are not limited to: disease resistance (common bacterial blight (5,70,95–98), halo blight (95), white mold (99–102), anthracnose (71,103–113), rust (77,96,114–118), bean common mosaic virus (5,70,95,119), *Fusarium* root rot (78,120)); seed bruchid resistance (56,70); growth habit (65,91,121); seed weight (65,122); seed color (7,123–128); yield (90); canning quality (129); and seed composition (62,63,130).

Generally these are low-density maps, but they have allowed breeders to gain an understanding of the inheritance of disease resistance—for example, in different tissues or genetic backgrounds or when challenged with different strains. Mapping analyses have shown that the major genes for domestication are concentrated on three to four linkage groups (65). As in other species, genes for disease resistance appear to be clustered, and some 15 putative clusters have been identified (76). Resistance gene analogs (RGAs) have been identified, but so far they overlap only partially with the disease resistance gene clusters (8), suggesting that additional disease resistance genes and RGAs need to be mapped.

Molecular markers have also been used to characterize the genetic diversity of organisms associated with the bean host. A parallel distribution of genetic diversity in the host and the organism suggest a possible coevolution as shown for the angular leaf spot pathogen (131,132). Indeed, there are two major gene pools for this pathogen, which are generally more virulent on the host from the same geographical origin, i.e., Mesoamerica and the Andes. Similar observations have been made for other pathogens, such as anthracnose, rust, and common bacterial blight. This coevolution provides excellent opportunities to study coevolution at the molecular level and to follow the evolution of disease resistance genes and their diversification process. For anthracnose, a cluster on linkage group B4 includes both Mesoamerican and Andean resistance specificities, suggesting that this cluster preexisted the divergence from the ancestral gene pool in Ecuador and northern Peru (109,110).

Bacterial Artificial Chromosome Libraries

BAC libraries are large-insert libraries that have been very useful in cloning large portions of genomes in common bean (133), pearl millet (134), sugarbeet (135), cotton (136,137), sunflower (138), and maize (137–139). Insert sizes are generally from 100 to 150 kb. Availability of such libraries allows researchers to develop a physical map of the genome of choice by developing contigs of overlapping clones (140). In turn, this physical map can be related to the genetic map by locating existing genetic markers on the contigs and, conversely, locating markers obtained from analysis of the contigs onto the genetic map (141). Contigs become a starting point for positional cloning of specific genes (142,143).

A correlation of physical and genetic maps can accelerate the discovery of genes underlying phenotypes of agronomic interest (142). It can also help in identifying more closely linked markers for marker-assisted selection (MAS). Furthermore, BAC clones can be used as probes in *in situ* hybridizations of chromosomes (144). In common bean (*P. vulgaris*), BAC libraries have been developed previously by Vanhouten and Mackenzie (133) for an Andean cultivar Sprite FR, and by Kami and Gepts (145) for the Mesoamerican breeding line BAT 93. In common bean, initial *in situ* hybridizations have been performed to kayotype the common bean genome and relate the cytological, genetic, and physical maps (146).

A compelling aspect of the biology of common bean is the existence of information regarding its ancestry and phylogeny (Fig. 4.3) (24,147). Domesticated common bean consists of two major geographic gene pools, Andean and Mesoamerican,

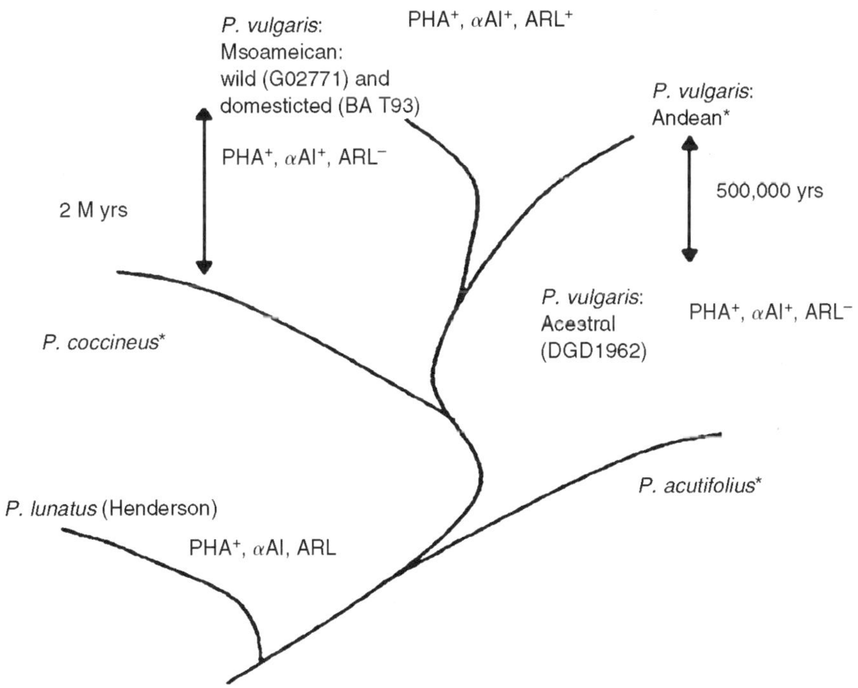

Figure 4.3. Phylogenetic and genealogical relationships among domesticated Phaseolus species (summarized from 161,162). BAC libraries have been developed in *P. vulgaris* genotypes G02771, BAT93, and DGD1962 and in *P. lunatus* Henderson. +PHA: phytohaemagglutinin; αAI: α-amylase inhibitor; ARL: arcelin. *Libraries in these taxa exist or should be developed to complete the model.

originating from independent domestications in the southern Andes and Mesoamerica. In turn, the wild common bean gene pools in these two areas are derived from a common ancestor located in the Andes Mountains from Ecuador and northern Peru (147). Several *Phaseolus* species, some of which have also been domesticated, are closely related to common bean (Fig. 4.1). These include the runner bean (*P. coccineus*), the year bean (*P. polyanthus,* a hybrid between a proto-*P. vulgaris* and *P. coccineus*), and the tepary bean (*P. acutifolius*). All the preceding species belong to the *P. vulgaris* clade or lineage within the genus *Phaseolus*. A more distantly related species is lima bean (*P. lunatus*), which belongs to a different clade altogether (24).

To analyze microevolutionary changes in genome structure, four BAC libraries are currently being developed in four different *Phaseolus* genotypes. Specifically, we want to follow the evolution of the APA complex locus on linkage group B4 (75). APA is a family of closely related seed proteins: α-amylase inhibitor, phytohaemagglutinin, and arcelin. Phytohaemagglutinin (PHA) proteins are widespread among all legumes, including *Phaseolus*. In the *P. vulgaris* lineage, the α-amylase inhibitor subfamily (αAI) appeared by duplication and divergence. Finally, in the Mesoamerican branch of *P. vulgaris*, some wild beans show a third subfamily, that of the arcelins (ARL). Consequently, the four BAC libraries are being developed in selected genotypes that encompass the entire evolution of the APA proteins in the *Phaseolus* genus. The four genotypes include *P. lunatus* cv. Henderson ($PHA^+\alpha AI^-ARL^-$), *P. vulgaris* wild DGD1962 ($PHA^+\alpha AI^+ARL^-$), *P. vulgaris* Mesoamerican domesticated BAT93 ($PHA^+\alpha AI^+ARL^-$), and *P. vulgaris* Mesoamerican wild G-2771 ($PHA^+\alpha AI^+ARL^+$). These complement the BAC library developed in cv. Sprite of Andean origin (133).

For each of the libraries, high molecular weight DNA was obtained after nuclei isolation (148). Following partial digestion with *Hind*III and electroelution (149), the DNA was ligated into the pIndigoBAC5 vector (Epicenter Technologies) and transformed into electrocompetent *E. coli* cells (Strain DH10B ElectroMax cells, Life Technologies). Clones have been distributed in 384-well plates and arrayed on high-density membranes. The characteristics of the libraries are presented in Table 4.2. Based on a conservatively large genome size of 637 Mb

TABLE 4.2
Main Characteristics of *P. vulgaris* BAC Libraries Developed at UC–Davis

Library	Number of clones	Size	Genome equivalents	Empties %
		Phaseolus vulgaris		
BAT93-A	36,864	~110	5.7	11
BAT93/*Hind*III	110,592	125	20.8	<0.5
DGD1962	52,608	105	8.7	1.40
G02771	55,296	139.4	12.1	<0.5
		Phaseolus lunatus		
"Henderson"	55,296	~130	11.3	<0.5

(150,151), the genome coverage ranges from 6× to 12×. There was generally a low frequency of empty and chloroplast DNA clones.

Transformation

Common bean has been transformed by several methods. These include electric particle discharge (152), *Agrobacterium* (153–155), and particle bombardment (156,157). Recent improvements in the transformation efficiency in *P. acutifolius* (153) suggest that improvements in *P. vulgaris* can be expected as well either by direct improvements in the method as applied to *P. vulgaris* or by sexual transfer of the transformability trait to *P. vulgaris* (A. Mejía, CIAT, Cali, Colombia, personal communication). An alternative is to use heterologous systems. Examples include phaseolin and arcelin regulatory sequences in tobacco and Arabidopsis (158,159).

Summary

An international consortium was formed during a meeting held at Cuernavaca, Mexico, 3–4 March, 2001. The consortium is called *Phaseomics* (160). The long term goal of the consortium is to coordinate the development of genomic resources that can be used in an applied setting to develop the next generation of beans that are higher yielding while also providing tolerance to disease or environmental stresses. The resources will include gene sequence data, protein profiles, global expression patterns, large insert libraries, transformation protocols, and information management.

Some of the earliest *Phaseomics* efforts are EST library sequencing. Hernandez *et al.* (161) are analyzing over 3,000 ESTs developed from *P. vulgaris* root nodules induced by Rhizobium infection. An anthracnose resistant *P. vulgaris* genotype was used to generate EST clones that were later sequenced. Currently, 735 unique sequences have been identified (162). The largest number

Chloroplast		APA	Phaseolin.
#	%	#	#
	Phaseolus vulgaris		
21	0.05	5	6
50	0.04	16	13
200	0.4	10	11
49	0.08	38	14
	Phaseolus lunatus		
16	0.03	ND	ND

(21,000) of ESTs was generated from a project to study embryonic development in *P. coccineus* (25).

References

1. Economic Research Service, United States Department of Agriculture, www.ers.usda.gov/briefing/drybeans/ (2003).
2. Andersen, J.W., L. Story, B. Sieling, W.-J.L. Chen, and M.S. Petro, Hypocholesterolemic Effects of Oat-Bran or Bean Intake for Hypercholesterolemic Men, *Am. J. Clin. Nutr. 40*:1146–1155 (1984).
3. Andersen, J.W., and N.J. Gustafson, Hypocholesterolemic Effect of Oat and Bean Products, *Michigan Dry Bean Digest 13*:2–5 (1989).
4. Bennett, M.D., and I.J. Leitch, Nuclear DNA Amounts in Angiosperms, *Ann. Bot. 76*: 113–176 (1995).
5. Vallejos, E.C., N.S. Sakiyama, and C.D. Chase, A Molecular Marker-Based Linkage Map of *Phaseolus vulgaris* L., *Genetics 131*:733–740 (1992).
6. Freyre, R., P. Skroch, V. Geffroy, A.F. Adam-Blondon, A. Shirmohamadali, W. Johnson, V. Llaca, R. Nodari, P. Pereira, S.M. Tsai, J. Tohme, M. Dron, J. Nienhuis, C. Vallejos, and P. Gepts, Towards an Integrated Linkage Map of Common Bean: 4. Development of a Core Map and Alignment of RFLP Maps, *Theor. Appl. Genet. 97*:847–856 (1998).
7. McClean, P., R. Lee, C. Otto, P. Gepts, and M. Bassett, Molecular and Phenotypic Mapping of Genes Controlling Seed Coat Pattern and Color in Common Bean (*Phaseolus vulgaris* L.), *J. Hered. 93*:148–152 (2002).
8. Rivkin, M., C. Vallejos, and P. McClean, Disease-Related Sequences in Common Bean, *Genome 42*:1–7 (1998).
9. Vallad, G., M. Rivkin, C.E. Vallejos, and P. McClean, Cloning and Homology Modeling of a *Pto*-Like Protein Kinase Family of Common Bean (*Phaseolus vulgaris* L.), *Theor. Appl. Genet. 103*:1046–1058 (2001).
10. Gepts, P., and F.A. Bliss, Phaseolin Variability Among Wild and Cultivated Common Beans (*Phaseolus vulgaris* L.) from Colombia, *Econ. Bot. 40*:469–478 (1986).
11. Gepts, P., Biochemical Evidence Bearing on the Domestication of *Phaseolus* (Fabaceae) Beans, *Econ. Bot. 44*:22–38 (1990).
12. Becerra Velásquez, V.L., and P. Gepts, RFLP Diversity in Common Bean (*Phaseolus vulgaris* L.), *Genome 37*:256–263 (1994).
13. Koenig, R., and P. Gepts, Allozyme Diversity in Wild *Phaseolus vulgaris*: Further Evidence for Two Major Centers of Diversity, *Theor. Appl. Genet. 78*:809–817 (1989).
14. Gepts, P, and D. Debouck, Origin, Domestication, and Evolution of Common Bean (*Phaseolus vulgaris* L.), in *Common Beans: Research for Crop Improvement,* edited by A. van Schoonhoven and O. Voyest, C.A.B. Intl., Wallingford and CIAT, Cali, 1991, pp. 7–53.
15. Debouck, D.G., O. Toro, O.M. Paredes, W.C. Johnson, and P. Gepts, Genetic Diversity and Ecological Distribution of *Phaseolus vulgaris* in Northwestern South America, *Econ. Bot. 47*:408–423 (1993).
16. Kami, J., B. Becerra Velásquez, D.G. Debouck, and P. Gepts, Identification of Presumed Ancestral DNA Sequences of Phaseolin in *Phaseolus vulgaris*, *Proc. Natl. Acad. Sci. USA 92*:1101–1104 (1995).

17. Singh, S.P., P. Gepts, and D.G. Debouck, Races of Common Bean (*Phaseolus vulgaris* L., Fabaceae), *Econ. Bot. 45*:379–396 (1991).
18. Miklas, P.N., R. Delorme, R. Hannan, and M.H. Dickson, Using a Subsample of the Core Collection to Identify New Sources of Resistance to White Mold in Common Bean, *Crop Sci. 39*:569–573 (1999).
19. Tohme, J., D.O. González, S. Beebe, and M.C. Duque, AFLP Analysis of Gene Pools of a Wild Bean Core Collection, *Crop Sci. 36*:1375–1384 (1996).
20. Lewis, G.P., B.D. Schrire, B.A. Mackinder, and J.M. Lock, *Legumes of the World,* Royal Botanic Gardens, Kew, United Kingdom, 2003.
21. Wojciechowski, M.F., Reconstructing the Phylogeny of Legumes (Leguminosae): An Early 21st Century Perspective, in *Advances in Legume Systematics, Part 10: Higher Systematics,* edited by B.B. Klitgaar and A. Bruneau, Royal Botanic Garden, Kew, United Kingdom, 2003, pp. 5–35.
22. Doyle, J.J., J.A. Chappill, C.D. Bailey, and T. Kajita, Towards a Comprehensive Phylogeny of Legumes: Evidence from rbcL Sequences and Nonmolecular Data, in *Advances in Legume Systematics, Part 9,* edited by P. Herendeen and A. Bruneau, Royal Botanic Gardens, Kew, United Kingdom, 2000, pp. 249–276.
23. Kajita, T., H. Ohashi, Y. Tateishi, C.D. Bailey, and J.J. Doyle, rbcL and Legume Phylogeny with Particular Reference to *Phaseoleae, Millettieae,* and Allies, *Sys. Bot. 26*:515–536 (2001).
24. Delgado-Salinas, A., T. Turley, A. Richman, and M. Lavin, Phylogenetic Analysis of the Cultivated and Wild Species of *Phaseolus* (Fabaceae), *Syst. Bot. 24*:438–460 (1999).
25. Bui, A.Q., B.H. Le, K. Weterings, Y.P. Bi, J.S. Choi, K.E. McElroy, P.S. Choi, J.J. Harada, R.L. Fischer, and R.B. Goldberg, Gene Activity in Different Regions of Post-Fertilization Plant Embryo by EST Analysis, for example, *GenBank* Record AF533650 (2002).
26. Sato, S., Y. Nakamura, and S. Tabata, *Lotus japonicus* TAC End Sequences, for example, *GenBank* Record AG265785 (2002).
27. Shultz, J.L., K. Meksem, J. Shetty, C.D. Town, H. Koo, J. Potter, K. Wakefield, H. Zhang, C. Wu, and D.A. Lightfoot, End Sequencing of BACs Comprising a Minimal Tiling Path from a Fingerprint Physical Map of Soybean (*Glycine max*) Cultivar Forrest, for example, *GenBank* Record CG826126 (2003).
28. Murray, J., J. Larsen, T.E. Michaels, A. Schaafsma, C.E. Vallejos, and K.P. Pauls, Identification of Putative Genes in Bean (*Phaseolus vulgaris*) Genomic (Bng) RFLP Clones and Their Conversion to STSs, *Genome 45*:1013–1024 (2002).
29. Sun, S.M., J.L. Slightom, and T.C. Hall, Intervening Sequences in a Plant Gene—Comparison of the Partial Sequence of cDNA and Genomic DNA of French Bean Phaseolin, *Nature 289*:37–41 (1981).
30. Hall, T.C., M.B. Chandrasekharan, and G. Li, Phaseolin: Its Past, Properties, Regulation and Future, in *Seed Proteins,* edited by P.R. Shewry and R. Casey, Kluwer, Dordrecht, 1999, pp. 209–240.
31. Edwards, K., C.L. Cramer, G.P. Bolwell, R.A. Dixon, W. Schuch, and C.J. Lamb, Rapid Transient Induction of Phenylalanine Ammonia-Lyase mRNA in Elicitor-Treated Bean Cells, *Proc. Natl. Acad. Sci. USA 82*:6731–6735 (1985).
32. Broglie, K.E., J.J. Gaynor, and R.M. Broglie, Ethylene-Regulated Gene Expression: Molecular Cloning of the Genes Encoding an Endochitinase from *Phaseolus vulgaris*, *Proc. Natl. Acad. Sci. USA 83*:6820–6824 (1986).

33. Ryder, T.B., S.A. Hedrick, J.N. Bell, X.W. Liang, S.D. Clouse, and C.J. Lamb, Organization and Differential Activation of a Gene Family Encoding the Plant Defense Enzyme Chalcone Synthase in *Phaseolus vulgaris*, *Mol. Gen. Genet. 210*:219–233 (1987).
34. Blyden, E.R., P.W. Doerner, C.J. Lamb, and R.A. Dixon, Sequence Analysis of a Chalcone Isomerase cDNA of *Phaseolus vulgaris* L., *Plant Mol. Biol. 16*:167–169 (1991).
35. Yu, K., S.J. Park, and V. Poysa, Marker-Assisted Selection of Common Beans for Resistance to Common Bacterial Blight: Efficacy and Economics, *Plant Breed. 119*:411–415 (2000).
36. Yu, K., S. Park, V. Poysa, and P. Gepts, Integration of Simple Sequence Repeat (SSR) Markers into a Molecular Linkage Map of Common Bean (*Phaseolus vulgaris* L.), *J. Hered. 91*:429–434 (2000).
37. Yu, K.F., S.J. Park, and V. Poysa, Abundance and Variation of Microsatellite DNA Sequences in Beans (*Phaseolus* and *Vigna*), *Genome 42*:27–34 (1999).
38. Blair, M.W., F. Pedraza, H.F. Buendia, E. Gaitán-Solís, S.E. Beebe, P. Gepts, and J. Tohme, Development of a Genome-Wide Anchored Microsatellite Map for Common Bean (*Phaseolus vulgaris* L.), *Theor. Appl. Genet. 107*:1362–1374 (2003).
39. Ferrier-Cana, E., V. Geffroy, C. Macadre, F. Creusot, P. Imbert-Bollore, M. Sevignac, and T. Langin, Characterization of Expressed NBS–LRR Resistance Gene Candidates from Common Bean, *Theor. Appl. Genet. 106*:251–261 (2003).
40. McClean, P.E., R.K. Lee, and P.N. Miklas, Sequence Diversity Analysis of Dihydroflavanol Reductase Intron 1 in Common Bean, *Genome 47*:266–280 (2004).
41. Weeden, N.F., Distinguishing Among White-Seeded Bean Cultivars by Means of Allozyme Genotypes, *Euphytica 33*:199–208 (1984).
42. Weeden, N.F., Enzyme Loci Defined in *Phaseolus vulgaris*, *Annu. Rep. Bean Improv. Coop. 29*:53 (1986).
43. Singh, S.P., R. Nodari, and P. Gepts, Genetic Diversity in Cultivated Common Bean. I. Allozymes, *Crop Sci. 31*:19–23 (1991).
44. Singh, S.P., J.A. Gutiérrez, A. Molina, C. Urrea, and P. Gepts, Genetic Diversity in Cultivated Common Bean: II. Marker-Based Analysis of Morphological and Agronomic Traits, *Crop Sci. 31*:23–29 (1991).
45. Belletti, P., and S. Lotito, Identification of Runner Bean Genotypes (*Phaseolus coccineus* L.) by Isozyme Analysis, *J. Genet. Breed. 50*:185–190 (1996).
46. Piñeiro, D., and L. Eguiarte., The Origin and Biosystematic Status of *Phaseolus coccineus* ssp. *polyanthus*: Electrophoretic Evidence, *Euphytica 37*:199–203 (1988).
47. Wall, J.R., and S.W. Wall, Isozyme Polymorphisms in the Study of Evolution in the *Phaseolus vulgaris–P. coccineus* Complex of Mexico, in *Isozymes IV*, edited by C.L. Markert, Academic Press, New York, 1975.
48. Garvin, D.F., and N.F. Weeden, Isozyme Evidence Supporting a Single Geographic Origin for Domesticated Tepary Bean, *Crop Sci. 34*:1390–1395 (1994).
49. Schinkel, C., and P. Gepts, Allozyme Variability in the Tepary Bean, *Phaseolus acutifolius* A. Gray, *Plant Breed. 102*:182–195 (1989).
50. Bi Irie, Z., A. Maquet, and J.P. Baudoin, Population Genetic Structure of Wild *Phaseolus lunatus* (Fabaceae) with Special Reference to Population Sizes, *Am. J. Bot. 90*:897–904 (2003).

51. Esquivel, M., L. Castineiras, L. Lioi, and K. Hammer, The Domestication of Lima Bean (*Phaseolus lunatus* L.) in Cuba: Morphological and Biochemical Studies, *Plant Genet. Res. Newslet. 91/92*:21–22 (1993).
52. Lioi, L., C. Lotti, and I. Galasso, Isozyme Diversity, RFLP of the rDNA and Phylogenetic Affinities Among Cultivated Lima Beans, *Phaseolus lunatus* (Fabaceae), *Plant Syst. Evol. 213*:153–164 (1998).
53. Maquet, A., I. Zoro Bi, M. Delvaux, B. Wathelet, and J.P. Baudoin, Genetic Structure of a Lima Bean Base Collection Using Allozyme Markers, *Theor. Appl. Genet. 95*:980–991 (1997).
54. Zoro Bi, I., A. Maquet, and J.P. Baudoin, Spatial Patterns of Allozyme Variants within Three Wild Populations of *Phaseolus lunatus* L. from the Central Valley of Costa Rica, *Belg. J. Bot. 129*:149–155 (1997).
55. Hall, T.C., J.L. Slightom, D.R. Ersland, P. Scharf, R.F. Barker, M.G. Murray, J.W.S. Brown, and J.D. Kemp, The Phaseolin Family of Seed Protein Genes: Sequences and Properties, in *Advances in Gene Technology: Molecular Genetics of Plants and Animals,* edited by K. Downey, Academic Press, New York, 1984, pp. 349–367.
56. Osborn, T.C., T. Blake, P. Gepts, and F.A. Bliss, Bean Arcelin: 2. Genetic Variation, Inheritance and Linkage Relationships of a Novel Seed Protein of *Phaseolus vulgaris, Theor. Appl. Genet. 71*:847–855 (1986).
57. Slightom, J.L., R.F. Drong, R.C. Klassy, and L.M. Hoffman, Nucleotide Sequences from Phaseolin cDNA Clones: The Major Storage Proteins from *Phaseolus vulgaris* are Encoded by Two Unique Gene Families, *Nucleic Acids Res. 13*:6483–6498 (1985).
58. Talbot, D.R., M.J. Adang, J.L. Slightom, and T.C. Hall, Size and Organization of a Multigene Family Encoding Phaseolin, the Major Seed Storage Protein of *Phaseolus vulgaris* L., *Mol. Gen. Genet. 198*:42–49 (1984).
59. Gutiérrez Salgado, A., P. Gepts, and D. Debouck, Evidence for Two Gene Pools of the Lima Bean, *Phaseolus lunatus* L., in the Americas, *Genet. Resour. Crop Evol. 42*:15–22 (1995).
60. Schinkel, C., and P. Gepts, Phaseolin Diversity in the Tepary Bean, *Phaseolus acutifolius* A. Gray, *Plant Breed. 101*:292–301 (1988).
61. Gepts, P., Phaseolin as an Evolutionary Marker, in *Genetic Resources of Phaseolus Beans,* edited by P. Gepts, Kluwer, Dordrecht, The Netherlands, 1988, pp. 215–241.
62. Delaney, D.E., and F.A. Bliss, Selection for Increased Percentage of Phaseolin in Common Bean. 1. Comparison of Selection for Seed Protein Alleles and S1 Family Recurrent Selection, *Theor. Appl. Genet. 81*:301–305 (1991).
63. Delaney, D.E., and F.A. Bliss, Selection for Increased Percentage of Phaseolin in Common Bean. 2. Changes in Frequency of Seed Protein Alleles with S_1 Family Recurrent Selection, *Theor. Appl. Genet. 81*:306–311 (1991).
64. Johnson, W., C. Menéndez, R. Nodari, E. Koinange, S. Singh, and P. Gepts, Association of a Seed Weight Factor with the Phaseolin Seed Storage Protein Locus Across Genotypes, Environments, and Genomes in *Phaseolus–Vigna* spp., *J. Quant. Trait Loci 2:*Article 5, www.cabi-publishing.org/gateways/jag/papers96/paper596/index596.html (1996).
65. Koinange, E.M.K., S.P. Singh, and P. Gepts, Genetic Control of the Domestication Syndrome in Common Bean, *Crop Sci. 36*:1037–1045 (1996).
66. Mirkov, T.E., J.M. Wahlstrom, K. Hagiwara, F. Finardi-Filho, S. Kjemtrup, and M.J. Chrispeels, Evolutionary Relationships Among Proteins in the Phytohemagglutinin-arcelin-alpha-amylase Inhibitor Family of the Common Bean and Its Relatives, *Plant Mol. Biol. 26*:1103–1113 (1994).

67. Osborn, T.C., D.C. Alexander, S.S.M. Sun, C. Cardona, and F.A. Bliss, Insecticidal Activity and Lectin Homology of Arcelin Seed Protein, *Science 240*:207–210 (1988).
68. Kornegay, J., C. Cardona, and C.E. Posso, Inheritance of Resistance to Mexican Bean Weevil in Common Bean, Determined by Bioassay and Biochemical Tests, *Crop Sci. 33*:589–594 (1993).
69. Gepts, P., R. Nodari, R. Tsai, E.M.K. Koinange, V. Llaca, R. Gilbertson, and P. Guzmán, Linkage Mapping in Common Bean, *Annu. Rep. Bean Improv. Coop. 36*:24–38 (1993).
70. Nodari, R.O., S.M. Tsai, R.L. Gilbertson, and P. Gepts, Towards an Integrated Linkage Map of Common Bean. II. Development of an RFLP-Based Linkage Map, *Theor. Appl. Genet. 85*:513–520 (1993).
71. Adam-Blondon, A., M. Sévignac, and M. Dron, A Genetic Map of Common Bean to Localize Specific Resistance Genes Against Anthracnose, *Genome 37*:915–924 (1994).
72. Vallejos, C., P. Skroch, and J. Nienhuis, *Phaseolus vulgaris*—The Common Bean. Integration of RFLP and RAPD-Based Linkage Maps, in *DNA-Based Markers in Plants,* Kluwer, Dordrecht, The Netherlands, 2001, pp. 301–317.
73. Nodari, R.O., S.M. Tsai, P. Guzmán, R.L. Gilbertson, and P. Gepts, Towards an Integrated Linkage Map of Common Bean. 3. Mapping Genetic Factors Controlling Host–Bacteria Interactions, *Genetics 134*:341–350 (1993).
74. Yu, Z., R. Stall, and C. Vallejos, Detection of Genes for Resistance to Common Bacterial Blight of Beans, *Crop Sci. 38*:1290–1296 (1998).
75. Gepts, P., Development of an Integrated Genetic Linkage Map in Common Bean (*Phaseolus vulgaris* L.) and Its Use, in *Bean Breeding for the 21st Century,* edited by S. Singh, Kluwer, Dordrecht, The Netherlands, 1999, pp. 53–91 and 389–400.
76. Kelly, J.D., P. Gepts, P.N. Miklas, and D.P. Coyne, Tagging and Mapping of Genes and QTL, and Molecular Marker-Assisted Selection for Traits of Economic Importance in Bean and Cowpea, *Field Crops Res. 82*:135–154 (2003).
77. Correa, R.X., M.R. Costa, P.I. Good-God, V.A. Ragagnin, F.G. Faleiro, M.A. Moreira, and E.G. de Barros, Sequence Characterized Amplified Regions Linked to Rust Resistance Genes in the Common Bean, *Crop Sci. 40*:804–807 (2000).
78. Fall, A.L., P.F. Byrne, G. Jung, D.P. Coyne, M.A. Brick, and H.F. Schwartz, Detection and Mapping of a Major Locus for Fusarium Wilt Resistance in Common Bean, *Crop Sci. 41*:1494–1498 (2001).
79. Jung, G., S. Beebe, J. Nienhuis, S. Park, D. Coyne, J. Marita, F. Pedraza, and J. Tohme, Development of SCAR Markers Linked to Common Bacterial Blight Resistance Genes (QTL) in Common Bean, *Ann. Rep. Bean Improv. Coop. 41*:97–98 (1998).
80. Melotto, M., and J. Kelly, SCAR Markers Linked to Major Disease Resistance Genes in Common Bean, *Ann. Rep. Bean Improv. Coop. 41*:64–65 (1998).
81. Singh, S.P., F.J. Morales, P.N. Miklas, and H. Terán, Selection for Bean Golden Mosaic Resistance in Intra- and Interracial Bean Populations, *Crop Sci. 40*:1565–1572 (2000).
82. Beebe, S., P. Skroch, J. Tohme, M. Duque, F. Pedraza, and J. Nienhuis, Structure of Genetic Diversity among Common Bean Landraces of Middle-American Origin Based on Correspondence Analysis of RAPD, *Crop Sci. 40*:264–273 (2000).
83. Eichenberger, K., F. Gugerli, and J.J. Schneller, Morphological and Molecular Diversity of Swiss Common Bean Cultivars (*Phaseolus vulgaris* L., Fabaceae) and Their Origin, *Botanica Helvetica 110*:61–77 (2000).
84. Galvan, M.Z., M.B. Aulicino, S. Garcia Medina, and P.A. Balatti, Genetic Diversity Among Northwestern Argentinian Cultivars of Common Bean (*Phaseolus vulgaris* L.) as Revealed by RAPD Markers, *Genet. Resour. Crop Evol. 48*:251–260 (2001).

85. Mienie, C.M.S., L. Herselman, and R.E. Terhlanche, RAPD Analysis of Selected Common Bean (*Phaseolus vulgaris* L.) Cultivars from Three African Countries, *South African J. Plant and Soil 17*:93–94 (2000).
86. Freyre, R., R. Ríos, L. Guzmán, D. Debouck, and P. Gepts, Ecogeographic Distribution of *Phaseolus* spp. (Fabaceae) in Bolivia, *Econ. Bot. 50*:195–215 (1996).
87. Caicedo, A.L., E. Gaitan, M.C. Duque, O.T. Chica, D.G. Debouck, and J. Tohme, AFLP Fingerprinting of *Phaseolus lunatus* L. and Related Wild Species from South America, *Crop Sci. 39*:1497–1507 (1999).
88. Papa, R., and P. Gepts, Asymmetry of Gene Flow and Differential Geographical Structure of Molecular Diversity in Wild and Domesticated Common Bean (*Phaseolus vulgaris* L.) from Mesoamerica, *Theor. Appl. Genet. 106*:239–250 (2003).
89. Negri, V., and N. Tosti, Genetic Diversity Within a Common Bean Landrace of Potential Economic Value: Its Relevance for On-Farm Conservation and Product Certification, *J. Genet. Breed. 56*:113–118 (2002).
90. Johnson, W., and P. Gepts, The Role of Epistasis in Controlling Seed Yield and Other Agronomic Traits in an Andean × Mesoamerican Cross of Common Bean (*Phaseolus vulgaris* L.), *Euphytica 125*:69–79 (2002).
91. Tar'an, B., E. Michaels Thomas, and K.P. Pauls, Genetic Mapping of Agronomic Traits in Common Bean, *Crop Sci. 42*:544–556 (2002).
92. Masi, P., P.L. Spagnoletti Zeuli, and P. Donini, Development and Analysis of Multiplex Microsatellite Marker Sets in Common Bean (*Phaseolus vulgaris* L.), *Mol. Breed. 11*:303–313 (2003).
93. Gaitán-Solís, E., M.C. Duque, K.J. Edwards, and J. Tohme, Microsatellite Repeats in Common Bean (*Phaseolus vulgaris*): Isolation, Characterization, and Cross-Species Amplification in *Phaseolus* ssp., *Crop Sci. 42*:2128–2136 (2002).
94. Yaish, M.W.F., and M. Pérez de la Vega, Isolation of (GA)(n) Microsatellite Sequences and Description of a Predicted MADS-Box Sequence Isolated from Common Bean (*Phaseolus vulgaris* L.), *Genet. Mol. Biol. 26*:337–342 (2003).
95. Ariyarathne, H.M., D.P. Coyne, G. Jung, P.W. Skroch, A.K. Vidaver, J.R. Steadman, P.N. Miklas, and M.J. Bassett, Molecular Mapping of Disease Resistance Genes for Halo Blight, Common Bacterial Blight, and Bean Common Mosaic Virus in a Segregating Population of Common Bean, *J. Am. Soc. Hortic. Sci. 124*:654–662 (1999).
96. Jung, G., D. Coyne, P. Skroch, J. Nienhuis, E. Arnaud-Santana, J. Bokosi, H. Ariyarathne, J. Steadman, J. Beaver, and S. Kaeppler, Molecular Markers Associated with Plant Architecture and Resistance to Common Blight, Web Blight, and Rust in Common Beans, *J. Am. Soc. Hortic. Sci. 121*:794–803 (1996).
97. Jung, G., P. Skroch, D. Coyne, J. Nienhuis, H. Ariyarathne, S. Kaeppler, and M. Bassett, Molecular-Marker Based Genetic Analysis of Tepary Bean–Derived Common Bacterial Blight Resistance in Different Developmental Stages of Common Bean, *J. Am. Soc. Hortic. Sci. 122*:329–337 (1997).
98. Park, S.O., D.P. Coyne, N. Mutlu, G. Jung, and J.R. Steadman, Confirmation of Molecular Markers and Flower Color Associated with QTL for Resistance to Common Bacterial Blight in Common Beans, *J. Am. Soc. Hortic. Sci. 124*:519–526 (1999).
99. Kolkman, J.M., and J.D. Kelly, QTL Conferring Resistance and Avoidance to White Mold in Common Bean, *Crop Sci. 43*:539–548 (2003).
100. Miklas, P., W. Johnson, R. Delorme, and P. Gepts, QTL Conditioning Physiological Resistance and Avoidance to White Mold in Dry Bean, *Crop Sci. 41*:309–315 (2001).

101. Miklas, P.N., R. Delorme, and R. Riley, Identification of QTL Conditioning Resistance to White Mold in Snap Bean, *J. Am. Soc. Hort. Sci. 128*:564–570 (2003).
102. Park, S.O., D.P. Coyne, J.R. Steadman, and P.W. Skroch, Mapping of QTL for Resistance to White Mold Disease in Common Bean, *Crop Sci. 41*:1253–1262 (2001).
103. Adam-Blondon, A., M. Sévignac, H. Bannerot, and M. Dron, SCAR, RAPD, and RFLP Markers Tightly Linked to a Dominant Gene (*Are*) Conferring Resistance to Anthracnose in Common Bean, *Theor. Appl. Genet. 88*:865–870 (1994).
104. Alzate-Marin, A.L., H. Menarim, M. Chagas Jose, G. de Barros Everaldo, and M.A. Moreira, Identification of a RAPD Marker Linked to the *Co-6* Anthracnose Resistant Gene in Common Bean Cultivar AB136, *Genet. Mol. Biol. 23*:633–637 (2000).
105. Alzate-Marin, A.L., H. Menarim, G.S. Baia, T.J. Paula Junior, K.A. De Souza, M.R. D.A. Costa, E.G. De Barros, and M.A. Moreira, Inheritance of Anthracnose Resistance in the Common Bean Differential Cultivar G2333 and Identification of a New Molecular Marker Linked to the *Co-42* Gene, *J. Phytopathol. 149*:259–264 (2001).
106. Cardoso de Arruda, M.C., A.L. Alzate-Marin, J. Mauro Chagas, M.A. Moreira, and E. Goncalves de Barros, Identification of Random Amplified Polymorphic DNA Markers Linked to the Co-4 Resistance Gene to *Colletotrichum lindemuthianum* in Common Bean, *Phytopathology 90*:758–761 (2000).
107. Creusot, F., C. Macadre, E. Ferrier Cana, C. Riou, V. Geffroy, M. Sevignac, M. Dron, and T. Langin, Cloning and Molecular Characterization of Three Members of the NBS-LRR Subfamily Located in the Vicinity of the *Co-2* Locus for Anthracnose Resistance in *Phaseolus vulgaris*, *Genome 42*:254–264 (1999).
108. Geffroy, V., F. Creusot, J. Falquet, M. Sévignac, A.F. Adam-Blondon, H. Bannerot, P. Gepts, and M. Dron, A Family of LRR Sequences in the Vicinity of the *Co-2* Locus for Anthracnose Resistance in *Phaseolus vulgaris* and Its Potential Use in Marker-Assisted Selection, *Theor. Appl. Genet. 96*:494–502 (1998).
109. Geffroy, V., D. Sicard, J. de Oliveira, M. Sévignac, S. Cohen, P. Gepts, C. Neema, and M. Dron, Identification of an Ancestral Resistance Gene Cluster Involved in the Coevolution Process Between *Phaseolus vulgaris* and Its Fungal Pathogen *Colletotrichum lindemuthianum*, *Mol. Plant Microbe Interact. 12*:774–784 (1999).
110. Geffroy, V., M. Sévignac, J. De Oliveira, G. Fouilloux, P. Skroch, P. Thoquet, P. Gepts, T. Langin, and M. Dron, Inheritance of Partial Resistance Against *Colletotrichum lindemuthianum* in *Phaseolus vulgaris* and Colocalization of QTL with Genes Involved in Specific Resistance, *Mol. Plant Microbe Interact. 13*:287–296 (2000).
111. Mendoza, A., F. Hernandez, S. Hernandez, D. Ruiz, O. Martinez de la Vega, G. Mora, J. Acosta, and J. Simpson, Identification of Co-1 Anthracnose Resistance and Linked Molecular Markers in Common Bean Line A193, *Plant Dis. 85*:252–255 (2001).
112. Young, R., and J. Kelly, RAPD Markers Linked to Three Major Anthracnose Resistance Genes in Common Bean, *Crop Sci. 37*:940–946 (1997).
113. Young, R., M. Melotto, R. Nodari, and J. Kelly, Marker-Assisted Dissection of the Oligogenic Anthracnose Resistance in the Common Bean Cultivar 'G 2333', *Theor. Appl. Genet. 96*:87–94 (1998).

114. Gelape Faleiro, F., W. Santos Vinhadelli, A. Ragagnin Vilmar, X. Correa Ronan, A. Moreira Maurilio, and E. Goncalves de Barros, RAPD Markers Linked to a Block of Genes Conferring Rust Resistance to the Common Bean, *Genet. Mol. Biol. 23*:399–402 (2000).
115. Haley, S.D., P.N. Miklas, J.R. Stavely, J. Byrum, and J.D. Kelly, Identification of RAPD Markers Linked to a Major Rust Resistance Gene Block in Common Bean, *Theor. Appl. Genet. 86*:505–512 (1993).
116. Johnson, E., P. Miklas, J. Stavely, and J. Martínez-Cruzado, Coupling- and Repulsion-RAPDs for Marker-Assisted Selection of PI 181996 Rust Resistance in Common Bean, *Theor. Appl. Genet. 90*:659–664 (1995).
117. Miklas, P.N., J.R. Stavely, and J.D. Kelly, Identification and Potential Use of a Molecular Marker for Rust Resistance in Common Bean, *Theor. Appl. Genet. 85*:745–749 (1993).
118. Park S.O., D.P. Coyne, J.R. Steadman, and P.W. Skroch, Mapping of the *Ur-7* Gene for Specific Resistance to Rust in Common Bean, *Crop Sci. 43*:1470–1476 (2003).
119. Melotto, M., L. Afanador, and J. Kelly, Development of a SCAR Marker Linked to the *I* Gene in Common Bean, *Genome 39*:1216–1219 (1996).
120. Schneider, K., K. Grafton, and J. Kelly, QTL Analyses of Resistance to Fusarium Root Rot in Bean, *Crop Sci. 41*:535–542 (2001).
121. Beattie, A.D., J. Larsen, T.E. Michaels, and K.P. Pauls, Mapping Quantitative Trait Loci for a Common Bean (*Phaseolus vulgaris* L.) Ideotype, *Genome 46*:411–422 (2003).
122. Park, S., D.P. Coyne, G. Jung, P.W. Skroch, E. Arnaud-Santana, J. Steadman, H. Ariyarathne, and J. Nienhuis, Mapping of QTL for Seed Size and Shape Traits in Common Bean, *J. Am. Soc. Hortic. Sci. 125*:466–475 (2000).
123. Bassett, M., The Dark Corona Character in Seedcoats of Common Bean Cosegregates with the Pink Flower Allele *'vlae'*, *J. Am. Soc. Hortic. Sci. 120*:520–522 (1995).
124. Bassett, M., Tight Linkage Between the *'fin'* Locus for Plant Habit and the *'Z'* Locus for Partly Colored Seedcoat Patterns in Common Bean, *J. Am. Soc. Hortic. Sci. 122*:656–658 (1997).
125. Bassett, M., C. Shearon, and P. McClean, Allelism Between the Hilum Ring Locus 'D' and the Partly Colored Seedcoat Locus 'Z' in Common Bean, *J. Am. Soc. Hortic. Sci. 124*:649–653 (1999).
126. Bassett, M.J., R. Lee, C. Otto, and P.E. McClean, Classical and Molecular Genetic Studies of the Strong Greenish Yellow Seedcoat Color in 'Wagenaar' and 'Enola' Common Bean, *J. Am. Soc. Hortic. Sci. 127*:50–55 (2002).
127. Brady, L., M. Bassett, and P. McClean, Molecular Markers Associated with *T* and *Z*, Two Genes Controlling Partly Colored Seed Coat Patterns in Common Bean, *Crop Sci. 38*:1073–1075 (1998).
128. Erdmann, P.M., R.K. Lee, M.J. Bassett, and P.E. McClean, A Molecular Marker Tightly Linked to *P*, a Gene Required for Flower and Seedcoat Color in Common Bean (*Phaseolus vulgaris* L.), Contains the Ty3-gypsy Retrotransposon Tpv3g, *Genome 45*:728–736 (2002).
129. Posa-Macalincag, M.-C.T., G.L. Hosfield, K.F. Grafton, M.A. Uebersax, and J.D. Kelly, Quantitative Trait Loci (QTL) Analysis of Canning Quality Traits in Kidney Bean (*Phaseolus vulgaris* L.), *J. Am. Soc. Hortic. Sci. 127*:608–615 (2002).

130. Guzmán-Maldonado, S.H., O. Martínez, J.A. Acosta-Gallegos, F. Guevara-Lara, and O. Paredes-Lopez, Putative Quantitative Trait Loci for Physical and Chemical Components of Common Bean, *Crop Sci. 43*:1029–1035 (2003).
131. Gepts, P., and F.A. Bliss, F1 Hybrid Weakness in the Common Bean: Differential Geographic Origin Suggests Two Gene Pools in Cultivated Bean Germplasm, *J. Hered. 76*:447–450 (1985).
132. Guzmán, P., R.L. Gilbertson, R. Nodari, W.C. Johnson, S.R. Temple, D. Mandala, A.B.C. Mkandawire, and P. Gepts, Characterization of Variability in the Fungus *Phaeoisariopsis griseola* Suggests Coevolution with the Common Bean (*Phaseolus vulgaris*), *Phytopathology 85*:600–607 (1995).
133. Vanhouten, W., and S. Mackenzie, Construction and Characterization of a Common Bean Bacterial Artificial Chromosome Library, *Plant Mol. Biol. 40*:977–983 (1999).
134. Allouis, S., X. Qi, S. Lindup, M.D. Gale, and K.M. Devos, Construction of a BAC Library of Pearl Millet, *Pennisetum glaucum, Theor. Appl. Genet. 102*:1200–1205 (2001).
135. Gindullis, F., D. Dechyeva, and T. Schmidt, Construction and Characterization of a BAC Library for the Molecular Dissection of a Single Wild Beet Centromere and Sugar Beet (*Beta vulgaris*) Genome Analysis, *Genome 44*:846–855 (2001).
136. Tomkins, J.P., D.G. Peterson, T.J. Yang, D. Main, T.A. Wilkins, A.H. Paterson, and R.A. Wing, Development of Genomic Resources for Cotton (*Gossypium hirsutum* L.): BAC Library Construction, Preliminary STC Analysis, and Identification of Clones Associated with Fiber Development, *Mol. Breed. 8*:255–261 (2001).
137. Wang, W., W. Zhai, M. Luo, G. Jiang, X. Chen, X. Li, R.A. Wing, and L. Zhu, Chromosome Landing at the Bacterial Blight Resistance Gene *Xa4* Locus Using a Deep Coverage Rice BAC Library, *Mol. Genet. Genomics 265*:118–125 (2001).
138. Gentzbittel, L., A. Abbott, J.P. Galaud, L. Georgi, F. Fabre, T. Liboz, and G. Alibert, A Bacterial Artificial Chromosome (BAC) Library for Sunflower, and Identification of Clones Containing Genes for Putative Transmembrane Receptors, *Mol. Genet. Genomics 266*:979–987 (2002).
139. Tomkins, J.P., G. Davis, D. Main, Y. Yim, N. Duru, T. Musket, J.L. Goicoechea, D.A. Frisch, E H. Coe, Jr., and R.A. Wing, Construction and Characterization of a Deep-Coverage Bacterial Artificial Chromosome Library for Maize, *Crop Sci. 42*:928–933 (2002).
140. Tao, Q., Y.L. Chang, J. Wang, H. Chen, M.N. Islam-Faridi, C. Scheuring, B. Wang, D.M. Stelly, and H.B. Zhang, Bacterial Artificial Chromosome-Based Physical Map of the Rice Genome Constructed by Restriction Fingerprint Analysis, *Genetics 158*:1711–1724 (2001).
141. Lewers, K., R. Heinz, H. Beard, L. Marek, and B. Matthews, A Physical Map of a Gene-Dense Region in Soybean Linkage Group A2 Near the Black Seed Coat and *Rhg4* Loci, *Theor. Appl. Genet. 104*:254–260 (2002).
142. Liu, N., Y. Shan, F.P. Wang, C.G. Xu, K.M. Peng, X.H. Li, and Q. Zhang, Identification of an 85-kb DNA Fragment Containing *pms*1, a Locus for Photoperiod-Sensitive Genic Male-Sterility in Rice, *Mol. Genet. Genomics 266*:271–275 (2001).
143. Xu, M., J. Song, Z. Cheng, J. Jiang, and S.S. Korban, A Bacterial Artificial Chromosome (BAC) Library of *Malus floribunda* 821 and Contig Construction for Positional Cloning of the Apple Scab Resistance Gene *Vf, Genome 44*:1104–1113 (2001).

144. Kim, J.S., K.L. Childs, M.N. Islam-Faridi, M.A. Menz, R.R. Klein, P.E. Klein, H.J. Price, J.E. Mullet, and D.M. Stelly, Integrated Karyotyping of Sorghum by in situ Hybridization of Landed BACs, *Genome 45*:402–412 (2002).
145. Kami, J., and P. Gepts, Development of a BAC Library in Common Bean Genotype BAT93, *Ann. Rep. Bean Improv. Coop. 43*:208–209 (2000).
146. Pedrosa, A., C. Vallejos, A. Bachmair, and D. Schweizer, Integration of Common Bean (*Phaseolus vulgaris* L.) Linkage and Chromosomal Maps, *Theor. Appl. Genet. 106*:205–212 (2003).
147. Gepts, P., Origin and Evolution of Common Bean: Past Events and Recent Trends, *Hort. Sci. 33*:1124–1130 (1998).
148. Kuehl, R., Isolation of Plant Nuclei, *Z. Naturforsch. 198*:525–534 (1964).
149. Strong, S.J., Y. Ohta, G.W. Litman, and C.T. Amemiya, Marked Improvement of PAC and BAC Cloning Is Achieved Using Electroelution of Pulsed-Field Gel-Separated Digests of Genomic DNA, *Nucl. Acids Res. 25*:3959–3961 (1997).
150. Bennett, M.D., P. Bhandol, and I.J. Leitch, Nuclear DNA Amounts in Angiosperms and Their Modern Uses—807 New Estimates, *Ann. Bot. 86*:859–909 (2000).
151. Arumuganathan, K., and E.D. Earle, Nuclear DNA Content of Some Important Plant Species, *Plant Mol. Biol. 9*:208–218 (1991).
152. Russell, D.R., K.M. Wallace, J.H. Bathe, and B.J. Martinell, Stable Transformation of *Phaseolus vulgaris* via Electric Discharge Mediated Particle Acceleration, *Plant Cell Rep. 12*:165–169 (1993).
153. De Clercq, J., M. Zambre, M. Van Montagu, W. Dillen, and G. Angenon, An Optimized Agrobacterium-Mediated Transformation Procedure for *Phaseolus acutifolius* A. Gray, *Plant Cell Rep. 21*:333–340 (2002).
154. Dillen, W., J. De Clercq, A. Goossens, M. Van Montagu, and G. Angenon, Agrobacterium-Mediated Transformation of *Phaseolus acutifolius* A. Gray, *Theor. Appl. Genet. 94*:151–158 (1997).
155. McClean, P., P. Chee, B. Held, J. Simental, R.F. Drong, and J. Slightom, Susceptibility of Dry Bean (*Phaseolus vulgaris* L.) to Agrobacterium Infection: Transformation of Cotyledonary and Hypocotyl Tissues, *Plant Cell, Tissue, Organ Cult. 24*:131–138 (1991).
156. Aragao, F., L. Barros, A. Brasileiro, S. Ribeiro, F. Smith, J. Sanford, J. Faria, and E. Rech, Inheritance of Foreign Genes in Transgenic Bean (*Phaseolus vulgaris* L.) Cotransformed via Particle Bombardment, *Theor. Appl. Genet. 93*:142–150 (1996).
157. Aragao, F., S. Ribeiro, L. Barros, A. Brasileiro, D. Maxwell, E. Rech, and J. Faria, Transgenic Beans (*Phaseolus vulgaris* L.) Engineered to Express Viral Antisense RNAs Show Delayed and Attenuated Symptoms to Bean Golden Mosaic Geminivirus, *Mol. Breed. 4*:491–499 (1998).
158. Chandrasekharan, M.B., K.J. Bishop, and T.C. Hall, Module-Specific Regulation of the Beta-Phaseolin Promoter During Embryogenesis, *Plant J. 33*:853–866 (2003).
159. De Jaeger, G., S. Scheffer, A. Jacobs, M. Zambre, O. Zobell, A. Goossens, A. Depicker, and G. Angenon, Boosting Heterologous Protein Production in Transgenic Dicotyledonous Seeds Using *Phaseolus vulgaris* Regulatory Sequences, *Nat. Biotechnol. 20*:1265–1268 (2002).
160. Broughton, W.J., G. Hernandez, M. Blain, S. Beebe, P. Gepts, and J. Vanderleyden, Beans (*Phaseolus* spp.)—Model Food Legumes, *Plant Soil 252*:55–128 (2003).

161. Hernandez, G., M. Lara, M. Ramirez, C.P. Vance, and L. Blanco, Functional Genomics in Bean (*Phaseolus vulgaris*): The Phaseomic Consortium, Plant and Animal Genome XI Conference, 11–15 Jan., 2003. www.intl-pag.org/11/abstracts/W32_W230_XI.html (2003).
162. Melotto, M., A.G. Bruschi, B.C. Monteiro-Vitorello, and L.E.A. Camargo, The Bean EST (BEST) Project: Providing Resource for Bean Genomics, Plant and Animal Genome XI Conference, 11–15 Jan., 2003. www.intl-pag.org/11/abstracts/P01_P25_XI.html (2003).

Chapter 5

Genomics and Genetic Improvement in the Cool Season Pulse Crops Pea, Lentil, and Chickpea

N.F. Weeden[a] and F.J. Muehlbauer[b]

[a]Department of Plant Sciences and Plant Pathology, Montana State University, Bozeman, MT 59717; and

[b]USDA ARS, 303 Johnson Hall, Washington State University, Pullman, WA 99164-6434

Introduction

The potential for genomic approaches to enhance and streamline breeding programs has been recognized for some time, and the advent of genomic techniques was predicted over 20 years ago (1,2). In the 5th Edition of *A Dictionary of Genetics* (3), *genomics* as a separate concept is not even mentioned, yet much of what is included under genomics today was in practice in the 1990s. What has changed is a matter of scale. Within the past decade techniques or tools that allow us to examine the entire genome have become available, and in many cases their use is routine. The goal of genomics is to further our understanding of how an organism operates, and it has proven a highly efficient approach in bacterial and fungal systems. It is clear that genomics will also prove to be highly productive in plant systems. A precise definition of the word *genomics* will not be given here, but a broad definition would include the characterization, investigation, and utilization of genes and gene products from all organisms, as well as the development of the tools to perform such studies and applications.

Applying the new advances in genomics to breeding programs often requires considerable effort, either on the part of the breeder, who must acquire the equipment and knowledge necessary for such applications, or on the part of molecular geneticists, who must develop intermediate tools such as DNA markers, genomic libraries, and laboratory procedures specific to the crop. Indeed, one of the major impacts of the emergence of genomics has been to make plant breeding much more of an interdisciplinary effort in which numerous fields of science must be drawn upon if the program is to be competitive. The purpose of this chapter is to describe and assess the genomic tools currently available for cool season pulse crops and to identify the strengths and weaknesses of genomic applications in these crops.

Included among the cool season pulses are pea (*Pisum sativum* L.), lentil (*Lens culinaris* Medik.), chickpea or garbanzo bean (*Cicer arietinum* L.), fava bean or broadbean (*Vicia faba* L.), and grasspea (*Lathyrus sativus* L). The last two are very minor crops in the United States and are not particularly interesting at the moment for genomic studies. Thus, this chapter will focus primarily on pea, lentil, and chickpea.

The pea has been important as a model for genetic investigations since the pioneering work of Gregor Mendel in the 1860s (4). The significance of Mendel's work was not recognized for another 30 years, but even the experiments that led to the rediscovery of transmission genetics in diploids also involved pea as an experimental organism (5). Throughout the 20th century the pea has often been at the forefront of genetic research, and a wealth of genomic information has accumulated for this species. The general outline of the following discussion will thus be to treat pea first and then to compare the information available for lentil and chickpea with that for pea. It should become apparent that for many genomic investigations pea is not the ideal model system. For many applications other models such as *Medicago truncatula* Gaertner, *Lotus japonicus* (Regel) K. Larsen, or, within the pulses, chickpea, may serve as the more appropriate organism for a particular study. However, it should also become clear that parallel and complementary work in several model legumes and crop species has led to a synergistic interaction that generates increased productivity in both basic science and breeding.

The main advantage of studying pea is its vast genetic diversity, including a host of easily observed monogenic polymorphisms. Mendel took advantage of seven of the morphological polymorphisms in his studies. The many other morphological and physiological mutations identified during the early 20th century provided the basis for one of the first linkage maps for a plant species (6–8). This genetic diversity is also apparent at the molecular level. More allozyme polymorphism has been observed within *P. sativum* than has been reported for any other crop except maize (9). Similarly, DNA sequence polymorphisms either as RFLPs (10–12), RAPDs (13), AFLPs (14), or CAPs analysis of sequence tagged sites (15,16) have been relatively easy to identify within crosses. An excellent germplasm collection for pea exists at the USDA Western Regional Plant Introduction Station in Pullman, WA; other collections exist in England, Germany, Russia, India, and Poland. At least at Pullman and Norwich, England, the germplasm is well characterized and freely available to interested parties. The species is self-compatible, allowing for convenient maintenance of true-breeding lines, and the seeds are long-lived under appropriate storage conditions, keeping maintenance costs to a minimum.

Both lentil and chickpea have fewer morphological mutations and less genetic diversity as measured by molecular markers than does pea. However, interspecific crosses can be made in both genera, providing access to a broader germplasm base. The cultivated lentil can be crossed with wild species *L. orientalis* (Boiss.) Hand.-Mazz, and (with difficulty) *L. ervoides* (Brign.) Grande. Unfortunately, even between *L. culinaris* and *L. ervoides* there appears to be less genetic diversity than exists between relatively divergent but easily crossed *P. sativum* lines (17). The cultivated forms of lentil are relatively low in morphological variation, but cultivated chickpea displays much more polymorphism, including unifoliate types and marked differences in seed size and shape. Genetic maps were not available for either lentil or chickpea until DNA markers became readily available in the early 1990s. Only when microsatellite markers had been developed for chickpea was its linkage map completed (18). Because chickpea has a much smaller genome than pea, and the development of many molecular tools for genomics becomes increasingly complicated as the genome size increases, chickpea may be the more appropriate species to use as a model for genomic studies. As with pea, both lentil and chickpea are self-compatible and have long-lived seed.

Genomic Applications and Limitations in Cool Season Pulses

Plant models for genomic studies (*Arabidopsis thaliana* (L.) Heyn., *Oryza sativa* L., and *M. truncatula*) all have a number of characteristics in common. Most important,, all have extremely small genomes with about 200 to 500 Mb (million base pairs) relative to most flowering plants. Such a feature is of paramount importance for initial genomic studies because the smaller the genome, the easier to generate genomic libraries, isolate genes by genome walking, or obtain sequences of significant portions of the genome. These objectives represent important steps toward the understanding of how organisms operate. In addition, these model species have short generation times, are easily grown and crossed in the greenhouse, and are relatively easy to manipulate in tissue culture and to transform using either biolistics or *Agrobacterium*.

Pea and lentil have much larger genomes (4.2 to 4.4 billion base pairs) than the above-mentioned model species. The chickpea genome is only about tenfold larger than that of *A. thaliana* and only about three times the size of the *M. truncatula* genome. Thus, for the molecular biologist, *Cicer* would be the pulse of choice for the generation of genomic libraries or gene isolation. Furthermore, the additional DNA in pea and lentil (and grasspea and broadbean) is primarily interspersed repetitive DNA that seriously interferes with chromosome walking, positional cloning, and shotgun sequencing techniques. There have been no genes cloned in pea or lentil using the traditional map-based cloning approach, presumably because of the difficulties encountered with the large genome and high levels of repetitive DNA.

Another limitation associated with many pulse crops is that none are particularly convenient experimental organisms for tissue culture or transformation studies. Despite the clear demand for an efficient transformation system in legumes, most large-seeded legumes (pea, lentil, chickpea, common bean, soybean, cowpea, peanut, lupine, etc.) have been resistant to common methods of plant transformation. Schroeder first developed a reliable method for transforming pea (19), and this method has become routine in a number of laboratories but still requires considerable effort for the relatively low number of transformants obtained. Methods have also been worked out for lentil (20,21) and chickpea, but none of these give the transformation efficiencies obtained in *Medicago* or *Arabidopsis*.

A final set of limitations inherently associated with pulses in regard to their use as model organisms for genomic studies includes their moderate size, moderate generation time, and large seed size. This set of traits is predominantly associated with the large, nitrogen-rich seed that takes considerable resources to develop. Relatively few seeds are produced per pod (and thus per cross). Pea genotypes are available that are relatively small and bloom within a month of planting, but these produce few (2–10) seeds. Larger plants that produce 100 to 200 seeds take up more greenhouse or growth chamber space and require 3 to 4 months to mature. Although none of these characters represents a major limitation to the use of a pulse as a model organism in genomic studies, in combination they tend to encourage the use of other legume species in the molecular biology laboratory and have prevented pulse crops from being considered as appropriate models for the initial phase of genomic investigations.

Fortunately, both model legumes *M. truncatula* and *L. japonicus* selected for the initial genomic studies are closely related to the pulse crops. Both *Medicago* and *Lotus* are included in the temperate legumes (22), and both possess a chloroplast DNA lacking the inverted repeat, placing them in the same monophyletic lineage as *Pisum*, *Lens*, and *Cicer*. The currently recognized taxonomic relationships among the pulse crops and the two model legumes are shown in Fig. 5.1. Other legume crops, such as common bean, peanut, and soybean, are more distantly related to either of the model crops. This close relationship between the models and the cool season pulses permits researchers working with the pulses to rapidly exploit the tools and information developed in the models.

Recalling that the overall goal of genomics is to understand how an organism operates, not the development of genomic libraries or sequences, we must expect that as the problems addressed by biologists change, those organisms best suited for the topic may also change. Although the entire pea or lentil genome may never be sequenced, the sequencing of the *L. japonicus* or *M. truncatula* genome should provide sufficient information to permit genomic investigations to proceed in pulse crops without the need for many of the tools used for the initial genomic studies. Indeed, with the wealth of physiological and genetic information already available on pea, it may again become the preferred model organism for specific questions. Certainly questions dealing with characteristics specific to the crop will have to be

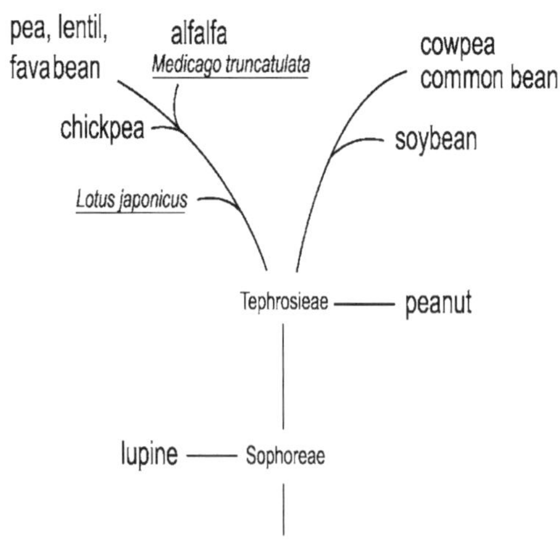

Figure 5.1. Phylogenetic relationships of the major seed legume crops. The central tribes of the papilionoid legumes are indicated by small print, the major crops by large print, and the two model legumes by underlining. The cool season pulses are clustered in the upper left corner of the diagram along with the two model species. Common bean, cowpea, and soybean are on a divergent lineage arising from the tribe Tephrosieae, and peanut is on a third lineage from this tribe. Lupine is much more distantly related to the other crops.

addressed through studies on that crop. The next sections will present the genomic tools currently available in pea, lentil, and chickpea and discuss current directions.

Genomics in Pea—Possibilities in Both Basic and Applied Research

Probably the first tool consulted for any genomic study is a genetic linkage map, i.e., a diagram of the arrangement of the known genes of an organism relative to each other. Usually such maps are presented as a set of vertical lines, each line corresponding to one of the chromosomes found in a sperm or egg cell. Not only do such maps allow one geneticist working on a species to communicate efficiently with another geneticist working on that same species, but also—and perhaps more important in this era of model species—to facilitate comparisons and information exchange among all species.

Pea is a diploid species with $2x = 2n = 14$ chromosomes in somatic tissue and seven linkage groups. One version of the pea map is presented in Fig. 5.2. This map does not show all the DNA markers available for pea; rather, it presents the location of those sequences that are expected or known to have homologues in the genomes of all other higher plants. Many of these genes have been located in *Lens*, *Cicer*, and *Medicago*, and comparison of their locations has identified several regions of conserved gene arrangement (17, 23, 24). Thus, these genes are particularly important for genetic extrapolations from model species.

The consensus linkage map developed for pea (25) includes most of the morphological "anchor markers" that had been placed on the classical map, as well as many of the DNA markers that have been developed during the previous decade. This map is consistent with most of the other linkage maps for pea that have been developed for specific populations (12,14,15) and are based primarily on RAPDs, AFLPs, or RFLPs. Interestingly, it is not consistent with many parts of the classical map (8), revealing the difficulty of assembling a map from many two- or three-factor crosses. The consensus map of pea was based on a population of F_{10} recombinant inbred lines (RILs) derived from a relatively wide cross within *P. sativum* (23). The parental lines involved in generating these populations display polymorphism (either CAPs or RFLP) for nearly every reasonably large (>1000 bp) DNA sequence that has been tested, allowing convenient mapping of new genes by examining segregation patterns of the gene within these "immortal" RIL populations.

The consensus map has an overall saturation of about one marker per centiMorgan (cM) and spans approximately 800 cM. The length of the linkage map converts to about 6 Mb per cM. This degree of saturation does not include the many AFLP, RAPD, and other markers that have formed the basis of full genomic or more focused regional maps in pea. Fine maps have been made around several genes (26,27), but few extensive genomic sequences are currently available in pea. From the number of markers that have been identified in the various mapping studies of pea, it is likely that a marker exists within 500 kb of any location in the pea genome. However, most of these markers contain or consist of repetitive DNA sequences,

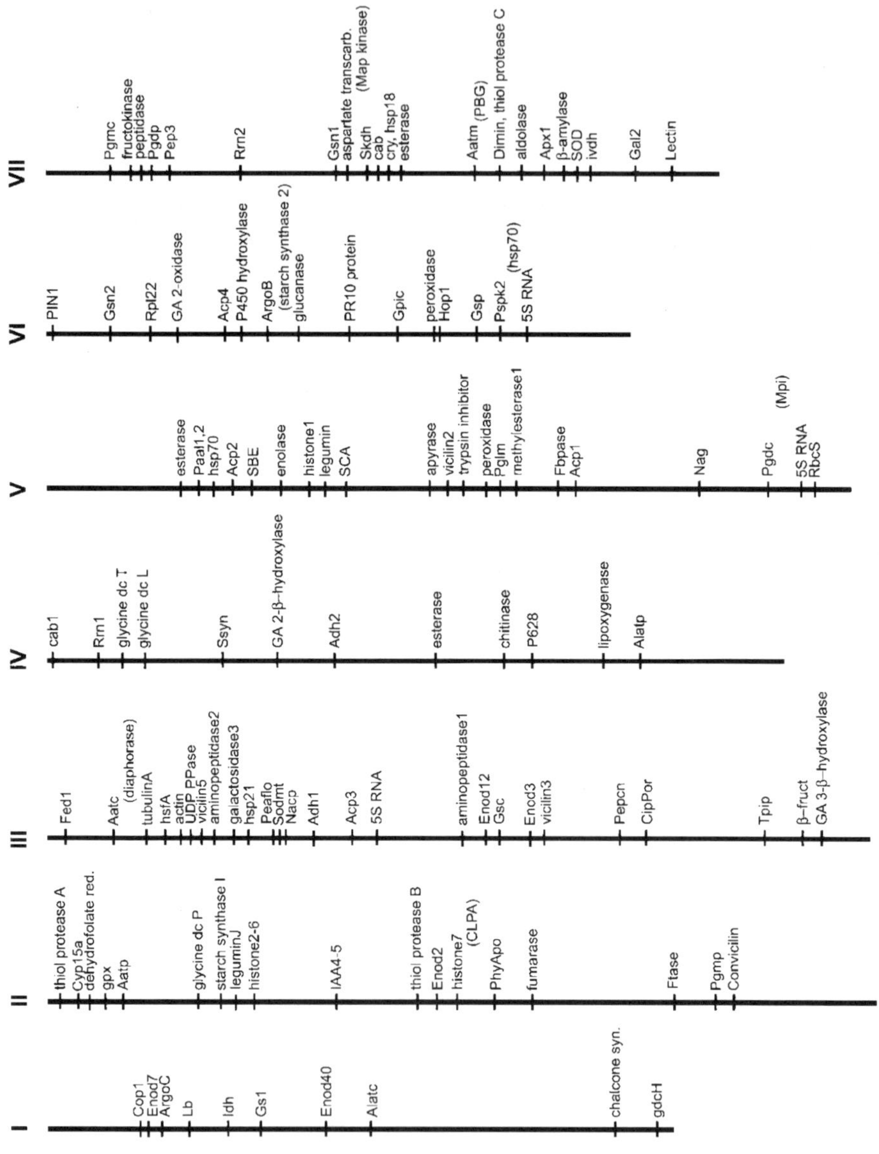

Figure 5.2. Linkage map for pea, showing the relative positions of expressed genes that are known also to be present in other legumes and most other plant species. Not shown are the many morphological mutations and DNA markers that have been mapped in pea but for which homologous genes have yet to be identified in other legumes. The positions of those genes given in parentheses are not precisely known. The scale on the left presents map distances in centimorgans (cM).

making it difficult or impossible to tag or isolate a unique portion of the pea genome (e.g., a bacterial artificial chromosome [BAC]) using the marker.

One of the more promising marker types for breeding and genetic studies within a species is the microsatellite or simple sequence repeat (SSR). A consortium was formed in 1999 that had as its objective development of a large number of PCR-based microsatellite markers in pea. The project is operating under a 5-year time frame and is expected to publish most of the sequences in 2004, providing an excellent set of additional anchor markers that should be particularly useful for breeding programs. A few of the SSR markers have already been used in germplasm diversity studies (28).

The prevalence of repetitive DNA sequences in the pea genome has been the primary reason genes have not been isolated in this crop by map-based cloning or chromosomal walking. A number of genomic libraries have been made for pea. Most of these have been in lambda phage. In addition, cDNA libraries have been made for several tissue types. Screening such libraries has isolated a number of recognized genes from pea (29–31), but the relatively small size of the genomic insert in a lambda genomic library, as well as the high probability that it will possess repetitive sequences, limits the utility of such genomic libraries. Two large insert (BAC) libraries also have been constructed for pea. The one most thoroughly tested consists of about 120,000 clones representing approximately a threefold coverage of the pea genome (32). It has a low (1%) frequency of plastid DNA inserts, and approximately half of the coding sequences used to probe the library identified homologous sequences (17). Contigs have yet to be assembled from this library, and no attempt has been made to begin ordering the BACs along the chromosomes. Yeast artificial chromosome (YAC) or transformation-competent artificial chromosome (TAC) genomic libraries have not been produced for pea.

The large amount of repetitive DNA in pea makes it a low-priority species for genomic sequencing. To generate a complete sequence of the pea genome would require an effort comparable to that of sequencing the human genome. However, much of the repetitive DNA in pea has been characterized (33,34). Thus, the genomic sequence of a model organism that has the same genes in approximately the same order as pea but lacked most of the repetitive DNA would be a very satisfactory substitute, as well as representing a much easier goal. *Medicago truncatula* appears to be such a model system. Many low copy sequences useful in pea as genetic markers have also been used in *M. truncatula* (16; D.J. Kim, Department Plant Pathology, University of California–Davis, personal communication). The colinearity of much of the *M. truncatula* linkage map with that of pea and other pulses has been documented in several studies (17,24; D.J. Kim, unpublished). Most important, an interesting mutation, *sym2*, first identified in pea, was identified and isolated from *M. truncatula* through the process described above (33), thereby validating this model-system approach.

Another approach that has permitted genomic level research to be performed in pea involves the concept of candidate genes. With the genome of *A. thaliana* having been sequenced and most of the coding sequences identified, and with gene sequences from many other species being contributed to publicly available databases, predicting the gene in pea that induces a specific mutation may be a much more efficient process than map-based cloning. Indeed, this approach has been successfully used to identify several genes that produce wrinkled seeds (35–37), control stem height (38–40), and influence flowering (41,42). The efficacy of this approach is more dependent on an individual's knowledge of biochemistry, physiology, and comparable mutations in other species than on the genomic tools available within pea. It appears that map-based cloning will not be a necessary procedure to develop for pea.

Other tools important for genomic studies are either less advanced than those developed for *Medicago* or are not available. Transformation by *A. tumifaciens* is now routine using cotyledonary node tissue (19), but the process does take more effort and time than the parallel process in *Medicago* or *Lotus*. Similarly, tissue culture procedures are available for pea, but remain tedious. Expressed sequence tag (EST) libraries for pea are not currently being constructed in the United States. Thus, key information on the relative level of expression of a particular gene in a specific tissue will be available only for a very few genes in the near future. Finally, the use of microarrays to study relative gene expression in pea is still in its infancy and is dependent on material developed for *Medicago*.

In summary, pea is considerably dependent on *Medicago*, *Lotus*, or *Arabidopsis* for providing sequence information for primer development or microarray applications. The pea BAC library probably will be ordered through a comparison with an ordered BAC library in *Medicago*. Identification of the genes responsible for interesting morphological or physiological mutations will be accomplished primarily by the second approach described above—that of identifying potential candidate genes and obtaining sequences for these genes from either *Medicago* or *Arabidopsis*. Pea still remains competitive with these models for research focused on understanding nitrogen fixation, apical dominance, and inflorescence development. Furthermore, with so many important mutants available, pea probably will be the model plant for unraveling the development of the compound leaf. There are also many problems specific to pea that cannot be investigated in a model system. Many of these will have to be addressed only after a more complete set of genomic tools is available in pulse crops.

Genomics in Lentil—Dependence on Other Genera as Models

Lentil ($2x = 2n = 14$ chromosomes) has a genome size similar to that of pea; however, the available genetic linkage map is still rudimentary, and unfortunately a consensus map has yet to be developed. Two crosses have been particularly useful in developing the current linkage map. One is a very wide interspecific cross between the cultivated lentil (*L. culinaris*) and a wild species (*L. ervoides*). This cross is difficult to make, requires embryo rescue, and displays considerable sterility and segregation distortion in the F_2 (43). However, populations derived from the *L. culinaris* × *L. ervoides* cross have been used by several groups for genetic mapping studies. These populations provide the highest level of segregating genetic polymorphism available within the genus. A second interspecific cross between the cultivated lentil and its immediate wild ancestor, *L. orientalis*, has also proven useful for mapping studies. It produces less sterility and segregation distortion in F_2 and RIL populations, but it also possesses less genetic diversity (44).

Zamir and Ladizinsky (45) initially reported two linkage groups from isozyme and morphological traits segregating in F_2 populations derived from *L. culinaris* × *L. orientalis* crosses. The map was extended to include additional isozyme loci and a translocation breakpoint in studies on a *L. culinaris* × *L. ervoides* F_2 population (46), and a number of RFLPs were mapped in the narrower cross (47). The *L. culinaris* × *L. ervoides* map was expanded to include many RFLPs and compared with the pea map (23). Although the number of loci compared were limited, it became clear that at least 40% of the linkage relationships among markers were conserved between lentil and pea. RAPDs were added to the *L. culinaris* × *L. orientalis* to produce a map of over 100 loci (48). The map has been further improved using recombinant inbred lines (49–51), and further comparisons with the pea linkage map have demonstrated relatively few rearrangements (17).

This map has been very important for the analysis of quantitatively inherited traits in lentil. Three quantitative trait loci (QTL) for winter hardiness in lentil have been reported (52), and one QTL was common over sites of evaluation in the United States and Turkey. Genes for resistance to ascochyta blight of lentil have been identified and mapped (53,54), and markers for use in selection programs were identified and are available for use in breeding programs.

Research is underway toward increasing marker density in the lentil linkage map using sequence tagged sites developed from ESTs in *M. truncatula*. In general, the position of the locus on the *Medicago* linkage map will serve as a good prediction of its location on the lentil map. This approach is expected to facilitate the use of genomic information from the model legume toward the isolation and characterization of genes of interest in lentil.

Genomics in Chickpea—a Potential Intermediate Model Species for Pulse Crops

Chickpea genetic mapping was initially hampered by a paucity of polymorphism in *C. arietinum* ($2x = 2n = 16$ chromosomes). However, the development of sequence-tagged microsatellite sites (STMS) made it possible to increase marker density (55). The STMS markers have shown a wealth of polymorphism within cultivated *C. arietinum* and have been successfully used to enhance the genetic map of the species. Prior to the development of STMS markers for chickpea, mapping efforts relied on progenies from interspecific crosses, particularly *C. arietinum* × *C. reticulatum*. In those crosses there appeared to be about a tenfold increase in polymorphism in the progenies when compared to progenies from intraspecific crosses within *C. arietinum*. Initial efforts toward genetic mapping of chickpea lead to the first linkage map that comprised 26 isozyme and three morphological trait loci (56). Several additions improved the coverage of the map (48,57) before an integrated map was established using several types of molecular markers including STMS, DAF (DNA amplification fingerprinting), AFLP, and RAPDs (18,58). The STMS markers are proving to be particularly useful, permitting researchers to place resistance gene analogs and other potentially valuable loci on the integrated map (59).

Interest in mapping the genes for Ascochyta blight resistance in chickpea lead Tekeoglu *et al.* (60) to incorporate STMS markers to a map developed from a recombinant inbred line population that had been phenotyped for blight resistance. The map was then used to identify three QTL for blight resistance. One of the QTLs is significantly robust to appear in data from Ascochyta blight evaluation sites in both the United States and Turkey.

Bacterial Artificial Chromosome libraries for chickpea have been developed for germplasm accessions with resistance to Fusarium wilt (G. Kahl, University of Frankfurt, Germany, personal communication) and Ascochyta blight (61). The latter library was calculated to have 3.8× coverage and has been used to identify clones containing markers linked to QTL for Ascochyta blight resistance resulting in the identification of six clones. Those clones are currently being "shot-gun" sequenced to determine the actual gene content and whether putative disease resistance genes are present. Genetic transformation experiments are also underway to confirm that the BAC clones in fact contain disease resistance genes.

There is great interest in utilizing the wild species of *Cicer* as a source of genes for resistance to Ascochyta blight and to *Heliocoverpa armigera* Hübner, the pod borer. Of particular interest are *C. echinospermum*, *C. bijugum*, *C. pinnatifidum*, and *C. anatolicum*, because these species appear to be highly resistant to Ascochyta blight. At least three research groups are currently working toward overcoming barriers to hybridization of these species to cultivated types. Thus far only the cross *C. echinospermum* × *C. arietinum* has been successfully made, and progenies are available for study. Of particular interest is the perennial *C. anatolicum* because of its near immunity to Ascochyta blight; however, crossing *C. arietinum* with this species has thus far not been successful. Histological studies of the pollen tube growth through the style indicate that double fertilization likely takes place; however, the developing ovules abort at about 8 days after pollination. Attempts are currently being made to develop embryo rescue techniques.

Summary

For genomic studies, all of the cool season pulse crops will be at least partially dependent on the information and tools being developed in the model species *A. thaliana*, *M. truncatula*, and *L. japonica*. In particular, *M. truncatula* should prove particularly useful because it is relatively close taxonomically to the pulses and, except for the amount of repetitive DNA present, has much of its genetic information in common with the pulses. Based on the cloning of the *sym2* homolog in *M. truncatula* and an isolation of a BAC clone containing *sym2* in pea, it is clear that information can flow in both directions and provides for a synergistic interaction for information exchange among different species. Once linkage maps and genomic libraries developed in the various species become sufficiently characterized, information transfer should become nearly immediate, and the cool season legumes could become a large model system for legume genetics.

Differences among the pulses will be exploited for further genomic studies. Pea, with its multitude of interesting mutations and wealth of physiological information, will be used to investigate the genetic basis of characters such as leaf, inflorescence, and nodule development. Chickpea, with its much smaller genome, may become the model system for traits such as development of a protein-rich seed and resistance to prevalent diseases. The sharing of knowledge and tools among researchers in the pulses should lead to a synergistic and rapid expansion in our understanding of the genetic basis for many important traits. Crop breeding should become more efficient in the near term, and novel pulse types should be anticipated as our ability to manipulate the pulse genome increases.

References

1. Moore, G.A., and G.B. Collins, New Challenges Confronting Plant Breeders, in *Isozymes in Plant Genetics and Breeding, Part A*, edited by S.D. Tanksley and T.J. Orton, Elsevier, Amsterdam, 1983, pp. 25–58.
2. Beckmann, J.S., Oligonucleotide Polymorphisms: A New Tool for Genomic Genetics, *Biotechnology 6*:1061–1064 (1989).
3. King, R.C., and W.D. Stansfield, *A Dictionary of Genetics*, 5th ed., Oxford University Press, New York, 1997.
5. Mendel, G., Versuche über Pflanzen-Hybriden, *Verhandlungen des Naturforschenden den Vereines in Brünn 4*:3–47 (1865).
6. Correns, C., G. Mendel's Regel über das Verhalten der Nachkommenshaft der Rassenbastarde, *Berichte der Deutschen Botanischen Gesellschaft 18*:158–168 (1900).
7. Lamprecht, H., Die Genenkarte von *Pisum* bei normaler Struktur der Chromosomen, *Agri Hort. Genetica 19*:360–401 (1961).
8. Blixt, S., The Pea, in *Handbook of Genetics*, edited by R.S. King, Plenum Press, 1975, Vol. 2, pp. 181–221.
9. Weeden, N.F., Polymorphic Isozyme Loci Identified in *Pisum, Pisum Genetics 20*:46–48.
10. Polans, N.O., N.F. Weeden, and W.F. Thompson, The Inheritance, Organization and Mapping of *rbcS* and *cab* Multigene Families in Pea, *Proc. Natl. Acad. Sci. USA 82*:5083–5087 (1985).
11. Polans, N.O., N.F. Weeden, and W.F. Thompson, The Distribution, Inheritance and Linkage Relationships of Ribosomal DNA Spacer Length Variants in Pea, *Theor. Appl. Genet. 72*:289–295 (1986).
12. Ellis, T.H.N., L. Turner, R.P. Hellens, D. Lee, C.L. Harker, C. Enard, C. Domoney, and D.R. Davies, Linkage Maps in Pea, *Genetics 130*:649–663 (1992).

13. Weeden, N.F., G.M. Timmerman, and J. Lu, Identifying and Mapping Genes of Economic Significance, *Euphytica 73*:191–198 (1994).
14. McCallum, J., G. Timmerman-Vaughan, T.J. Frew, and A.C. Russell, Biochemical and Genetic Linkage Analysis of Green Seed Color in Field Pea (*Pisum sativum* L.), *J. Am. Soc. Hort. Sci. 122*:218–225 (1997).
15. Gilpin, B.J., J.A. McCallum, T.J. Frew, and G.M. Timmerman-Vaughan, A Linkage Map of the Pea (*Pisum sativum* L.) Genome Containing Cloned Sequences of Known Function and Expressed Sequence Tags (ESTs), *Theor. Appl. Genet. 95*:1289–1299.
16. Brauner, S.R., L. Murphy, J.G. Walling, J. Przyborowski, and N.F. Weeden, STS Markers for Comparative Mapping in Legumes, *J. Am. Soc. Hort. Sci. 127*:616–622 (2002).
17. Walling, J.G., Cytological and Molecular Investigations in *Lens* and *Pisum*, M.S. Thesis, Montana State University, Bozeman, 2003, pp.141.
18. Winter, P., A.M. Benko-Iseppon, B. Huttel, M. Ratnaparkhe, A. Tullu, G. Sonnante, T. Pfaff, M. Tekeoglu, D. Santra, V.J. Sant, P.N. Rajesh, G. Kahl, and F.J. Muehlbauer, A Linkage Map of the Chickpea (*Cicer arietinum* L.) Genome Based on Recombinant Inbred Lines from a *C. arietinum* × *C. reticulatum* Cross: Localization of Resistance Genes for Fusarium Wilt Races 4 and 5, *Theor. Appl. Genet, 101*:1155–1163 (2000).
19. Schroeder, H.E., A.H. Schotz, T. Wardley-Richardson, D. Spencer, and T.J.V. Higgins, Transformation and Regeneration of Two Cultivars of Pea (*Pisum sativum* L.), *Plant Physiol. 101*:951–957 (1993).
20. Mahroudian, M., M. Yucel, and H.A. Oktem, Transformation of Lentil (*Lens culinaris* M.) Cotyledonary Nodes by Vacuum Infiltration of *Agrobacterium tumefaciens, Plant Molec. Biol. Rept. 20*:251–257 (2002).
21. Gulati, A., P. Schryer, and A. McHugen, Production of Fertile Transgenic Lentil (*Lens culinaris* Medik) Plants Using Particle Bombardment, *In Vitro Cell. Devel. Biol.—Plant 38*:316–324 (2002).
22. Polhill, R.M., Papilionoideae, in *Advances in Legume Systematics*, edited by R.M Polhill and P.H. Raven, Royal Botanic Gardens, Kew, Vol. 1, 1981, pp. 191–208.
23. Weeden, N.F., F.J. Muehlbauer, and G. Ladizinsky, Extensive Conservation of Linkage Relationships Exists between Pea and Lentil Genetic Maps, *J. Hered. 83*:123–129 (1992).
24. Vincent, J.L., M.R. Knox, T.H.N. Ellis, P. Kalo, G.B. Kiss, and N.J. Brewin, Nodule-Expressed Cyp15a Cystine Protease Genes Map to Syntenic Genome Regions in *Pisum* and *Medicago* spp., *Mol. Plant Microbe Interact. 13*:715–723 (2000).
25. Weeden, N.F., T.H.N. Ellis, G.M. Timmerman-Vaughan, W.K. Swiecicki, S.M. Rozov, and V.A. Berdnikov, A Consensus Linkage Map for *Pisum sativum*, *Pisum Genet. 30*:1–4 (1998).
26. Kozik, A., M. Matvienko, B. Scheres, V. G. Paruvangada, T. Bisseling, A. Van Kammen, T.H.N. Ellis, T.A. LaRue, and N.F. Weeden, The Pea Early Nodulin Gene *Enod7* Maps in the Region of Linkage Group I Containing *sym2* and Leghemoglobin, *Plant Mol. Biol. 31*:149–156 (1996).
27. McClendon, M.T., D.A. Inglis, K.E. McPhee, and C.J. Coyne, DNA Markers Linked to Fusarium Race 1 Resistance in Pea, *J. Amer. Soc. Hort. Sci 127*:602–607 (2002).
28. Ford, R., K. Le Roux, C. Itman, J.B. Brower, and P.W.J. Taylor, Diversity Analysis and Genotyping in *Pisum* with Sequence Tagged Microsatellite Site (STMS) Primers, *Euphytica 124*:397–405 (2002).

29. Yang, Z., C.L. Cramer, and J.C. Watson, Protein Farnesyltransferase in Plants: Molecular Cloning and Expression of a Homolog of the Beta Subunit from the Garden Pea, *Plant Physiol. 101*:667–674 (1993).
30. Finan, P.M, I.R. White, S.H. Redpath, J.B.C. Findlay, and P.A. Pillner, Molecular Cloning, Sequence Determination and Heterologous Expression of Nucleoside Diphosphate Kinase from *Pisum sativum, Plant Mol. Biol. 25*:5967 (1994).
31. Burgess, D., A. Penton, P. Dunsmuir, and H. Dooner, Molecular Cloning and Characterization of ADP-Glucose Pyrophosphorylase cDNA Clones Isolated from Pea Cotyledons, *Plant Mol. Biol. 33*:431–444 (1997).
32. Coyne, C.J., K. Medsem, D.A. Lightfoot, K.E. Keller, R.R. Martin, M.T. McClendon, D.A. Inglis, E.W. Storlie, and K.E. McPhee, Construction of a Bacterial Artificial Chromosome Library for Pea (*Pisum sativum* L.), *Pisum Genet. 32*:23–26 (2000).
33. Turner, L., and T.H.N. Ellis, Repeated Sequence Markers in Pea (*Pisum sativum*), *New Phytol. 137*:35–41 (1997).
34. Neumann, P., M. Nouzova, and J. Macas, Molecular and Cytogenetic Analysis of Repetitive DNA in Pea (*Pisum sativum* L.), *Genome 44*:716–728 (2001).
35. Bhattacharyya, M.K., A.M. Smith, T.H.N. Ellis, C. Hedley, and C. Martin, The Wrinkled-Seed Character of Pea Described by Mendel is caused by a Transposon-like Insertion in a Gene Encoding Starch-Branching Enzyme, *Cell 60*:115–122 (1990).
36. Rochat, C., S. Wuilleme, J.P. Boutin, and C.L. Hedley, A Mutation at the Rb Gene, Lowering ADPGPPase Activity, Affects Storage Product Metabolism of Pea Seed Coats, *J. Expt. Bot. 46*:415–421 (1995).
37. Harrison, C.J., R.M. Mould, M.J. Leech, S.A. Johnson, L. Turner, S.L. Schreck, K.M. Baird, P.L. Jack, S. Rawsthorne, C.L. Hedley, and T.L. Wang, The *rug3* Locus of Pea Encodes Plastidial Phosphoglucomutase, *Plant Physiol. 122*:1187–1192 (2000)
38. Martin, D.N., W.M. Proebsting, and P. Hedden, Mendel's Dwarfing Gene: cDNAs from the *Le* alleles and Function of the Expressed Proteins, *Proc. Natl. Acad. Sci. USA 94*:8907–8911 (1997).
39. Martin, D.N., W.M. Proebsting, and P. Hedden, The *SLENDER* Gene of Pea Encodes a Gibberellin 2-Oxidase, *Plant Physiol. 121*:775–781 (1999).
40. Davidson, S.E., R.C. Elliott, C.A. Helliwell, A.T. Poole, and J.B. Reid, The Pea Gene *NA* Encodes Ent-Kaurenoic Acid Oxidase, *Plant Physiol. 131*:335–344 (2003).
41. Hofer, J., L. Turner, R. Hellens, M. Ambrose, P. Matthews, A. Michael, and N. Ellis, *UNI-FOLIATA* Regulates Leaf and Flower Morphogenesis in Pea, *Curr. Biol. 7*:581–587 (1997).
42. Taylor, S., J. Hofer, and I. Murfet, *Stamina pistilloida*, the Pea Ortholog of Fim and UFO, is Required for Normal Development of Flowers, Inflorescences and Leaves, *Plant Cell 13*:31–46 (2001).
43. Ladizinsky, G., D. Cohen, and F.J. Muehlbauer, Hybridization in the Genus *Lens* by Means of Embryo Culture, *Theor. Appl. Genet. 70*:97–101 (1985).
44. Pinkas, R., D. Zamir, and G. Ladizinsky, Allozyme Divergence and Evolution in the Genus *Lens*, *Plant Syst. Evol. 151*:131–140 (1985).
45. Zamir, D., and G. Ladizinsky, Genetics of Allozyme Variants and Linkage Groups in Lentil, *Euphytica 33*:329–336 (1984).
46. Tadmor, Y., D. Zamir, and G. Ladizinsky, Genetic Mapping of an Ancient Translocation in the Genus *Lens*, *Theor. Appl. Genet. 73*:883–892 (1987).

47. Havey, M.J., and F.J. Muehlbauer, Linkages Between Restriction Fragment Length, Isozyme and Morphological Markers, *Theor. Appl. Genet. 77*: 395–401 (1989).
48. Simon, C.J., and F.J. Muehlbauer, Construction of a Chickpea Linkage Map and Its Comparison with Maps of Pea and Lentil, *J. Hered. 88*:115–119 (1997).
49. Tahir, M., and F.J. Muehlbauer, Gene Mapping in Lentil with Recombinant Inbred Lines, *J. Hered. 85*:306–310 (1994).
50. Tahir, M., F.J. Muehlbauer, and S.C. Spaeth, Association of Isozyme Markers with Quantitative Trait Loci in Random Single Seed Descent Derived Lines of Lentil (*Lens culinaris* Medik.), *Euphytica 75*:111–119 (1994).
51. Eujayl, I., M. Baum, W. Erskine, E. Pehu, and F.J. Muehlbauer, The Use of RAPD Markers for Lentil Genetic Mapping and the Evaluation of Distorted F-2 Segregation, *Euphytica 96*:405–412 (1997).
52. Kahraman, A., I. Kusmenoglu, A. Aydin, A. Aydogan, W. Erskine, and F.J. Muehlbauer, QTL Mapping of Winter Hardiness Genes in Lentil, *Crop Sci. 44*:13–22 (2004)
53. Ford, R., E.C.K. Pang, and P.W.J. Taylor, Genetics of Resistance to Ascochyta Blight (*Ascochyta lentis*) of Lentil and the Identification of Closely Linked RAPD Markers, *Theor. Appl. Genet. 98*:93–98 (1999).
54. Tullu, A., L. Buchwaldt, T. Warkentin, B. Taran, and A. Vandenberg, Genetics of Resistance to Anthracnose and Identification of AFLP and RAPD Markers Linked to the Resistance Gene in PI 320937 Germplasm of Lentil (*Lens culinaris* Medikus), *Theor. Appl. Genet. 106*:428–434 (2003).
55. Huttel, B., P. Winter, H. Weising, W. Choumane, F. Weigand, and G. Kahl, Sequence-Tagged Microsatellite Site Markers for Chickpea (*Cicer arietinum* L.), *Genome 42*:210–217 (1999).
56. Gaur, P.M., and A.E. Slinkard, Inheritance and Linkage of Isozyme Coding Genes in Chickpea, *J. Hered. 81*:455–46 (1990).
57. Kazan, K., F.J. Muehlbauer, N.F. Weeden, and G. Ladizinsky, Inheritance and Linkage Relationships of Morphological and Isozyme Loci in Chickpea (*Cicer arietinum* L.), *Theor. Appl. Genet. 86*:417–426 (1993).
58. Santra, D.K., M. Tekeoglu, M. Ratnaparkhe, W.J. Kaiser, and F.J. Muehlbauer, Identification and Mapping of QTLs Conferring Resistance to Ascochyta Blight in Chickpea, *Crop Sci. 40*:1606–1612 (2000).
59. Flandez-Galve, H., R. Ford, E.C.K. Pang, and P.W.J. Taylor, An Intraspecifc Linkage Map of the Chickpea (*Cicer arietinm* L.) Genome Based on Sequence Tagged Microsatellite Site and Resistance Gene Analog Markers, *Theor. Appl. Genet. 106*:1447–1456 (2003).
60. Tekeoglu, M., P.N. Rajesh, and F.J. Muehlbauer, Integration of Sequence Tagged Microsatellite Sites to the Chickpea Genetic Map, *Theor. Appl. Genet. 105*:847–854 (2002).
61. Rajesh, P.N., C. Coyne, K. Meksem, K.D. Sharma, V.S. Gupta, and F.J. Muehlbauer, Construction of a *Hin*dIII Bacterial Artificial Chromosome Library and Its Use in Identification of Clones Associated with Disease Resistance in Chickpea, *Theor. Appl. Genet.* (*in press*) (2004).

Chapter 6

Genomics and Genetic Enhancement of Peanut

Andrew H. Paterson,[a] H. Thomas Stalker,[b] Maria Gallo-Meagher,[c] Mark D. Burow,[d] Sangam L. Dwivedi,[e] Jonathan H. Crouch,[e] and Emma S. Mace[e]

[a]University of Georgia, 111 Riverbend Road, Athens, GA 30602;

[b]Department of Crop Sciences, Box 7620, North Carolina State University, Raleigh, NC;

[c]Department of Agronomy, PO Box 110300, University of Florida, Gainesville, FL;

[d]Texas A&M University Research and Extension Center, Lubbock, TX 29403; and

[e]International Crops Research Institute for the Semi Arid Tropics (ICRISAT), Hyderabad, India

Introduction

Cultivated peanut, also known as groundnut (*Arachis hypogaea* L.), is grown on 25.5 million hectares between latitudes 40° N and 40° S with a total global production of 35 million tons (Mt)(1). Peanuts originated in South America; however, the vast majority of peanut is produced in Asia (66.8%: 23.4 Mt) and Africa (24.6%: 8.6 Mt). The remaining 8.6% (3 Mt) comes from North America, the Caribbean, Europe, and Oceania.

Peanut is among the top five oilseeds grown in the world with 65% of world production used by India, China, and the United States. The cake remaining after peanut oil extraction can be used in human food or incorporated into animal feeds (2). Wild *Arachis* species (*A. pintoi* Krapov. and W.C. Gregory and *A. glabrata* Benth.) have nutritive values higher than those of most commercially important tropical forage legumes, and have been used for pasture improvement in North America, Central America, South America, and Australia (3). The greater adaptability of rhizomatous perennial peanut (*A. glabrata*) to the tropical environment and its high yield when harvested for hay give it the potential of becoming one of the most important forages in the tropics (4).

In the early 1900s, Dr. George Washington Carver of the Tuskegee Institute in Alabama developed more than 300 uses for cultivated peanut. Largely because of his research efforts, peanut became the second-largest row crop in the Southern United States (after cotton). During the 1990s, the U.S. peanut crop had an average value of over $1 billion, ranking it as our second most important seed legume, after soybean. Per capita, Americans consume more than 3 kg of peanuts and peanut products per year, ranking it among our most popular foods.

Peanut offers numerous human health benefits. It contains mostly unsaturated fat, which has been shown to lower LDL-cholesterol levels in the blood, and resveratrol, which leads to improved cardiovascular health. Peanut is also a good source of folic acid, which helps prevent neural tube defects, and it contains nearly half of the 13 essential vitamins and 35% of the essential minerals. Because of its high nutritional value, the peanut is being widely investigated as a key food source for astronauts during extended space missions.

Unique Botanical Features of Peanut and Their Impact on Crop Productivity

Peanut has one of the most intriguing reproductive systems in the plant kingdom. Peanuts produce flowers above the ground, but their fruits and seeds develop below the ground. The temporal and spatial separation of fertilization and fruit development is probably an adaptation to drought in the Amazon basin that constitutes peanut's center of diversity (5).

The reproductive mechanism of peanut offers new avenues of inquiry that should lead to a better understanding of plant reproductive and developmental biology. Following fertilization, peanut ovules advance to the 8- to 16-cell stage and then become quiescent for up to three weeks. Discovering what controls this discrete cessation of ovule development may lead to future strategies that could be employed in other crops for the production of seedless fruits. Subsequent stages of peanut reproduction, such as the meristematic activity of the ovary, the gravitropic response of the developing peg, and the dark and Ca^{2+} requirements for pod and seed development, also reflect meaningful changes to the standard reproductive program for legumes and most higher plants. They may provide insight into mechanisms at work in other plant biological processes which involve meristematic activity, gravitropism, light regulation, and Ca^{2+} regulation.

Peanut's reproductive mechanism also has major implications for crop productivity and genetic diversity. A primary repercussion of its reproductive strategy is a low success rate for completion of reproduction. After pollination, not all flowers produce pegs, and not all pegs grow sufficiently long to reach the soil. Consequently, a high percentage of pegs yield nothing. Depending on the variety, the reproductive success rate ranges from 11% (6) to 68% (7). Although a number of factors ultimately affect seed set, it is clear from breeding work that peanut reproductive success can be genetically improved. There is a high likelihood that peanut yields could be significantly enhanced if one were able to improve peanut's ability to form pegs that reach the soil and complete development. Additionally, a better understanding of peanut reproduction could substantially increase the extremely limited genetic diversity in cultivated peanut by improving the very low success rate of interspecific hybridization (8).

Genetic Diversity and Chromosomal Relationships of Cultivated and Wild *Arachis*

Unlike many other natural polyploid species for which multiple polyploidization events have been identified and that permit exchange of moderate levels of genetic variability among polyploids (9, 10), cultivated peanut, *A. hypogaea* is a tetraploid ($2n = 4x = 40$) believed to have originated recently from a single hybridization event (11). Archaeological evidence from excavations in Peru place the origin of *A. hypogaea* at least 3,500 years ago (12). Cytological studies of *A. hypogaea* observed 20 chromosome bivalents at meiosis in 88 to 98% of cells; the exceptions were rare univalents, trivalents, and quadrivalents, which suggested limited homoeologous pairing between the A and B genomes that compose the tetraploid (13,14).

Despite the paucity of DNA-level variability among cultivated genotypes, a wealth of diversity exists in other *Arachis* species. In section *Arachis* of the genus there are 27 species representing three genomes, A, B, and D. Additionally, 42 congeners have been identified in the other eight sections of the genus (22). Abundant DNA marker polymorphisms have been detected between wild species in section *Arachis* (16,23,24) and also within some species such as *A. duranensis* Krapov. and W.C. Gregory (25). This large amount of variation supports the hypothesis that *A. hypogaea* likely originated from a single hybridization event followed by chromosome doubling, with very little subsequent introgression from related diploid species (26). Kochert *et al.* (11) concluded that *A. duranensis* and *A. ipaensis* Krapov. and W. C. Gregory are the most likely diploid progenitors of the cultivated peanut, and *A. duranensis* was probably the female parent.

Only one other species in section *Arachis*, *A. monticola* Krapov. and Rigoni, is tetraploid and readily crossable with *A. hypogaea*, but it is indistinguishable from *A. hypogaea* based on DNA markers (15,16). Thus, *A. hypogaea* and *A. monticola* can be considered the same biological species.

Conventional Genetic Improvement of Peanut

Over 276 peanut cultivars were released between 1920 and 2000 for cultivation in various countries in Asia, Africa, and the Americas. Each has specific adaptation to its respective region of production and cropping system (27–29). A yearly genetic gain of nearly 15 kg/ha (per hectare) has been reported for large-seeded Virginia type cultivars released from the 1950s to the 1970s reaching a maximum yield of 4.6 t/ha in the United States (30). However, since the 1970s there has been increased emphasis on improving pest resistance and quality traits. The result of this emphasis is that the yield potential of cultivars released since that time has not surpassed those of the highest-yielding cultivars released during the 1970s.

Many breeding programs, including the one at ICRISAT, made substantial progress toward improving the yield potential of cultivars adapted to rain-fed or irrigated high-input situations. Some cultivars have produced pod yield of over 9 t/ha in Zimbabwe (31) and China (32). However, there is a wide gap between realized yields at farm level (world average yield of 1.4 t/ha) when compared to the highest average yields in China (3.1 t/ha) and the United States (3.4 t/ha) (1). There is therefore a need to incorporate multiple resistances to biotic and abiotic stresses into improved genetic backgrounds, even if it requires some sacrifice in yield potential, as experienced by U.S. peanut breeders, to narrow the gap between realized and potential yields (33).

Plant introductions have had increasing importance to peanut improvement, particularly in Virginia-type cultivars which have increased resistance to tomato spotted wilt virus; but relatively few accessions are in the pedigrees of cultivars grown in the U.S. (28). Within the cultivated and wild species gene pools of peanut there is a large amount of genetic variation for many agronomically important traits (34,35). Continued evaluation of the U.S. germplasm collection, consisting of more than 8,000 plant introductions (36), and the collection maintained by the ICRISAT with more than 14,000 accessions (37), will be critical for genetic improvement of cultivated peanut.

Recently developed cultivars have reduced vegetative mass, shorter main stem length, and greater reproductive allocation (partition more of their daily assimilate to fruit) than those developed previously, as predicted by Duncan *et al.* (38). High yield in recently released cultivars appears to be related more to total flower production than to reproductive efficiency. Therefore, future increases in seed yield might be accomplished by developing cultivars with a combination of high reproductive efficiency, harvest index, and total flower count (39).

Consistent with its predominantly inbreeding nature, the most commonly used breeding methods in peanut are pedigree selection, bulk-pedigree selection, and single-seed descent. Backcross breeding has not been used extensively because most of the economically important traits in peanut are quantitatively inherited (40,41). Breeders often make single crosses to generate variability. However, with increased emphasis on multiple resistance breeding, emphasis is now focused on complex crosses followed by intercrossing of segregants to bring the desired improvement into breeding populations. While selection for resistance to insect pests and diseases is practiced in early generations, selection for yield and yield component traits is delayed until later generations. Recurrent selection has also been used for continued genetic enhancement in peanut (42,43). As the application of molecular markers becomes routine, it is likely that greater emphasis will be placed on backcross breeding, particularly to assist the rapid introgression of disease resistances from wild *Arachis* species and exotic *A. hypogaea* germplasm.

Genetic Mapping of Peanut

The paucity of DNA polymorphism in cultivated peanut (see above) posed a considerable obstacle to genetic mapping. The use of a synthetic amphidiploid, TxAG-6 (44,45) made possible the generation of the first molecular map representing the entire tetraploid genome of peanut, as well as study of the transmission genetics of chromatin from a synthetic amphidiploid into cultivated peanut. TxAG-6 was developed through the cross [*A. batizocoi* Krapov. and W. C. Gregory × (*A. cardenasii* Krapov. and W. C. Gregory × *A. diogoi* Hoehne)]^{4x}, where *A. cardenasii* and *A. diogoi* are A-genome diploids, and *A. batizocoi* was once considered to be a B-genome diploid ancestor (46–48). This cross has been used to introduce root-knot nematode resistance into cultivated peanut (49,50) and also harbors many other potentially valuable attributes.

The discovery of a high level of polymorphism between the cultivar Florunner and the parents of TxAG-6 by RAPD analysis (49) was followed by RFLP surveys that reported 83% polymorphism on a per-band basis between Florunner and TxAG-6 (8). Using 78 peanut BC_1 lines generated using TxAG-6 (45,49) as the donor parent and the *A. hypogaea* cultivar Florunner as recurrent parent, 220 cDNA probes were used to map 370 RFLP loci onto 23 linkage groups. A total of 917 bands was observed, for an average of 4.1 bands per probe. A mean 1.68 loci per probe were mapped. The total length of the tetraploid map, 2210 cM (centimorgans), was slightly greater than twice the length (1063 cM) of a map previously reported from a cross between two A-genome diploid species (51). Based on a small number of DNA markers mapped in both crosses, there is a high degree of colinearity between the diploid and tetraploid chromosomes (8).

As expected of a recently formed polyploid, many DNA markers mapped to two different linkage groups, and patterns of duplication shed light on the identities of homoeologous chromosomes. Eighty-nine probes produced markers on both homoeologous linkage group pair members. The 23 linkage groups were comprised of nine pairs of homoeologous linkage groups, one trio representing a homoeologous chromosome pair, one fragment consisting of two markers, and one linkage group that was possibly an artifact. Given that cultivated peanut is a disomic polyploid ($2n = 4x = 40$), 20 linkage groups were expected. By comparison of "alloalleles" in tetraploid peanut to the alleles present in the respective diploid progenitors of TxAG-6, the subgenomic affinities of the linkage groups were readily discerned. There was evidence for large structural differences in the LG1/LG11 pair, and weaker evidence for possible rearrangements in the LG7/LG17, LG4/LG14, and LG5/LG15 pairs. Other homoeologous linkage group pairs appeared to be collinear to the degree of resolution afforded by this experiment. This is in agreement with cytological observations (52,53) of incomplete pairing of chromosomes in *A. cardenasii* and *A. batizocoi* diploid F_1 hybrids resulting in five or more univalents per pollen mother cell.

A consensus molecular map of the peanut tetraploid genome based on SSR and AFLP markers is also being developed (Mace, ICRISAT, unpublished data). However, due to the lack of polymorphism observed between the parental genotypes, to date only approximately 70 SSR and AFLP markers have been mapped to 17 linkage groups covering approximately 420 cM. Future research efforts will aim to map the new SSR markers on mapping populations derived from selected diverse parental genotypes such as TxAG-6.

Marker-Assisted Selection

Marker-assisted selection (MAS) offers great promise for improving the efficiency of conventional plant breeding. Molecular markers are especially advantageous for traits where conventional phenotypic selection is difficult, expensive, or lacks accuracy or precision. This includes resistance to certain pathogens and insect pests plus tolerance to abiotic stresses, quality parameters, and complex agronomic traits with low heritabilities. The essential requirements for developing marker-assisted selection systems are (i) availability of germplasm with substantially contrasting phenotypes for the traits of interest, (ii) highly accurate and precise screening techniques for phenotyping mapping populations for the trait of interest, (iii) identification of flanking marker(s) closely associated with the loci of interest and the flanking regions on either side, and (iv) simple robust DNA marker technology to facilitate rapid and cost-effective screening of large populations.

In peanut, a few examples of traits that may justify the cost and time required to develop and apply DNA markers include early leafspot (*Cercospora arachidicola* Hori.), late leafspot (*Cercosporidium personatum* Berk. et Curt.) Deighton, nematodes (*Meloidogyne* spp.), leafminer, and *Spodoptera*, for which there are only low to moderate levels of resistance (or tolerance) available in cultivated peanut but high levels available in wild species. Traits associated with seed quality—as measured by oleic/linoleic (O/L) ratio, for which the higher the ratio the better the shelf-life of the peanut products—and drought tolerance (specific leaf area, total transpiration, water use efficiency, and partitioning), which are difficult to measure in large segregating generations and substantially influenced by genotype-by-environment interaction, may also benefit from MAS (54). MAS holds great promise for the improvement of such characters if markers can be identified for genes controlling components of these traits.

MAS within *A. hypogaea* is constrained by the low level of DNA polymorphism found in the cultivated gene pool. The RFLP-based tetraploid map developed by Burow *et al.* (8), based on an interspecific cross, is likely to be useful in terms of locating specific genes of interest in this interspecific cross and also provides valuable information about genome organization and evolution. However, the markers themselves are of less value in elite cultivated germplasm, in which very little polymorphism is found. Recent advances in the development of PCR-based marker protocols have opened new possibilities in the study of complex traits in crop plants. The hybridization-based co-dominant markers (RFLP) and PCR-based dominant markers (RAPD and AFLP) in many crops have been superseded by co-dominant PCR-based markers (SSR). The AFLP assay has been used frequently in diversity and mapping studies in many crop plants. However, efforts to convert AFLP markers into simple co-dominant PCR markers are laborious, expensive, and time consuming, and above all have met with only mixed success, particularly in complex polyploids (55).

Although MAS has been little used within *A. hypogaea*, even with the limitations afforded by present technologies it has much potential use for introgressing genes from closely related *Arachis* species into the cultivated genome. Garcia *et al.* (56) reported introgression of genes from *A. cardenasii* (an A-genome species) into *A. hypogaea* in 10 of 11 linkage groups on the diploid RFLP map developed by Halward *et al.* (51). Garcia *et al.* (57) then used RAPD and SCAR technologies to map two dominant genes conferring resistance to the nematodes (*M. arenaria* (Neal) Chitwood, Race 1) where a tetraploid breeding line derived an *A. hypogaea* × *A. cardenasii* cross was analyzed. Burow *et al.* (49) reported RFLP markers linked to *M. arenaria* resistance in another interspecific cross involving the species *A. hypogaea*,

A. batizocoi, *A. cardenasii*, and *A. diogoi*. Additional linkages of RAPD markers have been found for components of early leafspot and corn rootworm (*Diabrotica undecimpunctata howardi* Barber) resistances in (cultivar NC 7 × NC GP WS 1)^{4x} interspecific crosses. In crosses restricted to the *A. hypogaea* parents NC 7 and PI 109839, linkages of RAPD primers with genes conferring resistance to leafhopper (*Empoasca fabae* Harris), Cylindrocladium black rot and several components of early leafspot were found (58). Among diploids, several AFLP markers were linked to tomato spotted wilt virus resistance (59). Although the above linkages of resistance genes to different molecular markers may prove useful for selecting breeding lines with desirable traits, there have been only a few successes in peanut for utilizing these materials for cultivar development.

Clearly, there is a need to explore new assays with greater power to reveal polymorphisms in peanut, such as single nucleotide polymorphisms (SNP). Recent developments in SNP technology indicate that in the near future there may be additional options available for rapid identification of large numbers of polymorphic markers (60). SNPs comprise the largest set of sequence variants in most organisms (61, 62). SNPs are biallelic markers but occur very frequently within the genome, their mutation rate is low, and they are capable of high throughput genotyping and are often linked to genes (63). For example, a map containing 1.42 million SNPs distributed throughout the human genome has been constructed with an average density of one SNP every 1.9 kb (64). SNPs have also been reported in many crop plants and show much promise for MAS, but to date there have been no substantial explorations of the peanut genome.

Tools and Strategies for Physical Mapping of Peanut

The cultivated peanut genome is estimated to contain about 2.9 pg DNA per haploid genome equivalent (65). The A and B genomes are thought to contribute approximately equal amounts of genetic information to cultivated peanut, based on (1c) estimates of 1.34 pg DNA for *A. duranensis* and 1.42 pg DNA for *A. ipaensis*, respectively (66). Cot analysis of the genomic DNA of *A. duranensis* reveals genomic composition of 27% highly repetitive DNA, 37% middle-repetitive DNA, and 36% low-copy DNA (67). These parameters of size and repetitiveness are similar to those for the human genome, suggesting that a comprehensive physical map of peanut will require a much greater effort than was necessary for botanical models such as *Arabidopsis* or *Oryza*.

The cornerstone of a physical mapping strategy for peanut will be a comprehensive arrayed library of large-insert DNA clones. Early progress toward a bacterial artificial chromosome (BAC) library for the tetraploid peanut cultivar Florunner (68) has been advanced to include about 180,000 clones (roughly six genome-equivalent coverage). Hybridization-based anchoring of genetically-mapped DNA markers, together with fingerprinting and alignment of BACs to one another will help assemble tools suitable for many applications, including positional cloning anywhere in the peanut genome. Utilization of synteny information from other legumes or botanical models (see Chapter 9) may help identify gene probes that can be targeted to gaps in the peanut physical map and accelerate closure. Selected BAC end sequencing may help to extend contigs, although the abundance of repetitive DNA in peanut suggests that only 36% of peanut BAC ends will be locus-specific.

A fundamental complication in peanut physical mapping will be the presence of two duplicated copies of most genes, one from each of the two diploid progenitors. By hybridization-based approaches, it is likely to be very difficult to distinguish which BACs contain true homologs of a particular probe sequence, and which contain homoeologs, or corresponding sequences from the alternate subgenome. Methods have been established (69) to resolve the subgenome specificity of individual BACs in the tetraploid libraries. An alternative approach would be to produce additional BAC libraries for diploid progenitors of peanut, which would reduce the magnitude of the problem.

Tools and Strategies for Functional Genomics of Peanut

Fundamental to analysis of gene function in an organism is a detailed picture of the transcriptome (the population of DNA sequences that encode mRNA products). For peanut, this information is almost completely lacking but urgently needed. Other cultivated legumes such as *Glycine*, *Medicago*, and *Phaseolus* enjoy extensive resources of 359,130, 194,501, and 21,931 ESTs, respectively, representing diverse sets of tissues and in some cases multiple genotypes. However, as of 15 August 2003, peanut had only 1,540 public EST (gene) sequences. Generation of peanut ESTs for many diverse tissues is a key first step in associating peanut genes with their respective functions. In view of the basic and applied importance of peanut's unusual reproductive system, a particularly high priority is generation of ESTs for reproductive tissues. The addition of a peanut EST resource to the extensive resources for *Glycine* and *Medicago*, and emerging resources for *Phaseolus,* will empower a host of investigations of features that distinguish these four major legume lineages from one another, and indeed, which distinguish the legumes from other well-studied rosids such as the *Brassicales* (crucifers, including *Arabidopsis*) and *Malvales* (cotton), as well as asterids (lettuce and sunflower), core eudicots (ice plant and sugarbeet), and monocots. EST resources are a crucial first step toward most functional genomic approaches, in particular large-scale analysis of the timing and levels of expression of peanut genes.

Prospects for Sequencing the Peanut Genome

The large size and high fraction of repetitive DNA make the complete sequencing of the peanut genome a daunting task, one of similar cost and complexity to sequencing of the human genome. A clear first step is sequencing large populations of expressed sequences (ESTs) to rapidly gain information about many of the regularly expressed genes in peanut. The cost will be a tiny fraction of resources needed to sequence the entire genome. While this primary goal will provide much information, it will also leave important questions unanswered, such as the identities of key regulatory genes that are expressed only briefly and/or at low levels, and therefore tend not to be sampled by the EST approach. Further, EST sequencing will not provide information about regulatory sequences that determine timing and levels of gene expression, and which may be key to the molecular basis of many morphological changes that distinguish related plant taxa.

Once basic information such as EST resources for peanut is developed, a logical next step may be the exploration of the peanut genome by a recently described approach called Cot-based cloning and sequencing (CBCS) (70,71). CBCS is based on fractionation of total genomic DNA into components that are comprised of populations of DNA elements that occur in the genome in similar copy numbers. For example, the "highly repetitive" component typically contains many retrotransposons, centromeric repeats, and other highly-abundant elements often thought of as "junk DNA." The "middle-repetitive" component may contain lower-copy repetitive elements, ribosomal DNA, and some large multigene families. The "low-copy" component contains most genes. By first fractionating a genome into such components, then cloning each component and sequencing into the clone library to a depth that provides appropriate coverage of the sequence complexity of the respective components, the population of nonredundant DNA sequences in a genome can be obtained at substantially less time and cost than by approaches that are presently in use (71). CBCS offers a particularly appealing approach by which one might efficiently obtain not only the gene-rich sequences in peanut, but also quickly identify the repetitive DNA families that complicate peanut genomics to mitigate the impact of these elements on sequence assembly and annotation.

Summary

Despite its major economic importance both in the United States and internationally, peanut is badly underexplored at the genomic level. This deficiency hinders geneticists from being able to provide intrinsic low-cost and environmentally benign solutions to many challenges that increase the cost and risk of peanut production, and cause peanut to fall short of consumer needs and desires. Urgent needs include the development of large numbers of user-friendly genetic mapping tools; sequencing of substantial populations (100,000 or more) of expressed sequences from diverse tissues and genotypes; the assembly of a genetically anchored physical map and its alignment to the emerging sequences of small-genome legumes such as *Medicago*; and sampling the gene-rich regions to quantify the additional information that may be gained by further sequencing of peanut genomic DNA.

Acknowledgments

We thank many colleagues for valuable contributions, and the Texas Higher Education Coordinating Board, USDA National Research Initiative, Goldkist Inc., and the Georgia, Florida, North Carolina, and Texas Agricultural Experiment Stations for funding.

References

1. FAO, *FAO Production Year Book 55*:118–119 (2001).
2. Savage, G.P., and J.I. Keenan, The Composition and Nutritive Value of Groundnut Kernels, in *The Groundnut Crop: A Scientific Basis for Improvement*, edited by J. Smartt, Chapman and Hall, London, 1994, pp. 173–213.
3. Kerridge, P.C., and B. Hardy, *Biology and Agronomy of Forage* Arachis, CIAT, Cali, Colombia, 1994, No. 240, pp. 206.
4. Ruiz, T.M., R. Ramos-Santana, and A. Sotomayor-Rios, Dry Matter Yield of Rhizoma Perennial Peanut (*Arachis glabrata*) Harvested at Six, Nine and 12 Weeks at Two Semiarid Sites, *J. Agric. Univ. P.R. 84*:115–131 (2000).
5. Sauer, J.D. *Arachis*—Peanut, in *Historical Geography of Crop Plants*, Lewis Publishers, Boca Raton, FL, 1993, pp. 80–83.
6. Gregory, W.C., B.W. Smith, and J.A. Yarbrough, Morphology, Genetics, and Breeding, in *The Peanut—The Unpredictable Legume*, National Fertilizer Assoc., Washington, D.C., 1951, pp. 28–88.
7. Pattee, H.E., H.T. Stalker, and F.G. Giesbrecht, Reproductive Efficiency in Reciprocal Crosses of *Arachis monticola* with *A. hypogaea* Subspecies, *Peanut Sci. 25*:7–12 (1998).
8. Burow, M.D., C.E. Simpson, J.L. Starr, and A. H. Paterson, Transmission Genetics of Chromatin From a Synthetic Amphiploid in Cultivated Peanut (*A. hypogaea* L.): Broadening the Gene Pool of a Monophyletic Polyploid Species, *Genetics 159*:823–837 (2001).
9. Soltis, D.E., and P.S. Soltis, Molecular Data and the Dynamic Nature of Polyploidy, *Crit. Rev. Plant Sci. 12*:243–273 (1993).
10. Soltis, P.S., and D.E. Soltis, The Role of Genetic and Genomic Attributes in the Success of Polyploids, *Proc. Natl. Acad. Sci. USA 97*:7051–7057 (2000).
11. Kochert, G., H.T. Stalker, M. Gimenes, L. Galgaro, C.R. Lopes, and K. Moore, RFLP and Cytological Evidence on the Origin and Evolution of Allotetraploid Domesticated Peanut *Arachis hypogaea* (Leguminosae), *Am. J. Bot. 83*:1282–1291 (1996).
12. Singh, A.K., and C.E. Simpson, Biosystematics and Genetic Resources, in *The Groundnut Crop: A Scientific Basis for Improvement*, edited by J. Smartt, Chapman and Hall, London, 1994, pp. 24–42.
13. Singh, A.K., and J.P. Moss, Utilization of Wild Relatives in Genetic Improvement of *Arachis hypogaea* L. 2 Chromosome Complements of Species of Section *Arachis*, *Theor. Appl. Genet. 61*:305–314 (1982).
14. Wynne, J.C., and T. Halward, Cytogenetics and Genetics of *Arachis*, *Crit. Rev. Plant Sci. 8*:189–220 (1989).
15. Kochert, G., T. Halward, W.D. Branch, and C.E. Simpson, RFLP Variability in Peanut (*Arachis hypogaea* L.) Cultivars and Wild Species, *Theor. Appl. Genet. 81*:565–570 (1991).
16. Halward, T.M., H.T. Stalker, E. LaRue, and G. Kochert, Genetic Variation Detectable with Molecular Markers among Unadapted Germplasm Resources of Cultivated Peanut and Related Wild Species, *Genome 34*:1013–1020 (1991).
17. Hopkins, M.S., A.M. Casa, T. Wang, S.E. Michell, R.E. Dean, G.D. Kochert, and S. Kresovich, Discovery and Characterization of Polymorphic Simple Sequence Repeats (SSRs) in Peanut, *Crop Sci. 39*:1243–1247 (1999).
18. Dwivedi, S.L., S. Gurtu, S. Chandra, W. Yuejin, and S.N. Nigam, Assessment of Genetic Diversity among Selected Groundnut Germplasm I: RAPD analysis, *Plant Breed. 120*:345-349 (2001).

19. He, G., and C. Prakash, Evaluation of Genetic Relationships among Botanical Varieties of Cultivated Peanut (*Arachis hypogaea* L.) Using Markers, *Genet. Resour. Crop Evol. 48*:347–352 (2001).
20. Cifarelli, R.A., M. Gallitelli, and F. Cellini, Random Amplified Hybridization Microsatellites (RAHM): Isolation of a New Class of Microsatellite-Containing Clones, *Nucleic Acids Res. 23*:3802–3803 (1995).
21. Fischer, D., and K. Bachmann, Microsatellite Enrichment in Organisms with Large Genomes (*Allium cepa* L.), *BioTechniques 24*:796–802 (1998).
22. Krapovikas, A., and W.C. Gregory, Taxonomy of Genus *Arachis* (Leguminosae), *Bonplandia 8*:1–186 (1994).
23. Paik-Ro, O.G., R.L. Smith, and D.A. Knauft, Restriction Fragment Length Polymorphism Evaluation of Six Peanut Species within the *Arachis* Section, *Theor. Appl. Genet. 84*:201–208 (1992).
24. Lanham, P.G., S. Fennell, J.P. Moss, and W. Powell, Detection of Polymorphic Loci in *Arachis* Germplasm Using Random Amplified Polymorphic DNAs, *Genome 35*:885–889 (1992).
25. Stalker, H.T., J.S. Dhesi, and G. Kochert, Genetic Diversity within the Species *Arachis duranensis* Krapov. and W. C. Gregory, a Possible Progenitor of Cultivated Peanut, *Genome 38*:1201–1212 (1995).
26. Young, N.D., N.F. Weeden, and G. Kochert, Genome Mapping in Legumes (Fam. Fabaceae), in *Genome Mapping in Plants*, edited by A.H. Paterson, Landes Bioscience Press, Austin, TX, 1996, pp. 211–227.
27. Godoy, I. J., and E. H. Giandana, Groundnut Production and Research in South America, in *Groundnut—A Global Perspective*, edited by S.N. Nigam, ICRISAT, Patancheru, India, 1992, pp. 77–85.
28. Isleib, T.G., C.C. Holbrook, and D.W. Gorbet, Use of Plant Introductions in Peanut Cultivar Development, *Peanut Sci. 28*:96–113 (2001).
29. Isleib, T.G., J.C. Wynne, and S.N. Nigam, Groundnut Breeding, in *The Groundnut Crop: A Scientific Basis for Improvement*, edited by J. Smartt, Chapman and Hall, London, 1994, pp. 552–623.
30. Mozingo, R.W., T.A. Coffelt, and J.C. Wynne, Genetic Improvement in Large-Seeded Virginia-Type Peanut Cultivars Since 1944, *Crop Sci. 27*:228–231 (1987).
31. Smartt, J., Makulu Red—A 'Green Revolution' Groundnut Variety? *Euphytica 27*:605–608 (1978).
32. Yanahao, S., and W. Caibin, Factors Contributing to High Yields of Groundnut in Shandong, China, *Int. Arachis Newsletter 8*:7–9 (1990).
33. Nigam, S.N., Some Strategic Issues in Breeding for High and Stable Yield in Groundnut in India, *J. Oilseeds Res. 17*:1–10 (2000).
34. Stalker, H.T., and C.E. Simpson, Genetic Resources in *Arachis*, in *Advances in Peanut Science*, edited by H.E. Pattee and H.T. Stalker, American Peanut Research Education Society, Stillwater, OK, 1995, pp. 14–53.
35. Holbrook, C.C., and H.T. Stalker, Peanut Breeding and Genetic Resources, in *Plant Breeding Reviews*, edited by J. Janick, John Wiley & Sons, NJ, 2003, Vol. 22, pp. 297–356.
36. Pittman, R.N., Progress and Status of the U.S. Peanut Collection, *Proc. Amer. Peanut Res. Educ. Soc. 32*:49 (2000).
37. Upadhyaya, H.D., M.E. Ferguson, and P.J. Bramel, Status of the *Arachis* Germplasm Collection at ICRISAT, *Peanut Sci. 28*:89–96 (2001).

38. Duncan, W.G., D.E. McCloud, R.L. McGraw, and K.J. Boote, Physiological Aspects of Peanut Yield Improvement, *Crop Sci. 18*:1015–1020 (1978).
39. Coffelt, T.A., M.L. Seaton, and S.W. VanScoyoc, Reproductive Efficiency of 14 Virginia-Type Peanut Cultivars, *Crop Sci. 29*:1217–1220 (1989).
40. Wynne, J.C., and W.C. Gregory, Peanut breeding, *Adv. Agron. 34*:39–71 (1981).
41. Knauft, D.A., and J.C. Wynne, Peanut Breeding and Genetics, *Adv. Agron. 55*:393–445 (1995).
42. Guok, H.P., J.C. Wynne, and H.T. Stalker, Recurrent Selection within a Population from an Interspecific Peanut Cross, *Crop Sci. 26*:249—253 (1986).
43. Halward, T.M., J.C. Wynne, and H.T. Stalker, Recurrent Selection Progress in a Population Derived from an Interspecific Peanut Cross, *Euphytica 52*:9–84 (1991).
44. Simpson, C.E., Pathways for Introgression of Pest Resistance into *Arachis hypogaea* L., *Peanut Sci. 18*:22–26 (1991).
45. Simpson, C.E., S.C. Nelson, J.L. Starr, K.E. Woodard, and O.D. Smith, Registration of TxAG-6 and TxAG-7 Peanut Germplasm Lines, *Crop Sci. 33*:1418 (1993).
46. Husted, L., Cytological Studies of the Peanut *Arachis*. I. Chromosome Number and Morphology. *Cytologia 5*:109–117 (1933).
47. Husted, L., Cytological Studies of the Peanut *Arachis*. II. Chromosome Number, Number Morphology and Behavior and their Application to the Origin of Cultivated Forms, *Cytologia 7*:396–423 (1936).
48. Smartt, J., W.C. Gregory, and M.P. Gregory, The Genomes of *Arachis hypogaea. 2*. The Implications in Interspecific Breeding, *Euphytica 27*:677–680 (1978).
49. Burow, M.D., C.E. Simpson, A. H. Paterson, and J.L. Starr, Identification of Peanut (*Arachis hypogaea* L.) RAPD Markers Diagnostic of Root-Knot Nematode (*Meloidogyne arenaria* (Neal) Chitwood) Resistance, *Mol. Breed. 2*:369–379 (1996).
50. Simpson, C.E., and J.L. Starr, Registration of Coan Peanut, *Crop Sci. 41*:918 (2001).
51. Halward, T.M., H.T. Stalker, and G. Kochert, Development of an RFLP Map in Diploid Peanut Species, *Theor. Appl. Genet. 87*:379–384 (1993).
52. Smartt, J., W.C. Gregory, and M.P. Gregory, The Genomes of *Arachis hypogaea*. 1. Cytogenetic Studies of Putative Genome Donors, *Euphytica 27*:665–675 (1978).
53. Stalker, H.T., J.S. Dhesi, D.C. Parry, and J.H. Hahn, Cytological and Interfertility Relationships of *Arachis* Section *Arachis, Am. J. Bot. 78*:238–246 (1991).
54. Dwivedi, S. L., J.H. Crouch, S.N. Nigam, M.E. Ferguson, and A.H. Paterson, Molecular Breeding of Groundnut for Enhanced Productivity and Food Securities in the Semi-Arid Tropics: Opportunities and Challenges, *Adv. Agron. 80:*153–221 (2003) (in print).
55. Shan, X., T.K. Blake, and L.E. Talbert. Conversion of AFLP Markers to Sequence-specific PCR Markers in Barley and Wheat, *Theor. Appl. Genet. 98*:1072–1078 (1999).
56. Garcia, G.M., H.T. Stalker, and G. Kochert, Introgression Analysis of an Interspecific Hybrid Population in Peanuts (*Arachis hypogaea* L.) using RFLP and RAPD Markers, *Genome 38*:166–176 (1995).
57. Garcia, G.M., H.T. Stalker, E. Shroeder, and G. Kochert, Identification of RAPD, SCAR and RFLP Markers Tightly Linked to Nematode Resistance Genes Introgressed from *Arachis cardenasii* to *A. hypogaea*, *Genome 39*:836–845 (1996).
58. Stalker, H.T., and L.G. Mozingo, Molecular Markers of Arachis and Marker-Assisted Selection, *Peanut Sci. 28*:117–123 (2001).
59. Milla, S., Relationships and Utilization of *Arachis* Germplasm in Peanut Improvement, Ph.D. Thesis, North Carolina State University, Raleigh, 2003.

60. Kanazin, V., H. Talbert, D. See, P. DeCamp, E. Nevo, and T. Blake, Discovery and Assay of Single-Nucleotide Polymorphisms in Barley (*Hordeum vulgare*), *Plant Mol. Biol. 48*:529–537 (2002).
61. Kwok, P.Y., Q. Deng, H. Zakeri, S.L. Taylor, and D.A. Nickerson, Increasing the Information Content of STS-Based Genome Maps: Identifying Polymorphisms in the Mapped STSs, *Genomics 31*:123–126 (1996).
62. Kruglyak, L., The Use of a Genetic Map of Biallelic Markers in Linkage Studies, *Nature Genet. 17*:21–24 (1997).
63. Kwok, P.Y., and Z. Gu, SNP Libraries: Why and How Are We Building Them? *Mol. Medicine Today 12*:538–543 (1999).
64. The International SNP Map Working Group, A Map of Human Genome Sequence Variation Containing 1.42 Million Single Nucleotide Polymorphisms, *Nature 409*:928–933 (2001).
65. Arumunganathan, K., and E.D. Earle, Nuclear DNA Content of Some Important Plant Species, *Plant. Mol. Biol. Rptr. 9*:208–218 (1991).
66. Singh, K.P., S.N. Raina, and A.K. Singh, Variation in Chromosomal DNA Associated with the Evolution of *Arachis* Species, *Genome 39*:890–897 (1996).
67. Dhillon, S. S., A.V. Rake, and J.P. Miksche, Reassociation Kinetics and Cytophotometric Characterization of Peanut (*Arachis hypogaea* L.) DNA, *Plant Physiol. 65*:1121–1127 (1980).
68. Yuksel, B., D. Peterson, S. Lee, M. Burow, and A.H. Paterson, Progress Toward Construction and Characterization of a Peanut BAC library, *International Plant and Animal Genome Conference*, San Diego, CA, www.intlpag.org/pag/11/abstracts/P5f_P499_XI.html (2003).
69. Lin, Y.R., X. Draye, X. Qian, S. Ren, L. Zhu, J. Tomkins, R.A. Wing, Z. Li, and A.H. Paterson, Fine-Scale Mapping and Sequence-Ready Contig Assembly in Highly-Duplicated Genomes, using the BAC-RF Method, *Nucleic Acids Res. 2823*: www3.oup.co.uk/nar/methods/Volume_28/Issue_07/gnd023_gml.abs.html (2000).
70. Peterson, D.G., S.R. Schulze, E.B. Sciara, S.A. Lee, J.E. Bowers, A. Nagel, N. Jiang, D.C. Tibbitts, S.R. Wessler, and A.H. Paterson, Integration of Cot Analysis, DNA Cloning, and High-Throughput Sequencing Facilitates Genome Characterization and Gene Discovery, *Genome Res. 12*:795–807 (2002).
71. Peterson, D.G., S.R. Wessler, and A.H. Paterson, Efficient Capture of Sequence Complexity Using Cot-Based Cloning and Sequencing, *Trends Genet. 18*:547–550 (2002).

Chapter 7

Genomics Research in Alfalfa, *Medicago sativa* L.

E. Charles Brummer

Raymond F. Baker Center for Plant Breeding, Department of Agronomy, Iowa State University, Ames, IA 50011

Introduction

Alfalfa, *Medicago sativa* L., is one of the most important forage crops in the world and is the most important forage legume in the United States. Alfalfa is cultivated on 32 million ha worldwide, with U.S. production concentrated in the upper Midwest, the Great Plains, and the desert Southwest (1). Although alfalfa is primarily used for ruminant livestock feed as hay, silage, green chop, and pasture, a number of nontraditional uses of alfalfa are being investigated, including as a bioreactor for the production of industrial enzymes (2) or other products via transgenesis and as a bioenergy crop.

Alfalfa has a number of characteristics that make its cultivation appealing, including its high rate of atmospheric nitrogen fixation, which is estimated to be approximately 180 kg N/ha/year (3). Alfalfa has beneficial effects on soil tilth, reduces pest problems when incorporated into a cropping rotation, and produces more protein per hectare than any other crop cultivated in the United States. As a perennial plant, it covers the ground throughout the year, decreasing soil erosion and improving water quality compared with annual grain or oilseed crops such as maize (*Zea mays* L.) or soybean (*Glycine max* (L.) Merr.).

The genus *Medicago* includes annual and perennial herbs and one perennial shrub distributed in 56 (4) to 83 (5) species. Alfalfa belongs to the *M. sativa–falcata* complex, which includes several subspecies at the diploid ($2n = 2x = 16$) and tetraploid ($2n = 4x = 32$) levels as well as several closely related species that are completely interfertile (6). Ploidy can be determined in a high throughput manner using flow cytometry (7). Tetraploid representatives of the complex are generally regarded as tetrasomic tetraploids (8). A recent report suggesting that preferential pairing may be common (9) is likely incorrect; a reanalysis of the data shows that no preferential pairing occurs (10). Nevertheless, further investigation of the actual pairing relationships, particularly among subspecies, is warranted.

Within the genus, alfalfa has historically received the most breeding and genetic attention, but improved cultivars of several of the annual species, including *M. polymorpha* L., *M. truncatula* Gaertn., and *M. scutellata* (L.) Miller, have also been produced. Alfalfa has served as a model system to investigate nitrogen fixation, secondary metabolite biochemistry, and polyploid genetics. Over the past decade, *M.*

truncatula has been developed into a model species more tractable than alfalfa to investigate legume-specific traits. Research on *M. truncatula* will have a substantial impact on alfalfa improvement, particularly because many tools, such as SSR markers, can be directly transferred from the model to the crop with little loss of power. The gene-space sequence of *M. truncatula* is currently being determined, serving as a backbone for genome investigations in all Medicago species.

Alfalfa breeding and genetics began in the early 1900s with selection for winter hardiness and adaptation. Over the past century, alfalfa populations have been selected for numerous traits, including biomass production, nutritive value, stress tolerance, and disease and pest resistance (11). Alfalfa is an allogamous perennial cultivated primarily as a tetraploid, although diploid germplasm exists under natural conditions. Inbreeding depression is usually considered to be severe in alfalfa (12), and as a consequence, most cultivars are synthetic populations, consisting of a heterogeneous mixture of heterozygous genotypes. Hybrid cultivars have been recurrently proposed in the literature, and the presence of substantial levels of heterosis between some germplasms (13) suggests they are worth investigating, but methods to produce them efficiently need to be developed more fully.

Genomics efforts in alfalfa have been underway for 20 years. Molecular genetic markers have been developed and used to differentiate among populations and to create genetic maps. Alfalfa tissue culture, regeneration, and transformation are well characterized. Nascent programs in gene, protein, and metabolite profiling are underway, all leaning heavily on *M. truncatula* for support. The objective of this review is to comprehensively summarize the genomic research in alfalfa, to highlight some of the community needs relative to genomic infrastructure, and to briefly consider the role of *M. truncatula* in alfalfa improvement.

Genetic Markers

Most major genetic marker systems have been applied to alfalfa. This section provides a catalog of experiments using various types of markers; subsequent sections elaborate on the biological significance of the results.

Isozymes were first applied to alfalfa over 20 years ago (14) and have subsequently been used to identify alfalfa plants and cultivars (15), to document multiple alleles at a locus (16), to assess relationships among members of the *M. sativa–falcata* complex (6), and to provide evidence for gene flow between cultivated and wild alfalfa populations (17).

Restriction fragment length polymorphism (RFLP) markers have been used widely to assess levels of diversity within and among populations of alfalfa (18–22), for genetic map construction (23–28), to map agronomically important traits (29–33), to assess the approach to homozygosity under inbreeding (34,35), and to study relationships of heterozygosity and forage yield (36–39).

Randomly amplified polymorphic DNA (RAPD) markers have been used to estimate relationships among alfalfa genotypes and/or populations (40–47), to map alfalfa

(48–50), to document interspecific *Medicago* crosses (51), to analyze and map meiotic mutants (52–55), to estimate somoclonal variation (56), to assess whether progeny resulted from cross- or self-pollination (57), to assess risk of transgene movement (58), and to estimate relationships among annual *Medicago* species (59).

Simple sequence repeat (SSR) markers have been developed for alfalfa using both alfalfa and *M. truncatula* sequences. Based on screening an alfalfa genomic library, approximately 19,000 CA + AT + CT + ATT repeats exist within the alfalfa genome (60). SSR markers have been identified by screening this genomic library and by searching GenBank sequences, and they appear to be easily transferred across species within the genus (60,61). SSR markers have also been mapped (62–66). Inter–simple sequence repeat (ISSR) markers have been applied to genetic mapping (55), and anchored microsatellite priming (AMSP) markers have been applied to estimation of genetic variation (46,47).

Amplified fragment length polymorphism (AFLP) markers have been applied to alfalfa mapping (55,65,67), to assessments of diversity among germplasms (68,69), to relationships with heterosis (70), and to identifying associations with disease resistance (71). Variations of the AFLP technique, including sequence-specific amplification polymorphism (S-SAP), selective amplification of microsatellite polymorphic loci (SAMPL) (72), and microsatellite-AFLP (73) have also been applied to linkage mapping. In addition to nuclear DNA markers, variation among accessions and/or species in the chloroplast and mitochondrial genome has also been reported (74–76).

Genetic Diversity

Because alfalfa is an allogamous species, populations consist of many different heterozygous genotypes. Methods to discriminate among populations have been developed based on genotyping individuals within populations (40) or on bulked samples (43). The issue of bulking samples for population differentiation has been evaluated by several authors. Kidwell *et al.* (19) assessed individual plants as well as bulks of up to 12 plants. They determined that increasing the number of plants in a bulk more clearly resolved the germplasms they were evaluating; what appeared to be indistinct clusters of individuals resolved into clear groups as the number of plants per bulk increased to six. In contrast, an analysis of four populations using either five individual plants or five-plant bulks with RAPD markers showed that the genetic distances between cultivars remained similar with either method (43). Bulks of seven individuals appeared sufficient to represent the diversity of the accession (43).

In general, accession-specific markers are rare unless one germplasm is clearly distinct, such as *M. sativa* subsp. *falcata*, and typically rare alleles are only present in a subset of plants within an accession (20,40). Because these fragments tend to disappear upon bulking, the use of bulking will eliminate the ability to fingerprint a population based on unique alleles, but it will not eliminate the ability to differentiate populations based on allele frequency (19,20,69). Fragments present in fewer than 25% of individuals were not detected in bulks analyzed by AFLP (68) and were unlikely to be detected by RFLP (20). Thus, bulking removes some of the variability between populations because rare alleles present in one population but not the other are lost.

Clearly, accurate estimation of shifts in allele frequencies needs to be evaluated by analysis of individual plants (40,77). However, the successful separation of populations based on bulked samples (69) shows that analysis of individual plants is not necessary if precise estimates of allele frequencies are not of interest. Segovia-Lerma *et al.* (69) found an average of 40% polymorphism across 34 primer pairs when amplifying bulk samples consisting of 30 plants of each of nine germplasms. Bulks were not able to differentiate between two Italian ecotypes, but a reanalysis of individual plants failed to find a significant difference either (68).

The first experiment to use DNA markers to assess variation in alfalfa germplasm was by Brummer *et al.* (18), who used RFLP markers on diploid germplasm and on three tetraploid cultivars. Clustering of diploid genotypes based on presence and absence of DNA fragments showed a clear separation between *M. sativa* subsp. *falcata* and *M. sativa* subsp. *caerulea*. *Medicago sativa* subsp. *caerulea* is purple flowered and represents the diploid form of subsp. *sativa*. A diploidized population of tetraploid alfalfa, Cultivated Alfalfa at the Diploid Level (CADL) (78), also clustered with the subsp. *caerulea* entries. At the tetraploid level, the 20 plants from each of three cultivars mainly clustered into discrete groups; the most distinct cultivar, Spredor II, which contains subsp. *falcata* germplasm, was completely separated from the others. From this experiment, the differences between subsp. *falcata* and subsp. *sativa*, which differ based on morphological characteristics, were also evident at the DNA level. Finally, large amounts of genetic variation were shown to be present within populations.

Alfalfa introduced into the United States has been documented to derive from nine historical sources (79). Evaluation of these nine germplasm classes using 180 RFLP marker alleles showed that they are generally similar, except for two of the nine (subsp. *falcata* and Peruvian), which clearly separated from the others (19). These results are clouded by the history of the sampled germplasms, which may have undergone cross-pollination with other groups. Thus, a reinvestigation of these nine groups was undertaken by Segovia-Lerma *et al.* (69) using recreated germplasms of all nine that are likely to represent the actual introductions more closely. This analysis was conducted on bulked samples of 30 plants per accession, which were scored for 1,541 polymorphic AFLP marker fragments. Again, *M. sativa* subsp. *falcata* was the most genetically dissimilar germplasm, but Peruvian did not appear to be as widely different from the others, as the RFLP data suggested (19). The AFLP experiment was exhaustive in its methodology and likely represents the best estimate of relationships among the progenitors of U.S. alfalfa germplasm that it is possible to obtain.

An assessment of a wide range of breeding material and wild germplasm from Spain showed that *M. sativa* subsp. *falcata* clusters apart from most Flemish and Mediterranean cultivars and germplasm (41), even though two Spanish ecotypes cluster with *falcata* based on morphological analyses (80). Two *sativa* cultivars with moderate levels of variegated flower color and a Tunisian landrace clustered close to the two *falcata* wild populations, though not particularly closely related to either. Evidence for gene flow between cultivated and wild populations undoubtedly will serve to obscure the relationships among these taxa (17). One of these *falcata* germplasms, Maron, was also evaluated in another experiment using RAPD markers and was clearly distinguished from a wild *sativa* population and six European *sativa* cultivars (40).

A number of other experiments evaluating differences among alfalfa germplasm have been conducted. An analysis of Australian cultivars based on individual plant samples from 10 cultivars found high intrapopulation variation based on RAPD markers, which precluded differentiating among cultivars. The exception was an *M. sativa* subsp. *falcata* population included as a control, which showed clear distinctiveness to the cultivated *sativa* germplasm (44). Finally, the cultivar Anik, a pure *falcata* population, was clearly distinct from populations derived from *sativa* or *sativa* × *falcata* intercrossing (43).

Most evidence, then, suggests that *M. sativa* subsp. *falcata* represents a distinct source of germplasm for alfalfa improvement (13,81). However, most of the experiments conducted to date have evaluated primarily cultivated alfalfa populations that have undergone some artificial selection. Very little research is available exploring the relationships among wild *sativa* and *falcata* populations. Some recent results show a clear delineation of diploid *falcata* and *caerulea* (E.C. Brummer, unpub. data), but the tetraploid wild species do not show such clear clustering (82, E.C. Brummer, unpub. obs.). Based on hypervariable chloroplast DNA markers, no clear differentiation between *M. falcata* and other *sativa* accessions, including both wild and cultivated, were observed (74). In fact, although some clustering of particular geographic groups (e.g., Morocco) was observed, many of the wild accessions of both *sativa* and *falcata* appeared to form no clear groups at all. Further exploration of the relationships among wild populations is warranted.

Several experiments have attempted to resolve relationships of Italian ecotypes and populations derived from them. Analysis of molecular variance (AMOVA) showed that 75 to 90% of total variance resides within populations, a number typical of outcrossing plant populations, and populations within ecotypes or cultivars accounted for most of the rest (42,68). Different populations derived from the ecotypes or cultivars appeared to diverge both morphologically and based on markers, which made differentiating the ecotypes difficult. Supporting evidence for this conclusion was provided by an analysis of chloroplast variation (76). A similar experiment examined six Iranian ecotypes with a single RAPD primer and was unable to resolve all of them into distinct groups.

Genetic Mapping

Map Development

A number of genetic maps have been published for alfalfa (Table 7.1). Most of the initial mapping work focused on diploid populations; although a tentative consensus group was developed using RAPD markers in a tetraploid alfalfa population (49), the completed map was never published. Although the F_1 maps constructed by Tavoletti *et al.* (28) were relatively sparse, they provided a first look at synteny between *M. sativa* subsp. *falcata* and *sativa*; markers in the two subspecies appeared to be entirely collinear.

Segregation distortion has been observed in all maps constructed to date. The most severe distortion has been observed in diploid F_2 populations, with as many as 50% of markers showing distortion, generally toward excess heterozygosity (23,26). In backcross populations, the distortion dropped to 34% (83), and only 8.8% of loci exhibited distorted segregation ratios in an F_1 population (28). Inbreeding depression has been repeatedly shown to be severe in alfalfa (12), and the approach to homozygosity upon inbreeding is much slower than expected (i.e., heterozygotes survive disproportionately) (35). Thus, a lower amount of distortion in less inbred populations is expected. In a backcross tetraploid population, between 4 and 9% of the loci exhibited segregation distortion, as expected based on the greater ability of tetraploids to buffer deleterious alleles (24). Some of the distorted markers in the tetraploid population fall on the same linkage group as some of the most severely distorted markers in an F_2 population (23) (LG 8 and 4, respectively). In contrast, a map constructed in a tetraploid F_1 population had 35% of AFLP and 26% of SSR loci exhibiting distortion, which was spread throughout the genome (65).

Linkage mapping in polyploids is complex. Initial attempts were based on the use of single-dose alleles (e.g., Aaaa) that are present in one parent but absent in the other (84). This approach, though useful, has a number of drawbacks, not least of which is the loss of substantial amounts of information from markers segregating as double dose or present in both parents. The map of Brouwer and Osborn (24) was constructed using single-dose alleles in conjunction with duplex alleles, based on formulae presented by Yu and Pauls (49). Subsequently, more robust methods that can analyze complex allelic configurations have been developed (85–90), and a computer program to integrate information across homologues and present a fully integrated map has been published (91). This procedure is being used by several groups mapping alfalfa (33,65). Finally, a method to map QTL in autopolyploids based on bivalent formation has been developed and applied to alfalfa mapping (9,10).

A solid reference map for alfalfa does not exist. Although some comparisons among maps are possible, based on markers mapped in common (24,65,92), clear comparisons among maps are not possible. Thus, references to linkage groups refer to the particular population under consideration rather than to a uniform nomenclature. Tying the *M. sativa* map to *M. truncatula* is difficult due to the lack of a comprehensive reference map for *M. truncatula*. A preliminary comparison of the *M. truncatula* and *M. sativa* genomes suggests that they are essentially syntenous (65). A major need for the alfalfa community is the development and maintenance of thorough reference maps in both the crop and the model species.

TABLE 7.1
Characteristics of Alfalfa Genetic Maps

Reference	Population type	Marker types	No. markers	Linkage groups	Length (cM)
Brummer *et al.* (23)	F_2 diploid	RFLP	108	10	467.5
Echt *et al.* (83)	BC diploid	RFLP, RAPD	86; 61	10, 7	553; 603
Kiss *et al.* (26)	F_2 diploid	RFLP, RAPD, isozyme, morph.	89	8	659
Tavoletti *et al.* (28)	F_1 diploid	RFLP	50; 55	10, 8	234; 261
Brouwer and Osborn (24)	BC tetraploid	RFLP	88	7	443
Barcaccia *et al.* (67)	F_1 diploid	AFLP, RAPD, RFLP	67	19	368.6
Kaló *et al.* (27)	F_2 diploid	as Kiss *et al.* (26) plus many known genes	868	8	754
Porceddu *et al.* (72)	F_1 diploid	as Barcaccia *et al.* (67) plus SAMPL, S-SAP	180	14	568.5
Julier *et al.* (65)	F_1 tetraploid	AFLP, SSR	339, 350 107[a]	8, 8, 8	662, 761, 709[a]
Robins *et al.* (33)	F_1 tetraploid	RFLP, SSR	101/78	8, 8	723, 666

[a]Consensus map of both parents based only on SSR markers.

Mapping Traits

During map development, six isozymes and 80 markers representing genes of known function were mapped in one diploid population (27), and several genes were mapped in another (83). Several simply inherited morphological traits have been mapped in alfalfa. The unifoliate leaf/cauliflower head mutation, *uni*, has been tightly linked to RFLP markers Hg2a7, Vg2c3, and Vg1e9 (29), which map to Linkage Group 4 (28,83). Anthocyanin and xanthophyll pigments were both mapped to Linkage Group 1 (27). Two other morphological traits, dwarf and sticky leaf, were mapped to Linkage Groups 2 and 6, respectively, in the same population.

The most elegant genetic mapping reported to date localized the genetic locus *nn1*, responsible for the nonnodulating phenotype in tetraploid alfalfa (93). The mapping work led directly to the map-based cloning of nn1 (94). Mapping was initiated in tetraploid alfalfa populations using bulked segregant analysis (BSA). The two bulks, representing the Nod^- and Nod^+ phenotypes, were screened with more than 500 RAPD markers; four primers were confirmed to produce fragments in the Nod^+ but not in the Nod^- bulk. The RAPD fragments were isolated and used as RFLP to map their location to a region on Linkage Group 5 on a saturated diploid genetic map (27), a location confirmed by careful mapping of selected markers in the tetraploid populations. Fine mapping of the region was then conducted based on molecular marker genotypes in the diploid population and ultimately in a larger F_2 tetraploid population of 2576 individuals. The gene was subsequently cloned by constructing a BAC contig in *M. truncatula*. This experiment demonstrated that despite the complexity of the cultivated alfalfa genome, molecular genetics tools can be successfully applied to carefully and thoughtfully designed experiments.

Few experiments have identified quantitative trait loci (QTLs) for complex agronomic traits. A RAPD marker allele was associated with somatic embryogenesis in a small tetraploid F_1 population, but the recombination fraction, 36.3%, and the lack of a clear genomic location for the trait make the potential for further manipulation low (95). Two QTLs for aluminum tolerance were identified in a diploid F_2 population and subsequently confirmed in diploid backcross populations (32). The QTLs were located on Linkage Groups 1 and 8 (32). These QTLs were identified based on an assay of callus growth, but a field study confirmed their value in imparting aluminum tolerance.

Winter hardiness and autumn biomass production are two traits commonly antagonistic to one another, yet independent manipulation to improve biomass production and winter survival concurrently would be ideal for alfalfa production (96). With analysis of variance, genetic mapping of these traits and the related trait of freezing injury measured by electrolyte leakage identified several QTLs for each trait in each of 2 years (30). Individual QTLs generally explained less than 10% of the variation, but multilocus models explained from 6 to 52% of the total variation for the traits. As with most QTL studies, the magnitude of effects of particular markers changed between years. For autumn growth, alleles from the nondormant Peruvian parent tended to increase plant height and winter/freezing injury, whereas alleles from the semidormant Blazer parent often acted in the opposite manner. The locations of the QTLs suggested that breeders could use markers to dissociate these traits. A subsequent reanalysis of these data using a mixture model algorithm detected a nearly nonoverlapping set of QTLs for these traits (9), but this appears to be due to a computational error (10). Other research to associate winter injury, biomass production, and physiological constituents related to these traits with markers has identified several QTLs for all traits (33,97).

Investigations are ongoing to map the locations of genes controlling various meiotic mutants involved with apomeiosis and parthenogenesis (98). The *jp* gene controls the development of jumbo pollen; homozygous individuals produce 100% 4*x* pollen (31). Using two diploid populations, *jp* has been localized close to marker Vg1G1b on Linkage Group 6 (31). Further, several QTLs for multinucleate microspore formation have also been identified, one of which is in the same region as *jp*, which may or may not be active in determining microspore nucleus number. RFLP markers have proven useful for half-tetrad analysis, which identified the mode of 2n egg formation in the diploid genotype PG-F9 to be predominantly a second-division restitution mechanism (99). Using BSA, QTLs for the production of these eggs have been located, with one RAPD marker explained 43% of the variation for seed set in interploidy crosses (55), although the authors suggest that a minimum of five genes is involved.

Finally, AFLP markers were associated with downy mildew resistance by comparing allele frequencies between a population selected for mildew resistance and the unselected base population (71). Only 36 plants from each population were screened, but two AFLP markers were associated with the trait, suggesting that marker-based methods can identify the location of loci for agronomically important traits in contrasting breeding populations (71). Further explication of the methods used to associate markers with traits in these types of population is provided in Skinner *et al.* (77).

Genetic Distance and Trait Expression

Alfalfa is known to express severe inbreeding depression, particularly at the tetraploid level (12). The cause of inbreeding depression is commonly assumed to be the loss of complementary dominant alleles in repulsion-phase linkages (100). Upon hybridization, heterosis is often expressed, especially in crosses between subspecies *sativa* and *falcata* (13,101). The phrase "maximizing heterozygosity" has been used to connote the maximization of complementary linkage blocks and/or of the number of alleles at any given locus in order to explain the heterotic effect (102).

The decline in heterozygosity at molecular marker loci in diploid alfalfa is much lower than expectations (35). Observed heterozygosity was 30–40% in two different diploid families, far in excess of expectation of 6%. This suggests that homozygous individuals, or those that lose one of the two complementary linkage blocks upon inbreeding, are less fit than heterozygotes, and that the latter survive preferentially each inbreeding cycle. In tetraploid alfalfa, heterozygosity loss more closely tracked expectations, although it was still less than expected (34). These authors suggest that the selection for heterozygotes during inbreeding may occur at a different generation depending on the germplasm; in fact, selection to increase the frequency of desirable alleles and/or linkage blocks, as was done with this material, may mean that heterozygosity can decline without adverse effects (34). More analysis of the relationship of heterozygosity and inbreeding is needed, particularly when examined at loci relevant to fitness traits.

Marker diversity and heterozygosity may also be related to yield, with more genetically dissimilar parents leading to progeny with higher levels of heterozygosity (or complementary linkage blocks) and hence, higher yield. This is precisely what happened in tetraploid alfalfa, but a similar relationship did not hold in isogenic diploids (38). The cause of this difference may be that the tetraploids exhibited more complementary linkage blocks because up to four alleles per locus were possible, and in fact, the diploid progenies were about 25% less heterozygous than the tetraploids derived from isogenic parental genotypes (36).

Using marker diversity estimates to select parents for synthetics has been used in two experiments with disappointing results. Marker diversity was examined among a set of 93 commercial alfalfa clones selected for use in cultivar development (37). Based on marker data, synthetics of 2 to 24 parents were selected for both similarity and dissimilarity. Little difference was noted among the similar or dissimilar synthetics, a fact attributed to linkage equilibrium in the population and the use of markers that were not necessarily associated with yield. In a second experiment, development of alfalfa synthetics based on genetic diversity assessed with AFLP markers was conducted to determine whether small parental numbers could still maintain pest resistance, fall dormancy, and improve yield, but no differences among synthetics were identified (39). Further work on this subject is warranted.

Particular combinations of alfalfa genotypes have been shown to produce substantial amounts of heterosis (13,101). Prediction of heterosis based on genetic markers would enable a more targeted search for potential hybrid combinations than creating and evaluating a host of test cross progenies. However, genetic distance did not predict heterosis in a limited study of 14 genotypes (13). In this and other experiments, the use of markers known to be linked to the traits of interest would presumably make the relationship with genetic distance stronger (103).

Gene Expression Analyses

Large-scale gene-expression profiling is in its infancy in alfalfa. To date, most expression studies have been the routine analysis of one to few genes via Northern analysis (e.g., 104,105–108). Several alfalfa expressed sequence tag (EST) libraries and corresponding cDNA arrays have been developed, the most extensive of which consists of 10,000 clones from cold-acclimated alfalfa crowns (109). Arrays of cDNA clones isolated from cold-acclimated alfalfa plants were used to identify genes that were up or down regulated under cold acclimation (110) and retroelement-containing cDNAs that were strongly expressed under cold acclimation (111). An EST library from plants with and without glandular hairs was developed using cDNA subtraction, which identified a number of genes not represented in other databases at the time (112). Using differential display, ESTs potentially involved with apomeiosis (113) and cold tolerance (J. Volenec, Purdue University, personal communication) have been isolated.

The development of large EST libraries for *M. truncatula* has limited the drive for alfalfa-specific libraries, given the similarity in sequence between the two species and the ability to assay alfalfa gene-expression patterns using arrays developed from *M. truncatula* (I. Ray, New Mexico State University, personal communication). Several experiments of alfalfa gene expression using *M. truncatula* oligonucleotide arrays are underway (I. Ray, G. May, S.R. Noble Foundation, personal communication).

Proteomics and Metabolomics

Even less developed than large-scale gene-expression studies are evaluations of the proteome and metabolome of alfalfa. Two experiments have been conducted in alfalfa to profile metabolites (114,115). Fifteen alfalfa and 27 *M. truncatula* saponins were identified using HPLC coupled to an electrospray ion-trap mass spectrometer (114). An assessment of the differences in phenolic metabolites between transgenic plants with modified lignin biosynthesis and their respective untransformed controls was conducted using HPLC (115). In the latter experiment, analysis of the metabolic profiles with principal component analysis enabled the separation of the transgenic and nontransgenic plants. This approach is useful to assess the broad effects of transgenes on many metabolites simultaneously. As these technologies become more routine in the model system, their use in alfalfa will increase.

Transformation

Alfalfa is fortunately both easily regenerable and easily transformable. Tissue culture experiments beginning over 30 years ago demonstrated that alfalfa could be regenerated from tissue culture but that some genotypes are better than others (116). Further, selection can successfully improve the ability of alfalfa populations to regenerate from callus tissue (117). Regeneration from suspension culture and from protoplasts has also been demonstrated (118). Most of the genotypes used in transformation experiments (Table 7.1) have been selected for their regeneration ability. Regeneration of alfalfa can be effected through somatic embryogenesis, either directly from leaf explants or indirectly from callus or suspension cells derived from explants (119). A comparison of the two methods showed that with direct somatic embryogenesis, only 5–10% of the explants produced embryos and, hence, transformed plants (120). In contrast, with indirect somatic embryogenesis, about 50% of the explants produced transformed calli, but only 5–10% regenerated plantlets, making the two systems essentially equivalent in efficiency for producing plants. The advantage of direct somatic embryogenesis is that it can be faster than indirect methods to attain transformed plants (120). The methods for production of somatic embryos and for regenerating alfalfa are summarized in McKersie and Brown (119).

Alfalfa is readily transformed using *Agrobacterium tumefaciens*, *A. rhizogenes*, and particle bombardment (Table 7.2). Stem, petiole, and leaf explants; somatic embryos; protoplasts; and suspension cultures have all been transformed successfully by at least one method. *A. tumefaciens*–mediated transformation has received the most attention; *A. rhizogenes*–mediated transformation works well, but it has numerous morphological effects on the transformed plants (121). The tissues used for *A. tumefaciens*–mediated transformation have been mainly stems and leaves. Somatic embryos have been successfully transformed but at a low frequency (122). In comparative experiments, leaves appear to be the most useful tissue, at least for some genotypes (123). A cocultivation time of 4 days appeared to be better than shorter times (123); a longer cocultivation of 6 to 8 days was optimum for strain LBA4404, but efficiency was best at 6 days for strain AGL1 (124). Kanamycin resistance is widely used to test for the presence of the selectable marker gene neophosphotransferase II, *nptII*; 50 μg/mL appears to be optimum to select for transformants without killing the tissue (123,125,126). The use of the *bar* gene, with selection of transformants using phosphinothricin (PPT), resulted in superior transformation efficiency of elite Australian cultivars than did *nptII* constructs (127).

The length of time from transformation until a regenerated plant is formed has been reported to be as short as 10 weeks (2,120). Transformation efficiency varies widely across experiments, ranging from near 0% (122) to 100% (128). Transformation efficiency is a nebulous metric, however, with some reports calculating efficiency based on kanamycin-resistant callus and others calculating it based on regenerated transformed plants. For example, Ziegelhoffer *et al.* (129) reported 99% of explants produced transgenic callus, based on kanamycin resistance, but only 57% of them could be regenerated, and only 8% of the resulting plants expressed the transgenic cellulase protein. In any case, plant genotype, *Agrobacterium* strain, selectable marker, and length of cocultivation affect the ability to produce transformed, regenerated plants, and the optimization of all parameters is necessary when considering a change in any of the variables.

A variety of genes has been transformed into alfalfa, designed to make the plants more disease resistant, nutritious, or stress tolerant or to use alfalfa to produce novel compounds (Table 7.2). Commercialization of transgenic alfalfa should occur in the near future, with Roundup Ready alfalfa being the first transgene to be marketed (130). Considerable interest in manipulating cell walls and protein to improve ruminant nutrition, eliminate bloat, and impart new pest resistances is driving commercial development of other transgenic products (M. McCaslin, Forage Genetics, International, personal communication).

The promoter most widely used in alfalfa transformation has been the cauliflower mosaic virus (CaMV) 35S promoter. Interestingly, this promoter does not perform as well in alfalfa as it does in other species (2,131), and it is likely not an optimum choice for constitutive expression (132). An attempt to increase expression of 35S by adding a second enhancer element was unsuccessful (133). Several experiments have used the light-regulated promoter of ribulose bisphosphate carboxylase small subunit genes, and the expression of genes using this promoter has been better than that observed using 35S (127). The use of the TR2' promoter from the Ti plasmid was less effective than 35S (134). Three *Arabidopsis* promoters involved in cell division, *cdc2a*, *CycB1*, and *CycA2*, all had better activity than 35S in actively dividing tissues; *cdc2a* had the strongest activity and broadest tissue expression of the three (120). The promoter of the salt-inducible gene *Ppt1* showed lower expression than 35S (133). The promoter of isoflavone reductase, IFR, had better expression in response to an elicitor or a fungus at inducing GUS expression than did 35S (135). The cryptic promotor *EntCUP*Δ has been used to show equivalent or higher

TABLE 7.2
Transformation of Alfalfa[a]

Gene	Promoter	Marker(s)	Strain	Tissue
Agrobacterium tumefaciens–mediated transformation				
nptII	nopaline synthase (nos)	——	bo42	Stem
nptII	nos	——	LBA4404	Stem
aphII	nos	——	A281, LBA4404	Leaf, petiole, stem
bar	CaMV 35S or TR2′	*nptII; hpt*	C58C1Rif[R]	Stem, petiole
Alfalfa mosaic virus (AMV) coat protein	355	*nptII*	LBA4404	Leaf
Chicken ovalbumin	35S	*nptII*	?	Leaf
GUS	35S	*nptII*	LBA4404	Stem
β-phaseolin seed storage protein	35S	*nptII*	A208	Leaf
Tomato chymotrypsin Inhibitor I gene	35S	*nptII*	?	Cotyledon
Mn-superoxide dismutase (SOD)	35S	*nptII*	C58C1Rif[R]	Stem, petiole
GUS, SOD	35S	*nptII*	A281, GV3101, C58-R1000, C58	Petiole, upper stem
Anti-elastase proteinase inhibitor	35S; salt inducible *Ppc1*	*nptII*	LBA4404	Leaf, petiole
GUS	35S; IFL (isoflavone reductase)	*nptII*	LBA4404	Leaf
A. rhizogenes rol genes	native promotors; 35S in one case	*nptII*	LBA4404	Leaf, petiole
nptII	nos	——	A281, LBA4404, C58	Leaf
GUS	35S	*nptII*	A281	Somatic embryos
Fungal Mn-dependent lignin peroxidase; bacterial α-amylase	Mac (35S derivative)	*nptII*	LBA4404	Leaf

Germplasm	Results	Ref.
M. varia genotype A2	Regenerated plants had kanamycin resistance.	(155)
CUF101	Regenerated plants from ~20% of stem slices	(142)
A2	Leaflets, strain A281, 4 day cocultivation resulted in 72% transformation efficiency.	(123)
Regen-S genotype	Variation among lines.	(134)
RA3	Efficiency of transformed callus ~50%.	
cv. Rambler, Roamer, Regen-S, Rangelander, Heinrichs; RA3	18% explants produced Kan resistant calli; 3% explants regenerated. Progeny ranged in expression of AMV coat protein but best genotype was resistant to AMV.	(149)
Rangelander, Siriver, Trifecta	Rangelander transformed ~20-100x better than commercial cultivars.	(143)
Ladak 65, Spredor II, Rangelander	Used PCR to identify true transformants.	(156)
Rangelander	Accumulated protein in seeds only.	(157)
93–21; 93–13	Expression substantially lower in alfalfa than in tobacco or nightshade.	(131)
RA3	Varying levels of expression of primary transformants; stable inheritance.	(158)
A70-34 from Rangelander; RA3; C2-4 elite genotype	20% C2-4 explants regenerated plants with strain A281; no regenerants otherwise.	(125)
V-154	Proteinase expressing plants may have lower thrip injury.	(133)
Regen-SY	Trangenic plants produced GUS in leaves in response to fungus.	(135)
Rangelander	30–40% Kan resistant callus; regenerated plants had increased root numbers.	(159)
Elite genotypes 1.5, 8.8, 11.9 selected for regeneration	Transformation efficiency as high as 85% based on Kan-resistant calli. Genotype 1.5 produced transgenic calli but few embryos, apparently losing embryogenic ability during transformation procedure.	(140)
cv. Zaječarska 83	Regenerated transformed plants at low frequency (2.8%).	(122)
Regen SY genotype RSY27	Mn-P had deleterious effects on plant growth; α-amylase did not. Transformation efficiency ~16% for aa; 13% for Mn-P (plants with detectable expression/total explants).	(160)

TABLE 7.2
Transformation of Alfalfa[a] *(Cont'd)*

Gene	Promoter	Marker(s)	Strain	Tissue
Sunflower albumin	35S; rbcS	*nptII, bar*	AGL1	Leaf
GUS	35S	*nptII*	LBA4404, ABI, AGL1	Cotyledon, stem
Glucanase and chitinase	35S or 35S e nhancer	*nptII*		
GUS	*Blec4* gene from pea	*nptII*	LBA4404	?
GUS	Enod12A; Srglb3	*nptII*	EHA105, GV3101	Leaf
Two bacterial cellulases	Mac (modified CaMV 35S)	*nptII*	LBA4404	Leaf
Alfalfa lectins MsLEC1, MsLEC2	35S?	*nptII*	LBA4404	Leaf
Sense and antisense IOMT	35S	*nptII*	LBA4404	Leaf
GUS	CaMV 35S, *Arabidopsis cdc2a, CycB1, CycA2*	*nptII*	LBA4404, EHA101	Leaf
Resveratrol synthase	35S	*nptII*	LBA4404	Leaf
barstar, barnase	pSoylec, pTA29	?	?	Stem, petiole
Three enzymes for PHB synthesis	35S	*nptII*	LBA4404	Leaf
Fungal phytase *phyA*	35S	*nptII*	LBA4404	Leaf
nptII	35S or *EntCUPΔ*	——	GV3101	Petioles, stems, leaves
Mn-SOD; Fe-SOD	35S	*nptII*	C58C1	Petiole
Anthocyanin regulatory genes, *Lc, B-Peru, C1*	35S, 35S with enhancer, rbcS (pea)	*nptII*	C58C1	Stem, petiole
Particle bombardment				
bar, nptII, hygromycin resistance	?	——	——	Callus
GUS or *nptII*	35S; 35S with enhancer	*nptII*	——	Petioles, callus
GUS	?	?	——	Pollen

Germplasm	Results	Ref.
cv. Aquarius, Siriver	*bar* constructs gave better transformation of cultivars.	(127)
Nine germplasms plus Rangelander	Compared *Agrobacterium* strains for tumorigenesis, transformation ability in 10 germplasms.	(124)
Regen-SY	Enzymes expressed and active.	(161)
Regen-SY	Targets epidermal layer in shoot apices.	(138)
PI564263 diploid *M. falcata*	Transformation and regeneration efficiency of 100%.	(128)
RSY27	99% explants produced Kan-resistant calli, 57% produced embryos, and only ~8% produced plants expressing protein.	(129)
Regen SY	Antisense, but not sense, constructs affected plant growth and development.	(162, 163)
Regen SY	IOMT (isoflavone 7-*O*-methyltransferase) overexpression leads to resistance to *Phoma medicaginis.*	(104)
M. falcata lines 47/1–5 and 47–150	Compared direct vs. indirect somatic embryogenesis, the former is faster but both produce similar numbers of plants.	(120)
Regen SY	Reduced sporulation of *Phoma medicaginis.*	(164)
RA3	Little pollen production; prevented self-fertility.	(139)
Regen SY genotype	Low levels of polyhydroxybutyrate (PHB) produced; range of expression among regenerants.	(132)
RSY27	Transgenic plants had a 13x increase in phytase and 17x increase in phytase-specific activity in leaves compared to controls. Phytase quality is acceptable for use as industrial enzyme.	(165, 166)
N4 breeding line	*EntCUPΔ* promotor had equal or better transformation efficiency than 35S.	(136)
N4	SOD overexpressed in organelles. Evidence for posttranslational control.	(167)
A01 (derived from N4); A04	Unknown stress-responsive factor needed to stimulate anthocyanin response, but only Lc responded.	(168)
?	Transformed plants obtained	(169)
C2-4 breeding line	Very few transformants (2,035 explants gave 7 transformed plants all from same event; no GUS activity; fragment lost.	(126)
?	Pollen transformation successful but some transformed plants lost GUS activity after several generations	(170)

TABLE 7.2
Transformation of Alfalfa[a] *(Cont'd)*

Gene	Promoter	Marker(s)	Strain	Tissue
Agrobacterium rhizogenes				
None	None	pRi/nos	NCPPB 1855	Stem inoculation
None	None	pRi/nos	A4(Rif^R)	Suspension culture
γ-zein	35S; rbcS	*nptII*	NCPPB 1855	Stem inoculation
γ-zein	*rbcS*	*nptII*	NCPPB 1855	Stem inoculation
Other				
GUS	—	—	—	Dry somatic embryos
—	—	nos	—	Protoplasts

[a]With a few exceptions, all transformation experiments resulted in regenerated plants containing the transgene.

transformation efficiencies than 35S, although expression levels in regenerated plants is not known (136). A comparison of the ability of several constitutive promoters to drive GUS expression in alfalfa has recently been published (137). Relatively little attention has focused on tissue-specific expression, although a promoter of the pea (*Pisum sativum*) gene *Blec4*, targeting the epidermis in actively growing tissue, has been used successfully (138), as has the tapetum-specific TA29 promoter driving the *barnase* gene (139).

Most *Agrobacterium tumefaciens*–mediated transformation experiments have used strain LBA4404 (Table 7.2). A direct comparison between LBA4404 and other strains showed the former to have better transformation efficiency in most cases (120,124,140), but it was only half as efficient as A281 in another (123). Virulent strains of *Agrobacterium,* including A208, A348, and A281, were more tumorigenic on nondormant than dormant alfalfa under greenhouse conditions following hypocotyl and stem inoculation, but even nondormant populations had resistance levels near 70% (124), and alfalfa genotype A2 did not express tumors following infection with A281 (123). A field-isolated *A. tumefaciens* strain, 1D1609, is highly virulent on alfalfa, especially nondormant CUF101, making it a possible future choice for effective alfalfa transformation (141).

The alfalfa germplasm used in transformation greatly affects success, with substantial variation observed among populations for both transformation and regeneration ability (124). Successful transformation of plants from germplasm representing the

Germplasm	Results	Ref.
cv. Europe, Adriana; P1 cell line from Adriana	Hairy roots occurred in most infected plants, but hairy root explants only regenerated from P1. 50% of regenerants produced agropine.	(121)
?	70% of callus produced agropine but no regenerants produced.	(171)
P1 cell line from cv. Adriana	No rbcS plants recovered; no significant accumulation of γ-zein in tissue possibly because of posttranscriptional effects.	(172)
?	0.026% of total soluble protein in leaves.	(173)
?	Transient expression noted, suggesting this is an avenue to direct DNA uptake	(174)
cell suspension line A6 from cv. Rambler	Microinjection into nuclei of plasmids from *A. tumefaciens* strains C58, LBA4404, and two unnamed. 15–26% calli transformed; no plants recovered.	(175)

spectrum of autumn dormancy has been reported (124). Dormant germplasm containing large amounts of *M. sativa* subsp. *falcata* appears to be best suited to transformation, probably because it is more likely to regenerate successfully. However, stable transformants have been regenerated from nondormant cultivars, including CUF101 (142). The dormant cultivar Rangelander was transformed 20 to 100 times better than commercial Australian cultivars, but the latter were transformable and presumably resulted in more agronomically desirable plants (143).

A number of alfalfa germplasms have been developed for use in transformation experiments, based largely on their ability to regenerate into adult plants following tissue culture. The most commonly used germplasms are Regen-S (144), Regen-Y (117), and Regen-SY (145), all of which were selected for their regeneration ability. Regen-S is derived from the semidormant cultivars DuPuit and Saranac, Regen-Y from dormant Canadian cultivars including Rambler and Roamer, and Regen-SY from a cross of the former two. The agronomic characteristics of these germplasms are less than desirable, so that even if they are easily transformed and regenerated, backcrossing into elite germplasm is necessary. A diploid population, Regen-2X, was developed by selecting for high regeneration ability (146). Two diploid populations, WY-RF1 and WY-RF2, derived from *M. sativa* subsp. *falcata* have also been developed for regeneration ability (147), although WY-RF2 has been shown to be tetraploid (148).

No relationship between autumn dormancy and GUS expression was noted on callus cultures of a diverse set of germplasm, but the germplasms did differ for transformation ability (124). Three *Agrobacterium* strains were used to transform all ten germplasms; LBA4404 was uniformly successful and always equal or superior to the strains AGL1 (a disarmed version of A281) and ABI. The latter two were unsuccessful at producing GUS-expressing calli in several of the germplasms. Plants regenerated from a clone from Regen-SY indicated that LBA4404 was most successful at producing transformants (up to 58%); AGL1 was intermediate, and no transformants were identified from ABI (124).

Expression of proteins in alfalfa has often been lower than expected or, at least, than desired. Expression of a chicken ovalbumin gene averaged 0.005% of total soluble protein (143). Similarly, a sunflower albumin gene high in sulfur-containing amino acids accumulated only to 0.11% of total soluble leaf protein in the best transformant (127). In both cases, this level of expression is not sufficient to affect animal nutrition; in the latter example, the best case is 25 to 50 times less than needed to improve wool growth in sheep (127). Expression of genes producing a biodegradable plastic was below levels seen in model plants and was insufficient for economical production (132). In all cases, the CaMV 35S promoter was used. The use of the promoter of ribulose bisphosphate carboxylase small subunit genes had better expression than 35S, but it still did not produce plants with adequate protein levels (127). In the best transformants, expression of alfalfa mosaic virus (AMV) coat protein under the control of 35S was at levels comparable to those observed in other experiments on tobacco (149).

Alfalfa is ready for large-scale testing of transgenic constructs, given its relative ease of transformation and regeneration. Several areas of emphasis at this time appear to be the identification of a constitutive promoter that results in stronger expression than CaMV 35S, the identification of promoters that target specific tissues, and the broadening of the germplasm that can be routinely transformed.

As with most transgenic experiments, variation in expression among regenerated plants is observed in many experiments in alfalfa (119,132). Thus, selection for the best-expressing events will be required prior to introducing the gene into a desirable cultivar. Breeding transgenic cultivars requires more thought than many crops because pure lines are not available. Several possible methods to breed transgenic cultivars have been reported (119,150,151). The most likely seems to be that of McCaslin *et al.* (150), which relies on the independent introduction of two or more copies of the transgenic construct so that the multiple loci will segregate in such a way that virtually all individuals in the resulting synthetic population will express the transgene. This method appears to be susceptible to transgene silencing problems, but to date, none has been identified (S. Temple, Forage Genetics International, personal communication).

Prospects

Although a rather substantial amount of research is devoted to alfalfa, coverage of each individual topic is quite limited. Based on the survey of genomic research above, I suggest a number of areas that need to receive more attention. First, despite the plethora of diversity experiments, very little is known about the evolution of the *M. sativa–falcata* complex. Understanding how tetraploids arose from diploids and how accessions of various taxa relate to one another based on DNA sequence or marker information would be useful to breeders attempting to use diverse or wild germplasm in selection programs. Thousands of accessions collected from the wild are in germplasm repositories around the world, yet characterization of them as to ploidy and subspecies status is often incorrect. Similarly, virtually nothing is known about the genetic structure of alfalfa breeding populations or cultivars. The number of alleles present in populations, the allele frequencies, and the extent of random mating and linkage equilibria need to be understood more clearly. The suggestion that genetic distance did not relate to synthetic performance because of linkage equilibrium in the breeding population under investigation (37) needs empirical support.

Obviously, a much more vigorous examination of agronomically important traits needs to be undertaken to identify QTLs controlling them. Linking QTL information with results from gene expression studies, DNA sequence, or candidate genes identified via comparative mapping will help provide a better understanding of how agronomically important traits are controlled and how marker-assisted breeding procedures can be used. Better genetic markers, particularly those associated with functional differences between alleles (152), are needed to make MAS a reality in a complex polyploid plant. High-throughput SNP (single nucleotide polymorphisms) markers would be a step toward that direction (153), until genes responsible for traits, and the functional differences between alleles, are identified.

On transgenics, new promoters need to be identified and adopted to improve transgene expression levels and to target transgene products to particular tissues under particular conditions. Further refinements in transformation and regeneration protocols could result in more efficient and rapid production of transgenic plants. A wealth of new genes is arriving from *M. truncatula* and other alfalfa programs, and their phenotypic functions need to be assessed in transgenic plants of various germplasms. Simple methods to alter gene expression, such as viral induced gene silencing (VIGS)(154), need to be developed to enable high-throughput phenotypic screening of many genes identified from gene expression experiments.

The development of a substantial research community working with *M. truncatula* is a boon for alfalfa researchers. Numerous tools developed for the model system will work just as well in alfalfa. For example, the fact that almost all SSR markers developed from *M. truncatula* EST sequences work in alfalfa (61) suggests high levels of sequence similarity between the species. Further, oligonucleotide arrays based on *M. truncatula* EST sequences work well for alfalfa expression experiments (I. Ray, personal communication). Genetic mapping experiments show that marker orders are collinear between the species as well (65). Thus, the tools from the model are directly applicable to alfalfa.

For alfalfa scientists to make the most of *M. truncatula*, however, will require more than just using the tools. The salient differences between the two species is striking; the inbred lines, diploidy, self-pollination, and annual life cycle of *M. truncatula* are obviously advantageous in terms of rapid cycling of materials, especially when compared to the outcrossing populations, tetraploidy, inbreeding depression, and perenniality of alfalfa. How well the model can be used as a proxy for alfalfa in dissecting complex traits like regrowth, winter hardiness, dormancy, persistence, and others remains to be seen. Nevertheless, alfalfa scientists would be well served by considering parallel research programs in both species to facilitate crossing from one to the other.

Mapping QTLs in *M. truncatula* would be much easier than in alfalfa, provided the trait being measured corresponds to the one of interest in alfalfa. Further investigation is needed to see how far the model can be pushed in service of alfalfa. In addition, alfalfa offers a number of alternatives itself to the problems of cultivated alfalfa, including a broad selection of diploid germplasm. A model example of the creative use of alfalfa and its relatives is shown in the cloning of *nn1* from alfalfa using a map-based approach (94). Among the resources that need to be developed in *M. truncatula* if it is to be useful to alfalfa breeders interested in complex traits are large, segregating, recombinant inbred populations from a variety of ecotypes and a solid reference map created in an immortalized population with genotype data freely available.

Genomics technologies will aid alfalfa improvement efforts in three major ways. First, they will undoubtedly lead to the identification of a suite of new genes that can add novel traits to alfalfa (129,132). These will be easily commercialized and, as such, will increase the value of alfalfa to the breeding companies, to the consumer, and (hopefully) to the farmer. Second, and more importantly for long-term genetic gain in alfalfa, genomics will help probe the genetic control of complex phenotypes that are difficult to measure accurately and in a timely fashion. Yield, long term persistence under grazing or hay production, winter hardiness, salt and drought tolerance, and others are difficult to evaluate in short-term, greenhouse environments and require four- to five-year field nurseries before selection can occur. The promise of genomic technologies lies in the fact that they may enable more rapid selection to decrease cycle time and improve genetic gain. Similarly, manipulation of entire germplasms via transgenesis will help to adapt new germplasm more rapidly, thereby improving existing germplasm more quickly. Third, genomics may result in new methods to develop cultivars, such as improved male sterility systems (139), that may capitalize on heterosis (81) or provide other benefits over current cultivar-development methods.

Unless genomics programs are intimately tied to alfalfa breeding programs, their promise will not be realized. Priority areas of investigation should be set by breeders, in conjunction with farmers, consumers, and other users of the products, rather than by genomics programs. What can be done with genomics may not reflect the needs of alfalfa improvement programs. Simple breeding has and will continue to lead to improved cultivars. Genomics technologies that enhance the breeding programs are those that should be emphasized. Finally, genomics research programs should look more broadly at the entire cropping system to consider whether engineering one crop at great expense is more or less effective at producing long-term sustainability and productivity of the nation's cropping systems than conducting elementary breeding and genetics work on another species.

Summary

Many current genomics technologies are being applied to alfalfa, albeit on a limited scale. Genetic mapping has progressed through a formative stage in which few markers were readily available to the current time when abundant SSR markers can be quickly mapped in any population on a high throughput basis. This should greatly expedite future mapping and genetic mapping experiments. With concurrent phenotype assessment, the identification of QTLs should increase, leading to the potential benefits of marker-assisted selection for complex traits. Transformation and regeneration systems are routine in alfalfa, and though improvements in efficiency and in gene expression would be desirable, inserting novel genes is possible and commercializing transgenic cultivars is a reality. The development of *M. truncatula* resources will assist the dissection and understanding of the alfalfa genome and will aid in future improvement efforts.

References

1. Michaud, R., W.F. Lehman, and M.D. Rumbaugh, World Distribution and Historical Development, in *Alfalfa and Alfalfa Improvement*, edited by A.A. Hanson, D.K. Barnes, and R.R. Hill, Jr., Amer. Soc. Agron.-Crop Sci. Soc. Amer.-Soil Sci. Soc. Amer., Madison, Wisconsin, 1988, pp. 25–91.
2. Austin, S., and E.T. Bingham, The Potential Use of Transgenic Alfalfa as a Bioreactor for the Production of Industrial Enzymes, in *Biotechnology and the Improvement of Forage Legumes*, edited by B.D. McKersie and D.C.W. Brown, CAB International, Oxon, United Kingdom, 1997, pp. 409–424.
3. Vance, C.P., Nitrogen Fixation Capacity, in *Biotechnology and the Improvement of Forage Legumes*, edited by B.D. McKersie and D.C.W. Brown, CAB International, Oxon, United Kingdom, 1997, pp. 375–407.
4. Lesins, K., and C.B. Gillies, Taxonomy and Cytogenetics of *Medicago*, in *Alfalfa Science and Technology*, edited by C.H. Hanson, Amer. Soc. Agron.-Crop Sci. Soc. Amer.-Soil Sci. Soc. Amer., Madison, Wisconsin, 1972, pp. 53–86.
5. Small, E., and M. Jomphe, A Synopsis of the Genus *Medicago* (Leguminosae), *Can. J. Bot. 67*:3260–3294 (1989).
6. Quiros, C.F., and G.R. Bauchan, The Genus *Medicago* and the Origin of the *Medicago sativa* Complex, in *Alfalfa and Alfalfa Improvement*, edited by A.A. Hanson, D.K. Barnes, and R.R. Hill, Jr., Amer. Soc. Agron.-Crop Sci. Soc. Amer.-Soil Sci. Soc. Amer., Madison, Wisconsin, 1988, pp. 93–124.
7. Brummer, E.C., P.M. Cazcarro, and D. Luth, Ploidy Determination of Alfalfa Germplasm Accessions Using Flow Cytometry, *Crop Sci. 39*:1202–1207 (1999).
8. Busbice, T.H., R.R. Hill, Jr., and H.L. Carnahan, Genetics and Breeding Procedures, in *Alfalfa Science and Technology*, edited by C.H. Hanson, Amer. Soc. Agron.-Crop Sci. Soc. Amer.-Soil Sci. Soc. Amer., Madison, Wisconsin, 1972, pp. 283–318.

9. Ma, C.X., G. Casella, Z.J. Shen, T.C. Osborn, and R.L. Wu, A Unified Framework for Mapping Quantitative Trait Loci in Bivalent Tetraploids Using Single-Dose Restriction Fragments: A Case Study from Alfalfa, *Genome Res. 12*:1974–1981 (2002).
10. Cao, D., T.C. Osborn, and R.W. Doerge, Correct Estimation of Preferential Chromosome Pairing in Autotetraploids, *Genome Res. 14*:459–462 (2004).
11. Barnes, D.K., B.P. Golpen, and J.E. Baylor, Highlights in the USA and Canada, in *Alfalfa and Alfalfa Improvement*, edited by A.A. Hanson, D.K. Barnes, and R.R. Hill, Jr., Amer. Soc. Agron.-Crop Sci. Soc. Amer.-Soil Sci. Soc. Amer., Madison, Wisconsin, 1988, pp. 1–24.
12. Jones, J.S., and E.T. Bingham, Inbreeding Depression in Alfalfa and Cross-Pollinated Crops, in *Plant Breed. Rev.*, edited by J. Janick, John Wiley & Sons, New York, 1995, pp. 209–233.
13. Riday, H., and E.C. Brummer, Forage Yield Heterosis in Alfalfa, *Crop Sci. 42*:716–723 (2002).
14. Quiros, C.F., Starch Gel Electrophoresis Technique Used with Alfalfa and Other *Medicago* Species, *Can. J. Plant Sci. 61*:745–749 (1981).
15. Corts, M.R.M., and M.C.C. Martinez, Variation of Pgm and Idh Isozymes for Identification of Alfalfa Varieties, *Euphytica 112*:137–143 (2000).
16. Quiros, C.F., Tetrasomic Inheritance for Multiple Alleles in Alfalfa, *Genetics 101*:117–127 (1982).
17. Jenczewski, E., J.-M. Prosperi, and J. Ronfort, Evidence for Gene Flow between Wild and Cultivated *Medicago sativa* (Leguminosae) Based on Allozyme Markers and Quantitative Traits, *Am. J. Bot. 86*:677–687 (1999).
18. Brummer, E.C., G. Kochert, and J.H. Bouton, RFLP Variation in Diploid and Tetraploid Alfalfa, *Theor. Appl. Genet. 83*:89–96 (1991).
19. Kidwell, K.K., D.F. Austin, and T.C. Osborn, RFLP Evaluation of Nine *Medicago* Accessions Representing the Original Germplasm Sources for North American Alfalfa Cultivars, *Crop Sci. 34*:230–236 (1994).
20. Pupilli, F., S. Businelli, F. Paolocci, C. Scotti, F. Damiani, and S. Arcioni, Extent of RFLP Variability in Tetraploid Populations of Alfalfa, *Medicago sativa*, *Plant Breed. 115*:106–112 (1996).
21. Pupilli, F., P. Labombarda, C. Scotti, and S. Arcioni, RFLP Analysis Allows for the Identification of Alfalfa Ecotypes, *Plant Breed. 119*:271–276 (2000).
22. Valizadeh, M., K.K. Kang, A. Kanno, and T. Kameya, Analysis of Genetic Distance among Nine *Medicago* Species by Using DNA Polymorphisms, *Breed. Sci. 46*:7–10 (1996).
23. Brummer, E.C., J.H. Bouton, and G. Kochert, Development of an RFLP Map in Diploid Alfalfa, *Theor. Appl. Genet. 86*:329–332 (1993).
24. Brouwer, D.J., and T.C. Osborn, A Molecular Marker Linkage Map of Tetraploid Alfalfa (*Medicago sativa* L.), *Theor. Appl. Genet. 99*:1194–1200 (1999).
25. Echt, C.S., K.K. Kidwell, S.J. Knapp, T.C. Osborn, and T.J. McCoy, Linkage Mapping in Diploid Alfalfa (*Medicago sativa* L.), *Genome 37*:61–71 (1994).
26. Kiss, G.B., G. Csanádi, K. Kálmán, P. Kaló, and L. Ökrész, Construction of a Basic Genetic Map for Alfalfa Using RFLP, RAPD, Isozyme, and Morphological Markers, *Mol. Gen. Genet. 238*:129–137 (1993).

27. Kaló, P., G. Endre, L. Zimányi, G. Csanádi, and G.B. Kiss, Construction of an Improved Linkage Map of Diploid Alfalfa (*Medicago sativa*), *Theor. Appl. Genet. 100*:641–657 (2000).
28. Tavoletti, S., F. Veronesi, and T.C. Osborn, RFLP Linkage Map of an Alfalfa Meiotic Mutant Based on an F1 Population, *J. Hered. 87*:167–170 (1996).
29. Brouwer, D.J. and T.C. Osborn, Identification of RFLP Markers Linked to the Unifoliate Leaf, Cauliflower Head Mutation in Alfalfa, *J. Hered. 88*:150–152 (1997).
30. Brouwer, D.J., S.H. Duke, and T.C. Osborn, Mapping Genetic Factors Associated with Winter Hardiness, Fall Growth, and Freezing Injury in Autotetraploid Alfalfa, *Crop Sci. 40*:1387–1396 (2000).
31. Tavoletti, S., P. Pesaresi, G. Barcaccia, E. Albertini, and F. Veronesi, Mapping the Jp (Jumbo Pollen) Gene and QTLs Involved in Multinucleate Microspore Formation in Diploid Alfalfa, *Theor. Appl. Genet. 101*:372–378 (2000).
32. Sledge, M.K., J.H. Bouton, M. Dall'Agnoll, W.A. Parrott, and G. Kochert, Identification and Confirmation of Aluminum Tolerance QTL in Diploid *Medicago sativa* Subsp. coerulea, *Crop Sci. 42*:1121–1128 (2002).
33. Robins, J.G., D.R. Viands, T.A. Campbell, D. Luth, and E.C. Brummer, Mapping Biomass Yield in an Intersubspecific Cross of Alfalfa, in *Proc. Molecular Breeding of Forage and Turf, 3rd Int'l Symposium*, Dallas, Texas, and Ardmore, Oklahoma, 2003, p. 133.
34. Scotti, C., F. Pupilli, S. Salvi, and S. Arcioni, Variation in Vigour and in RFLP-Estimated Heterozygosity by Selfing Tetraploid Alfalfa: New Perspectives for the Use of Selfing in Alfalfa Breeding, *Theor. Appl. Genet. 101*:120–125 (2000).
35. Brouwer, D.J., and T.C. Osborn, Molecular Marker Analysis of the Approach to Homozygosity by Selfing in Diploid Alfalfa, *Crop Sci. 37*:1326–1330 (1997).
36. Kidwell, K.K., E.T. Bingham, D.R. Woodfield, and T.C. Osborn, Relationships among Genetic Distance, Forage Yield, and Heterozygosity in Isogenic Diploid and Tetraploid Alfalfa Populations, *Theor. Appl. Genet. 89*:323–328 (1994).
37. Kidwell, K.K., L.M. Hartweck, B.S. Yandell, P.M. Crump, J.E. Brummer, J. Moutray, and T.C. Osborn, Forage Yields of Alfalfa Populations Derived from Parents Selected on the Basis of Molecular Marker Diversity, *Crop Sci. 39*:223–227 (1999).
38. Kidwell, K.K., D.R. Woodfield, E.T. Bingham, and T.C. Osborn, Molecular Marker Diversity and Yield of Isogenic 2× and 4× Single-Crosses of Alfalfa, *Crop Sci. 34*:784–788 (1994).
39. Sledge, M.K., J.H. Bouton, and G. Kochert, Shifts in Pest Resistance, Fall Dormancy, and Tolerant Synthetics Derived Yield in 12-, 24-, and 120-Parent Grazing from CUF 101 Alfalfa, *Crop Sci. 43*:1736–1740 (2003).
40. Ghérardi, M., B. Mangin, B. Goffinet, D. Bonnet, and T. Huguet, A Method to Measure Genetic Distance between Allogamous Populations of Alfalfa (*Medicago sativa*) Using RAPD Molecular Markers, *Theor. Appl. Genet. 96*:406–412 (1998).
41. Crochemore, M.-L., C. Huyghe, M.C. Kerlan, F. Durand, and B. Julier, Partitioning and Distribution of RAPD Variation in a Set of Populations of the *Medicago Sativa* Complex, *Agronomie 16*:421–432 (1996).
42. Mengoni, A., A. Gori, and M. Bazzicalupo, Use of RAPD and Microsatellite (SSR) Variation to Assess Genetic Relationships among Populations of Tetraploid Alfalfa, *Medicago sativa*, *Plant Breed. 119*:311–317 (2000).
43. Yu, K.F., and K.P. Pauls, Rapid Estimation of Genetic Relatedness among Heterogeneous Populations of Alfalfa by Random Amplification of Bulked Genomic DNA Samples, *Theor. Appl. Genet. 86*:788–794 (1993).

44. Musial, J.M., K.E. Basford, and J.A.G. Irwin, Analysis of Genetic Diversity within Australian Lucerne Cultivars and Implications for Future Genetic Improvement, *Aust. J. Agric. Res. 53*:629–636 (2002).
45. Dehghan-Shoar, M., J.G. Hampton, and S.E. Gardiner, Genetic Analysis among and within Populations Forming Ecotypes and Cultivars of Lucerne, *Medicago sativa* (Leguminosae), Using RAPD Fragments, *Plant Syst. Evol. 208*:107–119 (1997).
46. Campbell, T.A., Molecular Analysis of Genetic Variation among Alfalfa (*Medicago sativa* L.) and *Medicago ruthenica* Clones, *Can. J. Plant Sci. 80*:773–779 (2000).
47. Campbell, T.A., Molecular Analysis of Genetic Relatedness among Alfalfa Clones Differing in Levels of Self-Incompatibility, *Can. J. Plant Sci. 80*:559–564 (2000).
48. Echt, C.S., L.A. Erdahl, and T.J. McCoy, Genetic Segregation of Random Amplified Polymorphic DNA in Diploid Cultivated Alfalfa, *Genome 35*:84–87 (1992).
49. Yu, K.F., and K.P. Pauls, Segregation of Random Amplified Polymorphic DNA Markers and Strategies for Molecular Mapping in Tetraploid Alfalfa, *Genome 36*:844–851 (1993).
50. Yu, K.F., and K.P. Pauls, Identification of an RAPD Marker Associated with Somatic Embryogenesis in Alfalfa, *Plant Mol. Biol. 22*:269–277 (1993).
51. McCoy, T.J., C.S. Echt, and L.C. Mancino, Segregation of Molecular Markers Supports an Allotetraploid Structure for *Medicago sativa* × *Medicago papillosa* Interspecific Hybrid, *Genome 34*:574–578 (1991).
52. Barcaccia, G., N. Tosti, E. Falistocco, and F. Veronesi, Cytological, Morphological, and Molecular Analyses of Controlled Progenies from Meiotic Mutants of Alfalfa Producing Unreduced Gametes, *Theor. Appl. Genet. 91*:1008–1015 (1995).
53. Barcaccia, G., S. Tavoletti, M. Pezzotti, M. Falcinelli, and F. Veronesi, Fingerprinting of Alfalfa Meiotic Mutants Using RAPD Markers, *Euphytica 80*:19–25 (1994).
54. Barcaccia, G., S. Tavoletti, M. Falcinelli, and F. Veronesi, Verification of the Parthenogenetic Capability of Unreduced Eggs in an Alfalfa Mutant by a Progeny Test Based on Morphological and Molecular Markers, *Plant Breed. 116*:475–479 (1997).
55. Barcaccia, G., E. Albertini, D. Rosellini, S. Tavoletti, and F. Veronesi, Inheritance and Mapping of 2n-Egg Production in Diploid Alfalfa, *Genome 43*:528–537 (2000).
56. Piccioni, E., G. Barcaccia, M. Falcinelli, and A. Standardi, Estimating Alfalfa Somaclonal Variation in Axillary Branching Propagation and Indirect Somatic Embryogenesis by RAPD Fingerprinting, *Intern. J. Plant Sci. 158*:556–562 (1997).
57. Gjuric, R., and S.R. Smith, Identification of Cross-Pollinated and Self-Pollinated Progeny in Alfalfa through RAPD Nulliplex Loci Analysis, *Crop Sci. 36*:389–393 (1996).
58. St. Amand, P.C., D.Z. Skinner, and R.N. Peaden. Risk of Alfalfa Transgene Dissemination and Scale Dependent Effects, in *Proc. 36th North Amer. Alf. Impr. Conf.*, Bozeman, Montana, 1998, NAAIC, Beltsville, Maryland.
59. Brummer, E.C., J.H. Bouton, and G. Kochert, Analysis of Annual *Medicago* Species Using RAPD Markers, *Genome 38*:362–367 (1995).
60. Diwan, N., A.A. Bhagwat, G.B. Bauchan, and P.B. Cregan, Simple Sequence Repeat DNA Markers in Alfalfa and Perennial and Annual *Medicago* Species, *Genome 40*:887–895 (1997).

61. Eujayl, I., M.K. Sledge, L. Wang, G.D. May, K. Chekhovskiy, J.C. Zwonitzer, and M.A.R. Mian, *Medicago truncatula* EST-SSRs Reveal Cross-Species Genetic Markers for *Medicago* spp., *Theor. Appl. Genet. 108*:414–422 (2003).
62. Campbell, T.A., E.C. Brummer, D. Luth, G.R. Bauchan, Z.L. Xia, and C. He, Screening Autotetraploid Alfalfa and *Medicago truncatula* SSR Markers for Potential Use in Linkage Mapping and Genotype Separation in Cultivated Alfalfa, *Agron. Abstracts [CD-ROM]*, Amer. Soc. Agron.-Crop Sci. Soc. Amer.-Soil Sci. Soc. Amer., Madison, Wisconsin (2003).
63. Jiang, G.M., M.K. Sledge, M.A.R. Mian, and I. Ray, A Genetic Map of Autotetraploid Alfalfa Using Single-Dose EST-SSSRs from M. Truncatula, *Agron. Abstracts [CD-ROM]*, Amer. Soc Agron.-Crop Sci. Soc. Amer.-Soil Sci. Soc. Amer., Madison, Wisconsin (2003).
64. Robins, J.G., M. Santra, D. Luth, T.A. Campbell, and E.C. Brummer, Mapping SSRs in Tetraploid Alfalfa, *Agron. Abstracts [CD ROM]*, Amer. Soc Agron.-Crop Sci. Soc. Amer.-Soil Sci. Soc. Amer., Madison, Wisconsin (2003).
65. Julier, B., S. Flajoulot, P. Barre, G. Cardinet, S. Santoni, T. Huguet, and C. Huyghe, Construction of Two Genetic Linkage Maps in Cultivated Tetraploid Alfalfa (*Medicago sativa*) Using Microsatellite and AFLP Markers, *BMC Plant Biology*, www.biomed-central.com/1471-2229/3/9 (2003).
66. Diwan, N., J.H. Bouton, G. Kochert, and P.B. Cregan, Mapping of Simple Sequence Repeat (SSR) DNA Markers in Diploid and Tetraploid Alfalfa, *Theor. Appl. Genet. 101*:165–172 (2000).
67. Barcaccia, G., E. Albertini, S. Tavoletti, M. Falcinelli, and F. Veronesi, AFLP Fingerprinting in *Medicago* spp.: Its Development and Application in Linkage Mapping, *Plant Breed. 118*:335–340 (1999).
68. Zaccardelli, M., S. Gnocchi, M. Carelli, and C. Scotti, Variation among and within Italian Alfalfa Ecotypes by Means of Bio-Agronomic Characters and Amplified Fragment Length Polymorphism Analyses, *Plant Breed. 122*:61–65 (2003).
69. Segovia-Lerma, A., R.G. Cantrell, J.M. Conway, and I.M. Ray, AFLP-Based Assessment of Genetic Diversity among Nine Alfalfa Germplasms Using Bulk DNA Templates, *Genome 46*:51–58 (2003).
70. Riday, H., E.C. Brummer, T.A. Campbell, D. Luth, and P.M. Cazcarro, Comparisons of Genetic and Morphological Distance with Heterosis between *Medicago sativa* subsp. *sativa* and subsp. *falcata*, *Euphytica 131*:37–45 (2003).
71. Obert, D.E., D.Z. Skinner, and D.L. Stuteville, Association of AFLP Markers with Downy Mildew Resistance in Autotetraploid Alfalfa, *Mol. Breed. 6*:287–294 (2000).
72. Porceddu, A., E. Albertini, G. Barcaccia, G. Marconi, F.B. Bertoli, and F. Veronesi, Development of S-SAP Markers Based on an Ltr-Like Sequence from *Medicago sativa* L., *Mol. Genet. Genomics 267*:107–114 (2002).
73. Albertini, E., A. Porceddu, G. Marconi, G. Barcaccia, L. Pallottini, and M. Falcinelli, Microsatellite-AFLP for Genetic Mapping of Complex Polyploids, *Genome 46*:824–832 (2003).
74. Skinner, D.Z., Non-Random Chloroplast DNA Hypervariability in *Medicago sativa*, *Theor. Appl. Genet. 101*:1242–1249 (2000).
75. Campbell, T.A. and G.R. Bauchan, Organelle Based Molecular Analyses of the Genetic Relatedness of Cultivated Alfalfa (*Medicago sativa* L) to *Medicago edgeworthii* Sirjaev, and *Medicago ruthenica* (L.) Ledebour, *Euphytica 125*:51–58 (2002).

76. Mengoni, A., C. Ruggini, G.G. Vendramin, and M. Bazzicalupo, Chloroplast Microsatellite Variations in Tetraploid Alfalfa, *Plant Breed. 119*:509–512 (2000).
77. Skinner, D.Z., T. Loughin, and D.E. Obert, Segregation and Conditional Probability Association of Molecular Markers with Traits in Autotetraploid Alfalfa, *Mol. Breed. 6*:295–306 (2000).
78. Bingham, E.T., and T.J. McCoy, Cultivated Alfalfa at the Diploid Level: Origin, Reproductive Stability, and Yield of Seed and Forage, *Crop Sci. 19*:97–100 (1979).
79. Barnes, D.K., E.T. Bingham, R.P. Murphy, O.J. Hunt, D.F. Beard, W.H. Skrdla, and L.R. Teuber, *Alfalfa Germplasm in the United States: Genetic Vulnerability, Use, Improvement, and Maintenance*, USDA-ARS, Washington, D.C. (1977).
80. Crochemore, M.-L., C. Huyghe, C. Ecalle, and B. Julier, Structuration of Alfalfa Genetic Diversity Using Agronomic and Morphological Characteristics. Relationship with RAPD Markers, *Agronomie 18*:79–94 (1998).
81. Brummer, E.C., Capturing Heterosis in Forage Crop Cultivar Development, *Crop Sci. 39*:943–954 (1999).
82. Cazcarro-Ruiz, P., Differentiating *M. sativa* subsp. *sativa* and subsp. *falcata* Using Molecular Markers, M.S. Thesis, Iowa State Univ., Ames, IA, 2000, 115 pp.
83. Echt, C.S., K.K. Kidwell, S.J. Knapp, T.C. Osborn, and T.J. McCoy, Linkage Mapping in Diploid Alfalfa (*Medicago sativa* L.), *Genome 106*:123–137 (1994).
84. Wu, K.K., W. Burnquist, M.E. Sorrells, T.L. Tew, P.H. Moore, and S.D. Tanksley, The Detection and Estimation of Linkage in Polyploids Using Single-Dose Restriction Fragments, *Theor. Appl. Genet. 83*:294–300 (1992).
85. Hackett, C.A., J.E. Bradshaw, and J.W. McNicol, Interval Mapping of Quantitative Trait Loci in Autotetraploid Species, *Genetics 159*:1819–1832 (2001).
86. Hackett, C.A., B. Pande, and G.J. Bryan, Constructing Linkage Maps in Autotetraploid Species Using Simulated Annealing, *Theor. Appl. Genet. 106*:1107–1115 (2003).
87. Luo, Z.W., C.A. Hackett, J.E. Bradshaw, J.W. McNicol, and D. Milbourne, Predicting Parental Genotypes and Gene Segregation for Tetrasomic Inheritance, *Theor. Appl. Genet. 100*:1067–1073 (2000).
88. Luo, Z.W., C.A. Hackett, J.E. Bradshaw, J.W. McNicol, and D. Milbourne, Construction of a Genetic Linkage Map in Tetraploid Species Using Molecular Markers, *Genetics 157*:1369–1385 (2001).
89. Wu, S.S., R.L. Wu, C.X. Ma, Z.B. Zeng, M.C. Yang, and G. Casella, A Multivalent Pairing Model of Linkage Analysis in Autotetraploids, *Genetics 159*:1339–1350 (2001).
90. Wu, R., M. Gallo-Meagher, R.C. Littell, and Z.-B. Zeng, A General Polyploid Model for Analyzing Gene Segregation in Outcrossing Tetraploid Species, *Genetics 159*:869–882 (2001).
91. Hackett, C.A., and Z.W. Luo, Tetraploidmap: Construction of a Linkage Map in Autotetraploid Species, *J. Hered. 94*:358–359 (2003).
92. Brummer, E.C., J.H. Bouton, M. Sledge, and G. Kochert, Molecular Mapping in Alfalfa and Related Species, in *DNA-Based Markers in Plants*, 2nd Ed., edited by I.K. Vasil and R. Phillips, Kluwer, Dordrecht, The Netherlands, 2001, pp. 169–180.
93. Endre, G., P. Kalo, Z. Kevei, P. Kiss, S. Mihacea, B. Szakal, A. Kereszt, and G.B. Kiss, Genetic Mapping of the Non-Nodulation Phenotype of the Mutant MN-1008

in Tetraploid Alfalfa (*Medicago sativa*), *Mol. Genet. Genomics 266*:1012–1019 (2002).

94. Endre, G., A. Kereszt, Z. Kevei, S. Mihacea, P. Kalo, and G.B. Kiss, A Receptor Kinase Gene Regulating Symbiotic Nodule Development, *Nature 417*:962–966 (2002).
95. Yu, K.F., and K.P. Pauls, Identification of a RAPD Marker Associated with Somatic Embryogenesis in Alfalfa, *Plant Mol. Biol. 22*:269–277 (1993).
96. Brummer, E.C., M.M. Shah, and D. Luth, Re-Examining the Relationship between Fall Dormancy and Winterhardiness in Alfalfa, *Crop Sci. 40*:971–977 (2000).
97. Alarcón-Zúñiga, B., E.C. Brummer, P. Scott, K. Moore, and D. Luth, Quantitative Trait Loci Mapping of Winter Hardiness Metabolites in Autotetraploid Alfalfa, in *Molecular Breeding of Forage and Turf*, edited by Z.-Y. Wang, Kluwer, Dordrecht, The Netherlands, 2003, pp. 97–104.
98. Barcaccia, G., S. Tavoletti, A. Mariani, and F. Veronesi, Occurrence, Inheritance, and Use of Reproductive Mutants in Alfalfa Improvement, *Euphytica 133*:37–56 (2003).
99. Tavoletti, S., E.T. Bingham, B.S. Yandell, F. Veronesi, and T.C. Osborn, Half Tetrad Analysis in Alfalfa Using Multiple Restriction Fragment Length Polymorphism in Alfalfa, *Proc. Natl. Acad. Sci. USA 93*:10918–10922 (1996).
100. Bingham, E.T., R.W. Groose, D.R. Woodfield, and K.K. Kidwell, Complementary Gene Interactions in Alfalfa Are Greater in Autotetraploids Than Diploids, *Crop Sci. 34*:823–829 (1994).
101. Riday, H., Heterosis in a Broad Range of Alfalfa Germplasm, *Crop Sci. 44*:in press (2004).
102. Bingham, E.T., Maximizing Hybrid Vigor in Autotetraploid Alfalfa, in *Better Crops for Food, Ciba Found. Symp. 97*, Pitman, London, 1983, pp. 130–143.
103. Bernardo, R., Relationship between Single-Cross Performance and Molecular Marker Heterozygosity, *Theor. Appl. Genet. 83*:628–634 (1992).
104. He, X.Z., and R.A. Dixon, Genetic Manipulation of Isoflavone 7-O-Methyltransferase Enhances Biosynthesis of 4'-O-Methylated Isoflavonoid Phytoalexins and Disease Resistance in Alfalfa, *Plant Cell 12*:1689–1702 (2000).
105. Hirt, H., M. Mink, M. Pfosser, L. Bogre, J. Gyorgyey, C. Jonak, A. Gartner, D. Dudits, and E. Heberlebors, Alfalfa Cyclins—Differential Expression During the Cell-Cycle and in Plant Organs, *Plant Cell 4*:1531–1538 (1992).
106. Junghans, H., K. Dalkin, and R.A. Dixon, Stress Responses in Alfalfa (*Medicago sativa* L) .15. Characterization and Expression Patterns of Members of a Subset of the Chalcone Synthase Multigene Family, *Plant Mol. Biol. 22*:239–253 (1993).
107. Cunningham, S.M., J.A. Gana, J.J. Volenec, and L.R. Teuber, Winter Hardiness, Root Physiology, and Gene Expression in Successive Fall Dormancy Selections from 'Mesilla' and 'CUF 101' Alfalfa, *Crop Sci. 41*:1091–1098 (2001).
108. Monroy, A.F., Y. Castonguay, S. Laberge, F. Sarhan, L.P. Vézina, and R.S. Dhindsa, A New Cold-Induced Alfalfa Gene Is Associated with Enhanced Hardening at Subzero Temperature, *Plant Physiol. 102*:873–879 (1993).
109. Laberge, S., Genomic Analysis of Cold Tolerance in Alfalfa, in *First International Conference on Legume Genomics and Genetics: Translation to Crop Improvement*, Minneapolis-St. Paul, MN, 2002.
110. Ivashuta, S., K. Uchiyama, M. Gau, and Y. Shimamoto, Linear Amplification Coupled with Controlled Extension as a Means of Probe Amplification in a cDNA Array and

Gene Expression Analysis During Cold Acclimation in Alfalfa (*Medicago sativa* L.), *J. Exp. Bot. 53*:351–359 (2002).

111. Ivashuta, S., M. Naumkina, M. Gau, K. Uchiyama, S. Isobe, Y. Mizukami, and Y. Shimamoto, Genotype-Dependent Transcriptional Activation of Novel Repetitive Elements During Cold Acclimation of Alfalfa (*Medicago sativa*), *Plant J. 31*:615–627 (2002).
112. Hays, D.B. and D.Z. Skinner, Development of an Expressed Sequence Tag (EST) Library for *Medicago sativa*, *Plant Sci. 161*:517–526 (2001).
113. Barcaccia, G., S. Varotto, S. Meneghetti, E. Albertini, A. Porceddu, P. Parrini, and M. Lucchin, Analysis of Gene Expression During Flowering in Apomeiotic Mutants of *Medicago* spp.: Cloning of ESTs and Candidate Genes for 2n Eggs, *Sex. Plant Reprod. 14*:233–238 (2001).
114. Huhman, D.V., and L.W. Sumner, Metabolic Profiling of Saponins in *Medicago sativa* and *Medicago truncatula* Using HPLC Coupled to an Electrospray Ion-Trap Mass Spectrometer, *Phytochemistry 59*:347–360 (2002).
115. Chen, F., A.L. Duran, J.W. Blount, L.W. Sumner, and R.A. Dixon, Profiling Phenolic Metabolites in Transgenic Alfalfa Modified in Lignin Biosynthesis, *Phytochemistry 64*:1013–1021 (2003).
116. Saunders, J.W., and E.T. Bingham, Production of Alfalfa Plants from Callus Tissue, *Crop Sci. 12*:804–808 (1972).
117. Bingham, E.T., L.V. Hurley, D.M. Kaatz, and J.W. Saunders, Breeding Alfalfa Which Regenerates from Callus Tissue in Culture, *Crop Sci. 15*:719–721 (1975).
118. Atanassov, A., and D.C.W. Brown, Plant Regeneration from Suspension Culture and Mesophyll Protoplasts of *Medicago sativa* L., *Plant Cell Tissue and Organ Culture 3*:149–162 (1984).
119. McKersie, B.D., and D.C.W. Brown, Somatic Embryogenesis and Artificial Seeds, in *Biotechnology and the Improvement of Forage Legumes*, edited by B.D. McKersie and D.C.W. Brown, CAB International, Oxon, United Kingdom, 1997, pp. 111–143.
120. Shao, C.Y., E. Russinova, A. Iantcheva, A. Atanassov, A. McCormac, D.F. Chen, M.C. Elliott, and A. Slater, Rapid Transformation and Regeneration of Alfalfa (*Medicago falcata* L.) Via Direct Somatic Embryogenesis, *Plant Growth Regul. 31*:155–166 (2000).
121. Spano, L., D. Mariotti, M. Pezzotti, F. Damiani, and S. Arcioni, Hairy Root Transformation in Alfalfa (*Medicago sativa* L), *Theor. Appl. Genet. 73*:523–530 (1987).
122. Ninkovic, S., J. Miljusdjukic, and M. Neskovic, Genetic-Transformation of Alfalfa Somatic Embryos and Their Clonal Propagation through Repetitive Somatic Embryogenesis, *Plant Cell Tissue and Organ Culture 42*:255–260 (1995).
123. Chabaud, M., J.E. Passiatore, F. Cannon, and V. Buchananwollaston, Parameters Affecting the Frequency of Kanamycin Resistant Alfalfa Obtained by *Agrobacterium tumefaciens* Mediated Transformation, *Plant Cell Rep. 7*:512–516 (1988).
124. Samac, D.A., Strain Specificity in Transformation of Alfalfa by *Agrobacterium tumefaciens*, *Plant Cell Tissue and Organ Culture 43*:271–277 (1995).
125. Du, S., L. Erickson, and S. Bowley, Effect of Plant Genotype on the Transformation of Cultivated Alfalfa (*Medicago sativa*) by *Agrobacterium tumefaciens*, *Plant Cell Rep. 13*:330–334 (1994).

126. Pereira, L.F., and L. Erickson, Stable Transformation of Alfalfa (*Medicago sativa* L) by Particle Bombardment, *Plant Cell Rep. 14*:290–293 (1995).
127. Tabe, L.M., T. Wardleyrichardson, A. Ceriotti, A. Aryan, W. McNabb, A. Moore, and T.J.V. Higgins, A Biotechnological Approach to Improving the Nutritive-Value of Alfalfa, *J. Anim. Sci. 73*:2752–2759 (1995).
128. Trinh, T.H., P. Ratet, E. Kondorosi, P. Durand, K. Kamate, P. Bauer, and A. Kondorosi, Rapid and Efficient Transformation of Diploid *Medicago truncatula* and *Medicago sativa* ssp. *falcata* Lines Improved in Somatic Embryogenesis, *Plant Cell Rep. 17*:345–355 (1998).
129. Ziegelhoffer, T., J. Will, and S. Austin-Phillips, Expression of Bacterial Cellulase Genes in Transgenic Alfalfa (*Medicago sativa* L.), Potato (*Solanum tuberosum* L.), and Tobacco (*Nicotiana tabacum* L.), *Mol. Breed. 5*:309–319 (1999).
130. McCaslin, M., An Update on the Development of Roundup Ready Alfalfa, in 2002 *California Alfalfa & Forage Symposium*, 2002. http://alfalfa.ucdavis.edu/subpages/2002Symposium/pdf/McCaslin%20-%20Alfalfa%20Update.pdf (verified February 10, 2004).
131. Narváez-Vásquez, J., M.L. Orozco-Cárdenas, and C.A. Ryan, Differential Expression of a Chimeric Camv-Tomato Proteinase Inhibitor-I Gene in Leaves of Transformed Nightshade, Tobacco, and Alfalfa Plants, *Plant Mol. Biol. 20*:1149–1157 (1992).
132. Saruul, P., F. Srienc, D.A. Somers, and D.A. Samac, Production of a Biodegradable Plastic Polymer, Poly-Beta-Hydroxybutyrate, in Transgenic Alfalfa, *Crop Sci. 42*:919–927 (2002).
133. Thomas, J.C., C.C. Wasmann, C. Echt, R.L. Dunn, H.J. Bohnert, and T.J. McCoy, Introduction and Expression of an Insect Proteinase-Inhibitor in Alfalfa (*Medicago sativa* L), *Plant Cell Rep. 14*:31–36 (1994).
134. D'Halluin, K., J. Botterman, and W. De Greef, Engineering of Herbicide-Resistant Alfalfa and Evaluation under Field Conditions, *Crop Sci. 30*:866–871 (1990).
135. Oommen, A., R.A. Dixon, and N.L. Paiva, The Elicitor-Inducible Alfalfa Isoflavone Reductase Promoter Confers Different Patterns of Developmental Expression in Homologous and Heterologous Transgenic Plants, *Plant Cell 6*:1789–1803 (1994).
136. Tian, L., H. Wang, K. Wu, M. Latoszek-Green, M. Hu, B. Miki, and D.C.W. Brown, Efficient Recovery of Transgenic Plants through Organogenesis and Embryogenesis Using a Cryptic Promoter to Drive Marker Gene Expression, *Plant Cell Rep. 20*:1181–1187 (2002).
137. Samac, D.A., M. Dornbusch, M. Tesfaye, S. Purev, and S.J. Temple, A Comparison of Constitutive Promoters for Expression of Transgenes in Alfalfa (*Medicago sativa*), *Transgenic Res. 5*:in press (2004).
138. Mandaci, S., and M.S. Dobres, A Promoter Directing Epidermal Expression in Transgenic Alfalfa, *Plant Mol. Biol. 34*:961–965 (1997).
139. Rosellini, D., M. Pezzotti, and F. Veronesi, Characterization of Transgenic Male Sterility in Alfalfa, *Euphytica 118*:313–319 (2001).
140. Desgagnés, R., S. Laberge, G. Allard, H. Khoudi, Y. Castonguay, J. Lapointe, R. Michaud, and L.P. Vézina, Genetic-Transformation of Commercial Breeding Lines of Alfalfa (*Medicago sativa*), *Plant Cell Tissue and Organ Culture 42*:129–140 (1995).
141. Palumbo, J.D., D.A. Phillips, and C.I. Kado, Characterization of a New *Agrobacterium tumefaciens* Strain from Alfalfa (*Medicago sativa* L.), *Arch. Microbiol. 169*:381–386 (1998).

142. Shahin, E.A., A. Spielmann, K. Sukhapinda, R.B. Simpson, and M. Yashar, Transformation of Cultivated Alfalfa Using Disarmed *Agrobacterium tumefaciens*, *Crop Sci. 26*:1235–1239 (1986).
143. Schroeder, H.E., M.R.I. Khan, W.R. Knibb, D. Spencer, and T.J.V. Higgins, Expression of a Chicken Ovalbumin Gene in 3 Lucerne Cultivars, *Aust. J. Plant Physiol. 18*:495–505 (1991).
144. Bingham, E.T., Registration of Regen-S Alfalfa Germplasm Useful in Tissue Culture and Transformation Research, *Crop Sci. 29*:1095–1096 (1989).
145. Bingham,. E.T., Registration of Alfalfa Hybrid Regen-Sy Germplasm for Tissue Culture and Transformation Research, *Crop Sci. 31*:1098–1098 (1991).
146. Ray, I.M., and E.T. Bingham, Breeding Diploid Alfalfa for Regeneration from Tissue Culture, *Crop Sci. 29*:1545–1548 (1989).
147. Groose, R.W., and Y.G. Li, Registration of Diploid Alfalfa (*Medicago sativa* spp. *falcata*) Germplasms for Biotechnology Research and Development, *Crop Sci. 33*:1116–1116 (1993).
148. Blondon, F., and S. Brown, Ploidy of Embryogenic *Medicago sativa* subsp. *falcata* Germplasms, *Agronomie 19*:69–72 (1999).
149. Hill, K.K., N. Jarviseagan, E.L. Halk, K.J. Krahn, L.W. Liao, R.S. Mathewson, D.J. Merlo, S.E. Nelson, K.E. Rashka, and L.S. Loeschfries, The Development of Virus-Resistant Alfalfa, *Medicago sativa* L., *Bio-Technology 9*:373–377 (1991).
150. McCaslin, M.H., S.J. Temple, and J.E. Tofte, Methods for Maximizing Expression of Transgenic Traits in Autopolyploid Plants. U.S. Patent Pending, Application No. 20020042928 (2001).
151. Woodfield, D.R., and E.C. Brummer, Integrating Molecular Techniques to Maximise the Genetic Potential of Forage Legumes, in *Proc. 2nd Int'l Symp., Molecular Breeding of Forage Crops, Lorne and Hamilton, Victoria, Australia, Nov. 19–24, 2000*, edited by G. Spangenberg, Kluwer, Dordrecht, The Netherlands, 2001, pp. 51–65.
152. Andersen, J.R., and T. Lübberstedt, Functional Markers in Plants, *Trends Plant Sci. 8*:554–560 (2003).
153. Forster, J.W., E.S. Jones, J. Batley, and K.F. Smith, Molecular Marker-Based Genetic Analysis of Pasture and Turf Grasses, in *Molecular Breeding of Forage and Turf*, edited by Z.Y. Wang, Kluwer, Dordrecht, The Netherlands, 2003, pp. 197–238.
154. Lu, R., A.M. Martin-Hernandez, J.R. Peart, I. Malcuit, and D.C. Baulcombe, Virus-Induced Gene Silencing in Plants, *Methods 30*:296–303 (2003).
155. Deak, M., G.B. Kiss, C. Koncz, and D. Dudits, Transformation of *Medicago* by *Agrobacterium* Mediated Gene Transfer, *Plant Cell Rep. 5*:97–100 (1986).
156. Blake, N.K., R.L. Ditterline, and R.G. Stout, Polymerase Chain Reaction Used for Monitoring Multiple Gene Integration in *Agrobacterium* Mediated Transformation, *Crop Sci. 31*:1686–1688 (1991).
157. Bagga, S., D. Sutton, J.D. Kemp, and C. Senguptagopalan, Constitutive Expression of the Beta-Phaseolin Gene in Different Tissues of Transgenic Alfalfa Does Not Ensure Phaseolin Accumulation in Nonseed Tissue, *Plant Mol. Biol. 19*:951–958 (1992).
158. McKersie, B.D., Y. Chen, M. de Beus, S.R. Bowley, C. Bowler, D. Inzé, K. D'Halluin, and J. Botterman, Superoxide Dismutase Enhances Tolerance of Freezing Stress in Transgenic Alfalfa (*Medicago sativa* L.), *Plant Physiol. 103*:1155–1163 (1993).

159. Frugis, G., S. Caretto, L. Santini, and D. Mariotti, *Agrobacterium rhizogenes* Rol Genes Induce Productivity-Related Phenotypical Modifications in Creeping Rooted Alfalfa Types, *Plant Cell Rep. 14*:488–492 (1995).
160. Austin, S., E.T. Bingham, D.E. Mathews, M.N. Shahan, J. Will, and R.R. Burgess, Production and Field Performance of Transgenic Alfalfa (*Medicago sativa* L.) Expressing Alpha-Amylase and Manganese-Dependent Lignin Peroxidase, *Euphytica 85*:381–393 (1995).
161. Masoud, S.A., Q. Zhu, C. Lamb, and R.A. Dixon, Constitutive Expression of an Inducible Beta-1,3-Glucanase in Alfalfa Reduces Disease Severity Caused by the Oomycete Pathogen *Phytophthora megasperma* f. sp. *medicaginis*, but Does Not Reduce Disease Severity of Chitin-Containing Fungi, *Transgenic Res. 5*:313–323 (1996).
162. Brill, L.M., and A.M. Hirsch, Stable Transformation of T1 and T2 Transgenic Alfalfa with Antisense-Lectin Constructs, *Symbiosis 27*:17–31 (1999).
163. Brill, L.M., C.J. Evans, and A.M. Hirsch, Expression of Mslec1- and Mslec2-Antisense Genes in Alfalfa Plant Lines Causes Severe Embryogenic, Developmental, and Reproductive Abnormalities, *Plant J. 25*:453–461 (2001).
164. Hipskind, J.D., and N.L. Paiva, Constitutive Expression of a Resveratrol-Glucoside in Transgenic Alfalfa Increases Resistance to *Phoma medicaginis*, *Mol. Plant-Microbe Interact. 13*:551–562 (2000).
165. Ullah, A.H.J., K. Sethumadhavan, E.J. Mullaney, T. Ziegelhoffer, and S. Austin-Phillips, Cloned and Expressed Fungal *phya* Gene in Alfalfa Produces a Stable Phytase, *Biochem. Biophys. Res. Commun. 290*:1343–1348 (2002).
166. Ullah, A.H.J., K. Sethumadhavan, E.J. Mullaney, T. Ziegelhoffer, and S. Austin-Phillips, Characterization of Recombinant Fungal Phytase (*phya*) Expressed in Tobacco Leaves, *Biochem. Biophys. Res. Commun. 264*:201–206 (1999).
167. Rubio, M.C., J. Ramos, K.J. Webb, F.R. Minchin, E. González, C. Arese-Igor, and M. Becana, Expression Studies of Superoxide Dismutases in Nodules and Leaves of Transgenic Alfalfa Reveal Abundance of Iron Containing Isozymes, Posttranslational Regulation, and Compensation of Isozyme Activities, *Mol. Plant-Microbe Interact. 14*:1178–1188 (2001).
168. Ray, H., M. Yu, P. Auser, L. Blahut-Beatty, B. McKersie, S. Bowley, N. Westcott, B. Coulman, A. Lloyd, and M.Y. Gruber, Expression of Anthocyanins and Proanthocyanidins after Transformation of Alfalfa with Maize Lc, *Plant Physiol. 132*:1448–1463 (2003).
169. Ortiz, J.P., M.I. Reggiardo, M.A. Spitteler, and R.H. Vallejos, Stable Transformation of Alfalfa (*Medicago sativa*) Employing Accelerated Microparticles, *Anales De La Asociacion Quimica Argentina 80*:159–166 (1992).
170. Ramaiah, S.M., and D.Z. Skinner, Particle Bombardment: A Simple and Efficient Method of Alfalfa (*Medicago sativa* L.) Pollen Transformation, *Curr. Sci. 73*:674–682 (1997).
171. Xu, Z.Q., J.F. Jia, and Z.D. Hu, Enhancement by Osmotic Treatment of Hairy Root Transformation of Alfalfa Suspension Cultures, and Chromosomal Variation in the Transformed Tissues, *Aust. J. Plant Physiol. 24*:345–351 (1997).
172. Bellucci, M., B. Lazzari, A. Viotti, and S. Arcioni, Differential Expression of a Gamma-Zein Gene in *Medicago sativa*, *Lotus corniculatus* and *Nicotiana tabacum*, *Plant Sci. 127*:161–169 (1997).

173. Bellucci, M., A. Alpini, and S. Arcioni, Zein Accumulation in Forage Species (*Lotus corniculatus* and *Medicago sativa*) and Co-Expression of the Gamma-Zein: KDEL and Beta-Zein:KDEL Polypeptides in Tobacco Leaf, *Plant Cell Rep. 20*:848–856 (2002).
174. Senaratna, T., B.D. McKersie, K.J. Kasha, and J.D. Procunier, Direct DNA Uptake During the Imbibition of Dry Cells, *Plant Sci. 79*:223–228 (1991).
175. Reich, T.J., V.N. Iyer, and B.L. Miki, Efficient Transformation of Alfalfa Protoplasts by the Intranuclear Microinjection of Ti Plasmids, *Bio-Technology 4*:1001–1004 (1986).

Chapter 8

Medicago truncatula as a Model Legume

Sergey Ivashuta,[a] Steven Gantt,[a] and Carroll Vance[b]

[a]Plant Biology Department; and [b]USDA/ARS, Plant Science Research, Agronomy and Plant Genetics, University of Minnesota, St. Paul, MN 55108

Introduction

Legumes, broadly defined by their unusual flower structure, podded fruit, and symbiotic N_2-fixing root nodules, are second only to the Gramineae in their importance to humans (1). The 670 to 750 genera and about 19,000 species of legumes include important grain, pasture, and agroforestry species (2). Domestication of legumes has been reported at various sites dating as early as 8,000 to 3,000 BP (3,4). Grain and forage legumes are grown on some 180 million ha, or about 15% of the earth's arable land (http://apps.fao.org/page/collections). They account for 27% of the world's primary crop production, with grain legumes alone contributing 33% of the dietary protein nitrogen needs of humans (1,5). Legumes, predominantly soybean and peanut, provide more than 35% of the world's processed vegetable oil. The political, social, and economic importance of the Leguminoseae necessitate a comprehensive genetic understanding of this family.

Complete genome sequencing, large-scale expressed sequence tag (EST) projects, high-density physical maps of linkage groups, facile transformation, and a large number of readily defined mutants have made *Arabidopsis* and rice (*Oryza sativa* L. ssp. *japonica*) model genetic systems for most dicot and monocot plants, respectively (6,7). However, their lack of or poorly defined symbiotic/pathogenic associations, unusual root traits, and limited geographical distribution necessitate that other species be invoked as models to study root-microbe interaction (rhizobial, mycorrhizal, pathogenic) and root development. Having widespread agricultural importance would make alternative models even more attractive.

Medicago truncatula L. Gaertner (*Mt*) (Fig. 8.1) has many attributes that make it a valuable alternative model species for genomics and genetics (8–10). Key characteristics that make *Mt* a valuable model system for legume genomics and genetics include small diploid genome (500 Mb) (Table 8.1), rapid generation time, self-fertilization, highly abundant seed production, numerous ecotypes, readily transformed with *Agrobacterium tumefaciens* and *A. rhizogenes*, efficient nodulation with *Sinorhizobium meliloti*, ready colonization by *Glomus* spp., a range of susceptibility to root pathogens including nematodes, extensive EST sequencing from a number of organs at various

Figure 8.1. *Medicago truncatula* is a model legume for genome sequencing and functional genomics. This photo depicts leaves, stems, flowers, pods, and root nodules. The sequencing of thousands of ESTs from various organs has been facilitated by the construction of some 40 cDNA libraries.

developmental stages, high-density genetic linkage map of *Mt's* eight linkage groups, robust evidence for macro- and microsynteny with other legume species, extensive array of mutants, and agricultural importance. The acceptance of *Mt* as a model legume is evidenced by the National Science Foundation's recent award of $10.8 million for full genome sequencing.

TABLE 8.1
Genome Sequencing Model Legumes

Organism	Genome Size (Mb)	Chromosome Number Haploid	ESTs Sequenced (1,000)
Medicago truncatula[a]	500	8	190[b]
Lotus japonicus[a]	470	6	36
Glycine max	1,100	20	300[b]
Phaseolus vulgaris	530	11	20

[a]Shotgun sequencing: ≈ 25% of genome has been accomplished.
[b]*Medicago*, approximately 17,000 TCs and 19,000 singletons; *glycine*, approximately 29,000 TCs and 36,000 singletons.

Genetic Maps and Synteny

Genome size in the legume family ranges from about 370 megabases (Mb) to greater than 13,000 Mb (11,12). Although independent of genome size, 1N chromosome number also varies in the Leguminoseae (www.rbgkew.org.uk/cval/homepage.html). The wide range in genome size is thought to be due to differences in number and amount of retroelements and large-scale gene duplications (12,13). Such differences necessitate that any model legume used for genomics/genetics has a small, compact, and relatively simple genome. The *Mt* genome fits these criteria.

Because of the out-crossing tetraploid nature of cultivated alfalfa, inbred lines are very difficult to establish (14). Moreover, the complex segregation patterns encountered in tetraploid inheritance of alfalfa make genetic studies difficult. To overcome this difficulty, geneticists frequently use diploid alfalfa species and subspecies for genetic studies. One of the first steps in developing *Mt* as a useful model system would be to develop a well-defined linkage map of diploid ($2n = 2x = 16$, 8 linkage groups) alfalfa. Such a map would bridge tetraploid cultivated alfalfa with diploid *Mt*. Kaló *et al.* (15) have constructed an improved linkage map of diploid alfalfa. They analyzed the inheritance of 868 genetic markers in an F_2 population of plants derived from a self-pollinated F_1 hybrid individual of the cross *Medicago sativa* ssp. *quasifalcata* × *Medicago sativa* ssp. *coerula*. Of the markers, 80 corresponded to known genes, of which two had been previously cytologically localized. The genetic map covers 754 centimorgans (cM) with an average marker density of 0.8/cM. The 868 genetic markers were distributed into eight linkage groups. The correlation between physical and genetic distances is approximately 1,150 kb/cM. The improved genetic map of diploid *M. sativa* lays the foundation for map-based cloning of genes controlling important agronomic traits.

Verification that *Mt* can serve as a model system for alfalfa and related species is dependent upon whether colinearity of genes and significant macro/microsynteny occurs between homologous linkage groups. To assess whether colinearity and broad-scale macrosynteny occur between *Mt* and diploid alfalfa, Thoquet *et al.* (16) constructed a molecular genetic linkage map of *Mt*. They mapped 289 markers in 124 individuals of an F2 segregating population derived from a cross between *Mt* lines Jemalong 6 and DZA315.16. These two lines are polymorphic for several traits, including rhizobial specificity, pod shape and coiling, and leaf anthocyanins. The resulting map showed *Mt* to have eight linkage groups. More than 90% of the markers were located within an interval of less than 10 cM from adjacent markers. The physical to genetic distance in *Mt* is 470 kb/cM as compared to 1,150 kb/cM for diploid *M. sativa*, 3,000 kb/cM for pea, and 263 kb/cM for Arabidopsis. In companion studies, Thoquet *et al.* selected two or more distant markers per chromosome from the diploid alfalfa genetic map and positioned them on the *Mt* genetic map to identify homologous linkage groups between the two species. Their results showed extensive colinearity and suggest good macrosynteny between the two species. To confirm the usefulness of the *Mt* genetic map, they showed that the traits inefficient symbiosis (*Mt sym6*) and pod coiling (*SPC*) mapped to linkage groups 8 and 7, respectively.

Choi *et al.* (17) have also constructed a sequence-based genetic map of *Mt* and assessed colinearity with diploid *M. sativa*. They analyzed the segregation of 288 sequence-characterized genetic markers (141 ESTs, 80 BAC end sequence-tags, and 67 resistance gene analogs) in an F_2 population composed of 93 individuals derived from a cross between *Mt* lines *A17* and *A20*. Similar to Thoquet *et al.* (16), they found eight genetically distinct linkage groups that readily align with the eight linkage groups of diploid *M. sativa*. Intron-targeted PCR with 60 EST markers identified single nucleotide polymorphisms (SNPs) occurring every 142 base pairs (bp) in introns versus 1 SNP every 509 bp in exons. Fluorescent *in situ* hybridization experiments localized 16 markers on pachytene chromosomes and showed colinearity on chromosomes 4 and 8. Marker alignment between *Mt* and diploid *M. sativa* showed extremely high synteny between the two species.

Inclusively, Thoquet *et al.* (16) and Choi *et al.* (17) have developed high-resolution genetic maps of *Mt*. In addition, they have shown colinearity of genomes and striking synteny between *Mt* and diploid *M. sativa*. The studies provide strong rationale for using *Mt* as a model for *M. sativa* and closely related genera.

Kulikova *et al.* (18) constructed a molecular cytogenetic map of *Mt* on the basis of a pachytene DAPI karyogram. In addition, fluorescence *in situ* hybridization (FISH) was used to determine the relative position of 20 BACs and three repetitive sequences on pachytene chromosomes, thus enabling one to correlate genetic linkage groups with chromosomes. They found *Mt* chromosomes to have a fairly simple organization with solid blocks of heterochromatin flanking the centromeres and long stretches of euchromatin in the distal arms. Chromosome length ranged from 29 to 68 μm, with a total combined length of the eight chromosomes approaching 406 μm. Assuming an average degree of condensation in the euchromatic regions of 300 kb/μm (as measured for chromosome 5), they calculated the total euchromatic fraction of *Mt* DNA to be 105 Mb. Because the *Mt* genome is 470 to 500 Mb, some 365 to 395 Mb, or 80%, is estimated to occupy heterochromatic regions.

To assign individual linkage groups to chromosomes, 20 BACs (19) representing each linkage group were used as probes for FISH mapping to pachytene chromosomes. Five or more sequence-characterized genetic markers developed from BAC end sequencing were also positioned (15,19) to provide a frame of reference to the genetic map. Most of the linkage group–specific BACs gave distinct FISH signals in euchromatic parts of chromosomes. Little to no background staining was detected, supporting the interpretation that interstitial segments of *Mt* chromosomes contain a low number and frequency of repeated sequences. Inclusively, Kulikova *et al.* (18) interpret their finding to mean the arms of *Mt* chromosomes are rich in transcribed genes and repeat sequences will be under-represented in the intergenic regions. The simplicity of the genome will facilitate map-based cloning strategies for genes located in euchromatic regions because high-density repeated sequences will not interfere.

Comparative genetic mapping of closely related species shows that on a broad scale gene order and content are frequently conserved (20). Synteny studies in legumes, although limited, have shown that within the Phaseoleae and Vicieae large linkage blocks are conserved (21,22). However, macrosyntenic comparisons do not provide sufficient detail needed for microstructural comparison required for cross-species gene prediction and positional gene cloning. To assess whether the *Mt* genome could serve as a physical and genetic model, microsyntenic gene order and content must be compared to that of diverse legumes. To assess whether the *Mt* genetic map could be used for positional cloning of pea symbiotic genes, Gualtieri *et al.* (23) evaluated the level of microsynteny between the *SYM2* region of pea and the orthologous region in *Mt*. Using a set of genetic markers that are widely distributed on pea linkage group I and *Mt* linkage group V, they showed that the two linkage groups are syntenic. Closer inspection of the region around *SYM2* using nine markers that are physically linked in *Mt*, spanning some 480 kb (0.5 to 1.0 cM), showed that they are also clustered in pea. More importantly, the order of the gene in pea is identical to that in *Mt*, demonstrating the gene order and content of pea and *Mt* genomes are conserved in that region.

Map-based cloning in phylogenetically distant related legumes having large genomes could be greatly facilitated using *Mt* gene position as a reference if microsynteny occurred at the BAC sequence data (contig) level. Yan *et al.* (24) recently estimated microsynteny at the BAC contig level between *Mt* and soybean. They performed comparative positional analysis of 144 low-copy soybean BAC-derived probes along with 43 RFLP probes derived from soybean contigs to their location on *Mt* BACs. Hybridization data indicated that between 30 and 54% of the soybean probes showed microsynteny with *Mt*. The authors note that microsyntenic relationships will make it possible to use the *Mt* genome as a reference model for isolating genes in soybean and other legume species having large genomes, abundant repetitive DNA, and extensive duplication. Comparative sequencing of homologous *Mt* and soybean BACs is now in progress to define the extent of gene duplication and rearrangement and to determine the nature of intergenic sequences. The macro and microsyntenic relationship among *Mt*, alfalfa, and pea has recently been verified and utilized for map-based cloning of two genes involved in recognition of and infection by *S. meliloti* (25,26).

Functional Genomics

The Medicago truncatula *Transcriptome*

Large-scale, in-depth, single-pass sequencing of cDNAs, ESTs, from a variety of organs and tissues exposed to a range of treatments (P, N, H_2O, shading, pathogens, etc.) gives not only an overview of the plant transcriptomes but also a quantitative estimate of individual transcript abundance within a given organ (27). Knowledge of the abundance of specific transcripts provides insight into gene expression networks related to many processes, including but not limited to metabolism, organ development, function, adaptation to abiotic stress, response to pathogenic and symbiotic agents, senescence, and signaling. The *Mt* EST collection currently contains some 190,000 sequences obtained from 40 libraries representing a wide range of organs and conditions; it can be accessed at both the Institute for Genome Research (TIGR) *Medicago truncatula* Gene Index (Mt GI, www.tigr.org/tdb/mtgi) and University of Minnesota *Medicago truncatula* Data Base 2.0 (www.medicago.org/MtDB). Individual overlapping ESTs are clustered to form tentative consensus sequences (TCs) that putatively represent a unique transcript from a single gene (28). The current version of MtGI is composed of some 17,000 TCs and 19,000 singletons, giving a complement of 36,000 unique genes. Similar numbers for *Mt* TCs and singletons are found in MtDB. This complement of unique genes will change as more TCs or singletons are identified through deeper sequencing. Both MtDB and MtGI can be queried through the language of Boolean formalism to identify TCs composed of ESTs that are either derived solely from or highly expressed in specific tissues and treatments. For example, Table 8.2 shows the number of TCs having 50–100% of ESTs derived from a specific organ. As one might expect, a query with 100% of the ESTs in TCs derived from a specific organ results in far fewer TCs than a query that requests TCs with 50% of the ESTs derived from a corresponding tissue. Secondary queries (filters) can be superimposed upon primary queries to refine the data set. Lamblin *et al.* (29) have described the MtDB relational database that integrates *Mt* transcriptome data and provides a wide range of user-defined data-mining operations.

Because a primary justification for developing legumes as model systems for genomics has relied upon the importance of rhizobial and mycorrhizal symbioses (8,9,30), it is not surprising that transcriptome studies of *Mt* have focused on symbiosis. Because root hair cells are the first cells to respond to rhizobial inoculation, Covitz *et al.* (31) sequenced 899 ESTs derived from an *Mt* root hair cDNA library. The most abundant mRNAs detected in this limited assessment were methionine synthase, β-glucosidase, and tonoplast intrinsic protein. The first analysis of *Mt* symbiotic root nodule ESTs was reported by Györgyey *et al.* (32), who sequenced 389 cDNA clones derived from emerging nodules. The most abundant transcripts were cdc 48, pectin acetylesterase, ascorbate peroxidase, WRKY3, Sec12p, several

TABLE 8.2
Medicago truncatula TCs Having Varying Proportions of ESTs from Specific Organs

Organ	Number of TCs with: (%)		
	100	80	50
Nodule	411	484	1,198
Root	2,081	2,578	8,149
Leaf	421	543	2,598
Stem	299	338	1,714
Pod	42	49	284

Data derived from MtGI, www.tigr.org/tdb/mtgi.

early nodulins, leghemoglobins, carbonic anhydrase, lysophosphatidic acyltransferase, and calmodulin-binding protein.

With the advent of international genomics projects having large-scale EST databases, it has become possible to conduct genome-wide *in silico* (electronic) analysis of transcripts either specific to or significantly enhanced in symbiotic interactions. Nodule-specific TCs occurring in Release 4.0 of MtGI were identified by Fedorova *et al.* (33). Version 4.0 of MtGI contained 141,500 ESTs, of which 20,347 (or 14%) were derived from nodules. Upon clustering, some 340 TCs were found to be nodule-specific. Another 1,867 ESTs were identified as nodule-specific TCs. Several of the most abundant nodule-specific TCs are shown in Table 8.3. The expression pattern of nodule-specific TCs was validated with high-density macroarrays and RNA blots. Statistical analysis of the frequency of ESTs within a TC revealed that, when a TC was composed of four or more ESTs, physical measurement of transcript abundance correlated well with enhanced expression. More recent releases of MtGI, Releases 5.0 and 6.0, show that, as depth of nodule sequencing increases, the number of nodule-specific TCs increases. Currently, Release 6.0 of MtGI contains 411 nodule-specific TCs derived from about 30,000 nodule ESTs.

Although several functionally important nodule-specific TCs were identified by Fedorova *et al.* (33), two unusual highly expressed groups were very conspicuous. One group comprised five TCs clustered from 44 ESTs corresponding to calmodulin-like proteins. The deduced amino acid sequence of these proteins share 41–93% identity with typical constitutive calmodulin proteins. However, unlike typical calmodulins, nodule-specific calmodulin-like proteins contain one to three EF hands and an N-terminal signal sequence. The signal sequence predicts targeting of the calmodulin-like proteins to the secretory system of nodule cells, perhaps the symbiosome. Although the function of the calmodulin-like proteins is unknown, their abundance and specificity suggest that they play a significant role in nodules. Interestingly, putative orthologs of calmodulin-like genes could not be found in the Arabidopsis and rice genomes.

TABLE 8.3
Abundant Nodule-Specific TCs

Transport	Metabolism	Signal Transduction	Unknown Function
Purine permease (14)[a]	Carbonic anhydrase (7)	Calmodulin-like-5 TCs (44)	Cysteine cluster proteins (CCPs)-114 TC (~300+ESTs)
Copper transport (5)	Stearoyl acyl carrier desaturase (4)	IRE-protein kinase (6)	Bark agglutin precursor (40)
Hexose transport (4)	Lysine desaturase (4)	Zinc finger-protein (5)	Embryo specific protein (10)
Sulfate transport (4)		Receptor protein kinase (4)	Hypothetical protein (8)

[a] (#) ESTs in TC (33).

A second group of unusual nodule-specific TCs identified by Fedorova *et al.* (33) comprised 114 TCs clustered from about 300 ESTs, corresponding to cysteine-cluster proteins (CCPs). The striking abundance of CCP TCs is noteworthy; they account for 33% of the nodule-specific TCs in Release 4.0 of MtGI, and their abundance makes them highly provocative. The CCP deduced proteins are small, ranging in Mr from 6 to 9 kDa. The CCP proteins also have an N-terminal signal sequence that is predicted to target the protein to the secretory system but does not resemble the calmodulin-like signal sequence. Mergaert *et al.* (34) have extended understanding of the nodule-specific CCP gene family (they refer to CCPs as nodule-specific, cysteine-rich proteins, NCRs). They analyzed Release 5.0 of MtGI, which contains 164,441 ESTs, and found 311 CCP (NCR) TCs. Through *in silico* searches and DNA blots they concluded that CCPs (NCRs) occurred only in galegoid legumes that form indeterminate nodules. The authors suggest that CCPs (NCRs) are legume-specific genes. Using CCP (NCR) chimeric reporter gene constructs, they showed CCP (NCR) proteins were directed to the secretory pathway in onion epidermal cells. Moreover, *in situ* hybridization showed that most CCP (NCR) transcripts were localized in infected cells. They proposed that CCPs (NCRs) function in cell-to-cell signaling and/or cellular defense.

Both rhizobial and mycorrhizal symbioses were considered in the analysis of *Mt* ESTs conducted by Journet *et al.* (35). *In silico* analysis of various *Mt* cDNA libraries accompanied by a refined R test to determine statistical significance of transcript abundance data derived from electron northerns enabled the authors to obtain a global view of the genetic programs involved in root symbioses. Similar to Fedorova *et al.* (33), Journet *et al.* (35) found 331 TCs that were predicted to be upregulated in root nodules. Among these were 92 that were enhanced in young nodules. Searching all available EST databases, they concluded that 141 nodule-specific TCs were legume specific. Moreover, they found 51 TCs that had enhanced expression in mycorrhizal roots. Cross-comparison (Boolean language) of TCs from both nodulated and mycorrhizal roots showed 19 TCs to have upregulation common to both symbioses. Of these common TCs, seven were involved in protein synthesis and processing, four have unknown functions, two have no homology in the database, two are involved in RNA metabolism, and four have miscellaneous functions. The authors have created a searchable database that can be accessed at http://medicago.toulouse.inra.fr/EST.

The observations that CCP (NCR) genes appear legume specific (33,34) and that as many as 141 TCs are found only in legumes (35) drive the question, "Are any plant genes specific to legumes?" With the large EST database available for many plant species (www.tigr.org/tdb/tgi/agi/, Arabidopsis; www.tigr.org/tdb/tgi/hvgi/, barley; www.tigr.org/tdb/tgi/gmgi/, soybean; www.tigr.org/tdb/tgi/ogi/, rice; and www.tigr.org/tdb/tgi/zmgi/, maize) and with the Arabidopsis and rice genomes sequenced (6,7), it is possible to address this question. M. Graham, K. Silverstein, and K. VandenBosch (University of Minnesota, St. Paul, personal communication) have queried the existing databases for legume-specific genes (i.e., those genes having no sequence homology, below a specific threshold, to any publicly available sequence of non-legumes). They identified 2,525 legume-specific genes, 861 from *Mt*, 1,572 from soybean, and 92 from *Lotus*. Less than 3% of these sequences had homology to previously characterized genes or gene products. Some 300 correspond to CCP (NCR) genes. It will be interesting to determine the specific function of these genes as related to processes occurring in legumes.

Three recent reports document expression profiling of plant genes at distinct developmental stages of the *Mt–Glomus* spp. mycorrhizal symbiosis. Suppression subtractive hybridization (SSH) approaches by Brechenmacher *et al.* (36) and Wulf *et al.* (37) were successful in generating 641 and 1,805 ESTs, respectively. Both reports confirm, through reverse Northern cDNA hybridizations, that a subset of the ESTs generated by SSH was highly expressed at distinct stages of symbiosis. Collectively, they demonstrated that some 30 plant genes had enhanced expression during the symbiotic process. A more comprehensive approach by Liu *et al.* (38) involved sequencing of 7,351 ESTs from a cDNA library representing several stages of symbiotic development (inoculated roots collected at 10, 17, 22, 31, and 38 days of inoculation). *In silico* comparison of the *Mt* mycorrhizal symbiotic ESTs to those in MtGI showed 888 unique ESTs not found in any other *Mt* libraries. The abundance of transcript representing each EST was evaluated by probing high-density macroarrays (on nylon membranes) with radioactive cDNA made from RNA isolated at each of the five points after inoculation. Ninety-two unique genes having enhanced expression at one or more stages of the symbiosis were discovered. Liu *et al.* (38) developed chimeric constructs for two of the genes and demonstrated that reporter activity in cells containing arbuscules of transformed *Mt* roots.

Other than abstracts from meetings, there are no published reports on expression profiling of *Mt* transcriptome using microarrays. However, D.A. Samac, S. Penuela, and colleagues (USDA-ARS, St. Paul MN, personal communication) have laid the foundation for genome-wide transcript profiling experiments by developing methods for hybridization and data analysis with microarrays containing 1,152 ESTs printed in triplicate on a slide. They evaluated changes in gene expression of *Mt* during interactions with foliar and leaf pathogens. Differential expression was detected for 54 genes, 19 in response to leaf pathogens and 35 related to root infection. Cluster analysis of 48 genes representing disease defense–response indicators revealed that 19 were upregulated in all experiments. They have also shown that gene expression in pathogenic interactions was not only related to the pathogen, but was also a function of disease development. Currently, 6 K and 8 K EST–based microarrays are being used and a 16 K oligonucleotide microarray has been brought on line for use in transcriptome profiling by the *Mt* research community. By the time that this volume is published, high-density *Mt* EST and oligonucleotide-based microarrays will be in standard use.

Proteomics

Although transcriptome analysis gives information about the presence, absence, and abundance of RNA involved in processes, it gives no information about the ultimate gene product, the protein (39). Knowledge of if, when, where, and how much protein corresponding to a gene is expressed is critical to understanding functional importance. In addition, evaluation of post-translational events occurring in proteins provides insight into regulation of activity and cellular targeting. Advances in protein separation, amino acid sequencing of proteins, and peptide mass fingerprinting have led to the integration of protein characterization with gene analysis, referred to as proteomics.

Again due to the importance of plant–microbe interaction in *Mt* particular symbiosis, initial proteome characterization focuses on roots and microbial interaction with roots. Mathesius *et al.* (40) established a root proteome reference map for *Mt*. They found more than 2,500 root proteins that could be reproducibly displayed via two-dimensional electrophoresis. Peptide mass fingerprinting was used to analyze 485 proteins, and 179 of those were identified by matching against *Mt* EST sequence in the MtGI database. The majority of proteins identified in this study were metabolic enzymes and stress response proteins. In a subsequent study Mathesius *et al.* (41) showed that *Mt* roots respond to nanomolar concentrations of the bacterial quorum-sensing signal N-acyl homoserine lactone (AHL). They found changes in the accumulation of more than 150 proteins, 99 of which were identified by peptide mass fingerprinting and matched against the MtGI database. Roots of *Mt* responded to AHLs from both the pathogen *Pseudomonas aeruginosa* and the symbiont *Sinorhizobium meliloti*. Surprisingly, about 75% of the root proteins responding to AHLs showed increases of greater than four-fold. The proteome map established in Mathesius *et al.* (40,41) is regularly updated and can be accessed at: semele.anu.edu.au/2d/2d.html.

Proteome analysis of *Mt* roots inoculated with either *S. meliloti* or *G. mosseae* identified changes in 107 proteins (42). In response to mycorrhizae, 12 proteins were upregulated, 2 were downregulated, and 41 proteins were newly induced. In response to nodulation, 4 proteins were upregulated, one was downregulated, and 46 were newly induced. Unlike EST and transcript analysis, no common protein was found to be induced in both mycorrhizal and rhizobial symbioses.

Organs of *Mt* other than roots have now been subjected to proteome analysis. Watson *et al.* (43) have developed protein reference maps for *Mt* leaves, stems, flowers, pods, cell suspension cultures, and roots. A total of 551 proteins, equating to some 90 proteins from each organ, was excised from gels and analyzed by mass spectrometry; 304 of these were identified. The majority of the proteins corresponded to highly expressed EST sequences. For example, 40% of leaf protein mass was composed of Rubisco large subunit, Rubisco small subunit, Rubisco activase, and oxygen-evolving enzyme. Several proteins involved in secondary metabolism and lignin biosynthesis were found in stems. In flowers, 21% of the identified proteins were involved in protein synthesis and processing. Roots had a preponderance of proteins related to disease and abiotic stress. Predominant proteins in seeds/pods were globulins, cupins, ABA-response proteins, and osmotin. Cell suspension culture proteins included those involved in primary metabolism, disease defense, and protein processing.

Gallardo *et al.* (44) have also evaluated proteins occurring during *Mt* seed development. Seeds were collected at 12, 14, 16, 18, and 20 days after pollination

(DAP), and proteins were separated by two-dimensional electrophoresis. One hundred twenty-eight proteins were analyzed by mass spectrometry and 84 identified by peptide mass fingerprinting. The identification of legumins, vicilius, convicilius, and lipoxygenases and their increase during seed development confirmed the validity of using *Mt* as a model for legume seed development. Several seed proteins were found to decrease between 12 and 20 DAP, including annexia, myo-inositol-phosphate synthase, adenine kinase, and ascorbate peroxidase. Kinetics of accumulation of seed proteins were used to develop model schemes for starch, fatty acids, and oil deposition during seed maturation.

Transformation, Insertional Mutagenesis, RNAi

Efficient gene discovery is a cornerstone for any model genetic system. Discovery of plant gene function can be approached through a variety of methods. One of the most direct is gene tagging by insertional mutagenesis (45–47). Insertional mutagenesis gene tagging requires an effective insertional delivery process such as endogenous transposable elements as developed for maize, snapdragon, and tomato. For recalcitrant species in which transposons are not effective, such as *Mt*, the delivery of either T-DNA or RiT-DNA via *A. tumefaciens* and *A. rhizogenes*, respectively, is a powerful tool for functional genomics, as evidenced by the huge T-DNA insertional mutant for Arabidopsis (www.salk.edu).

Transformation of *Mt* via *A. tumefaciens* and *A. rhizogenes* was reported by Thomas *et al.* (48) some years ago, but efficiency of transformation and recovery of transformed plants was low. Advances in recent years have made transformation and regeneration of *Mt* more tractable. Wang *et al.* (49), using a highly regenerable line of *Mt* Jemalong 2HA developed explicitly for transformation, achieved a transformation rate of 17%. It took 9 to 12 weeks for transformed embryogenic callus to develop versus 4 to 6 weeks for nontransformed embryogenic callus. Transformed plantlets could be obtained in 8 months. Within the same time frame Chabaud *et al.* (50) were also successful in obtaining fairly routine transformation of *Mt* via somatic embryogenesis. Both laboratories used *A. tumefaciens* strain LBA 4404 for transformation.

Efficiency of *Mt* transformation was increased (51) and time to explant production reduced (52), further increasing the attractiveness of *Mt* for insertional mutagenesis for genetic studies. Hoffmann *et al.* (51) developed *Mt* line R108-1 in which 60–70% of the explants produced embryogenic calli and up to 50% of the embryos regenerated to complete plants. Line R108-1 was further improved in regeneration capacity through three additional cycles of *in vitro* leaf explant culture selection to produce R108-1(c3). With this line, highly embryogenic calli could be obtained in 3 weeks and 60% of explants developed callus. Plantlet production was increased 10-fold, and plantlets were obtained in 12 weeks. Co-cultivation of explants with *A. tumefaciens* EHA105 or GV3101 and the use of an improved regeneration genotype gave 80% of transformed embryos developing into whole plants and a cycle of 12 to 16 weeks from explant to transformed whole plant.

A striking advance in *Mt* transformation procedures was announced by Trieu *et al.* (53) in which vacuum infiltration, similar to that used for Arabidopsis, was effective in flower and seedling transformation. Unfortunately, this advance has not been pursued by Trieu *et al.*, and several other laboratories have had difficulty in reproducing the results. Vacuum infiltration of flowers for transformation via *A. tumefaciens* would advance functional genomics in *Mt* to a stage equivalent to Arabidopsis. The work by Trieu *et al.* (53) deserves further attention.

The efficacy of T-DNA tagging in *Mt* for efficient gene discovery was solidified by Scholte *et al.* (54). They found three vectors from which T-DNA inserted as a single-copy T-DNA into the *Mt* genome. Analysis of the T-DNA borders gave evidence that several *Mt* genes were tagged.

A very promising arena for functional genomics of roots and nodule traits is gene construct delivery to roots via *A. rhizogenes*. The pathogen *A. rhizogenes* is a close relative of *A. tumefaciens*, but it infects roots. When roots become infected with *A. rhizogenes*, a portion of the Ri plasmid (T-DNA) is delivered to the root cell nucleus and hyperplastic tissue growth occurs, resulting in uncontrolled root growth, called hairy roots. Jensen *et al.* (55) first reported delivery of an alien gene to legumes via root transformation of *Lotus corniculatus* L. with *A. rhizogenes*. Interestingly, when plants are transformed with *A. rhizogenes* containing both an Ri-plasmid and a disabled Ti-plasmid, the hairy roots are often co-transformed with T-DNAs from both plasmids. If the disabled Ti-plasmid contains a selectable marker (e.g., kanamycin resistance), one can easily identify co-transformed roots because they are able to grow in the presence of the growth inhibitor; nontransformed root can be removed through excision. Boisson-Dernier *et al.* (56) described protocols for rapid and efficient transformation of *Mt* roots with *A. rhizogenes*. They developed composite plants with transformed hairy roots expressing β-glucuronidase (GUS) driven by the ENOD-11 gene promoter. Root nodules that formed showed high levels of GUS expression, reflecting the activation of the ENOD-11 : GUS transgene. Using this system it is possible to obtain genetically transformed roots within 3 weeks as compared to 3 months for *A. tumefaciens*–transformed plants. Moreover, this approach makes screening for root or nodule gene promoters very rapid. The efficacy of *A. rhizogenes*–mediated root transformation for complementation was recently shown by Endre *et al.* (25) as they rescued the nodulation phenotype of non-nodulation controlled by the nodulation receptor kinase (NORK) gene. Hairy roots can also be used for reverse genetic approaches to study gene function by silencing genes via RNA interference (RNAi). We have begun to use gene silencing to investigate the involvement of specific genes in nodule development and function. Verification of the effectiveness of gene silencing using *A. rhizogenes* has been demonstrated by silencing GUS expression in hairy roots of transgenic *Mt* plants that constitutively express the GUS gene (Fig. 8.2). Similar experiments have been conducted by Kumagai and Kouchi (57). Furthermore, Limpens *et al.* (26) used RNAi to silence a number of genes in the *SYM2*-like region of *Mt* and identified two, which encode receptor-like kinases, that appear to function in nodule development.

Although *A. tumefaciens* transformation and regeneration of *Mt* have been improved in both efficiency and time to recover transformed plants, the system is not amenable to high throughput for generation of the very high numbers of mutants needed for effective insertional mutagenesis studies. *Agrobacterium rhizogenes* transformation of roots is an exceptional tool for studying root nodule and mycorrhizal symbioses and root gene function, but this approach should be complemented by whole plant transformation with the RNAi-inducing T-DNA using *A. tumefaciens* and/or identification of classical genetic mutants in the targeted gene, perhaps by targeting induced local lesions in genomes (TILLING) (58). Further advances are

Figure 8.2. RNAi-induced silencing of 35S-GUS (β-glucuronidase) reporter transgene expression in *Medicago truncatula* hairy roots. The composite plant on the left has hairy roots co-transformed with T-DNA that expresses double-stranded RNA complementary to GUS mRNA, when transcribed. Note the lack of GUS staining in roots of the plant on the left, indicating that expression of GUS activity in roots is silenced. The composite control plant on the right has hairy roots co-transformed with T-DNA that, when transcribed, produces double-stranded RNA complementary to human myosin mRNA. Note GUS staining throughout the plant.

needed in the arena of legume insertional mutagenesis and RNAi delivery to enhance the *Mt* model paradigm and functional genomics of all legume species.

Whole Genome Sequencing

Whole genome sequencing of *Mt* has been a goal of the *Mt* research community since the species was invoked as a model. To that end, several advances have occurred in the last few years. Physical mapping of the *Mt* genome is being coordinated by D. J. Kim and D. Cook at the University of California–Davis (http://medicago.plantpath.ucdavis.edu/), where nearly 50,000 BACs corresponding to a 12X coverage of the genome have been fingerprinted. This provides genome coverage of 445 Mb with 1,308 contigs of 36,400 clones. To date BAC sequencing has progressed to the point where some 20% of the *Mt* genome has been sequenced in Bruce Roe's laboratory at the University of Oklahoma (www.genome.ou.edu/medicago.html) with funding from the Noble Foundation, Ardmore, OK (www.noble.org/medicago/index.html). Gene frequency appears to be one gene/6 kb (B. Roe, University of Oklahoma, personal communication; D. Cook, University of California–Davis, personal communication). The EST sequencing project has resulted in sequencing of 17,000 TCs and 18,000 singletons representing some 50 Mb.

A very exciting step toward the goal of complete sequencing of the *Mt* genome was announced in October 2003 by NSF. The National Science Foundation awarded $10.8 million to sequence the gene-rich euchromatic regions of the *Mt* genome (www.fastlane.nsf.gov/servlet/showaward?award=0321460). Under the direction of N. Young, University of Minnesota, and co-principal investigators (B. Roe, University of Oklahoma; D. J. Kim, University of California–Davis; E. F. Retzel, University of Minnesota; and C. D. Town, The Institute for Genome Research [TIGR]), six of *Mt's* eight chromosomes will be sequenced. The U.S. group will sequence chromosomes 1, 2, 4, 6, 7, and 8. Chromosomes 1, 4, 6, and 8 will be sequenced by B. Roe's Oklahoma group, while chromosomes 2 and 7 will be sequenced at TIGR by C. D. Town's group. The sequencing project is planned in close coordination with colleagues in Europe who plan to sequence chromosomes 3 and 5, with collaboration between the Sanger Institute, John Innes Center (United Kingdom), and Genescape/INRA (France). Therefore, it is likely that *Mt*'s genome will be sequenced by 2006.

Summary

Several attributes contribute to the use of *Mt* as a model plant for legume genomics: a small, simply organized genome; a fairly well-developed genetic and physical map; synteny with diploid alfalfa and soybean; prolific seed production and short generation time; an excellent population of mutants; and a large number of ESTs from more than 40 libraries derived from a wide range of organs and conditions. Moreover, *Mt* is readily transformed with both *A. tumefaciens* and *A. rhizogenes* and is susceptible to gene silencing through RNAi, thus facilitating functional genomics. These advances, coupled to transcript profiling and proteome analysis related to root symbiotic interaction with rhizobiae and mycorrhizae, place *Mt* in a strong position to define legume functions and processes. Adding substantial strength to the *Mt* model is the recent commitment by NSF to fund sequencing the euchromatic gene space.

Because gene sequencing, transcript profiling, proteome analysis, metabolic profiling, and mutant analysis progress at a logarithmic pace, we will be challenged to integrate information into coherent molecular models that define plant growth and development. This integration will require even more collaboration between geneticists, biochemists, and bioinformaticians than has occurred to date. Even more importantly, the information will need translation to crop improvement through plant breeders. The gen-, prote-, metabol-, transcript- omics umbrella is large and interdisciplinary (Fig. 8.3), and it will give unimaginable insight into crop plant production. However, to achieve these insights and improve human well-being through food security will require exceptional communication as well as exceptional science.

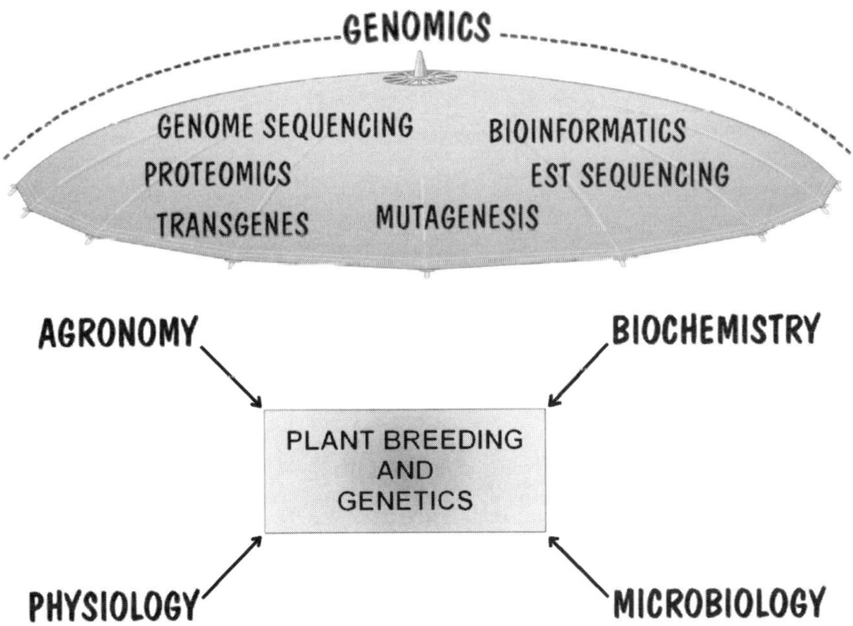

Figure 8.3. The broad umbrella of genomics covers a wide array of interdisciplinary approaches grounded in the disciplines of agronomy, physiology, biochemistry, and microbiology. Plant breeding and genetics are the foundational disciplines for translating genomics into public good.

Acknowledgments

Appreciation is extended to N. Young and D. Cook for sharing information about the *Mt* sequencing and mapping projects, respectively. Also, thank-you is extended to B. Bucciarelli for preparing the composite photo of *Mt* and G. Walker for assistance and proofreading on short notice.

References

1. Graham, P.H., and C.P. Vance, Legumes: Importance and Constraints to Greater Use, *Plant Physiol. 131*:872–877 (2003).
2. Polhill, R.M., P.H. Raven, and C.H. Stirton, Evolution and Systematics of the Leguminoseae, in *Advances in Legume Systematics*, Part 1, edited by R.M. Polhill and P.H. Raven, Royal Botanic Gardens, Kew, United Kingdom, 1981, pp. 1–26.
3. Cohen, M.N., The Food Crisis in Prehistory: Overpopulation and the Origins of Agriculture, Yale University Press, New Haven, 1977.
4. Hymowitz, T., and R.J. Singh, Taxonomy and Speciation, in *Soybeans: Improvement, Production and Use*, edited by J.R. Wilcox, Agronomy Monograph 16, American Society of Agronomy, Madison, Wisconsin, 1987, pp. 23–48.
5. Vance, C.P., Symbiotic Nitrogen Fixation and Phosphorus Acquisition: Plant Nutrition in a World of Declining Renewable Resources, *Plant Physiol. 127*:390–397 (2001).
6. Arabidopsis Genome Initiative: Analysis of the Genome Sequence of the Flowering Plant *Arabidopsis thaliana*, *Nature 408*:796–815 (2000).
7. Goff, S.A., D. Ricke, T.-H. Lan, G. Presting, R. Wang, *et al.*, A Draft Sequence of the Rice Genome (*Oryza sativa* L. ssp. *japonica*), *Science 296*:92–100 (2002).
8. Barker, D.G., S. Bianchi, F. Blondon, Y. Dattée, G. Duc, *et al.*, *Medicago truncatula*, a Model Plant for Studying the Molecular Genetics of the Rhizobium-Legume Symbiosis, *Plant Mol. Biol. Rep. 8*:40–49 (1990).
9. Cook, D.R., *Medicago truncatula*—A Model in the Making! *Current Opin. Plant Biol. 2*:301–304 (1999).
10. VandenBosch, K.A., and G. Stacey, Summaries of Legume Genomics Projects from around the Globe. Community Resources for Crops and Models, *Plant Physiol. 131*:840–865 (2003).
11. Broughton, W.J., G. Hernández, M. Blair, S. Beebe, P. Gepts, and J. Vanderleyden, Beans (*Phaseolus* spp.)—Model Food Legumes, *Plant & Soil 252*:55–128 (2003).
12. Young, N.D., J. Mudge, and T.H.N. Ellis, Legume Genomes: More than Peas in a Pod, *Curr. Opin. Plant Biol. 6*:199–204 (2003).
13. Vision, T.J., D.G. Brown, and S.D. Tanksley, The Origins of Genomic Duplication in Arabidopsis, *Science 290*:2114–2117 (2000).
14. Rumbaugh, M.D., J.L. Caddel, and D.E. Rowe, Breeding and Quantitative Genetics, in *Alfalfa* and *Alfalfa Improvement*, edited by A.A. Hanson, D.K. Barnes, and R.R. Hill, Agronomy Monograph 29, American Society of Agronomy, Madison, Wisconsin, 1988, pp. 777–808.
15. Kaló, P., G. Endre, L. Zimányi, G. Csanádi, and G.B. Kiss, Construction of an Improved Linkage Map of Diploid Alfalfa (*Medicago sativa*), *Theor. Appl. Genet. 100*:641–657 (2000).
16. Thoquet, P., M. Ghérardi, E.-P. Journet, A. Kereszt, J.-M. Ané, J.-M. Prosperi, and T. Huguet, The Molecular Genetic Linkage Map of the Model Legume *Medicago truncat-*

ula: An Essential Tool for Comparative Legume Genomics and the Isolation of Agronomically Important Genes, *Plant Biol. 2*:1–13 (2002).

17. Choi, H.-K., D. Kim, T. Uhm, E. Limpens, H. Lim, J.-H. Mun, P. Kaló, R.V. Penmetsa, A. Seres, O. Kulikova, B.A. Roe, T. Bisseling, G.B. Kiss, and D.R. Cook, A Sequence-Based Genetic Map of *Medicago truncatula* and Comparison of Marker Co-Linearity with *Medicago sativa*, *Genetics* (*in press*) (2003).
18. Kulikova, O., G. Gualtieri, R. Geurts, D.-J. Kim, D. Cook, T. Huguet, J.H. de Jong, P.R. Fransz, and T. Bisseling, Integration of the FISH Pachytene and Genetic Maps of *Medicago truncatula*, *Plant J. 27*:49–58 (2001).
19. Nam, Y.-W., R.V. Penmetsa, G. Entre, P. Uribe, D. Kim, and D.R. Cook, Construction of a Bacterial Artificial Chromosome Library of *Medicago truncatula* and Identification of Clones Containing Ethylene-Response Genes, *Theor. Appl. Genet. 98*:638–646 (1999).
20. Bennetzen, J.L., Comparative Sequence Analysis of Plant Nuclear Genomes: Microcolinearity and Its Many Exceptions, *Plant Cell 12*:1021–1029 (2000).
21. Boutin, S.R., N.D. Young, T.C. Olsen, Z.H. Yu, R.C. Shoemaker, and C.E. Vallejos, Genome Conservation among Three Legume Genera Detected with DNA Markers, *Genome 38*:928–937 (1995).
22. Weeden, N.F., F.J. Muehlbauer, and G. Ladizinsky, Extensive Conservation of Linkage Relationships between Pea and Lentil Genetic Maps, *J. Hered. 83*:123–129 (1992).
23. Gualtieri, G., O. Kulikova, E. Limpens, D.-J. Kim, D.R. Cook, T. Bisseling, and R. Geurts, Microsynteny between Pea and *Medicago truncatula* in the *SYM2* Region, *Plant Mol. Biol. 50*:225–235 (2002).
24. Yan, H.H., J. Mudge, D.-J. Kim, D. Larsen, R.C. Shoemaker, D.R. Cook, and N.D. Young, Estimates of Conserved Microsynteny among the Genomes of *Glycine max*, *Medicago truncatula* and *Arabidopsis thaliana*, *Theor. Appl. Genet. 106*:1256–1265 (2003).
25. Endre, G., A. Kereszt, Z. Kevei, S. Mihacea, P. Kaló, and G.B. Kiss, A Receptor Kinase Gene Regulating Symbiotic Nodule Development, *Nature 417*:962 (2002).
26. Limpens, E., C. Franken, P. Smit, J. Willemse, T. Bisseling, and R. Geurts, LysM Domain Receptor Kinases Regulating Rhizobial Nod Factor—Induced Infection, *Science 302*:630 (2003).
27. Ohlrogge, J., and C. Benning, Unraveling Plant Metabolism by EST Analysis, *Curr. Opin. Plant Biol. 3*:224–228 (2000).
28. Quackenbush, J., F. Liang, I. Holt, G. Pertea, and J. Upton, The TIGR Gene Indices: Reconstruction and Representation of Expressed Gene Sequences, *Nucleic Acids Res. 28*:141–145 (2000).
29. Lamblin, A.-F.J., J.A. Crow, J.E. Johnson, K.A.T. Silverstein, T.M. Kunau, A. Kilian, D. Benz, M. Stromvik, G. Endré, K.A. VandenBosch, D.R. Cook, N.D. Young, and E.F. Retzel, MtDB: A Database for Personalized Data Mining of the Model Legume *Medicago truncatula* Transcriptome, *Nucleic Acids Res. 31*:196–201 (2003).
30. Handberg, K., and J. Stougaard, *Lotus japonicus*, an Autogamous, Diploid Legume Species for Classical and Molecular Genetics, *Plant J. 2*:487–496 (1992).
31. Covitz, P.A., L.S. Smith, and S.R. Long, Expressed Sequence Tags from a Root-Hair-Enriched *Medicago truncatula* cDNA Library, *Plant Physiol. 117*:1325–1332 (1998).
32. Györgyey, J., D. Vaubert, J.I. Jiménez-Zurdo, C. Charon, L. Troussard, A. Kondorosi, and E. Kondorosi, Analysis of *Medicago truncatula* Nodule Expressed Sequence Tags, *Mol. Plant-Microbe Interact. 13*:62–71 (2000).

33. Fedorova, M., J. van de Mortel, P.A. Matsumoto, J. Cho, C.D. Town, K.A. VandenBosch, J.S. Gantt, and C.P. Vance, Genome-Wide Identification of Nodule-Specific Transcripts in the Model Legume *Medicago truncatula*, *Plant Physiol. 130*:519–537 (2002).
34. Mergaert, P., K. Nikovics, Z. Kelemen, N. Maunoury, D. Vaubert, A. Kondorosi, and E. Kondorosi, A Novel Family in *Medicago truncatula* Consisting of More Than 300 Nodule-Specific Genes Coding for Small, Secreted Polypeptides with Conserved Cysteine Motifs, *Plant Physiol. 132*:161–173 (2003).
35. Journet, E.-P., D. van Tuinen, J. Gouzy, H. Crespeau, V. Carreau, M.J. Farmer, A. Niebel, T. Schiex, O. Jaillon, O. Chatagnier, L. Godiard, F. Micheli, D. Kahn, V. Gianinazzi-Pearson, and P. Gamas, Exploring Root Symbiotic Programs in the Model Legume *Medicago truncatula* Using EST Analysis, *Nucleic Acids Res. 30*:5579–5592 (2002).
36. Brechenmacher, L., S. Weidmann, D. van Tuinen, O. Chatagnier, S. Gianinazzi, P. Franken, and V. Gianinazzi-Pearson, Expression Profiling of Up-Regulated Plant and Fungal Genes in Early and Late Stages of *Medicago truncatula-Glomus mosseae* Interactions, in *Mycorrhiza*, Springer-Verlag, 2003.
37. Wulf, A., K. Manthey, J. Doll, A.M. Perlick, B. Linke, T. Bekel, F. Meyer, P. Franken, H. Küster, and F. Krajinkski, Transcriptional Changes in Response to Arbuscular Mycorrhiza Development in the Model Plant *Medicago truncatula*, *Mol. Plant-Microbe Interact. 16*:306–314 (2003).
38. Liu, J., L.A. Blaylock, G. Endre, J. Cho, C.D. Town, K.A. VandenBosch, and M.J. Harrison, Transcript Profiling Coupled with Spatial Expression Analyses Reveals Genes Involved in Distinct Developmental Stages of an Arbuscular Mycorrhizal Symbiosis, *Plant Cell 15*:2106–2123 (2003).
39. Colebatch, G., B. Trevaskis, and M. Udvardi, Functional Genomics: Tools of the Trade, *New Phytol. 153*:27–36 (2002).
40. Mathesius, U., G. Keijzers, S.H.A. Natera, J.J. Weinman, M.A. Djordjevic, and B.G. Rolfe, Establishment of a Root Proteome Reference Map for the Model Legume *Medicago truncatula* Using the Expressed Sequence Tag Database for Peptide Mass Fingerprinting, *Proteomics 1*:1424–1440 (2001).
41. Mathesius, U., S. Mulders, M. Gao, M. Teplitski, G. Caetano-Anollés, B.G. Rolfe, and W.D. Bauer, Extensive and Specific Responses of a Eukaryote to Bacterial Quorum-Sensing Signals, *Proc. Natl. Acad. Sci. USA 100*:1444–1449 (2003).
42. Bestel-Corre, G., E. Dumas-Gaudot, V. Poinsot, M. Dieu, J.-F. Dierick, D. van Tuinen, J. Remacle, V. Gianinazzi-Pearson, and S. Gianinazzi, Proteome Analysis and Identification of Symbiosis-Related Proteins from *Medicago truncatula* Gaertn. by Two-Dimensional Electrophoresis and Mass Spectrometry, *Electrophoresis 23*:122–137 (2002).
43. Watson, B.S., V.S. Asirvatham, L. Wang, and L.W. Sumner, Mapping the Proteome of Barrel Medic (*Medicago truncatula*), *Plant Physiol. 131*:1104–1123 (2003).
44. Gallardo, K., C. Le Signor, J. Vandekerckhove, R.D. Thompson, and J. Burstin, Proteomics of *Medicago truncatula* Seed Development Establishes the Time Frame of Diverse Metabolic Processes Related to Reserve Accumulation, *Plant Physiol. 133*:664–682 (2003).
45. Bechtold, N., and G. Pelletier, *In planta Agrobacterium*-Mediated Transformation of Adult *Arabidopsis thaliana* Plants by Vacuum Infiltration, *Meth. Mol. Biol. 82*:259–266 (1998).
46. Meissner, R., V. Chague, Q. Zhu, E. Eyal, Y. Elkind, and A.A. Levy, A High Throughput System for Transposon Tagging and Promoter Trapping in Tomato, *Plant J. 13*:707–716 (2000).

47. Walbot, V., Genes, Genomes, Genomics. What Can Plant Biologists Expect from the 1998 National Science Foundation Plant Genome Research Program?, *Plant Physiol. 119*:1151–1155 (1999).
48. Thomas, M.R., R.J. Rose, and K.E. Nolan, Genetic Transformation of *Medicago truncatula* Using Agrobacterium with Genetically Modified Ri and Disarmed Ti Plasmids, *Plant Cell Rep. 11*:113–117 (1992).
49. Wang, J.H., R.J. Rose, and B.I. Donaldson, *Agrobacterium*-Mediated Transformation and Expression of Foreign Genes in *Medicago truncatula*, *Aust. J. Plant Physiol. 23*:265–270 (1996).
50. Chabaud, M., C. Larsonneau, C. Marmouget, and T. Huguet, Transformation of Barrel Medic (*Medicago truncatula* Gaertn.) by *Agrobacterium tumefaciens* and Regeneration via Somatic Embryogenesis of Transgenic Plants with the *MtENOD*12 Nodulin Promoter Fused to the *gus* Reporter Gene, *Plant Cell Rep. 15*:305–310 (1996).
51. Hoffman, B., T.H. Trinh, J. Leung, A. Kondorosi, and E. Kondorosi, A New *Medicago truncatula* Line with Superior *in vitro* Regeneration, Transformation, and Symbiotic Properties Isolated through Cell Culture Selection, *Mol. Plant-Microbe Interact. 10*:310–315 (1997).
52. Trinh, T.H., P. Ratet, E. Kondorosi, P. Durand, K. Kamaté, P. Bauer, and A. Kondorosi, Rapid and Efficient Transformation of Diploid *Medicago truncatula* and *Medicago sativa* spp. *falcata* Lines Improved in Somatic Embryogenesis, *Plant Cell Rep. 17*:345–355 (1998).
53. Trieu, A.T., S.H. Burleigh, I.V. Kardailsky, I.E. Maldonado-Mendoza, W.K. Versaw, L.A. Blaylock, H. Shin, T.-J. Chiou, H. Katagi, G.R. Dewbre, D. Weigel, and M.J. Harrison, Transformation of *Medicago truncatula* via Infiltration of Seedlings or Flowering Plants with *Agrobacterium*, *Plant J. 22*:531–541 (2000).
54. Scholte, M., I. d'Erfurth, S. Rippa, S. Monday, V. Cosson, P. Durand, C. Breda, H. Trinh, I. Rodriguez-Llorente, E. Kondorosi, M. Schultze, A. Kondorosi, and P. Ratet, T-DNA Tagging in the Model Legume *Medicago truncatula* Allows Efficient Gene Discovery, *Mol. Breed. 10*:203–215 (2002).
55. Jensen, J.S., K.A. Marcker, L. Otten, and J. Schell, Nodule-Specific Expression of a Chimeric Soybean Leghemoglobin Gene in Transgenic *Lotus corniculatus*, *Nature 321*:669–674 (1986).
56. Boisson-Dernier, A., M. Chabaud, F. Garcia, G. Bécard, C. Rosenberg, and D.G. Barker, *Agrobacterium rhizogenes*-Transformed Roots of *Medicago truncatula* for the Study of Nitrogen-Fixing and Endomycorrhizal Symbiotic Associations, *Mol. Plant-Microbe Interact. 14*:695–700 (2001).
57. Kumagai, H., and H. Kouchi, Gene Silencing by Expression of Hairpin RNA in *Lotus japonicus* Roots and Root Nodules, *Mol. Plant-Microbe Interact. 16*:663–668 (2003).
58. McCallum, C.M., L. Comai, E.A. Greene, and S. Henikoff, Targeting Induced Local Lesions in Genomes (TILLING) for Plant Functional Genomics, *Plant Physiol. 123*:439–442 (2000).

Chapter 9

Comparative Genomics of *Glycine max, Medicago truncatula,* Other Legumes, and *Arabidopsis thaliana*

Nevin Dale Young

Departments of Plant Pathology and Plant Biology, 495 Borlaug Hall, University of Minnesota, St. Paul, Minnesota 55108

Introduction

The legume family (Fabaceae) includes soybean, *Glycine max* L. Merr., and many other important crop species, such as bean, *Phaseolus vulgaris* L.; pea, *Pisum sativum* L.; alfalfa, *Medicago sativa* L.; and peanut, *Arachis hypogaea* L. Legumes are the third largest family of higher plants, second only to the Gramineae (grasses) in agricultural importance. Legumes range all the way from minute herbaceous plants to towering trees that dominate many tropical rainforests. Legumes form symbiotic relationships with Rhizobia bacteria (1), together producing biological nitrogen of immeasurable value to world agriculture and essential in global ecology. Legumes are a major source of organic fertilizer, especially in the developing world, and are an essential component of nearly all agricultural rotation schemes. Legumes also form symbiotic relationships with beneficial mycorrhizal fungi (as do many other plant families), though studies increasingly point to a common symbiotic pathway for both Rhizobium and mycorrhizae (2). This makes legumes an ideal system for studying the biology of symbiosis. In part because of their novel biological nitrogen metabolism, legumes are remarkably high in protein content, providing the largest single source of vegetable protein in human and livestock diets. Legumes also produce health-promoting secondary compounds, including chemicals effective in protecting against human cancers (3). Not surprisingly, legumes are one of the best-studied plant families in the world.

Legumes are typically divided into three subfamilies: Mimosoideae, Caesalpinoideae, and Papilionoideae (4). Most cultivated legumes, including soybean, are found within the Papilionoideae, the largest subfamily. Papilionoids are in turn divided into the tropical, or "phaseoloid," legumes (including the genera *Phaseolus*, *Vigna*, and *Cajanus*, along with soybean) and the temperate, or "galegoid," legumes (including the genera *Melilotus*, *Trifolium*, *Medicago*, *Pisum*, *Vicia*, *Lotus*, *Cicer*, and *Lens*). The prominent legume models, *Medicago truncatula* Gaertner and *Lotus japonicus* (Regel) K. Larsen, are both found within the temperate legumes. Molecular phylogenies based on *rbcL* sequence data (5,6) have recently clarified relationships among legume lineages. It is clear, for example, that the Caesalpinoideae are most basal and are thus expected to contain much of the genetic and evolutionary diversity present in the entire legume family. By contrast, cultivated legumes (including soybean) are within taxa that have diverged relatively recently. For this reason, the cultivated legume genera most often studied may not be good representatives of legume diversity generally, or of even the Papilionoideae in particular.

This review focuses on recent comparative genomics research demonstrating substantial macro- and microsynteny among legume genomes, and even with the model plant, *Arabidopsis thaliana* (L.) Heyn. (7–13). Genomic research is also adding to our knowledge about the duplicated nature of legume genomes, including soybean. Moreover, recent research provides strong evidence for nonhomogeneous genome organization in legumes involving gene-rich islands (9,10,13). These observations are important in the area of comparative genomics because efforts to analyze similarities among legume genomes and legume genome evolution require an appreciation of their duplicated and nonhomogeneous genome organization.

Genome Duplications in Legumes

Recent studies of the model dicot, Arabidopsis, highlight the critical role of whole-genome and large-scale segmental duplications in the evolution of higher plants (14,15). Indeed, large-scale genome duplications are clearly a common feature in eukaryotic genome evolution, with evidence for duplications in organisms all the way from corn (*Zea mays* L.) (16) to human (17). From a functional point of view, large-scale genome duplications create the potential for new diversity and the possibility of so-called subfunctionalization (18).

It was in soybean, however, where some of the first, and still the most convincing, data demonstrating wide-scale genome duplication were generated (19). In this study, Shoemaker and colleagues utilized multicopy (but still low copy) restriction fragment length polymorphism (RFLP) markers to infer the existence of syntenic and/or collinear runs of markers on two or more linkage groups. (Synteny refers to cases where two or more markers or genes are linked together in two different genomes or genome regions. Collinearity refers to cases where three or more markers or genes are linked and also have the same order in two different genomes or genome regions.) The RFLP results demonstrated that the soybean genome consists of many duplicated segments, with some present as many as six times. Frequent cases of nested duplications were observed, suggesting large-scale duplications occurred within regions that had themselves experienced earlier, more ancient duplications. Altogether these results indicated that soybean has a complex genome with a paleopolyploid evolutionary past.

Among duplicated regions showing synteny and collinearity, levels of DNA sequence similarity have recently been estimated. In a study of 10 bacterial artificial chromosome (BAC) contigs spanning one million base pairs (Mb) in a 10-centimorgan (cM) segment of soybean, Foster-Hartnett *et al.* (10) observed that duplicated BAC contigs from unlinked chromosomes exhibited remarkably similar restriction fragment fingerprint patterns. Homoeologous (duplicated) BACs also demonstrated 98% sequence identity in several short segments that were examined in detail. A similar study that focused on three groups of homoeologous soybean BAC contigs uncovered similar levels of sequence identity (11). Surprisingly, noncoding sequences in this study showed levels of similarity and microsynteny nearly as high as genic regions. In a genome-wide survey of intragenomic microsynteny in soybean, Yan *et al.* (12) determined that 67% of homoeologous sets of BAC contigs had at least one BAC contig pair with extensive microsynteny, with more than 90% of homoeologous BAC contig groups showing at least some microsynteny. From an evolutionary perspective, these results indicate at least one recent large-scale duplication event in soybean. From a practical perspective, the results suggest that genome-wide physical mapping, especially mapping based on restriction fragment fingerprinting, and shotgun genome sequencing are likely be very difficult in soybean.

Gene Clustering in Legume Genomes

The distribution of repetitive versus low copy sequences provides important insights into global genome organization—and by extension—the feasibility of targeting gene-rich regions for cost-effective genome sequencing. This is especially relevant in soybean, where there is great interest in genome sequencing but concern about the cost involved in complete sequencing of the entire genome. With this in mind, researchers have recently attempted to estimate the distribution of genes and global gene density in soybean and other legumes. In some cases, this has involved a combination of comparative physical and genetic mapping, and in others, targeted sequencing of genome regions up to 300 kilobase pairs (kb) in length.

In *Medicago*, for example, Kulikova *et al.* (13) compared the genetic linkage map with results of fluorescent *in situ* hybridization (FISH) of pachytene chromosomes using 20 previously mapped gene-rich BACs. Pachytene chromosomes are nearly 20 times longer than equivalent metaphase chromosomes typically used in FISH, and therefore provide impressive levels of optical resolution. The results demonstrated that the *Medicago* genome is organized into distinct regions with pericentromeric heterochromatin, rich in repeated sequences, and separate gene-rich euchromatic regions. Subsequent BAC sequencing suggests that *Medicago* gene density is approximately one gene every 6 to 10 kb (B. Roe, University of Oklahoma, personal communication), on par with gene density in *Arabidopsis*. Together these lines of evidence suggest that just 20–40% of the *Medicago* genome contains the majority of its genes. Although these estimates are too imprecise to predict the actual number of *Medicago* genes, they suggest that the total will be comparable to the 25,000 to 30,000 previously reported for Arabidopsis (14). Gene density in *L. japonicus* has also been estimated at 1 gene/9 kb (20). Thus, gene density in these two model legumes is clearly quite high in gene-rich regions.

In soybean, gene density has been examined using hypomethylated (and presumably genic) RFLPs as signatures of gene-containing regions (9). Of more than 2,000 BAC end sequences examined, RFLP-anchored BACs had 50% more genic sequences than simple sequence repeat (SSR)-anchored BACs. Moreover, RFLP-anchored BACs had just half as many repetitive sequences compared to SSR-anchored BACs. Based on one region of completely sequenced soybean genome more than 330 kb in length near the *rhg1* cyst nematode resistance gene, Young *et al.* (21) estimated gene density to be approximately 1 gene/8 kb. In summary, these estimates suggest a highly uneven gene density in soybean, with some regions containing higher densities than expected if genes were evenly distributed throughout the genomes.

Comparative Genomics of Soybean and Other Legumes

Early isozyme and RFLP-based genetic mapping in legumes were among the first to demonstrate macrosynteny (synteny at the level of entire linkage groups) among members of the same taxonomic family. For example, Weeden *et al.* (22) compared the molecular marker maps of lentil (*Lens*) and pea (*Pisum*) and found at least eight conserved genome regions constituting 40% of the lentil genome. Menancio-Hautea *et al.* (23) compared the genomes of mung bean (*Vigna radiata* (L.) Wilczek) and cowpea (*V. unguiculata* (L.) Walpers) and found multiple cases of linkage groups conserved between the species, as well as several smaller segments of collinearity.

Because of their great economic importance, a detailed comparison of *Vigna*, *Phaseolus*, and *Glycine* was carried out in (24). In this study, it was clear that the *P. vulgaris* RFLP map could be superimposed almost in its entirety on the *V. radiata* map (and vice versa). The average length of collinearity extended 37 cM, and in some cases more than 100 cM. By contrast, conserved segments between either *V. radiata* or *P. vulgaris* and soybean were generally much shorter, averaging just 12 to 13 cM. Widespread genome duplication and relatively low levels of DNA polymorphism frequently complicated efforts to map cross-species markers in soybean. For similar reasons, comparisons of the *Medicago* and soybean genomes have been difficult. However, syntenic relationships were examined through *Medicago* BAC-based markers with homology to soybean RFLPs. By this method, 60 loci could be mapped in common between the two genomes. Nonetheless, only eight relatively modest macrosyntenic blocks showing putative collinearity between *Medicago* and soybean were discovered (D. Cook, UC–Davis, personal communication).

Recently, levels of genome conservation between soybean and *Vigna*/*Phaseolus* have been reevaluated and found to be higher than originally believed (8). As an initial step, the authors first compared soybean and Arabidopsis. Surprisingly, cases of substantial macrosynteny between the two species were observed, though it is likely that macrosynteny between legumes and Arabidopsis is probably the exception, not the rule (12). Still, Grant *et al.* (7) were able to use comparisons with Arabidopsis to illuminate comparisons of soybean with *V. radiata* and *P. vulgaris* (8). To uncover higher levels of macrosynteny, it was necessary to consider the duplicated structure of the soybean genome and also to use Arabidopsis as a "bridging species." In effect, homoeologous regions of the soybean genome were first compared among themselves, and then with corresponding regions in both *P. vulgaris* and *V. radiata*. In this way, a network of macrosynteny among the legume taxa could be created and then compared to Arabidopsis. Yan *et al.* (11) used a similar strategy to characterize the underlying features of microsynteny between soybean, *Medicago*, and Arabidopsis (Fig. 9.1).

Several other examples of microsynteny between legumes and Arabidopsis have now been reported (Fig. 9.1). The first involved the 10-cM segment of soybean examined by Foster-Hartnett *et al.* (10) described earlier. Here, extensive microsynteny with Arabidopsis extending over the entire region was observed. However, genome segments in Arabidopsis showing microsynteny with soybean were multicopy, with only some of the conserved sequences found in any one of the Arabidopsis duplications. These results suggest that regions of inter-family microsynteny can be quite extensive—in this case, several centimorgans in soybean that correspond to regions nearly 2 Mb in length in Arabidopsis. Other examples of microsynteny between legumes and Arabidopsis have also been reported (9,25). Nevertheless, quantitative comparisons between soybean and Arabidopsis indicate that cases of conserved microsynteny may be relatively infrequent. In comparisons between more than 1,000 soybean BAC end sequences previously placed into mapped contigs and the published genome sequence of Arabidopsis, microsynteny between soybean and Arabidopsis was observed only 14% of the time (12).

Within legumes, microsynteny appears to be widespread. Based on the same 1,000 anchored soybean BAC ends just described, Yan *et al.* (12) found that more than half of all soybean BAC contig groups exhibit microsynteny with *Medicago*. Of these, more than 80% showed extensive microsynteny. To gain further insight into genomic relationships between soybean and *Medicago*, eight groups of BAC contigs, together spanning 2.6 Mb in soybean and 1.5 Mb in *Medicago*, were compared through high-resolution physical mapping and comparative sequencing (11). Cross-hybridization between soybean and *Medicago* contigs uncovered microsynteny in six of the contig groups and extensive microsynteny in three. (One example is shown in Fig. 9.1.) Sequence order was highly conserved between soybean duplicates and generally conserved between soybean and *Medicago*. These results also confirmed previous predictions of one recent genome-wide duplication in soybean and suggested that *Medicago* may have also experienced at least one ancient large-scale genome duplication.

In terms of gene discovery, microsynteny within legumes may be of important practical significance. Microsynteny was essential in the cloning of *NN1* from *M. sativa* and *SYM2* from *L. japonicus*, along with their orthologues *DMI2* in *M. truncatula* and *Sym19* in *P. sativum*, respectively (25,26). The gene sequences suggest the corresponding proteins are probably extracellular receptor kinases with roles in Nod-factor perception. Along with *Nin* (27), these interesting genes represent the first of what are likely to be many genes cloned through the use of model legumes. Given the high degree of microsynteny already discovered between soybean and *Medicago* (and Arabidopsis), this so-called genomic cross-referencing strategy may provide a valu-

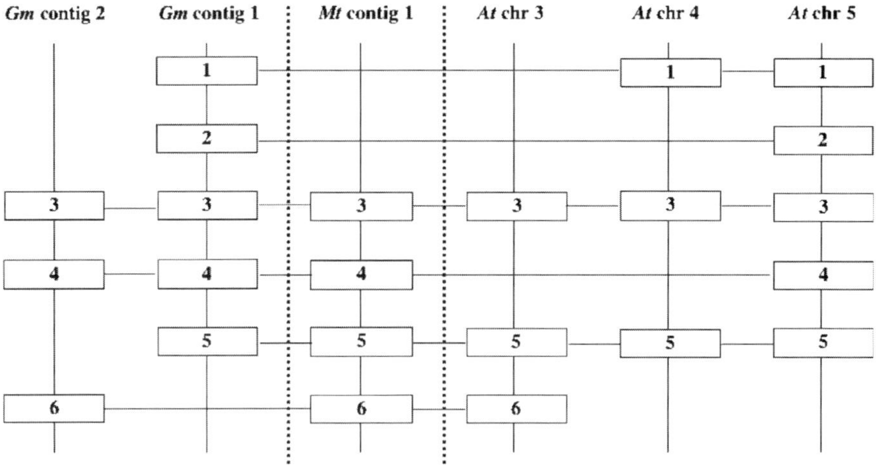

Figure 9.1. Extensive microsynteny among soybean, *Medicago*, and *Arabidopsis*. Graphical display of conserved sequences between two duplicated regions in soybean, a homologous region in *Medicago*, and three homologous regions in *Arabidopsis*. A soybean BAC contig, approximately 180 kb in size, was originally identified using RFLP K685 located on linkage group B2. Duplicate BAC contigs within soybean and homologous BAC contigs in *Medicago* were identified through cross-hybridization and restriction mapping. Homologous regions in *Arabidopsis* were identified through informatic comparison to the completed *Arabidopsis* genome sequence. Of 18 sequences analyzed and oriented on the two soybean contigs, 13 were conserved between the soybean duplicates, 9 were conserved in *Medicago*, and 6 were conserved in one of three genome regions of *Arabidopsis* (shown here). Adapted from (11). Legend: Gm, *Glycine max*; At, *Arabidopsis thaliana*; Mt, *Medicago truncatula*; chr, chromosome.

able method for discovering genes of interest in legumes by taking advantage of the rapidly increasing genome sequence resources available in model systems.

Summary

A growing body of data indicates extensive genome conservation among legumes, including soybean and other economically important crop species. Still, due to extensive and apparently recent genome duplication within soybean, cases of synteny and collinearity in this species tend to be difficult to discover. Moreover, genome-wide fingerprinting and shotgun sequencing will probably be difficult in soybean because of its highly duplicated genome organization. It is encouraging, therefore, that data point to the existence of soybean genome regions highly enriched in genes. The model systems *Medicago* and Arabidopsis are also beginning to provide bridges that help to connect soybean to other legumes. Detailed comparisons between soybean and model genomes are starting to reveal the details of microsynteny and microcollinearity, and this is likely to be helpful in the areas of gene discovery and positional gene cloning.

Acknowledgments

Yan Huihuang generously contributed to the research and design of Fig. 9.1, while Joann Mudge and Steve Cannon provided advice and data from ongoing research valuable in the preparation of this chapter. This paper is published as part of a series from the Minnesota Agricultural Experiment Station.

References

1. Schultze, M., and A. Kondorosi, Regulation of Symbiotic Root Nodule Development, *Ann. Rev. Genet. 32*:33–57 (1998).
2. Vanrhijn, P., Y. Fang, S. Galili, O. Shaul, N. Atzmon, S. Wininger, Y.Eshed, M. Lum, Y. Li, V. To, N. Fujishige, Y. Kapulnik, and A.M. Hirsch, Expression of Early Nodulin Genes in Alfalfa Mycorrhizae Indicates that Signal Transduction Pathways Used in Forming Arbuscular Mycorrhizae and Rhizobium-Induced Nodules May Be Conserved, *Proc. Natl. Acad. Sci. USA 94*:5467–5472 (1997).
3. Stark, A., and Z. Madar, Phytoestrogens: A Review of Recent Findings, *J. Ped. Endocrin. Metab. 15*:561–572 (2002).
4. Doyle, J.J., Leguminosae, in *Encyclopedia of Genetics*, edited by S. Brenner and J.H. Miller, Academic Press, San Diego, 2001, pp.1081–1085.
5. Doyle, J.J., J.L. Doyle, J.A. Ballenger, E.E. Dickson, T. Kajita, and H. Ohashi, A Phylogeny of the Chloroplast Gene *rbcL* in the Leguminosae—Taxonomic Correlations and Insights into the Evolution of Modulation, *Am. J. Bot. 84*:541–554 (1997).
6. Kajita, T., H. Ohashi, Y. Tateishi, C.D. Bailey, and J.J. Doyle, *RbcL* and Legume Phylogeny, with Particular Reference to Phaseoleae, Millettieae, and Allies, *Syst. Bot. 26*:515–536 (2001).
7. Grant, D., P. Cregan, and R.C. Shoemaker, Genome Organization in Dicots: Genome Duplication in *Arabidopsis* and Synteny between Soybean and *Arabidopsis*, *Proc. Natl. Acad. Sci. USA 97*:4168–4173 (2000).
8. Lee, J.M., D. Grant, C.E. Vallejos, and R.C. Shoemaker, Genome Organization in Dicots. II. Arabidopsis as a 'Bridging Species' to Resolve Genome Evolution Events among Legumes, *Theor. Appl. Genet. 103*:765–773 (2001).
9. Marek, L.F., J. Mudge, L. Darnielle, D. Grant, N. Hanson, M. Paz, Y. Huihuang, R. Denny, K. Larson, D. Foster-Hartnett, A. Cooper, D. Danesh, D. Larsen, T. Schmidt, R. Staggs, J.A. Crow, E. Retzel, N.D. Young, and R.C. Shoemaker, Soybean Genomic Survey: BAC-End Sequences Near RFLP and SSR Markers, *Genome 44*:472–581 (2001).
10. Foster-Hartnett, D., J. Mudge, D. Danesh, H. Yan, D. Larsen, R. Denny, and N.D. Young, Comparative Genomic Analysis of Sequences Sampled from a Small Region on Soybean Molecular Linkage Group 'G,' *Genome 45*:634–645 (2002).
11. Yan, H., J. Mudge, D.J. Kim, R.C. Shoemaker, D.R. Cook, and N.D. Young, Comparative Physical Mapping Reveals Features of Microsynteny Between *Glycine max*, *Medicago truncatula*, and *Arabidopsis thaliana*, *Genome* (*in press*) (2004).

12. Yan, H., J. Mudge, D.J. Kim, R.C. Shoemaker, D.R. Cook, and N.D. Young, Estimates of Conserved Microsynteny among the Genomes of *Glycine max*, *Medicago truncatula* and *Arabidopsis thaliana*, *Theor. Appl. Genet. 106*:1256–1265 (2003).
13. Kulikova, O., G. Gualtieri, R. Geurts, D.J. Kim, D. Cook, T. Huguet, J.H. de Jong, P.F. Fransz, and T. Bisseling, Integration of the FISH Pachytene and Genetic Maps of *Medicago truncatula*, *Plant J. 27*:49–58 (2001).
14. Arabidopsis Genome Initiative, Analysis of the Genome Sequence of the Flowering Plant *Arabidopsis thaliana*, *Nature 408*:796–815 (2000).
15. Vision, T.J., D.G. Brown, and S.D. Tanksley, The Origins of Genomic Duplications in Arabidopsis, *Science 290*:2114–2117 (2000).
16. Gaut, B.S., Patterns of Chromosomal Duplication in Maize and Their Implications for Comparative Maps of the Grasses, *Genome Res. 11*:55–66 (2001).
17. Li, W.H., Z. Gu, H. Wang, and A. Nekrutenko, Evolutionary Analyses of the Human Genome, *Nature 409*:847–849 (2001).
18. Lynch, M., and J.S. Conery, The Evolutionary Fate and Consequences of Duplicate Genes, *Science 290*:1151–1155 (2000).
19. Shoemaker, R.C., K. Polzin, J. Labate, J. Specht, E.C. Brummer, T. Olson, N.D. Young, V. Concibido, J. Wilcox, J.P. Tamulonis, G. Kochert, and H.R. Boerma, Genome Duplication in Soybean (*Glycine* subgenus *soja*), *Genetics 144*:329–338 (1996).
20. Nakamura, Y., T. Kaneki, E. Axamizu, T. Kato, S. Sato, and S. Tabata, Structural Analysis of a *Lotus japonicus* Genome. II. Sequence Features and Mapping of Sixty-Five TAC Clones Which Cover the 6.5 Mb Regions of the Genome, *DNA Res. 9*:63–70 (2002).
21. Young, N.D., J. Mudge, and T.H.N. Ellis, Legumes: More than Peas in a Pod, *Curr. Opin. Plant Biol. 6*:199–204 (2003).
22. Weeden, N.F., F.J. Muehlbauer, and G. Ladizinsky, Extensive Conservation of Linkage Relationships Between Pea and Lentil Genetic Maps, *J. Hered. 83*:123–129 (1992).
23. Menancio-Hautea, D., C.A. Fatokun, L. Kumar, D. Danesh, and N.D. Young, Comparative Genome Analysis of Mungbean (*Vigna radiata* (L.) Wilczek) and Cowpea (*V. unguiculata* (L.) Walpers) Using RFLP Mapping Data, *Theor. Appl. Genet. 86*:797–810 (1993).
24. Boutin, S.R., N.D. Young, T.C. Olsen, Z.H. Yu, R.C. Shoemaker, and C.E. Vallejos, Genome Conservation among Three Legume Genera Detected with DNA Markers, *Genome 38*:928–937 (1995).
25. Endre, G., A. Kereszt, Z. Kevei, S. Mihaçea, P. Kalo, and G.B. Kiss, A Receptor Kinase Gene Regulating Symbiotic Nodule Development, *Nature 417*:962–966 (2002).
26. Stracke, S., C. Kistner, S.Yoshida, L. Mulder, S. Sato, T. Kaneko, S. Tabata, N. Sandal, N.K. Stougaard, K. Szczyglowski, and M. Parniske, A Plant Receptor-Like Kinase Required for Both Bacterial and Fungal Symbiosis, *Nature 417*:959–962 (2002).
27. Schauser, L., A. Roussis, J. Stiller, and J. Stougaard, A Plant Regulator Controlling Development of Symbiotic Root Nodules, *Nature 402*:191–195 (1999).

Chapter 10

EST Projects for the Study of Genome Evolution in Legumes

J. Schlueter,[a] P. Dixon,[b] and R. Shoemaker[c]

[a]Department of Genetics, Development, and Cell Biology, Iowa State University, Ames, IA 50011;

[b]Department of Statistics, Iowa State University, Ames, IA 50011; and

[c]USDA-ARS, Iowa State University, Ames, IA 50011

Introduction

Gene and genome duplication have long been accepted as a driving force in the evolution and expansion of eukaryotic genomes (1,2). These phenomena are seen across numerous eukaryotic species and are highly prevalent within the plant kingdom. Polyploids are thought to have a greater genetic advantage over diploids and may be more resilient to environmental changes (3). Gene or genome duplication also acts as a mechanism for the creation of genetic diversity, genome expansion, gene silencing, and enhancement of new gene functions (4). Masterson (5) estimates that up to 80% of all angiosperms are of polyploid origin.

What often occurs after a polyploidy event is a divergence of duplicated regions at both the sequence and chromosomal levels (4). This process is referred to as diploidization (6), a concept originally envisioned by Clausen (7) in 1941. Diploidization allows for seeming diploid species to undergo multiple rounds of gene and/or genome duplication.

Analyses of duplicated genes allow us to gain an understanding of the processes of genome evolution and the evolutionary histories of various plant species. Previous studies have asked questions about the evolutionary history of plant species in the absence of whole genome sequence (8). A limitation to these analyses, however, has been the lack of large datasets and an underlying commonality across numerous plant species. Although whole genome sequences would be the ideal, current genome efforts have provided a wellspring of information in the expressed sequence tag (EST) datasets for a wide range of plant species (9–15). The sequences generated by these projects represent a largely untapped resource for dissection of gene families, construction of phylogenetic trees, and studies of evolutionary histories of genomes.

The purpose of this study was to utilize these large sequence sets to identify and further study duplicated genes and the evolutionary events that created them in four grasses, two solanaceae, and two legumes. Because we can determine the evolutionary distances between pairs of duplicated genes, we can estimate when these genome, or polyploidy, events may have occurred as well as what events are shared between species. We can gain insight as to the forces acting on these genomes over time and what the EST expression profiles of these duplicated pairs show us.

Materials and Methods

Identification of Potential Duplicates

TIGR tentative contigs (TCs) for *Glycine max* (L.) Merr. (soybean) and *Medicago truncatula* Gaertner (barrel medic) were obtained from the TIGR genome database (www.tigr.org/tdb/agi/). Each TC set was run through the program getorf, part of the EMBOSS bioinformatics package (16). The determined ORFs (open reading frames) are defined as regions of at least 30 bp (base pairs) between all possible stop codons. A Perl script selected the longest of all possible ORFs for each TC. For all species studied, the shortest open reading frame accepted was 120 bp.

A similarity search of each species-specific longest ORF TC set against itself was done using a tBLASTx search algorithm (17). The program MuSeqBox (18) was used to tabularize the BLAST output, and a Perl script parsed all potential paralogs. The criteria used were: (i) Query and subject coverage must be greater than or equal to 80%; (ii) the subject identified must not be the query itself (*a* identifying *a*); (iii) the subject and query were required to be reciprocal (if *a* identified *b*, then *b* must identify *a*); and (iv) only pairs that fit the criteria were considered (i.e., if a triplet fit all the criteria, it was not considered, etc.). Under these criteria, the average accepted E-value was 6×10^{-9}.

Nucleotide differences within a species, as a result of cultivar differences, were minimized by developing genotype-specific contigs. The most prominent genotype was used within each EST collection. For soybean, ESTs only from the cultivar Williams or Williams 82 were used. For *Medicago*, ESTs only from the genotype Jemalong were used. All ESTs derived from the acceptable genotype were retained for each TC, and unacceptable ESTs were removed. If a contig contained all acceptable genotypes, that contig and its subsequent consensus sequence were unchanged. Any contigs from which ESTs were removed were recontigged using CAP3 (19) with a 40-bp overlap and 95% similarity requirement. The consensus sequence for the genotype-specific contig was used in subsequent analysis instead of the consensus sequence obtained from TIGR. At this point, any contig containing fewer than two ESTs and its paralogous contig were removed.

Each genotype-specific consensus sequence was run through getorf, EMBOSS, and the previously mentioned Perl script to obtain the longest putative ORF. The shortest identified reading frame in all datasets was 120 bp. Alignments between identified pairs were performed using virtual translation and the Clustalx method with default parameters (MegAlign DNASTAR, Inc.). Alignments were cropped to the largest sequence overlap, or between start and stop codons. For any pairs that did not align, a BLASTx with the consensus sequence against the NCBI (nonredundant) database was used to determine the correct frame. That ORF was then used for virtual translation in MegAlign, and analysis proceeded as described for the other pairs.

Putative annotation of each gene pair was accomplished by searching against the NCBI nonredundant database using the tBLASTx algorithm and an E cutoff value of 1×10^{-10}. The top five queries identified were then parsed from the BLAST output and classified into the appropriate first level MIPS (Munich Information Center for Protein Sequences) categories (20). The additional category of nodulation/nitrogen fixation was added for legumes.

Evolutionary Distances and Mixtures of Normal Distributions

Synonymous (Ks) and nonsynonymous (Ka) distances as well as their respective standard deviations (sd) were calculated using the method of Goldman and Yang (21) as implemented in PAML (22). The sd of each Ks was not constant, larger values generally having larger sd's. Natural log (Ln) transforming the Ks values equalized, on average, the standard deviations. For determination of a mixture of normal distributions, Ks values less than 0.05, as well as any Ks values greater than 1.0, were not considered (22). Ks values greater than 1.0 were considered to reach a saturation point where the distance estimates may become less accurate.

The Ln Ks values were modeled as a mixture of k normal distributions (23). We considered models from $k = 1$ component through $k = 5$ components. Each component was allowed to have a different mean, a different variance, and a different probability of membership in that component. The membership probabilities were constrained to sum to 1. The model for k components has $3k - 1$ parameters.

Parameters were estimated by maximum likelihood, with the constraint that the variance of each component was larger than 0.0001. The log likelihood function has many false optima in which one component represents a single observation. The variance for that component converges to 0 and the log likelihood function converges to infinity. These false optima can be avoided by setting a lower bound on the component variance.

The log likelihood function was numerically maximized using Splus (MathSoft Inc., 1998). The reported maximum depended on the choice of starting values, especially for larger numbers of components. We used 100 different sets of starting values, randomly chosen from the space of reasonable parameter values. Solutions with a component that represented a single point were rejected. The maximum likelihood estimates were obtained from the nonrejected solution with the largest log likelihood. In general, the optimum for $k = 2$ or $k = 3$ components was reached from over 90% of the randomly chosen starting values. The log likelihood surface for $k = 4$ or $k = 5$ components had many local optima with slightly different log likelihood values. The reported optimum was reached from approximately 50% of the randomly chosen starting values.

The number of components was chosen by a series of log likelihood ratio tests, comparing models for k components to models for $k + 1$ components. The distribution of the log likelihood ratio statistic for this test is nonstandard because of the constraints on the component variances. For $k = 1$, 2, or 3 components, the bootstrap under the null hypothesis was used to estimate the distribution (24). The bootstrap distributions were generally consistent with a 5 d.f. chi-square distribution. Bootstrapping models with $k = 4$ and 5 components required excessive amounts of computer time. P-values for these tests were computed using the 5 d.f. chi-square distribution. The number of components chosen by hypothesis testing was generally consistent with those from AIC and BIC model selection criteria.

Genes were classified using the maximum posterior probability rule. For each gene, the probability of membership in each component was computed from the parameter estimates and the Ln Ks value. The classification of each gene was the component with the largest posterior probability.

Pearson Correlation

Clustering of libraries and contigs was done as described by Shoemaker *et al.* (14) and Ewing *et al.* (10). Libraries that were nonrandomly sampled or that contained a mixture of various tissues were removed from the dataset (14). All of the duplicated pairs were used in the analysis, as well as all of the available libraries. The original ESTs that composed the TIGR TCs were used to determine counts of ESTs for each duplicated contig. Data were normalized using percentage counts instead of raw EST data (14). Pearson correlation coefficients were calculated between each duplicated pair. All Pearson calculations were done using Microsoft Excel (Microsoft Corporation).

Results

Synonymous and nonsynonymous distances between potential duplicated genes (contigs) were obtained for both legume genera. Synonymous distances ranged from 0.680 ± 0.011 to 258.6 ± 1.65 for soybean and 0.690 ± 0.007 to 288.2 ± 1.70 for *M. truncatula*. Two hundred sixty-eight paralogs were identified for soybean and 293 were identified for *M. truncatula*. The average pairwise alignment length was 474 bp and 618 bp for soybean and *M. truncatula*, respectively. General characteristics of the identified duplicated genes are reported in Table 10.1.

We hypothesize that genes with similar synonymous distances cluster, and that these groupings represent duplication events that can be represented as a normal distribution. Multiple events are a mixture of normal distributions. Mixtures of normal distributions were determined for each species studied, and the most statistically significant number of distributions, or components, was determined with a statistical significance of $P < 0.05$.

Using this strategy, we found that across all studied species, there is evidence for more than one normal distribution. Soybean and *M. truncatula* both support a bimodal distribution with high statistical significance ($P = 0.005$ and $P = 0.005$, respectively).

TABLE 10.1
Characteristics of Identified Duplicated Genes

	Number of paralogs	Genotype[a]	Percent genotype[b]	Total ESTs[c]	Avg. Len. bp[d]	Avg. Syn. Distance[e]	Avg. Nonsyn. Distance[e]
Soybean	275	William+ Williams82	55%	236,461	473	1.105 (0.994)	0.101 (0.018)
Medicago	308	Jemalong	91%	164,304	619	1.342 (2.99)	0.125 (0.018)

[a]The genotype selected for in the identification of paralogs.
[b]Percent of a particular genotype within entire EST collection at time of analysis.
[c]Total ESTs used in TIGR TC development.
[d]Average alignment length of duplicate gene pair.
[e]Average synonymous or nonsynonymous distance between pairs. Standard error is in parentheses.

Coalescence Estimates

If we assume that silent, synonymous substitutions are selectively neutral and accumulate approximately linearly with respect to time, then we can infer the relative age -distribution of duplicated genes from the distribution of synonymous distances (25). Each normal distribution is assumed to be representative of gene pairs evolving at a similar rate, and it is further assumed that the rate is equal across all paralogs. An estimated synonymous substitution rate of 6.1×10^{-9} substitutions per synonymous site per year (25) was used to determine coalescence estimates.

The two statistically significant normal distributions, or genomic events, in soybean are estimated to have occurred approximately 14.5 and 41.6 million years ago (mya) (Fig. 10.1). A similar number of events is supported in *Medicago* and corresponds to an event at 20.2 and 54.6 mya (Fig. 10.2). If the distributions observed in both soybean and *M. truncatula* are overlaid (Fig. 10.3), events that occurred in concert should have very similar distributions. What we

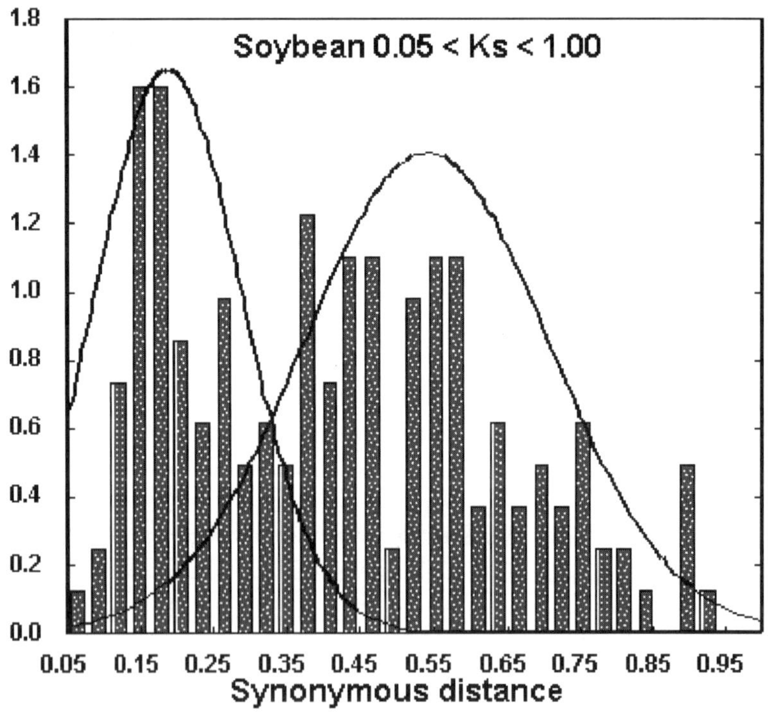

Figure 10.1. Graphical representation of determined normal distributions for soybean. The histograms represent the percent of duplicate gene pairs. The *y*-axis is the density function for normal distributions. Each distribution represents a cluster of genes.

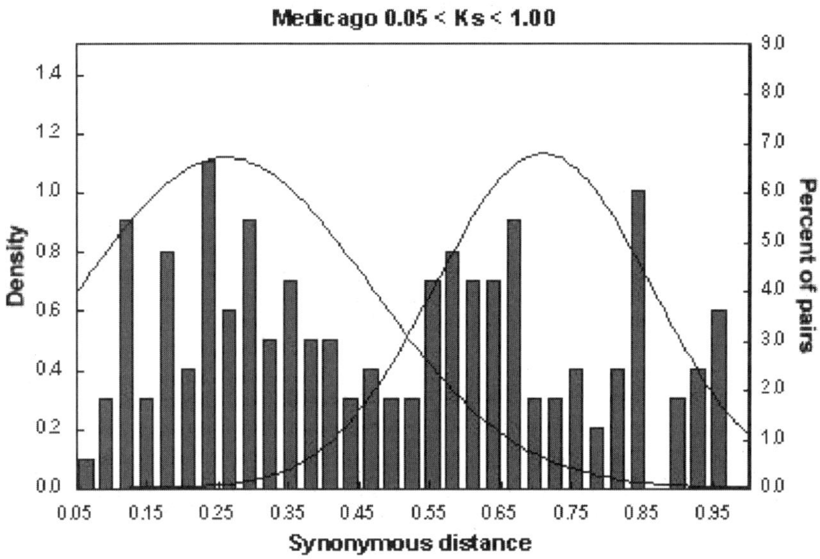

Figure 10.2. Graphical representation of determined normal distributions for *Medicago*. The histograms represent the percent of duplicate gene pairs. The *y*-axis is the density function for normal distributions. Each distribution represents a cluster of

observe is a shift in both distributions relative to the two species. Doyle and Luckow (26) estimated that soybean and *M. truncatula* had already diverged between 50 and 60 mya. Therefore, events predating the divergence of the two genera should coincide. Very probably, the two species had diverged prior to the observed duplication events.

Selection on Duplicated Genes

Utilizing the synonymous and nonsynonymous distance measures, it is possible to determine what type of selection pressure a pair of duplicated genes may be experiencing. If the ratio of nonsynonymous to synonymous substitutions (Kn/Ks) is greater than 1, it suggests that genes are under positive selection and that the gene is experiencing nucleotide changes that affect the resulting amino acid integrity. However, if the Kn/Ks is less than 1, it suggests that some force is acting upon the duplicated genes to preserve their structure (negative selection). Synonymous distances that are approximately equal to nonsynonymous distances are indicative of neutrality. Our results are consistent with those observed for a number of eukaryotic species, including *Homo sapiens* and *Drosophila melanogaster*, in that most all of our duplicated genes are undergoing purifying or negative selection (22,25). These findings are graphically represented in Fig. 10.4.

We observed evidence of possible positive selection. In *M. truncatula*, we found eight genes under positive selection. Five of these genes had no putative annotation,

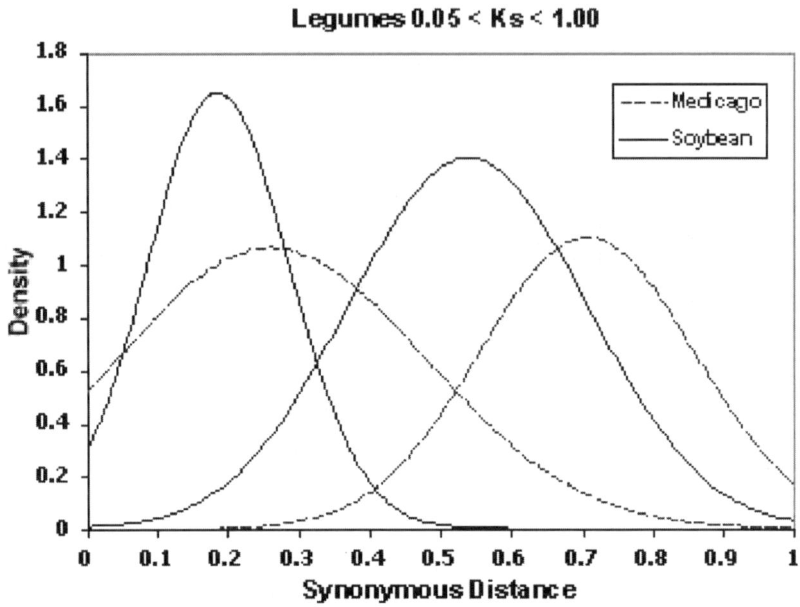

Figure 10.3. Diagrammatic representation of the normal distributions of soybean and *Medicago* overlaid. The *y*-axis is the density function for normal distributions. Each distribution represents a cluster of genes.

but two genes involved in nodulation and one gene involved in lipid transfer showed evidence of positive selection. Similarly, in soybean we also find a nonspecific lipid transfer protein as well as an unknown gene.

Duplicate Gene Expression

Duplication of a gene per se is inherently selectively neutral, as our results detailed previously support (25,27). However, the redundancy created by gene duplication results in a situation in which selective constraints are temporarily lessened on both duplicates (22). During this period, both genes accumulate mutations and both may acquire new functions. Thus, gene duplication is often considered to be a major source of genetic diversity.

To further understand how duplicated gene expression has changed over time, we utilized the EST libraries from which the genes were identified. Pearson correlation coefficients were determined for both soybean and *M. truncatula* for all genes originally identified. Both species have large EST collections from various tissues, and both are legumes, allowing for cross-family comparisons. If a gene pair shows a Pearson correlation coefficient of 1, the genes are showing precisely the same expression profile. A Pearson correlation of 0 equals no correlation at all, and the genes

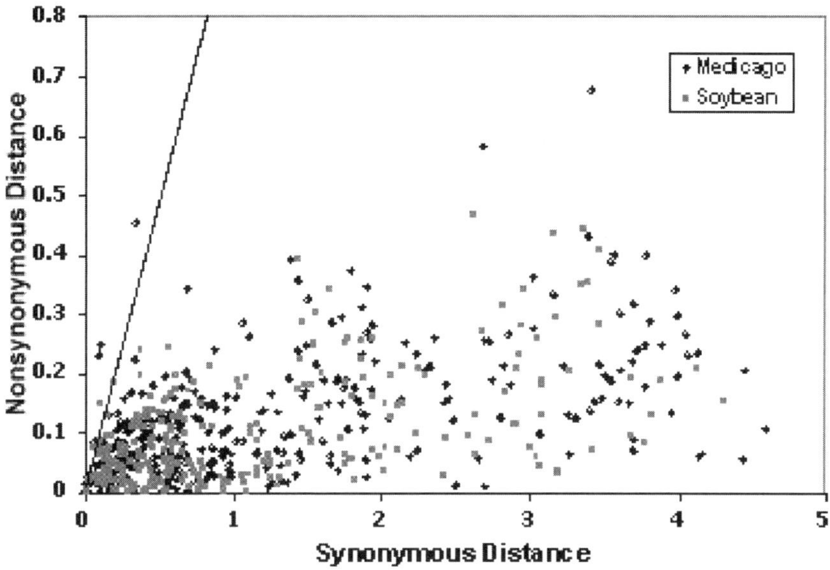

Figure 10.4. Scatterplot of synonymous distances against nonsynonymous distances. The line represents neutral evolution, where synonymous distances approximately equal nonsynonymous distances. All points below the line exhibit negative, purifying selection, where nonsynonymous distance is greater than synonymous distance. All points above the line exhibit positive selection, where synonymous distance is larger than nonsynonymous distance. Both soybean and *Medicago* data points are shown.

are showing basically random expression patterns. However, if a Pearson correlation of –1 is obtained, the genes are showing completely dichotomous expression profiles.

Across both soybean and *M. truncatula*, we expectedly find a wide range of Pearson correlation values. What is striking, however, is that in both soybean and *M. truncatula* we find Pearson correlations of very nearly 1. Those duplicated genes have not diverged remarkably in their expression profiles yet both copies have been retained. No Pearson correlations were found near –1, with –0.423 being the most negative correlation obtained.

Synonymous distances were compared with Pearson coefficients, and scatterplots depict these findings (Figs. 10.5 and 10.6). This comparison indicates that many duplicated genes probably undergo an almost immediate change in expression patterns. If we consider synonymous distance as a measure of time, then in both soybean and *M. truncatula* we see a general trend of correlations moving away from 1 toward 0. This reflects a change in control of gene expression over time, causing duplicated genes to no longer share the same expression profile.

Even more striking is the comparison of nonsynonymous distance measurements and Pearson coefficients (Figs. 10.7 and 10.8). Again, although the

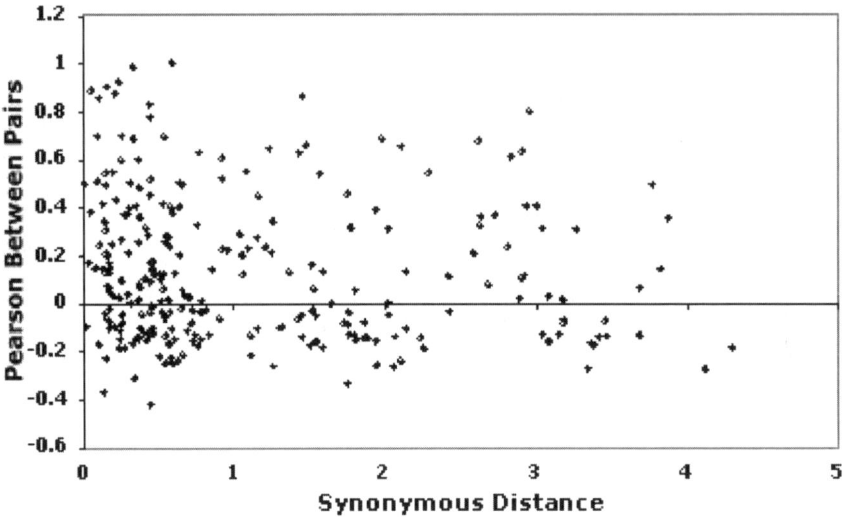

Figure 10.5. Scatterplot of synonymous distances against Pearson correlation coefficients for soybean. All data points are shown on this plot.

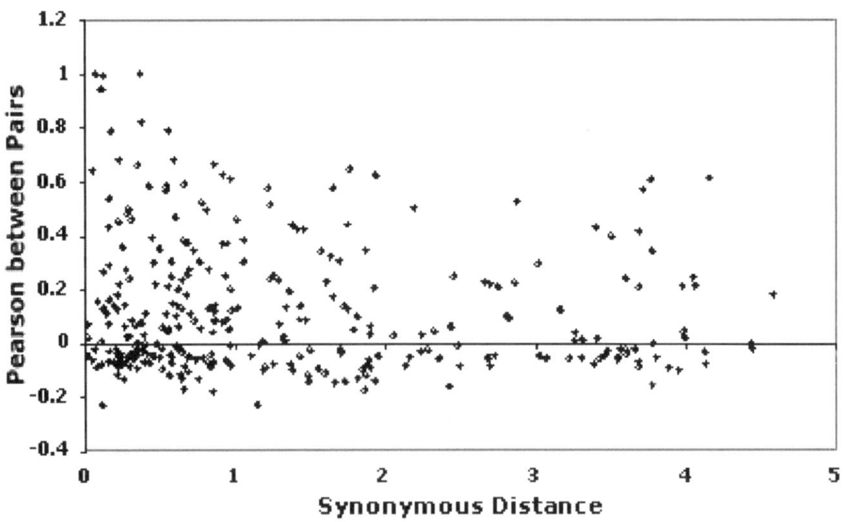

Figure 10.6. Scatterplot of synonymous distances against Pearson correlation coefficients for *Medicago*. All data points are shown on this plot.

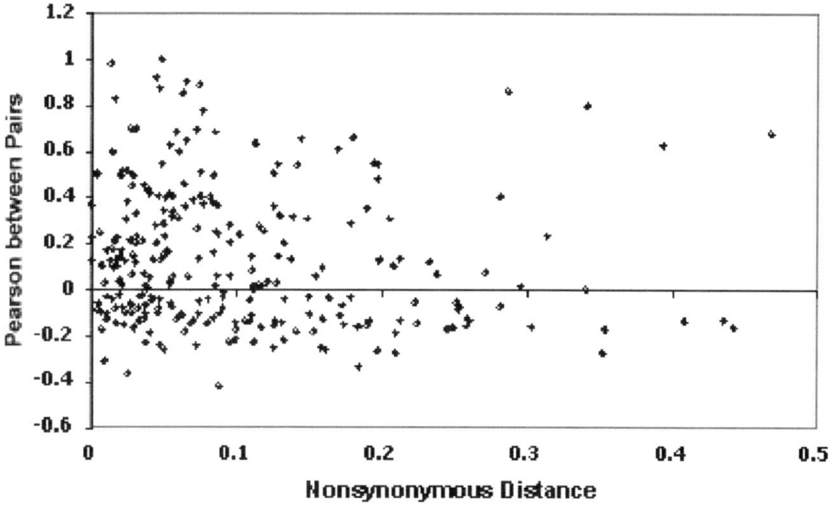

Figure 10.7. Scatterplot of nonsynonymous distances against Pearson correlation coefficients for soybean. All data points are shown on this plot.

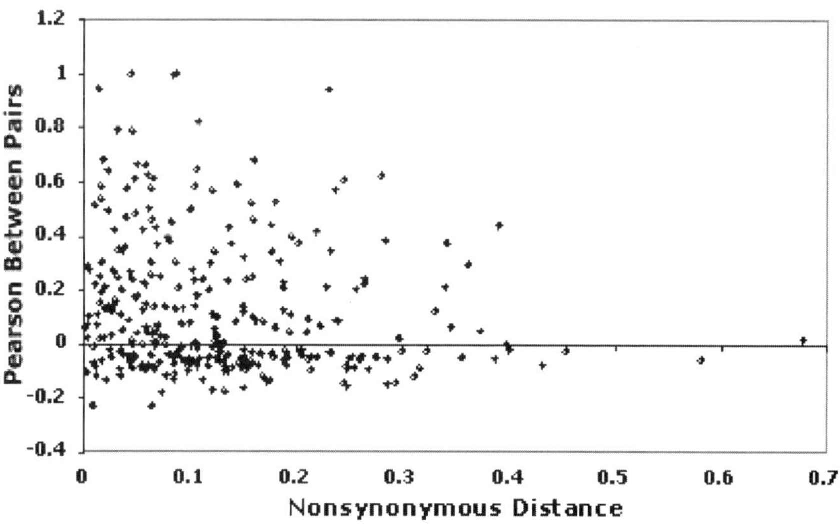

Figure 10.8. Scatterplot of nonsynonymous distances against Pearson correlation coefficients for *Medicago*. All data points are shown on this plot.

Pearson coefficients of many duplicated genes drop very quickly to zero, the distribution is skewed toward the positive. This indicates that even though the amino acid structure of duplicated genes is changing, expression patterns of many duplicates remain similar. Interestingly, in soybean there are four genes that seem to maintain a strong positive correlation of expression pattern with a large nonsynonymous distance. What this may reflect is that although these pairs have accumulated many nonsynonymous nucleotide changes, the most detrimental to protein integrity, and thereby function, may only be a few changes. Additionally, this may show that the promoter regions of these four genes have not undergone tremendous change.

When we examine the age distribution of duplicated genes within the distribution of nonsynonymous mutations, an interesting fact appears: Ancient duplications are intermixed with more recent duplications (Figs. 10.9 and 10.10). This coincides with our observation of negative selection among duplicated genes and also suggests that newly duplicated genes accumulate nonsynonymous substitutions very quickly and then seemingly stabilize. This observation agrees with that of Lynch and Conery (25) and Yang and Nielsen (22), and it suggests that duplicated genes experience a brief period of relaxed selection immediately after duplication. It is during this period that subfunctionalization likely occurs, after which purifying selection takes over. The duplicates surviving this initial process then remain relatively stable. It has

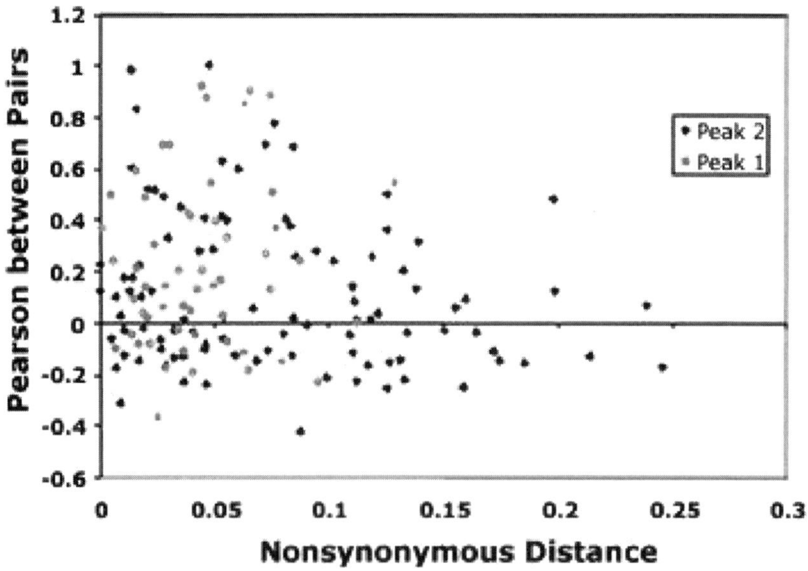

Figure 10.9. Scatterplot of nonsynonymous distances against Pearson correlation coefficients for soybean. The two different types of data points reference the determined distributions shown in Fig. 10.1.

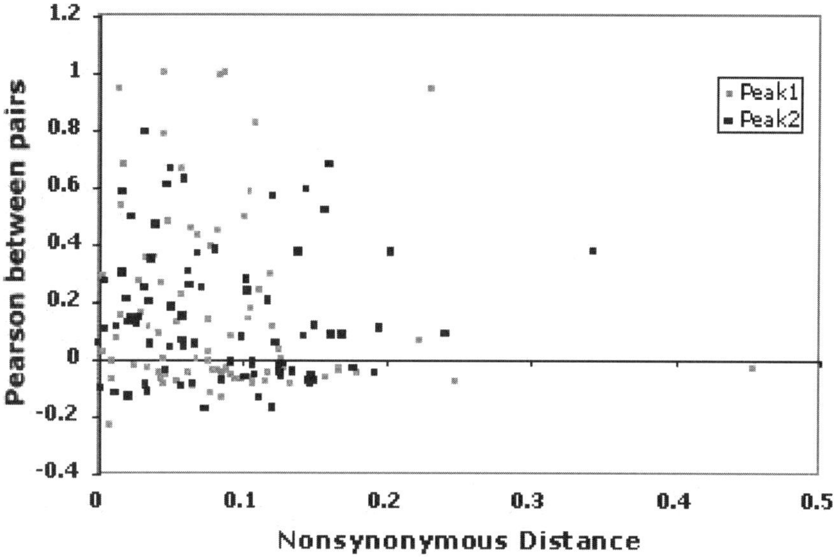

Figure 10.10. Scatterplot of nonsynonymous distances against Pearson correlation coefficients for *Medicago*. The two different types of data points reference the determined distributions shown in Fig. 10.2.

been estimated that the efficiency of purifying selection increases approximately tenfold after the initial relaxed selection (25).

Discussion

Legumes

Most genera of the Phaseolae have a genome complement of $2n = 22$, suggesting that soybean may have been derived from a diploid ancestor ($n = 11$) that underwent aneuploid loss to $n = 10$ and subsequent tetraploidization followed by diploidization (28). The detection of large, relatively well-maintained duplicated segments in the soybean linkage map (29) suggests that the youngest genome duplication is relatively recent. Study of relationships within gene families has long suggested that soybean is an ancient polyploid (30–34). Hybridization-based genetic maps are consistent with these findings (29,35,36). Combined data from nine different mapping populations uncovered extensive homeologous relationships among linage groups, with 90% of soybean RFLP probes detecting more than two fragments (29). In fact, nested duplications were observed, suggesting that one of the soybean genomes involved in the recent tetraploidization event may have undergone duplication prior to tetraploidization (29,36). Our results support these findings; soybean probably underwent two major genome events (14.5 mya and 41.6 mya) after the predicted divergence from *Medicago* (26).

Both soybean and barrel medic belong to the large flowering family Fabaceae and share the distinguishing ability to symbiotically fix nitrogen. The evolutionary history of *M. truncatula* is not as well studied as that of soybean. There is, however, evidence of gene duplication in *M. truncatula* because glutathione synthetase exists in two copies, most likely through tandem duplication (37). Additionally, *M. truncatula* chromosome 5 exhibits orthology with another region on the same chromosome showing evidence of segmental duplication (38). It has been suggested that *M. truncatula* has a simple genome structure, with only tandem and segmental duplications (39). However, prior to genome sequencing, the Arabidopsis genome was considered a true diploid with only 15% of the loci being duplicated (40). Now, whole genome duplications are well accepted as a driving force in the evolution of the Arabidopsis genome (41–44).

Our results support a hypothesis that *M. truncatula* probably underwent a major genome duplication event just after separation from soybean, at approximately 54.6 mya. We can explain the event at 20.2 mya in two ways: (i) Because it is represented as a somewhat broad peak, it may not actually represent a major event but the result of numerous chromosomal duplications spread across time; or (ii) that it does indeed represent a whole genome duplication occurring approximately 20.2 mya. The first hypothesis would be consistent with the general thought that *Medicago* has not undergone a recent whole-genome duplication event. However, in light of the 54.6 mya event, *M. truncatula* is most likely an ancient polyploidy. As the genome sequence of *M. truncatula* becomes available, it is possible that the same will be true as with Arabidopsis: the discovery that a simple diploid is not simple.

Distance Measures

A few caveats of distance measures should be addressed. First, an overestimation of time since divergence can occur due to asymmetrically bounded variables (nonelastic boundary at the present and an elastic boundary at the past) (45). Second, reduction in the precision of estimates of divergence can also be caused by variation in rates of sequence substitution (46), and rates of substitution may vary with functional differentiation, something often associated with gene duplication (4,47). Third, low numbers of ESTs in contigs can result in incomplete coverage of the coding region and can result in biased distance measures. Finally, ancient duplications become saturated at synonymous sites. This leads to potentially skewed estimates of the relative numbers of synonymous and nonsynonymous substitutions among duplicated genes in ancient duplications and may contribute to inaccuracies in the estimation of divergence times (47). Several of these experimental errors can be minimized. Utilizing EST contigs with large numbers of EST's increases the probability of obtaining the full coding region. Also, as the number of gene sequences increases, as is possible with analysis of large datasets, the accuracy of the divergence estimates also increases (48).

Selection and Expression of Duplicated Genes

Whole-genome duplications are also subject to varying levels of selective pressure. Ahn and Tanksley (49) suggested that after whole-genome duplication, chromosomal rearrangements accumulated in maize. In other words, genome duplication is followed by genome rearrangement. This was supported in a study involving recently synthesized polyploids in Brassica (3). Our results support these hypotheses by showing that most genes are not under positive selection after duplication but do indeed experience purifying selection.

After duplication, genes undergo one of several fates. Both may retain function, they may diverge to have different functions, or one may be lost while the remaining gene continues its original function (4). Maintenance of duplicated sequences within paleopolyploids suggests the involvement of negative selective pressures in their evolution. We observed this in our study.

If gene promoters were not undergoing purifying selection, we would expect to see a rapid progression of Pearson coefficients toward the zero (unrelated expression patterns) line. However, the maintenance of expression-pattern distribution over time is striking. This suggests that the duplicated genes surviving the initial relaxation in purifying selection have achieved an evolutionary advantage and have stabilized. This observation is consistent with the subfunctionalization hypothesis in which both copies of duplicated genes are required to achieve the function of the original ancestral gene.

As suggested by Lynch and Conery (25), genes experience a brief relaxation in purifying selective pressure immediately after duplication. During this period most copies are silenced, with the few remaining paralogs experiencing strong purifying, or negative, selection. Further, it has been suggested that during this relaxation in selection pressure, duplicated genes affect the fitness of the organism, leading to a temporary advantage for gene duplication (22). It has been hypothesized that the relaxation in purifying selection is the main driving force of duplicate gene evolution, whereas positive selection plays a less prominent role (22). Yang and Nielsen (22) also showed that among genes with similar levels of divergence, duplicated pairs evolved faster than single-copy genes.

Acknowledgments

The authors are grateful to the North Central Soybean Research Program for partial support of this research. Acknowledgment is given to Hilary Hagedorn for technical support in the preparation of this manuscript.

References

1. Stebbins, G.L., *Variation and Evolution in Plants*, Columbia University Press, New York, 1950.
2. Ohno, S., *Evolution by Gene Duplication*, Springer-Verlag, New York, 1970.
3. Song, K., P. Lu, K. Tang, and T.C. Osborn, Rapid Genome Change in Synthetic Polyploids of Brassica and Its Implications for Polyploidy Evolution, *Proc. Natl. Acad. Sci. USA 92*:7719–7723 (1995).

4. Pickett, F., and R. Meeks-Wagner, Seeing Double: Appreciating Genetic Redundancy, *Plant Cell 7*:1347–1356 (1995).
5. Masterson, J., Stomatal Size in Fossil Plants: Evidence for Polyploidy in Majority of Angiosperms, *Science 264*:421–424 (1994).
6. Stebbins, G.L., Chromosomal Variation and Evolution, *Science 152*:1463–1469 (1966).
7. Clausen, R.E., Polyploidy in Nicotiana, *Am. Nat. 75*:291–306 (1941).
8. Gaut, B.S., and J.F. Doebly, DNA Sequence Evidence for the Segmental Allotetraploid Origin of Maize, *Proc. Natl. Acad. Sci. USA 94*:6809–6814 (1997).
9. Lunde, C.F., D.J. Morrow, L.M. Roy, and V. Walbot, Progress in Maize Gene Discovery: A Project Update, *Funct. Integr. Genomics 3*:25–32 (2003).
10. Ewing, R.M., A.B. Kahal, O. Poirot, F. Lopez, S. Audic, and J.M. Claverie, Large-Scale Statistical Analysis of Rice ESTs Reveal Correlated Patterns of Gene Expression, *Genome Res. 9*:950–959 (1999).
11. Michalek, W., W. Weschke, K.P. Pleissner, and A. Graner, EST Analysis in Barley Defines a Unigene Set Comprising 4,000 Genes, *Theor. Appl. Genet. 104*:97–103 (2002).
12. Van der Hoeven, R., C. Ronning, J. Giovannoni, G. Martin, and S. Tanksley, Deductions About the Number, Organization, and Evolution of Genes in the Tomato Genome Based on Analysis of a Large Expressed Sequence Tag Collection and Selective Genomic Sequencing, *Plant Cell 14*:1441–1456 (2002).
13. Ronning, C.M., S.S. Stegalkina, R.A. Ascenzi, O. Bourgri, A.L. Hart, T.R. Utterbach, S.E. Vanaken, S.B. Riedmuller, J.A. White, J. Cho, G.M. Pertea, Y. Lee, S. Karamycheva, R. Sultana, J. Tsai, J. Quackenbush, H.M. Griffiths, S. Restrepo, C.D. Smart, W.E. Fry, R. Van der Hoeven, S. Tanksley, P. Zhang, H. Jin, M.L. Yamamoto, B.J. Baker, and C.R. Buell, Comparative Analyses of Potato Expressed Sequence Tag Libraries, *Plant Physiol. 131*:419–429 (2003).
14. Shoemaker, R.C., P. Keim, L. Vodkin, E. Retzel, S.W. Clifton, R. Waterston, D. Smoller, V. Coryell, A. Khanna, J. Erpelding, X.W. Gai, V. Brendel, C. Raph-Schmidt, E.J. Shoop, C.J. Vielweber, M. Schmatz, D. Pape, Y. Bowers, B. Theising, J. Martin, M. Dante, T. Wylie, and C. Granger, A Compilation of Soybean ESTs: Generation and Analysis, *Genome 45*:329–338 (2002).
15. Gyorgyey, J., D. Vauber, J.I. Jimenez-Zurdo, C. Charon, L. Troussard, A. Kondorosi, and E. Kondorosi, Analysis of *Medicago truncatula* Nodule Expressed Sequence Tags, *Mol. Plant Microbe Interact. 13*:62–71 (2000).
16. Rice, P., I. Longden, and A. Bleasby, EMBOSS: The European Molecular Biology Open Software Suite, *Trends Genet. 16*:276–277 (2000).
17. Altschul, S.F., W. Gish, W. Miller, E.W. Myers, and D.J. Lipman, Basic Local Alignment Search Tool, *J. Mol. Biol. 215*:403–410 (1990).
18. Xing, L., and V. Brendel, MuSeqBox: A Program for Multi-Query Sequence BLAST Output Examination, *Bioinformatics 17*:744–745 (2000).
19. Huang, X., and A. Madan, CAP3: A DNA Sequence Assembly Program, *Genome Res. 9*:868–877 (1999).
20. Mewes, H.W., D. Frishman, U. Guldener, G. Mannhaupt, K. Mayer, M. Mokrejs, B. Morgenstern, M. Munsterkotter, S. Rudd, and B. Weil, MIPS: A Database for Genomes and Protein Sequences, *Nucleic Acids Res. 30*:31–34 (2002).
21. Goldman, N., and Z. Yang, A Codon-Based Model of Nucleotide Substitution for Protein-Coding DNA Sequences, *Mol. Biol. Evol. 11*:725–736 (1994).
22. Yang, Z., and R. Nielsen, Estimating Synonymous and Nonsynonymous Substitution Rates under Realistic Evolutionary Models, *Mol. Biol. Evol. 17*:32–43 (2000).

23. McLachlan, G., and D. Peel, *Finite Mixture Models*, Wiley, New York (2000).
24. McLachlan, G., On Bootstrapping the Likelihood Ratio Test Statistic for the Number of Components in a Normal Mixture, *Applied Statistics 36*:318–324 (1987).
25. Lynch, M., and J.S. Conery, The Evolutionary Fate and Consequences of Duplicate Genes, *Science 290:*1151–1155 (2000).
26. Doyle, J.J., and M.A. Luckow, The Rest of the Iceberg. Legume Diversity and Evolution in a Phylogenetic Context, *Plant Physiol. 131*:900–910 (2003).
27. Force, A., M. Lynch, F.B. Pickett, A. Amores, Y. Yan, and J. Postlethwait, Preservation of Duplicate Genes by Complementary, Degenerative Mutations, *Genetics 151*:1531–1545 (1999).
28. Lackey, J.A., Chromosome Numbers in the Phaseoleae (Fabaceae:Faboideae) and Their Relation to Taxonomy, *Am. J. Bot. 67*:595–602 (1980).
29. Shoemaker, R., K. Polzin, J. Labate, J. Specht, E.C. Brummer, T. Olson, N. Young, V. Concibido, J. Wilcox, J. Tamulonis, G. Kochert, and H.R. Boerma, Genome Duplication in Soybean (*Glycine* Subgenus *soja*), *Genetics 144*:329–338 (1996).
30. Lee, J.S., and D.P.S. Verma, Structure and Chromosomal Arrangement of Leghemoglobin Genes in Kidney Bean Suggest Divergence in Soybean Leghemoglobin Gene Loci Following Tetraploidization, *EMBO J. 3*:2745–2752 (1984).
31. Hightower, R., and R. Meagher, Divergence and Differential Expression of Soybean Actin Genes, *EMBO J. 4*:1–8 (1985).
32. Grandbastien, M.A., S. Berry-Lowe, B.W. Shirley, and R. Meagher, Two Soybean Ribulose-1,5-Bisphosphate Carboxylase Small Subunit Genes Share Extensive Homology Even in Distant Flanking Sequences, *Plant Mol. Biol. 7*:451–465 (1986).
33. Nielsen, N.C., C. Dickinson, T.J. Cho, V.H. Thanh, B.J. Scallon, R.L. Rischer, T.L. Sims, G.N. Drews, and R.B. Goldberg, Characterization of the Glycinin Gene Family in Soybean, *Plant Cell 1*:313–328 (1989).
34. Granger, C., V. Coryell, A. Khanna, P. Keim, L. Vodkin, and R.C. Shoemaker, Identification, Structure, and Differential Expression of Members of a BURP Domain Containing Protein Family in Soybean, *Genome 45*:693–701 (2002).
35. Lee, J.M., A. Bush, J.E. Specht, and R. Shoemaker, Mapping Duplicate Genes in Soybean, *Genome 42*:829–836 (1999).
36. Lee, J.M., D. Grant, C.E. Vallejos, and R. Shoemaker, Genome Organization in Dicots. II. Arabidopsis as a Bridging Species to Resolve Genome Evolution Events Among Legumes, *Theor. Appl. Genet. 103*:765–773 (2001).
37. Frendo, P., M.J. Hernández, C. Mathieu, L. Duret, D. Gallesi, G. Van de Sype, D. Hárouart, and A. Puppo, A *Medicago truncatula* Homoglutathione Synthetase Is Derived from Glutathione Synthetase by Gene Duplication, *Plant Physiol. 126*:1706–1715 (2001).
38. Gualtieri, G., and T. Bisseling, Microsynteny Between the *Medicago truncatula SYM2*-Orthologous Genomic Region and Another Region Located on the Same Chromosome Arm, *Theor. Appl. Genet. 105*:771–779 (2002).
39. Kulikova, O., G. Gualtieri, R. Geurts, D.J. Kim, D.R. Cook, T. Huguet, H. De Jong, P.F. Franz, and T. Bisseling, Integration of the FISH Pachytene and Genetic Maps of *Medicago truncatula*, *Plant J. 27*:49–58 (2001).
40. McGrath, J.M., M.M. Jancso, and E. Pichersky, Duplicate Sequences with a Similarity to Expressed Genes in the Genome of *Arabidopsis thaliana*, *Theor. Appl. Genet. 86*:880–888 (1993).
41. The Arabidopsis Genome Initiative, Analysis of the Genome Sequence of the Flowering Plant *Arabidopsis thaliana*, *Nature 408*:796–815 (2000).

42. Vision, T.J., D.G. Brown, and S.D. Tanksley, The Origins of Genomic Duplications in Arabidopsis, *Science 290*:2114–2117 (2000).
43. Simillion, C., K. Vandepoele, M.C.E. Van Montagu, M. Zabeau, and Y. Van de Peer, The Hidden Duplication Past of *Arabidopsis thaliana*, *Proc. Natl. Acad. Sci. USA 99*:13627–13632 (2002).
44. Blanc, G., K. Hokamp, and K.H. Wolfe, A Recent Polyploidy Superimposed on Older Large-Scale Duplications in the Arabidopsis Genome, *Genome Res. 13*:137–144 (2003).
45. Rodriguez-Trelles, F., R. Tarrio, and F. Ayala, A Methodological Bias Toward Over-Estimation of Molecular Evolutionary Time Scales, *Proc. Natl. Acad. Sci. USA 9*:8112–8115 (2002).
46. Wray, G.A., Dating Branches on the Tree of Life Using DNA, *Genome Biol. 3*:1–7 (2001).
47. Ohta, T., Further Examples of Evolution by Gene Duplication Revealed Through DNA Sequence Comparisons, *Genetics 138*:1331–1337 (1994).
48. Nei, M., P. Xu, and G. Glazko, Estimation of Divergence Times from Multiprotein Sequences for a Few Mammalian Species and Several Distantly Related Organisms, *Proc. Natl. Acad. Sci. USA 98*:2497–2502 (2001).
49. Ahn, S., and S.D. Tanksley, Comparative Linkage Maps of the Rice and Maize Genomes, *Proc. Natl. Acad. Sci. USA 90*:7980–7984 (1993).

Chapter 11

The Gene Space of the Soybean Genome

Scott Jackson, Barbara Hass Jacobus, and Janice Pagel

Department of Agronomy & Plant Biology Program, Purdue University, 915 W. State St., West Lafayette, IN 47907

Introduction

Plant genomes are complex entities because of their high levels of repetitive DNA sequences and polyploidy. Soybean has a genome size of 1,100 million base pairs per haploid (Mb/C) (1), of which 40 to 60% is repetitive DNA based on Cot analyses (2,3). Soybean is likely an ancient diploidized tetraploid, as evidenced by chromosome number ($2n$ = 40) and multiple hybridizing restriction bands in RFLP mapping (4). Classical cytogenetic studies have shown that 35% of the fraction length of the pachytene chromosomes is heterochromatic and that six of the short arms are nearly completely heterochromatic (5).

The term *gene space* refers to the portion of nuclear DNA encoding active genes; it can also refer to the physical distribution of genes within a genome. Having a grasp of the gene space of soybean is important for an informed genomics approach to crop improvement and for the eventual sequencing of the soybean genome. For instance, given the high cost of sequencing an 1,100-Mb genome, an approach that selectively sequences the gene space would be attractive. Moreover, for functional genomics, including forward and reverse approaches, knowing what and where the genes are in the soybean genome is essential.

Composition of Plant Genomes

Polyploidy is a common theme in plant genomes. Estimates of the occurrence of polyploidy in the angiosperms range from 30 to 70%. Even many diploid plant genomes, including such species as maize (6) and *Arabidopsis thaliana* (L.) Heyn. with its minimalist 125-Mb genome (7), are actually "paleo"-polyploids. Polyploidy plays an integral role in the variation of plant genome sizes, but probably even more important in the determination of genome size are the repetitive sequences ubiquitous in plant genomes. Indeed, in species such as wheat, maize, pea, and many others, the repetitive fraction accounts for 75–90% of the total nuclear DNA.

In recent years, a picture of the organization of gene-encoding sequences within the morass of repetitive sequences has begun to evolve. Deletion mapping in wheat has shown that there are gene-rich regions located distally on its chromosomes and that these regions account for a high percentage of the observed recombination (8). It was also concluded that as many as 85% of the genes in wheat are located in less than 10% of the genomic DNA. At the molecular level, orthologous regions have been sequenced across several grass species (9), resulting in a vision of gene islands separated by clusters of nested retrotransposons (10) with increased repetitive content correlating with increasing genome size (11,12).

Approaches to Determining Gene Space in Complex Genomes

Several approaches have been successfully used to determine to some extent the gene space of eukaryotic organisms. These include techniques such as DNA:DNA renaturation, Cot-based sequencing, cDNA sequencing, methyl-filtration sequencing, and various physical mapping techniques such as contig assembly of genetically anchored large-insert clones, pachytene chromosome analyses, and cytogenetic mapping of gene-anchored bacterial artificial chromosomes (BACs) and repetitive sequences.

DNA renaturation studies in soybean indicate that 40–60% of the soybean genome consists of repetitive DNA sequences (2,3). These repeats may vary in complexity, frequency, and distribution within the soybean genome such that they are clustered rather than evenly interspersed.

Selective sequencing of Cot DNA fractions is a recent development that has been employed for species with complex genomes such as sorghum (13) and maize (14). Methyl-filtration sequencing is premised on the observation that repetitive sequences in eukaryotic genomes tend to be highly methylated, whereas genic sequences are undermethylated. Thus, using bacterial strains that are methyl-sensitive, one can selectively sequence unmethylated portions of complex genomes, thereby enriching the population of sequencing clones for genic sequences. This was shown to be a viable approach in maize, where the genic sequences were enriched fourfold (15). For further discussion of Cot- and methylfiltration-based sequencing in soybean, see Chapter 14 of this volume, by Gary Stacey.

The more traditional sequence-based approach used to study gene space is to sequence cDNA libraries derived from RNA recovered from expressed genes, resulting in expressed sequence tag (EST) libraries. This approach has been extremely useful and is generally preliminary to whole-genome sequencing, as ESTs are useful for genome annotation (16). Soybean has an extensive EST collection based on libraries from many tissue types (http://129.186.26.94/soybeanest.html). The EST collection gives an approximation of the number of genes in the gene space but never fully reaches saturation because of several confounding factors: Genes expressed at low levels may not be captured; genes expressed in specific tissues/cell types may be missed in library construction; and highly expressed genes tend to be over-represented. EST libraries, though an approximation of the gene content, reveal almost nothing of the physical organization of genes within the genome.

Physical mapping offers the best approximation of the organization of genic/repetitive sequences within a genome. Of course, the most complete physical map is an imperforate genome sequence, but in the absence of such, other physical mapping techniques can be employed to approximate the genome organization. Recently, BAC-based physical maps, constructed by ordering BAC clones into contigs based on shared restriction digest fragments (17,18), have proven useful as a basis for understanding the organization of genomes and as substrates for the cloning of genes and sequencing of genomes. Complete BAC-based physical maps are a necessary and fundamental prerequisite for assembling genome sequences, and they provide a resource for manipulating segments of the genome for efficient gene cloning.

The Soybean Genome

Soybean is hypothesized to be an ancient, diploidized polyploid species with a haploid chromosome number of $n = 20$. Previous work has shown that many RFLP mapping loci are present in two to three copies in the soybean genome (4), suggesting that there may have been at least one other segmental duplication event. Analyses of the soybean EST database have shown many genes to be present in two to four isoforms—indicating multiple loci (19). Moreover, an analysis of the incidence of specific isoforms in the various tissues represented in the EST libraries revealed that some isoforms are over-represented in certain libraries (19), indicating possible subfunctionalization (20).

Very little is known about the primary constituent of the soybean genome, that is, the repetitive DNA elements. In cereals, retroelements compose a large fraction of the repetitive sequences and are found clustered between blocks of genes (10,21). Currently, the only large-scale sequencing effort in legumes is that of *Medicago truncatula* Gaertner. The *M. truncatula* genome is 454 to 526 Mb, or roughly half the size of soybean. Cytological anlaysis of the *M. truncatula* genome has shown that gene-rich regions located in euchromatic regions of the chromosome arms are physically separated from repetitive DNA sequences that are located primarily in the pericentromeric regions (22). Preliminary evidence suggests that this may also be the case for soybean. Hymowitz and Singh (5) estimated that 35% of the pachytene chromosome length was heterochromatic and that this was concentrated in the pericentromeric regions and the short arms of a few chromosomes.

Recent work in our lab using genetically anchored BACs as probes for fluorescence *in situ* hybridization (FISH) has shown that pericentromeric repeats cover a large fraction of the metaphase chromosome length (Fig. 11.1). Gel blot analyses of these soybean centromere–derived BACs indicates that the majority of the sequences present on the BACs are high-copy in the soybean genome (data not shown). Moreover, several BACs confirmed by FISH to be centromeric have sequences that are not common (nonhybridizing bands: BACs 09M21, 76J21, and 36H16; see Fig. 11.1e and f). Also, centromere-specific repeats are excluded from low-copy, genic regions of the genome, as seen with BAC 123E07 (Fig. 11.1e and f, lane 1).

We are currently sequencing two centromere-derived BAC clones in order to identify specific DNA elements peculiar to centromeres and to understand the DNA structure of soybean centromeres. Since the pericentromeric regions occupy a large portion of the pachytene chromosomes of soybean (5), understanding the DNA

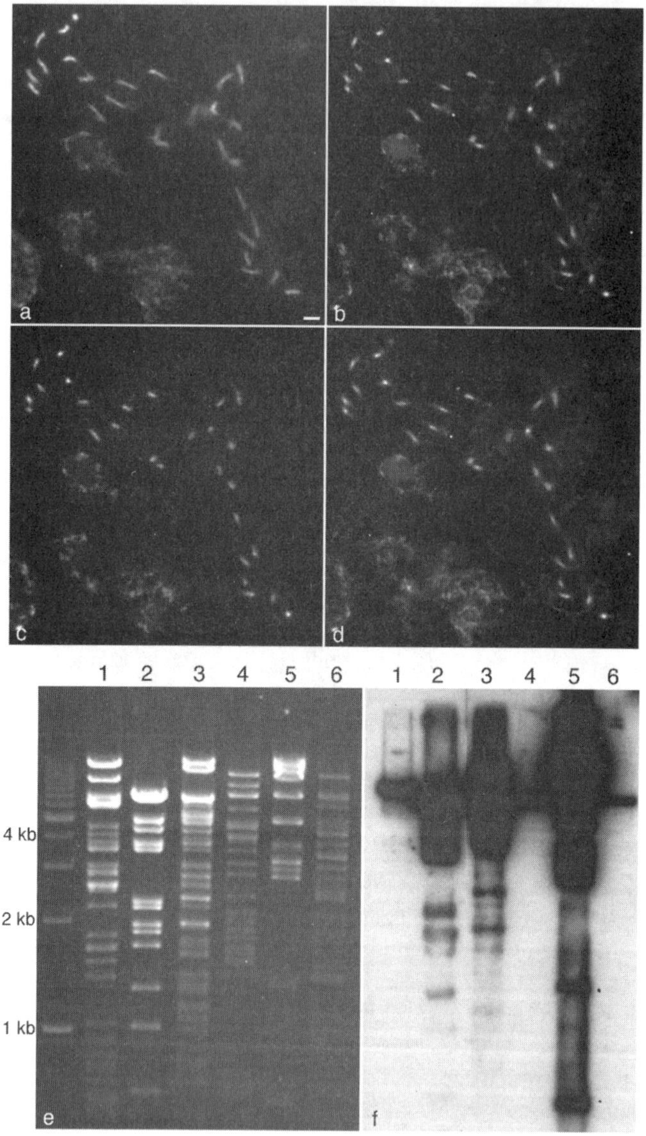

Figure 11.1. Molecular and cytological analysis of centromere-specific BACs of soybean. (a–d) FISH analysis of two centromere-specific BACs on soybean chromosomes: (a) DAPI stained chromosomes, (b) FITC detection of biotin-labeled BAC 09M21, (c) rhodamine detection of digoxigenin-labeled BAC 76J21, and (d) merged image. (Bar in panel a = 5 μm.) (e–f) Southern analysis of BAC 76J21 to restriction-digested BACs: (e) *HindIII* restriction digest of BACs, and (f) hybridization of radiolabeled BAC 76J21 to gel blot. Lane: (1) BAC 123E07, (2) BAC 09M21, (3) BAC 41A23, (4) BAC 55A20, (5) BAC 76J21, and (6) BAC 36H16.

sequence organization of centromeres will give us insights into the molecular organization of a sizable fraction of the soybean genome.

Only a few repetitive sequences from soybean have been characterized and described at the molecular level. Molecular and cytological characterization of SB92, a 92–base pair (bp) repeat within the soybean genome, showed that this repeat was homologous to about 0.9% of the soybean genome and was found at only four to five chromosomal locations (23). A few other soybean DNA repeats have been characterized, including SB192 (24) and two transposable elements: *Tgm* (25), and *Soymar1* (26). Recent work analyzing BAC end sequences from NCBI (National Center for Biotechnology Information) indicates that there may be as many as 100 different classes of repeats in soybean (Randy Shoemaker, USDA-ARS, Iowa State University, personal communication).

In soybean, there are two BAC-based physical mapping resources, neither of which meets the necessary qualifications for a large-scale genome project. One was developed from the cultivar Williams 82 (27), and the other from cultivar Forrest (28). Given the common acceptance of the Williams 82 genotype as the "standard" for soybean genomics, this will be the genetic background in which a complete, genetically anchored physical map will need to be developed.

Other physical mapping approaches have or are being used in soybean to estimate the distribution of the gene space. Our lab is using FISH of genetically anchored BACs and cloned repetitive sequences to describe the genome in terms of types and distribution of repetitive sequences relative to gene-rich regions defined by anchored BACs. We have found that the centromeric/pericentromeric regions are large and that some sequences are conserved in all 40 of the soybean centromeres. Evaluation of both the organization and conservation of repetitive sequences in the conspicuous centromeric regions is being followed up by selective BAC sequencing and comparative genomics within the Leguminosae.

Earlier cytological analyses of premeiotic chromosomes (pachytene), where heterochromatic and euchromatic regions can be more easily distinguished, showed that the bulk of the heterochromatin was indeed centromeric and that the chromosome arms were generally euchromatic (15). More recent studies using FISH of genetically anchored BACs to pachytene chromosomes in the related legume *M. truncatula* showed a similar genomic organization (22). This discovery led to a BAC-by-BAC sequencing approach starting with BACs anchored in euchromatic regions of the genome. What is not known is the prevalence of genes within the large blocks of pericentromeric heterochromatin. In general, few genes are found in large blocks of heterochromatin, and those that are found there may be epigenetically silenced through chromatin modifications as observed in position effect variegation (29,30). However, there are examples of important, expressed genes in close physical proximity to pericentromeric DNA (31). Thus, sequencing efforts directed only at gene-rich regions will be able to capture most of the gene-encoding sequences but will likely miss other genes outside of the gene-rich regions. Given the cost of sequencing the entire soybean genome at this time, however, a gene-rich targeted attack may be the most viable approach.

References

1. Arumuganathan, K., and E.D. Earle, Nuclear DNA Content of Some Important Plant Species, *Plant Mol. Biol. Rep. 9*:208–219 (1991).
2. Goldberg, R.B., DNA Sequence Organization in the Soybean Plant, *Biochem. Genet. 16*:45–68 (1978).
3. Gurley, W.B., A.G. Hepburn, and J.L Key, Sequence Organization of the Soybean Genome, *Biochem. Biophys. Acta. 561*:167–183 (1979).
4. Shoemaker, R.C., K. Polzin, J. Labate, J. Specht, E.C. Brummer, T. Olson, N. Young, V. Concibido, J. Wilcox, J.P. Tamulonis, G. Kochert, and H.R. Boerma, Genome Duplication in Soybean (*Glycine* subgenus *soja*), *Genetics 144*:329–338 (1996).
5. Hymowitz, T., and R. Singh, The Genomic Relationship between *Glycine max* and *G. soja* Sieb. and Zucc. as Revealed by Pachytene Chromosome Analysis, *Theor. Appl. Genet. 76*:705–711 (1988).
6. Gaut, B.S., and J.F. Doebley, DNA Sequence Evidence for the Segmental Allotetraploid Origin of Maize, *Proc. Natl. Acad. Sci. USA 94*:6809–6814 (1997).
7. Vision, T.J., D.G. Brown, and S.D. Tanksley, The Origins of Genomic Duplications in *Arabidopsis*, *Science 290*:2114–2117 (2000).
8. Gill, K.S., B.S. Gill, T.R. Endo, and T. Taylor, Identification and High-Density Mapping of Gene-Rich Regions in Chromosome Group 1 of Wheat, *Genetics 144*:1883–1891 (1996).
9. Feuillet, C., and B. Keller, High Gene Density Is Conserved at Syntenic Loci of Small and Large Grass Genomes. *Proc. Natl. Acad. Sci. USA 96*:8265–8270 (1999).
10. San Miguel, P., A.P. Tikhonov, Y.K. Jin, N. Motchoulskaia, and D. Zakharov, Nested Retrotransposons in the Intergenic Regions of the Maize Genome, *Science 274*:765–768 (1996).
11. Chen, M., P. San Miguel, and J. Bennetzen, Sequence Organization and Conservation in sh2/a1-Homologous Regions of Sorghum and Rice, *Genetics 148*:535–443 (1998).
12. Li, W., and B.S. Gill, The Colinearity of the Sh2/A1 Orthologous Region in Rice, Sorghum and Maize Is Interrupted and Accompanied by Genome Expansion in the Triticeae, *Genetics 160*:1153–1162 (2002).
13. Peterson, D.G., S.R. Schulze, E.B. Sciara, S.A. Lee, J.E. Bowers, A. Nagel, N. Jiang, D.C. Tibbitts, S.R. Wessler, and A.H. Paterson, Integration of Cot Analysis, DNA Cloning and High-Throughput Sequencing Facilitates Genome Characterization and Gene Discovery, *Genome Res. 12*:795–807 (2002).
14. Yuan, Y., P.J. San Miguel, and J.L. Bennetzen, High-Cot Sequence Analysis of the Maize Genome, *Plant J. 34*:249–260 (2003).
15. Rabinowicz, P.D., K. Shutz, N. Dedhia, C. Yordan, L.D. Parnell, L. Stein, W.R. McCombie, and R.A. Martienssen, Differential Methylation of Genes and Retrotransposons Facilitates Shotgun Sequencing of the Maize Genome, *Nat. Genet. 23*:305–308 (1999).
16. Yeh, R.F., L.P. Lim, and C.B. Burge, Computational Inference of Homologous Gene Structures in the Human Genome, *Genome Res. 5*:803–816 (2001).
17. Marra, M., T. Kucaba, M. Sekhon, L. Hillier, R. Martienssen, A. Chinwalla, J. Crockett, J. Fedele, H. Grover, C. Gund, W.R. McCombie, K. McDonald, J. McPherson, N. Mudd, L. Parnell, J. Schein, R. Seim, P. Shelby, R. Waterston, and R. Wilson, A Map for Sequence Analysis of the *Arabidopsis thaliana* Genome, *Nat. Genet. 22*:265–270 (1999).

18. Soderlund, C., S. Humphrey, A. Dunhum, and L. French, Contigs Built with Fingerprints, Markers and FPC V4.7, *Genome Res. 10*:1772–1787 (2000).
19. Schlueter, J.A., P. Dixon, C. Granger, and R.C. Shoemaker, Mining the EST Databases to Determine Evolutionary Events in the Legumes and Grasses, *23rd Stadler Genetics Symposium Proceedings*, in press. See also Chapter 10, this volume.
20. Lynch, M., M. O'Hely, B. Walsh, and A. Force, The Probability of Preservation of a Newly Arisen Gene Duplicate, *Genetics 159*:1789–1804 (2001).
21. Bennetzen, J.L., Transposable Element Contributions to Plant Gene and Genome Evolution, *Plant Mol. Biol. 42*:251–259 (2000).
22. Kulikova, O., G. Gualtieri, R. Geurts, D.J. Kim, D. Cook, T. Huguet, J.H. de Jong, P.F. Fransz, and T. Bisseling, Integration of the FISH Pachytene and Genetic Maps of *Medicago truncatula*, *Plant J. 27*:49–58 (2001).
23. Vahedian, M., L. Shi, T. Zhu, R. Okimoto, K. Danna, and P. Keim, Genomic Organization and Evolution of the Soybean SB92 Satellite Sequence, *Plant Mol. Biol. 29*:857–862 (1995).
24. Morgante, M., I. Jurman, L. Shi, T. Zhu, P. Keim, and J.A. Rafalski, The STR120 Satellite DNA of Soybean: Organization, Evolution and Chromosomal Specificity, *Chrom. Res. 5*:363–373 (1997).
25. Rhodes, P.R., and L.O. Vodkin, Organization of the Tgm Family of Transposable Elements in Soybean, *Genetics 120*:597–604 (1988).
26. Jarvik, T., and K.G. Lark, Characterization of Soymar1, a Mariner Element in Soybean, *Genetics 149*:1569–1574 (1998).
27. Marek, L., J. Mudge, L. Darnielle, D. Grant, N. Hanson M. Paz, Y. Huihuang, R. Denny, K. Larson, D. Foster-Hartnett, A. Cooper, D. Danesh, D. Larsen, T. Schmidt, R. Staggs, J.A. Crow, E. Retzel, N.D. Young, and R.C. Shoemaker, Soybean Genomic Survey: BAC-End Sequences near RFLP and SSR Markers, *Genome 44*:572–558 (2001).
28. Meksem, K., K. Zobrist, E. Ruben, D. Hyten, T. Quanzhou, H-B. Zhang, and D.A. Lightfoot, Two Large Insert Soybean Genomic Libraries Constructed in a Binary Vector: Applications in Chromosome Walking and Genome Wide Physical Mapping, *Theor. Appl. Genet. 101*:747–755 (2000).
29. Csink, A., A. Bounoutas, M.L. Griffith, J.F. Sabl, and B.T. Sage, Differential Gene Silencing by *Trans*-Heterochromatin in Drosophila Melanogaster, *Genetics 160*:257–269 (2001).
30. Reuter, G., and P. Spierer, Position Effect Variegation and Chromatin Proteins, *BioEssays 14*:605–612 (1992).
31. Bryan, G.T., K.S. Wu, L. Farrall, Y. Jia, H.P. Hershey, S.A. McAdams, K.N. Faulk, G.K. Donaldson, R. Tarchini, and B. Valent, A Single Amino Acid Difference Distinguishes Resistant and Susceptible Alleles of the Rice Blast Resistance Gene *Pi-ta, Plant Cell 12*:2033–2046 (2000).

Chapter 12

Targeting Induced Local Lesions in Genomes—TILLING

Rae Ritchie,[a,b] Hunt Wiley,[c] Dan Szymanski,[b] and Niels Nielsen[a,b]

[a]USDA-ARS, Crop Production and Pest Control Unit, West Lafayette, IN 47907;

[b]Department of Agronomy, Purdue University, West Lafayette, IN 47907; and

[c]Dairyland Seed Company, Inc., Otterbein, IN 47970

Introduction

New technology has been used to improve the human food supply for many centuries. Like most important grain crops, the early history of soybean predates written history. According to Piper and Morse (1), soybeans were referred to in the earliest Chinese writings from more than 5,000 years ago, where they were called *tchouan*. Later Chinese writings dating from the Christian Era referred to soybeans as *ta-teou* or *sou*. It is not difficult to imagine how the later ancient Chinese name for this crop evolved to soy, its modern name. The earliest, primitive efforts to domesticate soybean were dependent on empirical selection methodologies. Near the turn of the 20th century, the discovery of genetics augmented these efforts and greatly accelerated the crop improvement process. A number of studies have attempted to quantitate the magnitude of the developments that have occurred during recent history (2–5). The application of molecular genetic approaches beginning in the last two decades of the 20th century greatly enhanced these processes and brought with it the possibility of rapidly and specifically introducing or modifying individual genes to enhance food quantity and quality. A fact often lost or ignored because of the debate that is presently raging about the pros and cons of genetically modified organisms is that modern gene technology is also a powerful tool that can be used to identify natural or induced genetic variation in soybean populations and to aid in the selection of those recombinants that have inherited beneficial variation.

Even with the application of modern gene technology, several important features still impede the improvement of soybean. One is a lack of definitive knowledge about individual genes that encode or control important agronomic traits. For example, it is estimated that roughly 20,000 diverse genes are expressed in soybean seeds, 90% of which are expressed in all stages of development (6). The remaining 10% of genes are therefore differentially expressed and possibly involved in the development of soybean seeds; only a tiny percentage of these have been identified so far. Modern genomics approaches are rapidly evolving that will permit such genes to be identified and their interactions with other genes catalogued as they respond to various developmental and environmental stimuli. However, the ability to identify with accuracy those few genes among many that have a major influence on important agronomic features is still very tenuous. A second impediment concerns defining gene function and the discovery of allelic variation that can be applied to crop improvement. Gene insertion by T-DNA and TILLING (*T*argeting *I*nduced *L*ocal *L*esions *IN* *G*enomes) are two reverse genetic techniques that are presently available and can be applied to discovering gene function and allelic variation.

The identification of genes and their function requires versatile tools that interact in a complementary way. The remainder of this chapter will describe how some of these tools can be utilized together.

Subtraction Libraries Identify Important Genes

The period of seed fill during formation of soybean seeds is of critical importance to the agronomic value of the crop. It is during this interval that massive amounts of protein and oil are synthesized—the two storage components that make soybeans such a valuable commodity. Concerted efforts by many laboratories throughout the world have documented the genes that encode the prevalent 11S and 7S proteins that constitute the storage reserves (7,8). Other efforts have characterized the enzymes required to synthesize the storage triglycerides that are accumulated in soybeans (9). It has been established that the onset of expression of genes encoding the storage reserves is intimately associated with the appearance of a developmental process known as endoreduplication. Despite the enormous economic importance of events associated with this developmental phase change, almost nothing is known about the genetic events that control this process.

To understand the genetic events that control the onset of endoreduplication and seed fill, genes that are differentially expressed just before and just after initiation of the endocycle were identified. For this purpose cDNA libraries were constructed using mRNAs isolated from intact immature seeds of the Resnik soybean cultivar at 7 and 23 days after flowering (DAF) (10). At 7 DAF seeds from this cultivar are at the late heart stage of development, which is prior to initiation of endoreduplication. On the other hand, the 23-DAF seeds are near midmaturation, and when the DNA contents of nuclei are measured by flow cytometry, nuclei with 16- and 32S-ploidy are prominent, a marker of endoreduplication. Polymerase chain reaction (PCR) subtraction libraries were formed by suppressing the amplification of cDNAs common to both starting libraries, in order to identify differentially expressed cDNAs in the 7- and 23-DAF libraries (10). Mirror orientation selection was subsequently used to enhance the selection of genes expressed at the two time intervals (11). In addition, a pre-subtraction step was performed on the source 23-DAF library to remove cDNAs that encoded the prevalent seed proteins such as 7S and 11S subunits, protease inhibitors, and lipoxygenase. Approximately 3,500 clones each from the 7- and 23-DAF libraries were analyzed by a combination of cross-hybridization and single-pass DNA sequence analysis. So far, this effort has resulted in the identification of nearly 1,000 unique DNA sequences. These sequences were then compared to those found in the

GenBank nucleotide, protein, and EST databases. Information describing most of these clones can be found at http://soypr2.agry.purdue.edu/data/local-vb-sequences-2002/publish/daf7-diff-genes.htm and http://soypr2.agry.purdue.edu/data/local-vb-sequences-2002/publish/daf21-diff-genes.htm.

The types of proteins encoded by the genes in the subtraction libraries bear similarity to cell cycle genes, DNA binding and remodeling proteins, phosphatases, kinases, proteins involved in various metabolic pathways, and a cadre of proteins with unknown functions. The question arises as to what proportions of the clones in these libraries are unique and actually differentially expressed. To answer this question, a set of five clones was chosen randomly from the two subtraction libraries. Total RNA isolated at different stages of seed development was amplified by RT-PCR using primers derived from each of these sequences. The results obtained with two of them are shown in Fig. 12.1, which represents examples of the two types of results. The data revealed that four of the five clones were amplified differentially.

Clone O05, shown in Fig. 12.1, top panel, is 842 bases long. It is not detectable at 7 DAF but is easily detected by RT-PCR once seed-fill has commenced. When BLASTed against the NCBI (National Center for Biotechnology Information) protein databases (BLAST is a program developed at the NCBI), the protein sequence derived from O05 had similarity to the terminal flower-1 (TLF1) sequence from Arabidopsis (identity, 75%; positives, 89%) (12). The putative soybean TLF1 protein sequence also bears considerable similarity to the animal phosphotidylethanolamine binding proteins (13–15). The TLF1 family of proteins is involved in controlling stages of development throughout the lifecycle of plants. Although TLF1 has not been identified from soybean previously, the clone has 99% identity to five soybean EST sequences in GenBank and matches two contigs (overlapping sequences of genomic DNA typically containing several genes) from the Soybean Genomics Initiative database.

Clone B12, shown in Fig. 12.1, middle panel, is 390 base pairs in length. When analyzed using RT-PCR, it was present at 7 DAF but could not be detected at any of the soybean developmental stages in which endoreduplication was taking place. When BLASTed against the NCBI databases, it was found to have 84% identity to a putative MAP-kinase from apricot. There is about 16% similarity to the peptide domain of serine/threonine kinases. The clone is 99% identical to three soybean EST sequences and matches two contigs. Thus, the available data indicate that a study of clones from the suppression subtraction library should provide new information about genes involved in the control of cell development.

TILLING Permits the Function of Important Genes to Be Determined

It will be necessary to deduce the biological function of genes identified using the subtraction library. Two techniques that can be used for this purpose are knockout technology using T-DNA insertions and TILLING. The former approach involves the

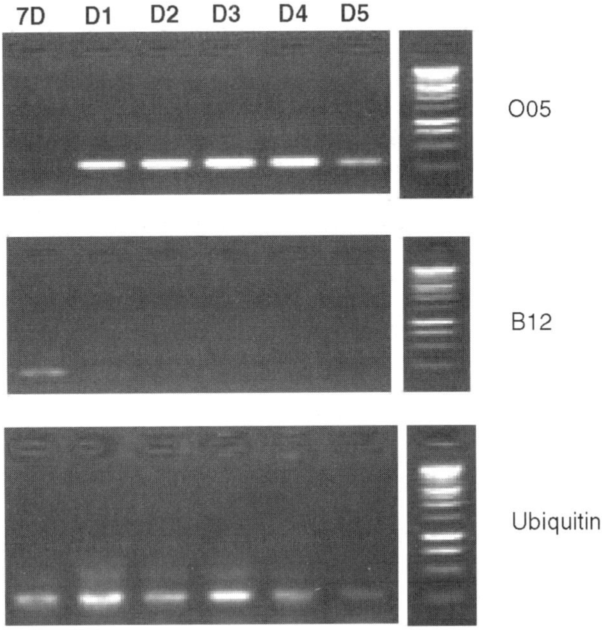

Figure 12.1. Total RNA was isolated from seeds at different stages of development (7 DAF (7D) and seed fill stages D1–D5). Primers unique to each clone were used to amplify the cloned sequences in each sample. Twenty-five cycles of PCR were run and the products loaded onto a 1.2% agarose gel. Ubiquitin was used for determining loading equivalents.

random insertion of T-DNA into genes, and it has been extensively applied to studies with Arabidopsis.

TILLING is a combination of traditional chemical mutagenesis and PCR-based screening for point mutations in specific genes of interest (16,17). The steps involved in this procedure are summarized in Fig. 12.2. With this reverse genetics technology, groups of mutants can be identified that contain mutations that affect a gene or part of a gene of interest. Four features of TILLING make it particularly attractive: (i) The screening is rapid and can be automated (18); (ii) chemical mutagenesis is reliable, and therefore the regions of genes most likely to be targeted can be determined; (iii) a wide range of mutants can be created and identified—the mutations can be silent and in the wobble position, and they can be either lethal or nonlethal; and (iv) any gene can be mutated irrespective of whether it is essential, and the mutation will be detected when either homozygous or heterozygous.

TILLING populations have been generated using three soybean cultivars. Production of the mutant DSR061 and DSR299 populations was started during summer 2002, and mutation of Williams 82 was carried out during summer 2003. In

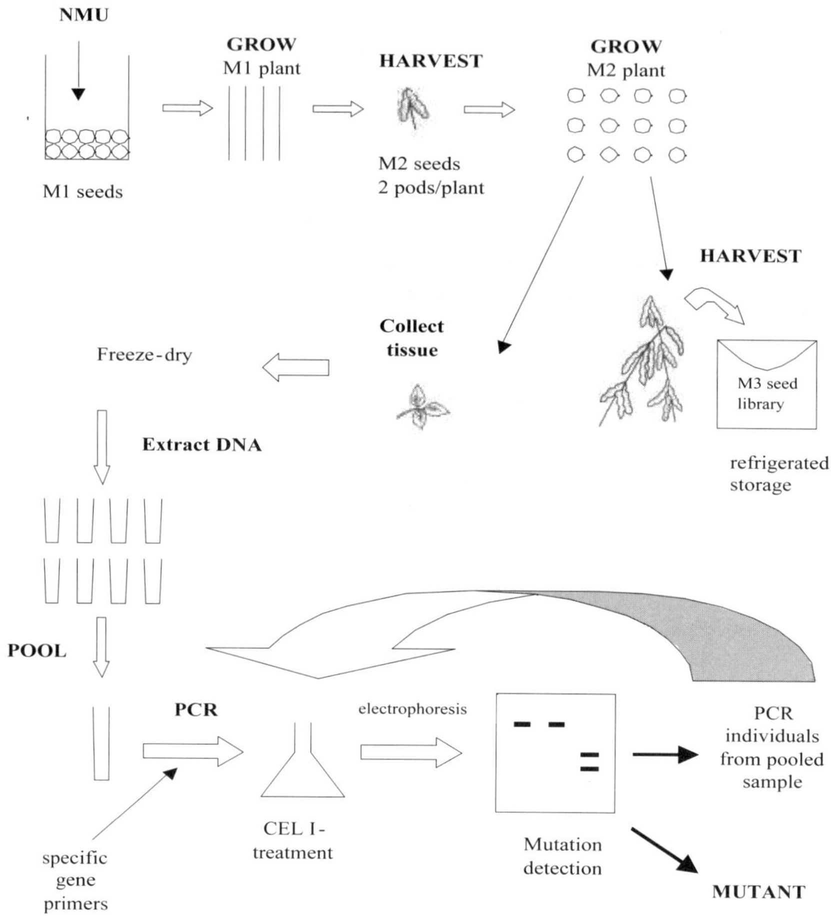

Figure 12.2. Outline of TILLING. Soybean seeds are treated with N-nitroso-N-methylurea (NMU). The M1 seeds are planted. Two pods from each M1 plant are hand harvested. The M2 seeds from one pod are planted in a hill, and 100 leaf discs are collected from each of the M2 plants into microtubes and stored at –80°C. After single plant harvest, the M3 seed libraries are stored refrigerated. The tissue samples are freeze-dried and the DNA extracted. Pools of DNA from 8 plants are put into 96-well plates. Specific primers for a gene of interest are used to amplify a portion of the gene (1,000 bases, an amplicon). After PCR, the amplicons are melted and then annealed. This step allows for annealing of wild type and mutant amplicons (if any). The products are treated with CEL I, an endonuclease specific for base mismatches. The amplicons are then separated by electrophoresis and examined for mutations. If present, the mutations will appear as smaller bands. If a pool of DNA contains a mutant, the individuals in the pool will be amplified individually and CEL I treated. The individual plant containing the mutant will be identified. Seed (M3) from this plant can be grown for further studies.

each case, seeds (M1) were treated with N-nitroso-N-methylurea (NMU) as described by Kerr and Sebastian (19). NMU alkylates the oxygens of nucleic acids (20), and one product, O^6-methylguanine, pairs efficiently with thymine (T). The pairing causes a transition from guanine (G) to adenine (A) and results in a base pair change from cytosine (C)/G to T/A. Seeds were air-dried following mutation treatment, just to the extent that they would not stick to one another and could be planted mechanically with a grain drill. Handling of control seeds during the mutation treatment decreased seed emergence by about 25%, and treatment with NMU approximately doubled that decrease. Two pods were collected from each of 8,000 plants in each of the DSR061 and DSR299 populations. M2 seeds that were all derived from one pod from each plant were planted together in hills that were arrayed in a matrix 30 inches on center. One or more plants per hill were obtained for just over 12,600 of the approximately 16,000 hills planted. One plant from each hill was assigned an accession number, and 100 leaf discs from that plant were collected into microtubes labeled with the same accession number. About 1,300 of the mutant plants that emerged either did not produce seed or were so small that it would not have been possible to collect leaf tissue for DNA purification. Tubes containing the leaf punches were stored at –80°C until they could be freeze-dried and used to prepare DNA (21). After single plant harvest, the M3 seed libraries were stored refrigerated.

Seeds per pod, from M1 plants, were scored to see if potential evidence of embryo lethality could be observed. Seeds per pod averaged 2.5 in the mutant population and 2.9 for the untreated controls. Approximately 1,000 seeds of the DSR299 M2 seed population were grown in the greenhouse during 2003. The emergence rate of seeds in the greenhouse was about 78% versus 78% in the field. Emergence of untreated controls was 100%. Approximately 15% of the plants grown in the greenhouse were phenotypically abnormal. It was not uncommon to observe seedlings that died about a week after emergence, which indicated that they probably contained lethal gene mutations. Other common phenotypic manifestations of NMU treatment were highly stunted growth patterns, male sterility, altered plant morphology, and altered pigmentation patterns. Characteristics of plants grown both in the field and in the greenhouse left little doubt that they were derived from mutant populations.

Presently, DNA from each of the accessions in the soybean mutant populations is being prepared. These are initially to be screened for accessions that contain mutations that specifically affect genes identified as being active during the transition from cell division to endoreduplication during seed development, like those described above. The careful selection of areas in genes of interest increases the likelihood of discovering a useful mutation. For example, choosing regions of genes that are conserved across species, and therefore likely to be required for gene function, will increase the likelihood that the mutation will affect gene function. Also, identifying regions where the C/G to T/A switch is likely to affect the protein sequence via truncation or missense mutations will increase the probability of finding mutations. A computer-based tool called CODDLE has been developed (22) to aid in the identification of these regions. The CODDLE software is available at www.proweb.org/coddle.

Considerations When Using TILLING as a Tool

An important issue to consider when establishing a TILLING population is the frequency of mutations in that population. It is essential to establish a balance between the potentially lethal effects of mutations and the average density of mutations in each plant. Clearly, the greater the number of mutations per plant, the fewer accessions must be searched during the discovery efforts. For this reason, soybean populations exposed to the same concentration of NMU, but for increasingly longer periods of time, have been generated and remain to be evaluated. Although scarce, some information about this is beginning to appear in the literature. The *Lotus japonicus* (Regel) K. Larsen project screens for phenotypes of interest and tests only selected phenotypes. For example, only plants that were non-nodulating, had few nodules, or had abnormalities in root development were screened for mutations in the SYMRK gene, which is required for formation of root symbioses (23). A total of 288 selected mutants were screened and 15 homozygous mutants (six alleles) with missense mutants were found. The *Medicago truncatula* Gaertner project has recovered an average of one single-nucleotide polymorphism (SNP) per 1 kb per 500 germlines (24), while the *Arabidopsis thaliana* project reported that with target fragments of 1 kb they found 3.6 mutations per 750 plants. This extrapolates to about 700 different mutations per plant (22).

A second important issue to be considered when evaluating soybean TILLING populations is that *Glycine max* (L.) Merr. has undergone substantial duplication in its genome during its evolution (25). The consequence is that there are multiple copies of most genes, and the likelihood of any one point mutation producing a phenotypic effect is thereby decreased. The presence of duplicate genes enables compensation for loss or reduction of activity by the mutated gene (26). This consideration implies that phenotypic manifestation of gene mutations will be reduced in soybean compared to plant genomes that do not include a similarly large complement of duplicated genes. Indeed, early attempts to mutate soybean with sodium azide described reduced phenotypic variation compared to barley (27). The presence of many duplicate genes also indicates the necessity of creating mutations in redundant genes to gain clues about their biological function.

Summary

It has been apparent for many years that a major impediment to efficient crop improvement is lack of knowledge about the many genes that control plant development and the acquisition of agronomically important traits. As knowledge about the many genes involved in plant growth and development increases, so too will our ability to direct crop improvement using conventional approaches of recombination and selection, directed molecular engineering, or a combination of the two. The experiments described briefly in this chapter illustrate several modern tools that can be applied to the problem of gene discovery and identification of gene function. Although only two of nearly 1,000 clones that encode proteins that are differentially expressed during seed development were described in this chapter, their DNA sequences conform closely with sequences in the soybean EST database. It was also possible to identify DNA sequences that perform orthologous functions. The results illustrate the value of the EST database to soybean researchers and justify the decision by the United Soybean Board to establish this important resource.

This chapter also briefly describes efforts to establish and characterize soybean TILLING populations. Theoretically, an allelic set of mutants that modify the activity of any gene of interest in the soybean genome can be identified from the TILLING populations. The mutant populations provide two opportunities for the soybean research community. The first is to provide tools by which to study the function of specific genes in the context of soybean. Although convincing arguments can be presented that many fundamental biological problems can be studied effectively in model organisms, such as *Arabidopsis,* that have shorter generation times than soybean, it can be anticipated that differences between the two species will have developed during evolution that will influence efforts to improve crop performance. For this reason it is important not to confine our attention to modeling plants during efforts to improve agronomically important crop species. The second important feature of TILLING populations is that they are a source of genetic variation that can be used for crop improvement. A maize TILLING population, currently being established at Purdue University, will function as a resource for the maize research community (Cliff Weil, Purdue University, personal communication). The driving force for generating the TILLING populations described in this chapter was to investigate the function of genes active during seed formation, but if supported appropriately by the soybean research community, the populations will become an important general resource for soybean researchers.

Acknowledgments

The efforts by members of Dairyland Seed, Inc. in Otterbein, Indiana, to mutate and generate the mutant populations is gratefully acknowledged. This work was supported in part by a grant from the United Soybean Board. Additional information about the Soybean Genomic Initiative can be accessed at http://soybean.ccgb.umn.edu/.

References

1. Piper, C.V., and W.J. Morse, *The Soybean,* McGraw-Hill, New York, 1923, pp. 35–54.
2. Wilcox, J.R., W.T. Schapaugh, R.L. Bernard, R.L. Cooper, W.R. Fehr, and M.H. Neihaus, Genetic Improvement of Soybean in the Midwest, *Crop Sci. 19*:803–805 (1979).
3. Boerma, R., Comparison of Past and Recently Developed Soybean Cultivars in Maturity Groups VI, VII and VIII, *Crop Sci. 19*:611–613 (1979).

4. Specht, J.E., and J.H. Williams, Contribution of Genetic Technology to Soybean Productivity—Retrospect and Prospect, in *Genetic Contributions to Yield Gains of Five Major Crops,* edited by W.R. Fehr, Crop Science Society of American and American Society of Agronomy, Madison, WI., 1980, Spec. Pub. 7, pp. 49–74.
5. Smith, K.J., and W. Huyser, World Distribution and Significance of Soybean, in *Soybeans: Improvement, Production and Uses,* edited by J.R. Wilcox, American Society of Agronomy, Madison, WI, 1987, Vol. 2, pp. 1–22.
6. Goldberg, R.B., S.J. Barker, and L. Perez-Grau, Regulation of Gene Expression during Plant Embryogenesis, *Cell 56*:149–160 (1989).
7. Nielsen, N.C., R. Bassuner, and T. Beaman, The Biochemistry and Cell Biology of Embryo Storage Proteins, in *Cellular and Molecular Biology of Plant Seed Development,* edited by B.A. Larkins and I.K. Vasil, Kluwer, Dordrecht, The Netherlands, 1997, pp. 151–220.
8. Beilinson, V., Z. Chen, R.C. Shoemaker, R.L. Fischer, R.B. Goldberg, and N.C. Nielsen, Genomic Organization of Glycinin Genes in Soybean, *Theor. Appl. Genet. 104*:1132–1140 (2002).
9. Ohlrogge, J.B., and J.G. Jaworski, Regulation of Fatty Acid Synthesis, *Ann. Rev. Plant Physiol. Plant Mol. Biol. 48*:109–136 (1997).
10. Beilinson, V., O. Moskalenko, and N.C. Nielsen, Differential Expression of the Genes During Soybean Seed Development in Plant Biology 2002 Final Program, American Society of Plant Biologists (Denver, CO, Aug. 3–7), p. 110, or online at http://abstracts.aspb.org/pb2002/public/P53/0888.html (2003).
11. Rebrikov, D.V., O.V. Britanova, N.G. Gurskaya, K.A. Lukyanov, V.S. Tarabykin, and S.A. Lukyanov, Mirror Orientation Selection (MOS): A Method for Eliminating False Positive Clones from Libraries Generated by Suppression Subtractive Hybridization, *Nucleic Acids Res. 28*:e90 (2000).
12. Levy, Y.Y., and C. Dean, The Transition to Flowering, *Plant Cell 10*:1973–1989 (1998).
13. Schoentgen, F., and J. Jolles, From Structure to Function: Possible Biological Roles of a New Widespread Protein Family Binding Hydrophobic Ligands and Displaying a Nucleotide Binding Site, *FEBS Lett. 369*:22–26 (1995).
14. Banfield, M.J., J.J. Barker, A.C.F. Perry, and R.L. Brady, Function from Structure? The Crystal Structure of Human Phosphatidylethanolamine-Binding Protein Suggests a Role in Membrane Signal Transduction, *Structure 6*:1245–1254 (1998).
15. Banfield, M.J., and R.L. Brady, The Structure of *Antrirrhinum* Centroradialais Protein (CEN) Suggests a Role as a Kinase Regulator, *J. Mol. Biol. 297*:1159–1170 (2000).
16. McCallum, C.M., L. Comai, E.A. Greene, and S. Henikoff, Targeted Screening for Induced Mutations, *Nature Biotech. 18*:455–457 (2000).
17. McCallum, C.M., L. Comai, E.A. Greene, and S. Henikoff, Targeting Induced Local Lesions in Genomes (TILLING) for Plant Functional Genomics, *Plant Physiol. 123*:439–442 (2000).
18. Colbert, T., B.J. Till, R. Tompa, S. Reynolds, M.N. Steine, A.T. Yeung, C.M. McCallum, L. Comai, and S. Henikoff, High-Throughput Screening for Induced Point Mutations, *Plant Physiol. 126*:480–484 (2001).
19. Kerr, P.S., and S.A. Sebastian, U.S. Patent 6,147,193 (2000).
20. Singer, B., and D. Grunberger, *Molecular Biology of Mutagens and Carcinogens,* Plenum, New York, 1983, pp. 45–96.
21. Preparation and Analysis of DNA, in *Current Protocols in Molecular Biology*, Volume I, edited by F.M. Ausubel, R. Brent, R.E. Kingston, D.D. Moore, J.G. Seidman, J.A. Smith, and K. Struhl, John Wiley & Sons, Hoboken, New Jersey (2003), unit 2.3.

22. Till, B.J., S.H. Reynolds, E.A. Greene, C.A. Codomo, L.C. Enns, J.E. Johnson, C. Burtner, A.R. Odden, K. Young, N.E. Taylor, J.G. Henikoff, L. Comai, and S. Henikoff, Large-Scale Discovery of Induced Point Mutations with High-Throughput TILLING, *Genome Res. 13*:524–530 (2003).
23. Perry, J.A., T.L. Wang, T.J. Welham, S. Gardner, J.M. Pike, S. Yoshida, and M. Parniske, A TILLING Reverse Genetics Tool and a Web-Accessible Collection of Mutants of the Legume *Lotus japonicus, Plant Physiol. 131*:866–871 (2003).
24. VandenBosch, K.A., and G. Stacey, Summaries of Legume Genomics Projects from around the Globe: Community Resources for Crops and Models, *Plant Physiol. 131*:840–865 (2003).
25. Shoemaker, R.C., K. Polzin, J. Labate, J. Specht, E.C. Brummer, T. Olson, N. Young, V. Concibido, J. Wilcox, J.P. Tamulonis, G. Kochert, and H.R. Boerma, Genome Duplication in Soybean (*Glycine* subgenus *soya*), *Genetics 144*:329–338 (1996).
26. Pickett, F.B., and D.R. Meeks-Wagner, Seeing Double: Appreciating Genetic Redundancy, *Plant Cell 7*:1347–1356 (1995).
27. Kleinhofs, A., R.L. Warner, F.J. Murhlbauer, and R.A. Nilan, Induction and Selection of Specific Gene Mutations in *Hordeum* and *Pisum, Mut. Res. 51*:29–35 (1978).

Chapter 13

Genetic Engineering of Soybean: Strategies and Utility

Tom E. Clemente[a] and Ted M. Klein[b]

[a]Center for Biotechnology/Plant Science Initiative/ Department of Agronomy and Horticulture, University of Nebraska, N308 Beadle Center, Lincoln, NE 68588-0665; and

[b]Pioneer Crop Genetics, Transformation Research, Stine-Haskell Research Center, P.O. Box 30, Building 614/103, Newark, DE 19714

Introduction

Plant molecular biology continues to advance at a stunning pace. The ability to characterize and understand the role and interaction of particular genes and proteins in the establishment of complex signaling and biochemical pathways has been facilitated by the application of genomic technologies. The practical use of this knowledge for the development of new traits for crop plants is the focus of many researchers in both academia and industry.

The first reports of the production of transgenic soybean occurred in the mid-1980s. During the ensuing years, the commercial application of biotechnology has significantly impacted soybean genetic improvement. In 2003, approximately 81% of the 74 million acres of soybeans planted in the United States carried a transgene that confers tolerance to the herbicide glyphosate. Glyphosate tolerance has enabled soybean growers to implement alternative and simplified weed-management programs. In 1999, this single trait helped soybean growers realize a $216 million savings in weed management and resulted in approximately 19 million fewer herbicide applications (1). Transgenic approaches for improving agronomic performance by enhancing tolerance to biotic and abiotic stresses will be an important complement to conventional soybean breeding programs. There are now a number of examples in which biotechnology has been used to manipulate pathways for the quality or quantity of protein, oil, and carbohydrates in soybean seed (2,3).

The ability to efficiently introduce and express foreign genes in plants is a necessary technology for crop improvement using the tools of biotechnology. There are numerous reports detailing methods for gene transfer for a wide range of crop plants (4,5). The two major routes for transformation of plant cells are by particle bombardment (also termed microprojectile bombardment or the biolistic approach) and *Agrobacterium*-mediated DNA delivery. The ability to utilize these processes for crop transformation differs among species and even among genotypes within species. For example, high rates of transformation are possible for several maize genotypes whereas transformation for most elite inbreds remains problematic. The same holds true for rice, wheat, and barley. On the other hand, an attractive feature of two model plant species, tobacco and *Arabidopsis*, is the ease with which high numbers of transgenic individuals can be obtained. Therefore, specific transformation protocols for particular crops must often be developed, and this holds true for soybean as well.

Plant transformation requires the occurrence of two independent events: the integration of foreign DNA into a cell and the subsequent development of the transformed cell into the lineage destined to contribute to a gamete. Aside from the floral-dip protocol used for *Agrobacterium*-mediated transformation of *Arabidopsis* (6,7), the ability to recover a heritable transgenic cell lineage requires *in vitro* culturing of cells. Therefore, the process of introducing foreign genes into plants requires the ability to propagate tissues in culture. Soybean transformation uses tissue culture approaches that rely on either organogenesis or embryogenesis to provide targets for gene transfer. These tissue culture approaches are generally coupled with *Agrobacterium*- or particle bombardment–mediated gene delivery, respectively. Several reviews of soybean transformation have been published (8,9). This review will focus on the major routes to produce transgenic soybean, the limitations of these systems, and the strategies being pursued to further enhance transformation frequencies of soybean. We will also discuss the utility of transformation as a tool to enhance the germplasm of soybean and to augment genomic studies of this most important crop plant.

Soybean Transformation Using Organogenic *In Vitro* Cultures

The first successful reports on transformation of soybean implementing organogenic culture regimes were published in 1988 (10,11). Hinchee and coworkers used *Agrobacterium* to infect axillary meristems that developed at the junction between cotyledon and hypocotyl (termed the cotyledonary node) of young seedlings (10). Another group of researchers used particle bombardment to deliver genes to the apical meristem of the embryonic axis isolated from mature seed (11).

The *in vitro* regeneration potential of the cotyledonary-node segments was first observed by Cheng *et al.* (12). In the presence of low levels of the cytokinin, benzylaminopurine (BAP), the axillary meristems proximal to the base of the cotyledon from young seedlings germinate and form multiple buds. These multiple buds can then proliferate following subculture of the excised cotyledonary-node segments on media supplemented with BAP. Transferring the developing buds to media containing reduced levels of BAP encourages shoot development, and roots are established on the shoots by culturing on media devoid of growth regulators.

By coupling the protocol developed by Cheng *et al.* (12) with a nopaline strain of *A. tumefaciens*, A208 (13), Hinchee and coworkers (10) were able to successfully recover soybean transformants and demonstrate faithful transmittance of the transgenes to the next generation. In this transformation system, soybean seeds were germinated for a period of 4 to 10 days on Gamborg's B5 medium (14). The cotyledons were then excised from the germinating seed, wounded at the axillary node region with a scalpel, and cocultivated with a *Agrobacterium* strain harboring a cointegrating plasmid containing the *npt*II gene (for resistance toward the aminoglycoside kanamycin). Following a 2-day cocultivation period, the explants were subcultured to shoot induction medium composed of Gamborg's B5 salts supplemented with 1.2 mg/L BAP, 200 to 300 mg/L kanamycin, 500 mg/L carbenicillin, and 100 mg/L cefotaxime. Shoot elongation was initiated by removal of BAP, and rooting was induced by reducing the basal salt concentration by 50%.

Histological analysis of developing buds from cotyledonary-node segments revealed that regeneration occurs via direct organogenesis (10). Direct organogenesis involves regeneration of plants without an intervening phase of unorganized tissue (callus) growth and therefore can be relatively fast. Because the duration of time that the tissues spend in culture is correlated with the accumulation of mutations, methods that minimize time in culture are advantageous.

Transformation using a direct organogenic regeneration regime requires the ability to efficiently select transformed cell lineages. The use of kanamycin in selecting transgenic cell lineages from cotyledonary-node explants is rather inefficient and results in the recovery of multiple escapes (nontransformed) plants (10,15). This may be related to the ability of the aminoglycoside to translocate efficiently within the explant combined with relatively high tolerance level of soybean to the antibiotic. It is noteworthy, however, that kanamycin can be effectively used to identify transgenic soybean callus lines (16), suggesting that the problem is more related to the translocation of the antibiotic during organogenesis.

Townsend and Thomas (17) identified a number of factors during the cocultivation step that influence the interaction of *A. tumefaciens* with the cotyledonary node explant. These include the pH of the medium, incubation temperatures, and the incorporation of acetosyringone in the medium (an activator of the two-component signal transduction pathway) (18). The positive impact of low pH and acetosyringone during the cocultivation of the *A. tumefaciens*/soybean interaction is in agreement with observations reported by Godwin *et al.* (19). The temperature effect detected by Townsend and Thomas (17) is undoubtedly related to enhanced *Vir* gene expression and stimulation of the T-DNA transfer apparatus, as compared to that observed at higher temperatures (20,21). This observation is consistent with the observation that formation of plant tumors by wild-type *A. tumefaciens* is temperature sensitive (22).

Clemente *et al.* (23) modified the cotyledonary node transformation protocol by further optimizing cocultivation conditions and improving the formulation for the medium used for shoot elongation. These modifications improved the quality of the cultures, which in turn enhanced transformation frequencies. This protocol used a gene (EPSP synthase) that confers resistance to glyphosate. The implementation of sublethal levels of glyphosate (0.05 to 0.15 μM) allowed efficient screening for the identification of transgenic soybean shoots.

The salient features of this revised cotyledonary node transformation protocol include a modified cocultivation medium composed of 1/10 Gamborg's salts, full strength vitamin formulation, 3% sucrose, 0.25 mg/L GA_3, 1.7 mg/L BAP, and

200 μM acetosyringone. The medium is buffered with 20 mM MES (pH 5.4). The elongation medium modifications include the use of Murashige and Skoog (MS) basal salts; Gamborg's B5 vitamins, supplemented with 1 mg/L zeatin riboside; 0.1 mg/L indole acetic acid; and 0.5 mg/L GA_3. Both the shoot induction and elongation media are buffered with 3 mM MES (pH 5.6). In addition to tolerance to glyphosate, genes encoding tolerance to the herbicide glufosinate (24,25) or the antibiotic hygromycin (26) have been used for the identification of transgenic shoots.

The refinements to the cotyledonary node transformation system that have been made over the past decade now permit the introduction of transgenes into advanced germplasm at reliable frequencies. However, the cotyledonary node transformation protocol, like other methods for plant transformation, has inherent variation in frequency among genotypes. Variations in transformation frequencies among genotypes can be attributed to two potentially interrelated factors. The capacity of a given genotype to form shoots and regenerate whole plants from the cotyledonary node explant is clearly a critical factor influencing transformation frequency. The susceptibility of the axillary meristematic cells to *Agrobacterium* also varies among genotypes. In addition, the regenerative capacity of some genotypes can be negatively impacted by the interaction with *Agrobacterium*. As mentioned, there are several elite soybean cultivars that are suitable for the cotyledonary node transformation system (23–25).

Many *Agrobacterium*-based transformation systems require physical wounding of the target tissue. Wounding can help to elicit the appropriate tissue culture response and also provide a means for the bacterium to come in contact with the appropriate cells. *In vitro* differentiation of the cotyledonary node explant is highly dependent on proper wounding of the axillary meristematic region. Insufficient wounding results in mere expansion of the axillary shoot, whereas excessive wounding decimates the meristem and curtails the ensuing *de novo* shoot differentiation about the nodal region. Three steps are associated with proper explant preparation, including: (i) A vertical slice is made approximately 5 mm below the cotyledonary node-hypocotyl junction, thereby excising the cotyledons with the associated embryonic axis from the root system; (ii) A second parallel slice is made through the attached hypocoytl/node junction, separating the cotyledons; and (iii) The axillary meristem is explosed by removing the embryonic axis, and subsequently by making 7 to 12 slices about the meristem, parallel with the axis. The slices should encompass a region approximately 5 mm^2 about the nodal region.

Following infection of the cotyledonary node, the growth of *Agrobacterium* must be controlled because, if left unchecked, bacterial growth would completely overwhelm the explant tissue. Antibiotics are added to the regeneration medium for this purpose. Typical antibiotics utilized in Agrobacterium-mediated transformation protocols to inhibit *A. tumefaciens* growth include, but are not limited to, carbenicillin, cefotaxime, ticarcillin, and vancomycin at levels of 50 mg/L up to 500 mg/L (26). Although these molecules do not have well-characterized physiological effects on plant cells, there are reports of some having growth regulator–type activity (27,28). There are also reports of antibiotic-induced methylation of plant genomes (29). Antibiotic supplements designed to control *Agrobacterium* growth also can negatively impact the ability of transgenic shoots to develop roots (30). Therefore, Clemente and coworkers (unpublished data) modified the antibiotic counter selection regime to maximize rooting efficiencies while maintaining effective counter selection against *Agrobacterium.* In this protocol, vancomycin, cefotaxime, and ticarcillin are added to the shoot initiation medium at 75 mg/L each for 2 weeks and then reduced to 50 mg/L each for the subsequent regeneration steps. At the rooting stage, the antibiotic counter-selection regime is completely withdrawn. This single modification within the system has resulted in rooting efficiency of transgenic shoots >90% (Clemente, unpublished data).

The second organogenic tissue culture regime that has provided an effective target for genetic engineering of soybean is the apical meristem of the zygotic embryo axis (11,31,32). Genes have been introduced to this explant by both microprojectile bombardment (11,32,33) and *Agrobacterium* (31,34,35). McCabe *et al.* (11) were the first to successfully exploit this explant in a gene transfer system using microprojectile bombardment (36,37). These researchers developed a custom designed gene gun that utilized capacitance discharge between two electrodes to produce a shock wave. This shock wave provided the force necessary to accelerate a mylar sheet, whose front surface was coated with DNA-bearing particles, into a metal screen. The screen stopped the mylar sheet and allowed the passage of the accelerated particles into the target tissue. Following bombardment, the meristems were placed on MS medium (38) supplemented with 3.0 mg/L BAP and 0.04 mg/L NAA. The tissue was incubated in the dark for up to 2 weeks. After the 2-week dark incubation period, the proliferating embryonic axes were subcultured to MS medium containing 0.4 mg/L BAP and 0.04 mg/L IBA. The cultures were then maintained under a 16-hour light regime until shoots from either the primary or axillary meristem reached 0.5 to 1.0 cm in length. This tissue culture regime permits the formation of from five to 10 shoots from each bombarded meristem. The visual marker gene, ß-glucuronidase (GUS), was used to screen for transgenic plantlets arising from bombarded meristems (11,33). Not all GUS-expressing shoots transmitted the transgene to the next generation. Many plants exhibited sectors of GUS expression, but not all of these sectors encompassed a cell lineage that gave rise to a gamete (39). After observing a large number of GUS-expressing shoots, these researchers were able to identify GUS expression patterns that could be used to predict the likelihood of inheritance of the transgene in the next generation. Plants with GUS sectors in the stem that extended from the epidermis through the cortex and pith usually produced seed carrying the transgene (39). This system has the advantage of minimal *in vitro* culture time, thus limiting the probability of generating tissue culture–derived mutations (40). The researchers also demonstrated the applicability of this system to a number of commercially important genotypes. However, the system is very inefficient and labor intensive.

Aragão *et al.* (32) incorporated a selection regime in the embryonic axes/microprojectile–bombardment protocol in an attempt to alleviate the labor involved in screening while using a visual selection system. The modified protocol used the imidazolinone resistance gene *ahas* (41) combined with the herbicidal agent imazapyr. Here, following bombardment, the embryonic axes were cultured on MS media supplemented with 5.0 mg/L BAP for 16 hours in the dark. Following this brief shoot-induction step, the explants were transferred to selection medium containing 500 nM imazapyr, with no growth regulator supplements, and were cultured under a 16-hour light regime. The transformation frequencies reported ranged from 3.9 to 20.1% (32). The authors suggested that these relatively high rates of transformation—as compared to the original system developed by McCabe *et al.* (11)—are due to the ability of imazapyr to rapidly translocate to meristematic cells and inhibit the division of nontransgenic cells. Therefore, a selective advantage is given to transgenic cell lineages, thereby providing an efficient means to identify transgenic plants (32).

The soybean embryonic axis also has been used as a target for gene transfer by *Agrobacterium*. Chee and coworkers (31) prepared explants by first imbibing seed for 24 hours and then excising one cotyledon to expose the embryonic axis and associated apical meristem. Inoculations were conducted by directly injecting *A. tumefaciens* (nopaline strain C58Z707) (42) in the nodal and epicotyl regions. The inoculated explants were placed in the dark for 4 hours and subsequently transferred to soil for whole-plant development (31). Although this protocol circumvents an *in vitro* culturing step, the transformation frequency is very low, with only 0.07% of the treated explants giving rise to a transgenic plant (34). Therefore, like the microprojectile bombardment protocol (11), the absence of a reliable selection system necessitates significant labor and space requirements to achieve success (34).

Martinell *et al.* (35) combined the attributes of the embryonic axis explant, *A. tumefaciens*, DNA delivery, and herbicide selection in a soybean transformation protocol. In this system, mature seeds were germinated in nutrient media in the dark. The apical meristem was subsequently exposed by removal of the embryonic leaves. The explants were then cultured overnight in the dark on modified MS base media supplemented with 3 mg/L BAP at 15°C. The precultured axes were then inoculated with *Agrobacterium* for 1 hour. In some experiments, the explants were wounded prior to inoculation with either a scalpel, needle, or sonication (43). Following the inoculation step, the explants were cocultivated on 1/10 Gamborg's B5 medium for 3 days. The explants were subsequently placed on regeneration medium consisting of a modified MS basal mixture supplemented with BAP for induction of multiple shoots and glyphosate or kanamycin as selective agents. With this protocol, Martinell *et al.* (35) reported transformation frequencies ranging from 0.9 to 3.3%. As was the case for bombardment of meristems, the inclusion of a selection system led to an enhanced rate of transformation.

Soybean Transformation Using Somatic Embryo *In Vitro* Cultures

Somatic embryos and embryogenic cultures often represent suitable targets for gene transfer. In somatic embryogenesis, embryo-like structures form from cells other than a gamete or the product of gamete fusion. Under the proper tissue culture conditions, these somatic embryos can proliferate by repetitive somatic embryogenesis to form embryogenic cultures. The first reports of somatic embryogenesis from soybean involved removing immature cotyledons from seed and placing them on semisolid media containing synthetic hormones such as 2,4-D and kinetin. Although conditions that permitted some proliferation of embryo-like structures were explored, these structures proved difficult to maintain and to regenerate into whole plants (44). A number of subsequent reports outline culture conditions that are important for development of soybean somatic embryos (45–48). In general, induction of somatic embryogenesis in soybean requires auxin (2,4-D or NAA), sucrose, and an organic nitrogen source such as adenine, asparagine, or glutamine. Ranch and coworkers (49) and Finer (50) found that relatively high levels of 2,4-D permitted the proliferation of secondary somatic embryogenesis from primary somatic embryos derived from immature cotyledons. Finer and Nagasawa (51) described a refined medium that permitted the development of proliferative embryogenic cultures. Other tissue culture parameters critical in establishment and maintenance of soybean somatic embryogenesis include wounding (52), pH (52), light (53), explant orientation (54), and solidifying agent (52).

Soybean embryogenic cultures can be initiated from immature embryos by culturing the explants on MS-based medium at pH 7.0 that are supplemented with 40 mg/L 2,4-D and 6% sucrose (51). The tissue is cultured in the dark at 25°C. Embryogenesis is induced within 3 to 4 weeks under these conditions. Maintenance and proliferation of embryogenic cultures is carried out by subculturing the tissue to induction medium at pH 5.8 that are supplemented with 20 mg/L 2,4-D and 3% sucrose (55). Methods for regenerating whole plants from soybean somatic embryos have been described. Bailey *et al.* (56) outlined a procedure for subsequent differentiation and maturation of soybean somatic embryos. Briefly, culturing on a hormone-free medium consisting of MS salts and Gamborg's vitamins, supplemented with 6% maltose and 0.5% activated charcoal, leads to differentiation of embryogenic tissue. The addition of activated charcoal at the differentiation step helps alleviate potential embryo developmental abnormalities presumably manifested by exposure to high levels of 2,4-D (57). Subsequent culturing of the somatic embryos individually on MS medium with Gamborg's vitamins containing 3% sucrose aids in the further maturation of the differentiated embryos. In general, the maturation steps take 4 weeks. The maturated embryos are then desiccated for several days and subsequently transferred to MS media to induce germination. Maturation of soybean somatic embryos can also be carried out in a liquid medium, which greatly decreases the time required for differentiation (58). The frequency of embryo germination using liquid media was improved by the addition of sorbitol as an osmoticum (59).

Soybean transformation using somatic embryogenic tissue as the recipient target was first achieved by microprojectile bombardment (60). Embryogenic material was cotransformed with plasmids harboring the visual marker gene ß-glucuronidase

(61) and the selectable marker gene *Aph*IV (62), which detoxifies the antibiotic hygromycin. Following DNA delivery, the tissue was cultured in liquid medium, supplemented with 50 mg/L hygromycin, for 6 to 8 weeks. Small yellowish-green clusters of hygromycin-resistant tissue were then removed from selection and allowed to grow for an additional 3 to 4 months. The subsequent differentiation, maturation, desiccation, and germination steps were conducted without further selection. Although fertile transgenic plants can be produced with the protocol outlined by Finer and McMullin (60), routine recovery of fertile plants proved problematic. The most likely explanation for this is the prolonged period of time the embryogenic tissue spent in culture prior to transformation. Tissue culture–induced mutations observed in regenerated plants are often correlated with the amount of time in culture, and this is clearly true for soybean (63). Bombardment of relatively young cultures (less than 6 months old) generally resulted in the recovery of few transformants (64). Repetitive subculturing over time may result in selection for tissue morphologies more receptive to stable transformation (64). Growth rates and morphologies of liquid embryogenic cultures were improved by optimizing the sugar and nitrogen composition of the medium (65). The improved growth rate of the target tissue permits the reproducible transformation of relatively young embryonic cultures by particle bombardment.

Research had been conducted to develop methods that circumvent the use of embryogenic cultures by bombarding immature cotyledons, but this has met with limited success. Santarem and Finer (66) used embryogenic cultures propagated on solid media and, following bombardment with a gene conferring resistance to hygromycin, performed the selection steps on solid media. This process proved slow and of relatively low efficiency.

Limited success has been achieved by attempting to couple *Agrobacterium*-mediated gene delivery with embryogenic cultures as targets. Ko and coworkers (67) infected immature cotyledons with *Agrobacterium* and induced somatic embryogenesis from the explants. They were able to recover transgenic plants at a frequency of about 1 to 2%. *Agrobacterium*-mediated transformation of embryogenic suspension cultures was attempted by first bombarding cultures with microprojectiles to produce "micro" wounds (68). This resulted in the recovery of transgenic cell lines, but recovery of transgenic plants was not reported. Sonication can also be used as a wounding mechanism (69,70), but only transgenic soybean cell lines were recovered.

Although drawbacks are associated with the original protocol outlined by Finer and McMullin (60), this system (with modifications made by other researchers) provided the basis for successful introduction of novel transgenes in soybean, including the Bt insecticidal protein cryIAc (71), a bovine ß-casein gene (72), a polypeptide derived from the coat protein of bean pod mottle virus (73), and a maize 15 kDa zein protein (74). As will be discussed, modifications have been made to seed storage lipids (75), amino acids (76), and proteins (77,78). Also of significance is the finding that a number of soybean cultivars can be transformed by bombardment of embryonic cultures (79,80).

Quality and Quantity of Transgenic Soybean

Two parameters to be considered when developing and optimizing transformation protocols are transformation frequency and the quality of the derived transformants. Transformation frequency represents the number of fertile, independent transgenic plants recovered from gene transfer to a given number of explants. Quality relates to the reliability with which genetically normal plants that consistently express the introduced transgene over multiple generations can be produced. Therefore, protocols that minimize tissue culture–induced mutations are important. Consistency and reliability of transgene expression is related to the complexity of the transgenic locus because transformants that harbor multicopy, rearranged and fragmented insertions are more likely to display instability of transgene expression. The relationship between expression instability and insertion complexity has been attributed to induction of epigenetic gene silencing (81–83). Hence, soybean transformation protocols that provide high frequency but have the tendency to generate complex transgene insertions are less desirable than protocols that produce high quality transformants at relatively low frequencies.

The interkingdom gene transfer of *A. tumefaciens* has been reviewed by others (84–86), and therefore a detailed description of the process will not be outlined herein. The beneficial features associated with *Agrobacterium*-mediated transformation include the transfer of a defined genetic element (T-DNA), resulting in the recovery of transgenic loci that generally posses low copy number. Most transformants produced by *Agrobacterium*-mediated transformation carry only transgenes defined by the borders of the T-DNA. However, in some instances the transgenic locus can harbor DNA derived from the plasmid backbone. The occurrence of vector backbone sequence in T-DNA insertion sites was observed in approximately 30% of rice transformants derived from *Agrobacterium*-mediated transformation (87). It is noteworthy that the probability of finding vector backbone sequences at a transgenic locus in rice was greater if the respective locus possessed multicopy T-DNA insertions (87). In a subset of T_3 progeny derived from nine randomly selected soybean transformants described by Buhr *et al.* (88), the offspring of only one line harbored *aadA*-hybridizing fragments (Clemente, unpublished data), indicating that a minority of transgenic soybean lines derived from *Agrobacterium*-mediated gene transfer will possess vector backbone sequences.

Olhoft *et al.* (26) conducted molecular characterization on 72 individual lines of transgenic soybean derived from *Agrobacterium*-mediated transformation. The data, revealing that 29 lines (40%) possessed hybridization patterns, are suggestive of complex T-DNA integration. The majority of the transgenic lines (48%) transmitted the transgenes as a single functional locus (26). Thus, complex loci containing multiple copies of the transgene can be inherited as a single locus. The observed segregation patterns are in agreement with the data reported by Clemente *et al.* (23).

Microprojectile bombardment (36,37) is a valuable tool for delivering transgenes to plant species. The major agriculture biotechnology products currently on the market today were derived from this technique, including RoundUp™ soybean (89) and Bt corn (90). Microprojectile bombardment has the advantage that multiple gene cassettes can be codelivered via separate molecules, which tend to be genetically linked in the resultant transformant. However, the downside of this DNA delivery technology is that it introduces the entire plasmid sequence and has the tendency to produce multicopy and fragmented inserts. The introduction of vector backbone DNA can be circumvented by coating particles with linearized DNA elements that contain only the gene(s) of interest. Transgenic rice produced by bombardment with these linearized gene cassettes had very simple patterns of transgene insertion (91). In addition, merely reducing the amount of DNA delivered per shot can lower insert complexity in microprojectile-derived soybean transformants (8). Thus, either *A. tumefaciens* or microprojectiles can be used to effectively generate soybean transformants that carry high quality transgenic alleles.

Transformation protocols often are limited to specific genotypes that are not of the highest agronomic quality. A number of avenues are being pursued to increase the efficiency of the soybean cotyledonary-node transformation protocol and to expand the range of transformable genotypes. Interaction of *A. tumefaciens* with plant cells can invoke a host response characteristic of programmed cell death (92). The resulting necrosis has a significant negative impact on the subsequent growth of tissue in culture. Tissue necrosis is clearly observed following infection of soybean cotyledonary-node with *A. tumefaciens* (93,94). Supplements added to the medium used during the cocultivation step of the soybean cotyledonary node transformation system, including L-cysteine, various thiol compounds, and chelators of copper and iron, significantly improved transformation frequencies (93,94). These additives appear to impact the plant cell's response to *A. tumefaciens* because they led to a decrease in tissue browning of the cotyledonary-node explants. These data suggest that strategies designed to alleviate cellular necrosis induced by *A. tumefaciens* can provide a means to enhance transformation efficiency without compromising the quality of the derived transformants.

Other approaches for optimizing *Agrobacterium*-mediated transformation protocols include modifications of the bacterium *per se* and/or the plant genotype. Results from transformation studies from other species indicate that several new approaches should be applicable to soybean.

Hansen (92) demonstrated that the introduction of genes that control programmed cell death, namely *p35* and *iap*, can negate the necrotic response of plant cells induced by *A. tumefaciens*. Both of these genes were first identified in baculoviruses, and they function to inhibit protease (caspase) activity in host cells. Caspases act in the degradation of cellular components and thus are important in the final stages of programmed cell death (95). In maize, about 80 to 90% of the immature embryos placed on tissue culture media designed to induce somatic embryogenesis respond appropriately. However, following inoculation with *A. tumefaciens*, embryogenesis is reduced to less than 10% (92). Stable transgenic maize lines expressing either *p35* or *iap*, inoculated with *A. tumefaciens*, displayed an embryogenic response similar to noninoculated control lines (92). Thus, a maize line expressing a negative regulator of programmed cell death can be used as a bridge to deliver novel transgenes into maize germplasm in a more efficient fashion. The newly introduced transgene(s) can then be segregated away from the genes that limit programmed cell death. Dickman *et al.* (96) demonstrated the utility of other negative regulators of programmed cell death that can be used to ameliorate tolerance toward necrotrophic fungi pathogenesis by blocking the necrotic response induced by these pathogens. To test the concept outlined by Hansen (92) in soybean, we recently introduced one of the antiapoptotic transgenes evaluated by Dickman *et al.* (96), *bcl-xl*, into the genotype Thorne, developed at Ohio State University. Expression levels have been monitored through T_3 generation in a subset of the lines, and seeds of homozygous individuals are currently being bulked up for testing in retransformation studies (Clemente, unpublished data).

Gordon-Kamm and coworkers (97) took an alternative approach to develop a genotype that can be used as an efficient bridge to introduce transgenes into a crop species by either *Agrobacterium*-mediated or microprojectile-derived transformation methods. The introduction of the *RepA* gene from the geminivirus wheat dwarf virus (WDV) was shown to stimulate cell divisions in both tobacco suspension cultures and maize embryogenic callus. This gene, and a number of other genes that control the cell cycle, limited programmed cell death and was used to increase rates of transformation in elite germplasm.

The concept of breeding for susceptibility to *A. tumefaciens* was explored by Mauro *et al.* (98). It was estimated that at least three genes controlled susceptibility to gall formation by wild-type strains of the bacterium. However, using a conventional breeding approach to derive a novel soybean genotype for genetic engineering has not proven to be an effective approach to date. On the other hand, recent studies have been reported that have identified plant genes that appear to be intimately involved in the *Agrobacterium*/plant cell interaction (99,100). These observations open the door to test novel strategies to enhance gene transfer protocols for crop plants. Tzfira *et al.* (101) identified an Arabidopsis gene, designated *VIP1*, that has homology to the bZIP family of transcription factors. *VIP1* has been shown to specifically interact with VirE2 and is postulated to assist in nuclear import of the T-complex (101). Expression of *VIP1* in tobacco rendered the transgenics more susceptible to *A. tumefaciens* than the wild-type counterpart (100). Transient expression of T-DNA elements, recovery of shoots under selection, and the frequency of recovery of transgenic whole plants were increased by the presence of the *VIP1* gene. These data suggest that *in planta* expression of *VIP1* can be a route to render a crop such as soybean more amendable to genetic engineering via *Agrobacterium*-mediated transformation.

As stated earlier, an alternative avenue for enhancing transformation frequencies using *A. tumefaciens* is to modify the virulence of the bacterium to soybean. The virulence of wild-type *A. tumefaciens* varies among strains (19,102). The hypervirulence of strain A281 exemplifies the synergistic interactions between the chromosomal background and resident Ti plasmid. The wild-type Dahlia strain designated Bo542 (L,L-succinamopine-type), carrying the resident Ti plasmid, pTiBo542, displays limited virulence to soybean. However, the presence of pTiBo542 in (nopaline-type) the chromosomal background of strain A136 manifests enhanced virulence on soybean and other plant species (103). In plant transformation systems, it is desirable to use *A. tumefaciens* strains that are disarmed. Disarming a wild-type *A. tumefaciens* strain involves deleting from the resident Ti plasmid the so-called oncogenes within the T-DNA element, while leaving the *vir* region intact. Torisky *et al.* (104) developed a partially disarmed Chry5 strain in order to exploit the hypervirulence of the Chry5 strain for use in soybean transformation. The resultant strain is designated KYRTI. Subsequent evaluation of KYRTI transconjugants (104) carrying the binary plasmid pBISN1 (105) revealed that it possessed increased virulence on soybean cotyledonary explants as compared to *A. tumefaciens* strains EHA105 (106) and GV3850 (107). However, the resident Ti plasmid in the Chry5 strain contains two T-DNA elements designated T-right (T_R) and T-left (T_L). Palanichelvam *et al.* (108) found that the partial disarming of KYRT1 was due to the deletion of only the RB region of the T_L T-DNA, whereas the T_R T-DNA was completely intact. Because the T_R and T_L T-DNAs can be processed independently, the T_R T-DNA is still transfer competent, explaining the weak oncogenic capacity of KYRT1 (104). Complete disarming of the Chry5 strain has been performed, and the resultant Ti plasmid is designated pKPSF2 (108).

Effecting expression patterns of various *vir* proteins has also been successful in modulating the virulence potential of *A. tumefaciens*. Constitutive expression or additional copies of *virG*, which regulates its own expression and that of others in the *vir* region, impacts the virulence potential of *A. tumefaciens* (109–111). As mentioned previously, strain A281 is hypervirulent on soybean and other plant species. The hypervirulence of this strain was found to be associated with an approximate 2.5 kb fragment of the *vir* region of the pTiBo542 (112), which contained *virG* and a segment of the *virB* operon. Moreover, the introduction of *virG* along with the entire *virB* operon of pTiBo542 imparted on other *A. tumefaciens* strains enhanced virulence (112). This led to the development of the so-called super binary system that is highly effective for transformation of monocot species, such as rice (113) and maize (114). The availability these novel *A. tumefaciens* strains will permit a broader evaluation of the interaction of *A. tumefaciens* with soybean and thereby allow further studies of the factors that limit the system.

Transformation for Functional Genomics

Considerable investments have been made in soybean genomics. This has resulted in a large repository of DNA sequence and expression information whose functional understanding will greatly improve our knowledge base of soybean biology. This in turn will help guide strategies to enhance soybean germplasm through genetic engineering and conventional breeding. A number of avenues are being employed to help elucidate the function of particular soybean genes, including bioinformatics, genetic mapping, and expression profiling. These tools strengthen our global understanding of gene function. This information, coupled with data acquired from reverse genetics approaches, serves as a strong complement to functional genomics programs.

The ability to directly regulate *in planta* gene expression is a powerful tool for functional genomics. Transformation can be used to either increase or decrease the expression of specific genes in plants. Down regulation can be triggered by the introduction of a transgene that can induce sequence specific RNA degradation through posttranscriptional gene silencing (82,83). Random mutagenesis by T-DNA tagging (87,115,116) or transposon mutagenesis (117–120) are other approaches to investigate the effect of decreasing the expression of specific genes. Nontransgenic mutagenic approaches using chemicals or UV irradiation also have been employed to modulate gene expression in plants. TILLING represents a powerful technique that couples chemical mutagenesis with methods to identify specific point mutations in a gene by denaturing HPLC (121). This technique permits the rapid identification of allelic variants of specific genes.

Rates of soybean transformation are now at sufficient levels to allow the modulation of gene expression by gene-silencing approaches (77,78,88,122). Staswick *et al.* (122) effectively down regulated the genes responsible for the production of soybean vegetative storage proteins by 50-fold by using a RNA antisense transgenic approach. Surprisingly this alteration in vegetative protein accumulation had no apparent phenotypic impact on the plant. Seed-specific attenuation of a Δ^{12} fatty acid desaturase, *FAD2-1*, resulted in substantial enhancement of the oleic acid component of the seed storage lipids (123), whereas dual seed-specific down-regulation of *FAD2-1* along with a palmitoyl preferred thioesterase, *fatB*, generated a high oleic and low palmitic fatty acid profile in the seed storage lipids (88). These reports clearly demonstrate the utility of exploiting posttranscriptional gene silencing as a means to modulate endogenous gene expression in soybean, which provides an enabling tool for soybean functional genomics programs.

One drawback associated with down-regulation via posttranscriptional gene silencing is that essential genes are difficult to characterize because a dominant lethal phenotype cannot be maintained in the heterozygous state. Therefore, additional means to modulate gene expression in soybean would be useful to augment posttranscriptional gene-silencing strategies.

Reverse genetic approaches using transposons (gene tagging) have been conducted for decades. Historically, transposons are classified into autonomous and nonautonomous categories. Autonomous transposons encode their own transposase and therefore are self-mobile. The nonautonomous members of a transposon family are typically deletions of the autonomous element and therefore lack coding capacity for a functional transposase. In fact, some nonautonomous transposons may resemble the corresponding autonomous member only at the termini.

Two maize transposon families have been effectively used for gene tagging in heterologus systems, *Activator/Dissociation* (*Ac/Ds*) (118,124,125) and

Enhancer/Suppressor-mutator (*En/Spm*) (126,127). Recently, d'Erfurth *et al.* (117) introduced the retrotransposable element *Tnt1* into the legume *Medicago truncatula.* The data revealed that *Tnt1* transposes early during the *in vitro* culturing of *M. truncatula* because some plants recovered through tissue culture (apparent escapes) harbored the retrotransposon element but lacked T-DNA integration. The number of *Tnt1* insertions varied from four to 30 per genome. Moreover, the retrotransposon element can be reactivated through additional *in vitro* culturing (117). Hence, this system may provide a valuable tool for saturated mutagenesis of this and possibly other leguminous species, such as soybean.

A potential shortcoming with random mutagenesis using a transposon-based system is that the disruption of a gene may not induce an observable phenotype. This can be avoided by coupling gene and enhancer-trap elements with a transposon tagging system (128). A gene trap element will activate the expression of a visual marker gene upon integration within a transcriptionally active region (129), whereas a visual marker gene in an enhancer trap construct will be activated when integrated proximal to an endogenous enhancer element (119). Importantly, expression patterns of the visual marker using these systems will reflect that of the tagged gene (128).

Both down-regulation and gene disruption approaches can serve as complementary tools for soybean functional genomics programs. However, the former can potentially down regulate an entire gene family, thereby confounding hypothesis testing, whereas the latter may disrupt a member of the family, but gene redundancy may mask the phenotype. A gain-of-function method can be an alternative method to help understand gene functionality without the potential complications associated with gene redundancy and/or influencing expression patterns of other genes within a family. Introducing a transgenic element that harbors enhancer regions oriented outward can generate misexpression mutants. This method has been successfully applied in *Arabidopsis* (116,130). Moreover, coupling a gain-of-function element with transposition capability (131) enhances the power of this methodology. A variation on a gain function element is to employ tissue-specific promoters to misexpress genes in a nonconstitutive manner (116). With respect to an oil seed crop such as soybean, by combining seed-specific promoters with gain-function elements that have transposition ability may provide a valuable tool to garner insight on metabolic pathways influencing oil and protein deposition in the developing embryo.

Transformation for Novel Trait Development

Enhancing germplasm for both input and output traits is the long-term mission of soybean breeding programs. Yield gains over the past few decades have been attributed to improved cultivars with superior agronomic performance and tolerance toward biotic and abiotic stresses, along with development of alternative management practices, including fertilization regimes and weed control. Advances in these areas will continue to play a major role in soybean production practices. The ability to genetically engineer soybean has provided the soybean breeder with the means to acquire novel traits that were inaccessible though conventional methods.

As mentioned earlier, soybean producers have capitalized on the cost-effective weed management offered by the use of RoundUp Ready™ soybean into integrated culture practices that has resulted in improved yields (1). However, significant barriers still hamper soybean production, including both biotic and abiotic stresses. The biotic agents limiting soybean production include fungi, nematodes, insects, and viruses. To date, enhanced tolerance toward only biotic stresses using transgenic approaches in soybean has been reported. These reports target the introduction of novel transgenes for imparted genetic resistance of white mold, lepidopteran insects, and two viral agents.

A fungal agent that is becoming a significant threat to production in North America is the necrotrophic pathogen *Sclerotinia sclerotiorum* (Lib.), the causal agent of white mold on soybean (132). A number of studies have revealed variability in susceptibility to this disease among different soybean genotypes and there is no known resistance (133–135). A primary determinant of pathogenesis in *S. sclerotiorum* is the production of oxalic acid (136). Donaldson *et al.* (137) introduced the wheat germin gene, designated *gf-2.8*, which encodes for oxalate oxygen oxioreductase (OxO) into soybean. The OxO catalyzes the reaction of oxalic acid with oxygen to produce carbon dioxide and hydrogen peroxide. Therefore, expression of OxO may impede pathogenesis of *S. sclerotiorum* in a two-pronged fashion, direct detoxification oxalic acid and generation of hydrogen peroxide, that functions in the defense response of plant cells (138). A 3-year field study of a transgenic line conducted in *S. sclerotinia* infested and noninfested soils revealed significant reductions in disease severity and incidence. Moreover, agronomic qualities of the transgenic line were not compromised (139).

A number of lepidopteran insect pests are capable of causing severe economic damage on soybean in southern growing regions of the United States (71,140). Management though chemical sprays of some of these pests may be cost prohibitive (141), and breeding efforts to develop elite soybean cultivars that harbor resistance has proven to be problematic (142,143). Stewart *et al.* (71) introduced a synthetic *Bt* gene designated *cry1Ac* into soybean in an attempt to deliver novel insect resistance into the germplasm. Insect-feeding assays with corn earworm, soybean looper, tobacco budworm, and velvetbean caterpillar on soybean tissue expressing *cryIAc* revealed significant reductions in both larvae weight and defoliation (71). Moreover, field evaluation on one of the transgenic *Bt* lines demonstrated the efficacy of this approach in controlling a number of lepidopteran pests (140). However, the insecticidal levels observed were not sufficient to limit the probability of developing *Bt* resistance buildup within the insect populations (71,140). To develop potentially more durable insect resistance within these *Bt* soybean lines, Walker *et al.* (144) pyramided a soybean QTL associated with corn earworm resistance with the *Bt* transgenic allele. BC_2F_3-derived populations were recovered employing marker-assisted selection that harbored either QTL alone or in combination with the *Bt* transgene. The combined effect of the dual mechanisms for insect control illustrates the potential of this approach for developing durable field resistance (144).

The first demonstration of a transgenic approach for disease control was reported in the mid-1980s. Tobacco transformants harboring the coat protein gene of tobacco mosaic virus displayed a delay in viral symptoms development upon challenge with TMV virions (145). This observation opened the door to a multitude of studies aimed at viral resistance through the introduction of viral genes into plants (146). This approach has been referred to as pathogen derived resistance (PDR) (147). PDR mechanistically operates by exploiting post transcriptional gene silencing (82,148). PDR has been successfully utilized in soybean toward the viral agents soybean mosaic virus (SMV) (149) and bean pod mottle virus (BPMV) (15,73). SMV is widespread in soybean-growing regions throughout the world, and infestations can lead to reduction in yields and seed quality (150,151). Wang *et al.* (149) introduced to soybean the coat protein gene of SMV strain N (152). A single soybean line was obtained that harbored four transgenic loci that segregated independently. Derived from this single transgenic line were four populations, which were subsequently challenged by mechanical inoculations with four strains of SMV. Two of the populations displayed resistance toward all SMV strains (149) under greenhouse and growth chamber conditions.

BPMV is a member of the comovirus group and is rather widespread throughout the United States. Soybean yield losses due to BPMV infection can reach greater than 50% (153,154). Reddy *et al.* (73) delivered the coat protein gene of BPMV into soybean. Subsequent challenge inoculations with BPMV revealed the ability of the virus to replicate in the inoculated leaf, but systemic movement was severely retarded in one of the 10 transgenic lines characterized (73). These reports reflect the utility of PDR as a means to introduce SMV and BPMV resistance into soybean germplasm for breeders to draw upon. Although these two traits may be of value individually, pyramiding SMV and BPMV transgenic resistance alleles would be ideal because soybean can be dually infected by both viruses, which leads to enhanced disease severity (155).

The major commodity products derived from soybean are meal and oil. Soybean seed consists of about 40% protein and 20% oil. The meal component (protein) is primarily utilized in animal feed supplements of which poultry and swine account for 66% of the consumption (www.unitedsoybean.org). In 2000, approximately 17.9 billion pounds of soybean oil was produced in the United States, with the vast majority being used for edible products, salad and cooking oil, baked products, and margarine (www.unitedsoybean.org). Implementing transformation technology for the introduction of novel output traits could provide a route to increased value of the commodity while delivering nutritional and functional benefits to the consumer without altering dietary habits.

The major storage proteins in soybean are globulins, glycinin and ß-conglycinin, which represent approximately 70% of the total seed proteins (156,157). Soybean seed storage proteins are rich in many of the essential amino acids, including lysine, isoleucine, and valine. However, soybean meal is deficient in the sulfur amino acids methionine and cysteine. Dinkins *et al.* (74) took a transgenic approach to express the methionine rich 15 kDa maize zein protein in a seed-specific fashion as a means to enhance methionine content of soybean meal. The data revealed a modest increase in methionine content, but it was still below applicable levels (74). Nonetheless this strategy appears to be viable option for enhancing methionine content of soybean meal.

Corn is typically used in cattle feed. However, due to corn meal's poor lysine content, soybean meal and/or direct lysine additives is often supplemented to corn-based animal feeds. Enhancing lysine content of seeds may provide a means to circumvent the need for direct lysine supplements to animal feeds. Lysine is produced through the aspartate-family amino acid biosynthetic pathway. Key regulatory enzymes within the pathway that control lysine production, and which themselves are feedback regulated by lysine, are aspartokinase (AK) and dihydrodipicolinic acid synthase (DHDPS). Falco *et al.* (76) introduced into soybean lysine feedback insensitive versions of DHDPS and AK, derived from *Corynebacterium* and *E. coli*, respectively. The two transgenes were placed under the control of the seed-specific ß-phaseolin promoter from common bean (158). In a subset of the transgenic soybean lines recovered, free lysine levels were observed up to 34% of the total amino acid pool within the seed. This reflects an approximate six-fold enhancement in free lysine content. Thus, nutritional improvements of soybean meal can be achieved by perturbing amino acid biosynthesis in the seed. However, it is noteworthy that transgenic lines that contained greater than 15% free lysine in the seed displayed negative agronomic phenotypes, including wrinkled seed and poor germination (76). Deregulating other branches within the aspartate-family biosynthetic pathway also generates negative phenotypes in plant systems (159), reflecting the potential pitfalls associated with modulating intracellular amino acid levels by perturbing cellular metabolism.

Genetic engineering has served as a tool to enhance the properties of soybean oil for both food and industrial applications. Commodity soybean oil is inherently oxidatively unstable due to a high proportion of polyunsaturated fatty acids. Therefore, the oil must be chemically treated to increase its shelf life and thereby limit the premature rancidity of food products. Soybean oil is typically hydrogenated in order to enhance its oxidative stability. Depending on application, varying degrees of hydrogenation are carried out. Soybean oil is partially hydrogenated to preferentially reduce the polyunsaturated fatty acids linoleic acid and linolenic acid for use in baked goods and frying applications (160). A more complete hydrogenation of soybean oil is typically conducted when end use is earmarked for shortenings and margarine (161). The downside of hydrogenating soybean oil is that a significant amount of trans fatty acids is produced (162). These trans fatty acids have been associated with heart disease (163,164). Moreover, soybean oil possesses approximately 17% saturated fatty acids, of which 13% are composed of palmitic acid, which has also been linked with heart disease (165). Therefore, soybean oil with reduced polyunsaturated fatty acids would possess increased oxidative stability and circumvent the need for hydrogenation, thereby circumventing the production of trans fatty acids. In addition, a concomitant reduction in palmitic acid levels can potentially augment the health attributes associated with non–trans fatty acid vegetable oil. To this end, genetic engineering approaches have been very successful in modifying fatty acid composition of soybean oil as a means to enhance oxidative stability (2,75,123,166–168).

Oleic acid (18:1) is converted to α-linoleic acid (18:2) via Δ^{12}-desaturase activity. In soybean, oleic acid is converted to α-linoleic in vegetative tissues by the *FAD2-2* gene, whereas *FAD2-2* and *FAD2-1* genes carry out this reaction during seed development (169). *FAD2-2* and *FAD2-1* are 73% identical at the DNA level; therefore, targeting the *FAD2-1* gene for PTGS (82,83) in a seed-specific fashion will down regulate both *FAD2-2* and *FAD2-1*. This in turn will lower polyunsaturated fatty acids and will raise oleic acid within the seed storage lipids to over 80% without compromising fatty acid profiles in vegetative tissues (168). Buhr *et al.* (88) successfully down-regulated Δ^{12} desaturase activity and palmitoyl-thioeserase activity (*FatB* gene) (170) in transgenic soybean, which resulted in elevated oleic acid and reduced palmitic acid in the seed storage lipids.

The enhanced oxidative stability of high oleic-acid soybean oil can be combined with elevated stearic acid for optimal functionality in solid-fat applications such as margarines (167). Pyramiding the Δ^{12} desaturase down-regulated line with a soybean line mutated a Δ^9 desaturase allele, resulting in a fatty acid profile in the resultant cross composed of approximately 30% stearic acid and 60% oleic acid (171). Importantly, the properties of high stearic/high oleic soybean oil mimic those of hydrogenated oil products (75). In addition to the health benefits associated with using high oleic acid and combined high oleic acid/high stearic acid soybean oils in food applications, these products may provide enhanced performance in biodiesel fuels (172).

Polyunsaturated fatty acids (PUFAs) possess a number of health attributes, and many of them have important commercial applications. Some of these novel PUFAs are produced only in a few plant species, for which cost-effective production is difficult. Two PUFAs that are of interest in both the pharmaceutical and nutraceutical industries are γ-linolenic acid (GLA) and stearidonic acid (STA) (173,174). GLA and STA are the Δ^6 desaturation products of linoleic acid and α-linolenic, respectively (175). Sato *et al.* (176) recently introduced *Borago officinalis* Δ^6 desaturse gene under the control of the seed-specific promoter ß-conglycinin, employing a two–T-DNA binary plasmid. A two–T-DNA binary plasmid simultaneously delivers two T-DNA elements, which is an effective means to derive marker-free transgenic plants (25,177). This strategy resulted in the recovery of marker-free transgenic soybean lines that harbored only the Δ^6 desaturase cassette. A T_3 homozygous, marker-free transgenic line grown under greenhouse conditions was evaluated in the T_4 generation under field conditions. Average GLA levels of T_3 seeds in the greenhouse ranged from 29.8 to 34.1%, and STA levels ranged from 3.7 to 5.1%. Under field conditions, at the T_4 generation, GLA levels averaged 23.7 to 30.4%, whereas average STA levels were observed at 2.1 to 2.9% (Clemente, unpublished data). These data reveal the power of genetic engineering to permit the production of these and other high-value molecules in a cost-efficient manner.

Summary

Although still somewhat technically demanding, soybean transformation can now be routinely performed in a number of laboratories around the world. Genetic transformation of soybean is being used to conduct basic studies of gene function. In addition, important agronomic traits from the tools of biotechnology can now be explored in soybean. Research will continue to improve soybean transformation, and as a result, the transformation technologies will become more robust.

References

1. Carpenter, J.E., and L.P. Gianessi, Agricultural Biotechnology: Updated Benefit Estimates, National Center for Food and Agricultural Policy, Washington, D.C., 2001.
2. Kinney, A.J., Engineering Soybeans for Food and Health, *AgBioForum 6*:1–5 (2003).
3. Galili, G., S. Galili, E. Lewinsohn, and Y. Tadmor, Genetic, Molecular, and Genomic Approaches to Improve the Value of Plant Foods and Feeds, *Crit. Rev. Plant Sci. 21*:167–204 (2002).
4. Veluthambi, K., A.K. Gupta, and A. Sharma, The Current Status of Plant Transformation Technologies, *Curr. Sci. 84*:368–380 (2003).
5. Twyman, R.M., P. Christou, and E. Stoger, Genetic Transformation of Plants and Their Cells, in *Biotechnology* and *Transgenic Plants*, edited by K.M. Oksman-Caldentey and W. Barz, Marcel Dekker, New York, 2002, pp. 111–141.
6. Bechtold, N., J. Ellis and G. Pelletier, In Planta *Agrobacterium*-Mediated Gene Transfer by Infiltration of Adult *Arabidopsis thaliana* Plants, *Crit. Rev. Acad. Sci., Paris 316*:1194–1199 (1993).
7. Desfeux, C., S.J. Clough, and A.F. Bent, Female Reproductive Tissues Are the Primary Target of Agrobacterium-Mediated Transformation by the Arabidopsis Floral-Dip Method, *Plant Physiol. 123*:895–904 (2000).
8. Simmonds, D., Molecular Methods of Plant Analysis, in *Genetic Transformation of Plants*, edited by J.F. Jackson and H.F. Linskus, Springer Verlag, New York, 2003, pp. 159–174.
9. Chee, P.P., and C.Y. Hu, Transgenic Soybean, in *Biotechnology in Agriculture and Forestry (Transgenic Crops)*, edited by Y.P.S. Bajaj, Springer, New York, 2000.
10. Hinchee, M.A., D.V. Connor-Ward, C.A. Newell, R.E. McDonnell, S.J. Sato, C.S. Gasser, D.A. Fischhoff, R.T. Fraley, and R.B. Horsch, Production of Transgenic Soybean Plants Using *Agrobacterium*-Mediated DNA Transfer, *Biotechnology* 6:915–922 (1988).
11. McCabe, D.E., W.F. Swain, B.J. Martinell, and P. Christou, Stable Transformation of Soybean (*Glycine max*) by Particle Acceleration, *Biotechnology* 6:923–926 (1988).

12. Cheng, T.Y., H. Saka, and T.H. Voqui-Dinh, Plant Regeneration from Soybean Cotyledonary Node Segments in Culture, *Plant Sci. Lett. 19*:91–99 (1980).
13. Sciaky, D., A.L. Montoya, and M.D. Chilton, Fingerprint of *Agrobacterium* Ti-Plasmids, *Plasmid 1*:238–253 (1978).
14. Gamborg, O.L., R.A. Miller, and K. Ojima, Nutrient Requirements of Suspension Cultures of Soybean Root Cells, *Exp. Cell Res. 50*:152–158 (1968).
15. Di, R., V. Purcell, G.B. Collins, and S.A. Ghabrial, Production of Transgenic Soybean Lines Expressing the Bean Pod Mottle Virus Coat Protein Precursor Gene, *Plant Cell Rep. 15*:746–750 (1996).
16. Luo, G., A. Hepburn, and J. Widholm, A Simple Procedure for the Expression of Genes in Transgenic Soybean Callus Tissue, *Plant Cell Rep. 13*:632–636 (1994).
17. Townsend, J.A., and L.A. Thomas, An Improved Method of *Agrobacterium*-Mediated Transformation of Cultured Soybean Cells, World Intellectual Property Organization, number *WO 94/02620* (1994).
18. Stachel, S.E., E. Messens, M. Van Montagu, and P. Zambryski, Identification of the Signal Molecules Produced by Wounded Plant Cells that Activate T-DNA Transfer in *Agrobacterium tumefaciens*, *Nature 318*:624–629 (1985).
19. Godwin, I., G. Todd, B. Ford-Lloys, and H.J. Newbury, The Effects of Acetosyringone and pH on *Agrobacterium*-Mediated Transformation Vary According to Plant Species, *Plant Cell Rep. 9*:671–675 (1991).
20. Fullner, K.J., and E.W. Nester, Temperature Affects the T-DNA Transfer Machinery of *Agrobacterium tumefaciens*, *J. Bacteriol. 178*:1498–1504 (1996).
21. Jin, S., Y.N. Song, W.Y. Deng, M.P. Gordon, and E.W. Nester, The Regulatory VirA Protein of *Agrobacterium tumefaciens* Does Not Function at Elevated Temperatures, *J. Bacteriol. 175*:6830–6835 (1993).
22. Riker, A.J., Studies on the Influence of Some Environmental Factors on the Development of Crown Gall, *J. Agric. Res. 32*:83–96 (1926).
23. Clemente, T., B.J. LaVallee, A.R. Howe, D. Conner-Ward, R. Rozman, P. Hunter, D.L. Broyles, D. Kasten, and M.A. Hinchee, Progeny Analysis of Glyphosate Selected Transgenic Soybeans Derived from *Agrobacterium*-Mediated Transformation, *Crop Sci. 40*:797–803 (2000).
24. Zhang, Z., A. Xing, P.E. Staswick, and T.E. Clemente, The Use of Glufosinate as a Selective Agent in Agrobacterium-Mediated Transformation of Soybean, *Plant Cell Tiss. Org. Cult. 56*:37–46 (1999).
25. Xing, A., Z. Zhang, S. Sato, P. Staswick, and T. Clemente, The Use of the Two T-DNA Binary System to Derive Marker-Free Transgenic Soybeans, *In Vitro Cell. Dev. Biol. 36*:456–463 (2000).
26. Olhoft, P.M., L.E. Flagel, C.M. Donovan, and D.A. Somers, Efficient Soybean Transformation Using Hygromycin-B Selection in the Cotyledonary-Node Method, *Planta 216*:723–735 (2003).
27. Mathias, R.J., and L.A. Boyd, Cefotaxime Stimulates Callus Growth Embryogenesis and Regeneration in Hexaploid Bread Wheat *Triticum aestivum* L., *Plant Sci. 46*:217–224 (1986).
28. Lin, J.-J., N. Assad-Garcia, and J. Kuo, Plant Hormone Effect of Antibodies on the Transformation Efficiency of Plant Tissues by *Agrobacterium tumefaciens* Cells, *Plant Sci. 109*:171–177 (1995).

29. Schmitt, F., E.J. Oakeley, and J.P. Jost, Antibiotics Induce Genome-Wide Hypermethylation in Cultured *Nicotiana tabacum* Plants, *J. Biol. Chem. 272*:1534–1540 (1997).
30. Nauerby, B., K. Billing, and R. Wyndaele, Influence of the Antibiotic Timentin on Plant Regeneration Compared to Carbenicillin and Cefotaxime in Concentrations Suitable for Elimination of *Agrobacterium tumefaciens*, *Plant Sci. 123*:169–177 (1997).
31. Chee, P.P., K.A. Fober, and J.L. Slighton, Transformation of Soybean (*Glycine Max*) by Infecting Germinating Seeds with *Agrobacterium tumefaciens*, *Plant Physiol. 91*:1212–1218 (1989).
32. Aragao, F.J., L. Sarokin, G.R. Vianna, and E.L. Rech, Selection of Transgenic Meristematic Cells Utilizing a Herbicidal Molecule Results in the Recovery of Fertile Transgenic Soybean [*Glycine max* (L.) Merrill] Plants at a High Frequency, *Theor. Appl. Genet. 101*:1–6 (2000).
33. Christou, P., W.F. Swain, N.S. Yang, and D.E. McCabe, Inheritance and Expression of Foreign Genes in Transgenic Soybean Plants, *Proc. Natl. Acad. Sci. USA 86*:7500–7504 (1989).
34. Chee, P.P., and J.L. Slighton, Transformation of Soybean (*Glycine max*) via *Agrobacterium tumefaciens* and Analysis of Transformed Plants, in *Methods in Molecular Biology*, edited by K.M.A. Gartland and M.R. Davey, Humana Press, Totowa, N.J., 1995, pp. 101–119.
35. Martinell, B.J., L.S. Julson, C.A. Emler, Y.S. Huang, D.E. McCabe, and E.J. Williams, U.S. Patent 6,384,301 (2002).
36. Klein, T.M., E.D. Wolf, R. Wu, and J.C. Sanford, High-Velocity Microprojectiles for Delivering Nucleic Acids into Living Cells, *Nature 327*:70–73 (1987).
37. Klein, T.M., M.E. Fromm, A.K. Weissinger, D. Tomes, D.J. Schaaf, M. Sletten, and J.C. Sanford, Transfer of Foreign Genes into Intact Maize Cells with High-Velocity Microprojectiles, *Proc. Natl. Acad. Sci. USA 85*:4305–4309 (1988).
38. Murashige, T., and F. Skoog, A Revised Medium for Rapid Growth and Bioassays with Tobacco Tissue Cultures, *Physiol. Plantarum 15*:473–497 (1962).
39. Christou, P., and D.E. McCabe, Prediction of Germ-Line Transformation Events in Chimeric R_0 Transgenic Soybean Plantlets Using Tissue-Specific Expression Patterns, *Plant J. 2*:283–290 (1992).
40. Barwale, U.B., and J.M. Widholm, Somaclonal Variation in Plants Regenerated from Cultures of Soybean, *Plant Cell Rep. 6*:365–368 (1987).
41. Sathasivan, K., G.W. Haughn, and N. Murai, Molecular Basis of Imidazolinone Herbicide Resistance in *Arabidopsis thaliana* Var. Columbia, *Plant Physiol.* 97:1044–1050 (1991).
42. Hepburn, A.G., J. White, L. Pearson, M.J. Maunders, L.E. Clarke, and A.G. Prescott, The Use of pNJ5000 as an Intermediate Vector for the Genetic Manipulation of *Agrobacterium* Ti-Plasmids, *J. Gen. Microbiol. 131*:2961–2969 (1985).
43. Trick, H.N., and J.J. Finer, SAAT: Sonication-Assisted *Agrobacterium*-Mediated Transformation, *Transgenic Res. 6*:329–336 (1997).
44. Phillips, G.C., and G.B. Collins, Induction and Development of Somatic Embryos from Cell Suspension Cultures of Soybean *Glycine max*, *Plant Cell Tiss. Org. Cult. 1*:123–130 (1981).

45. Christianson, M.L., D.A. Warnick, and P.S. Carlson, A Morphogenic Competent Soybean *Glycine max* Cultivar Mitchell Suspension Culture, *Science 222*:632–634 (1983).
46. Barwale, U.B., H.R. Kerns, and J.M. Widholm, Plant Regeneration from Callus Cultures of Several Soybean Genotypes Via Embryogenesis and Organogenesis, *Planta 167*:473–481 (1986).
47. Komatsuda, T., and K. Ohyama, Genotypes of High Competence for Somatic Embryogenesis and Plant Regeneration in Soybean *Glycine max*, *Theor. Appl. Genet. 75*:695–700 (1988).
48. Parrott, W.A., G. Dryden, S. Vogt, D.F. Hildebrand, G.B. Collins, and E.G. Williams, Optimization of Somatic Embryogenesis and Embryo Germination in Soybean, *In Vitro Cell. Dev. Biol. 24*:817–820 (1988).
49. Ranch, J.P., L. Oglesby, and A.C. Zielinski, Plant Regeneration from Embryo-Derived Tissue Cultures of Soybean by Somatic Embryogenesis, *In Vitro Cell. Dev. Biol. 21*:653–657 (1985).
50. Finer, J.J., Apical Proliferation of Embryogenic Tissue of Soybean (*Glycine max* (L.) Merrill), *Plant Cell Rep. 7*:238–241 (1988).
51. Finer, J.J., and A. Nagasawa, Development of an Embryogenic Suspension Culture of Soybean (*Glycine max* Merrill), *Plant Cell Tiss. Org. Cult. 15*:125–136 (1988).
52. Santarem, E.R., B. Pelissier, and J.J. Finer, Effect of Explant Orientation, pH, Solidifying Agent, and Wounding on Initiation of Soybean Somatic Embryos, *In Vitro Cell. Dev. Biol. 33*:13–19 (1997).
53. Lazzeri, P.A., D.F. Hildebrand, and G.B. Collins, Soybean Somatic Embryogenesis: Effects of Nutritional, Physical, and Chemical Factors, *Plant Cell Tiss. Org. Cult. 10*:209–220 (1987).
54. Hartweck, L.M., P.A. Lazzeri, D. Gui, G.B. Collins, and E.G. Williams, Auxin-Orientation Effects on Somatic Embryogenesis from Immature Soybean Cotyledons, *In Vitro Cell. Dev. Biol. 24*:821–828 (1988).
55. Wright, M.S., K.L. Launis, R. Novitzky, J.H. Duesing, and C.T. Harms, A Simple Method for the Recovery of Multiple Fertile Plants from Individual Somatic Embryos of Soybean (*Glycine max* (L.) Merrill), *In Vitro Cell. Dev. Biol. 27*:153–157 (1991).
56. Bailey, M.A., H.R. Boerma, and W.A. Parrott, Genotype Effects on Proliferative Embryogenesis and Plant Regeneration of Soybean, *In Vitro Cell. Dev. Biol. 29*:102–108 (1993).
57. Buchheim, J.A., S.M. Colburn, and J.P. Ranch, Maturation of Soybean Somatic Embryos and the Transition to Plantlet Growth, *Plant Physiol. 89*:768–775 (1989).
58. Samoylov, V.M., D.M. Tucker, F. Thibaud-Nissen, and W.A. Parrott, A Liquid-Medium–Based Protocol for Rapid Regeneration from Embryogenic Soybean Cultures, *Plant Cell Rep. 18*:49–54 (1998).
59. Walker, D.R., and W.A. Parrott, Effect of Polyethylene Glycol and Sugar Alchols on Soybean Somatic Embryo Germination and Conversion, *Plant Cell Tiss. Org. Cult. 64*:55–62 (2001).
60. Finer, J.J., and M.D. McMullen, Transformation of Soybean Via Particle Bombardment Embryogenic Suspension Culture Tissue, *In Vitro Cell. Dev. Biol. 27P*:175–182 (1991).
61. Jefferson, R.A., The GUS Reporter Gene System, *Nature 342*:837–838 (1989).

62. Gritz, L., and J. Davies, Plasmid-Encoded Hygromycin B Resistance: The Sequence of Hygromycin B Phosphotransferase Gene and Its Expression in *Escherichia coli* and *Saccaromyces cervisiae*, *Gene 26*:179–188 (1983).
63. Singh, R.J., T.M. Klein, C.J. Mauvais, S. Knowlton, T. Hymowitz, and C.M. Kostow, Cytological Characterization of Transgenic Soybean, *Theor. Appl. Genet. 96*:319–324 (1998).
64. Hazel, C.B., T.M. Klein, M. Anis, H.D. Wilde, and W.A. Parrott, Growth Characteristics and Transformability of Soybean Embryogenic Cultures, *Plant Cell Rep. 17*:765–772 (1998).
65. Samoylov, V.M., D.M. Tucker, and W.A. Parrott, Soybean (*Glycine max* (L.) Merrill) Embryogenic Cultures: The Role of Sucrose and Total Nitrogen Content on Proliferation, *In Vitro Cell. Dev. Biol. 34*:8–13 (1998).
66. Santarem, E.R., and J.J. Finer, Transformation of Soybean (*Glycine max* (L.) Merrill) Using Proliferative Embryogenic Tissue Maintained on Semi-Solid Medium, *In Vitro Cell. Dev. Biol. 35*:451–455 (1999).
67. Ko, T.S., S.B. Lee, S. Kransnyanski, and S.S. Korban, Two Critical Factors Are Required for Efficient Transformation of Multiple Soybean Cultivars: *Agrobacterium* Strain and Orientation of Immature Cotyledonary Explant, *Theor. Appl. Genet. 107*:439–447 (2003).
68. Droste, A., G. Pasquali, and M.H. Bodanese-Zanettini, Integrated Bombardment and *Agrobacterium* Transformation System: An Alternative Method for Soybean Transformation, *Plant Mol. Biol. Rep. 18*:51–59 (2000).
69. Trick, H.N., and J.J. Finer, Sonication-Assisted *Agrobacterium*-Mediated Transformation of Soybean (*Glycine max* (L.) Merrill) Embryogenic Suspension Culture Tissue, *Plant Cell Rep. 17*:482–488 (1998).
70. Santarem, E.R., H.N. Trick, J.S. Essig, and J.J. Finer, Sonication-Assisted *Agrobacterium*-Mediated Transformation of Soybean Immature Cotyledons: Optimization of Transient Expression, *Plant Cell Rep. 17*:752–759 (1998).
71. Stewart Jr., C.N., M.J. Adang, J.N. All, H.R. Boerma, G. Cardineau, D. Tucker, and W.A. Parrott, Genetic Transformation, Recovery, and Characterization of Fertile Soybean Transgenic for a Synthetic *Bacillus thuringiensis Crylac* Gene, *Plant Physiol. 112*:121–129 (1996).
72. Maughan, P.J., R. Philip, M.J. Cho, J.M. Widholm, and L.O. Vodkin, Biolistic Transformation, Expression, and Inheritance of Bovine ß-Casein in Soybean (*Glycine max*), *In Vitro Cell. Dev. Biol. 35*:344–349 (1999).
73. Reddy, M.S.S., S.A. Ghabrial, C.T. Redmond, R.D. Dinkins, and G.B. Collins, Resistance to Bean Pod Mottle Virus in Transgenic Soybean Lines Expressing the Capsid Polyprotein, *Phytopathology 91*:831–838 (2001).
74. Dinkins, R.D., M.S.S. Reddy, C.A. Meurer, B. Yan, H. Trick, N.F. Thibaud, J.J. Finer, W.A. Parrott, and G.B. Collins, Increased Sulfur Amino Acids in Soybean Plants Overexpressing the Maize 15 kDa Zein Protein, *In Vitro Cell. Dev. Biol. 37*:742–747 (2001).
75. Kinney, A.J., Plants as Industrial Chemical Factories—New Oils from Genetically Engineered Soybeans, *Fett/Lipid 100*:173–179 (1998).
76. Falco, S.C., T. Guida, M. Locke, J. Mauvais, C. Sanders, R.T. Ward, and P. Webber, Transgenic Canola and Soybean Seeds with Increased Lysine, *Biotechnology 13*:577–582 (1995).

77. Herman, E.M., R.M. Helm, R. Jung, and A.J. Kinney, Genetic Modification Removes an Immunodominant Allergen from Soybean, *Plant Physiol. 132*:36–43 (2003).
78. Kinney, A.J., R. Jung, and E.M. Herman, Cosuppression of the a Subunits of ß-Conglycinin in Transgenic Soybean Seeds Induces the Formation of Endoplasmic Reticulum-Derived Protein Bodies, *Plant Cell 13*:1165–1178 (2001).
79. Simmonds, D.H., and P.A. Donaldson, Genotype Screening for Proliferative Embryogenesis and Biolistic Transformation of Short-Season Soybean Genotypes, *Plant Cell Rep. 19*:485–490 (2000).
80. Droste, A., G. Pasquali, and M.H. Bodanese-Zanettini, Transgenic Fertile Plants of Soybean (*Glycine max* (L.) Merrill) Obtained from Bombarded Embryogenic Tissue, *Euphytica 127*:367–376 (2002).
81. Matzke, A.M., A.J.M. Matzke, and W.B. Eggleston, Paramutation and Transgene Silencing: A Common Response to Invasive DNA, *Trends Plant Sci. 1*:382–388 (1996).
82. Vance, V., and H. Vaucheret, RNA Silencing in Plants—Defense and Counterdefense, *Science 292*:2277–2280 (2001).
83. Cerutti, H., RNA Interference: Traveling in the Cell and Gaining Functions?, *Trends Genet. 19*:39–46 (2003).
84. Zupan, J., T.R. Muth, O. Draper, and P. Zambryski, The Transfer of DNA from *Agrobacterium tumefaciens* into Plants: A Feast of Fundamental Insights, *Plant J. 23*:11–28 (2000).
85. Zupan, J.R., and P. Zambryski, Transfer of T-DNA from *Agrobacterium* to the Plant Cell, *Plant Physiol. 107*:1041–1047 (1995).
86. Zhu, J., P.M. Oger, B. Schrammeijer, P.J.J. Hooykaas, S.K. Farrand, and S.C. Winans, The Bases of Crown Gall Tumorigenesis, *J. Bacteriol. 182*:3885–3895 (2000).
87. Sallaud, C., D. Meynard, J. Van Boxtel, C. Gay, M. Bès, J.P. Brizard, P. Larmande, D. Ortega, M. Raynal, M. Portefaix, P.B.F. Ouwerkerk, S. Rueb, M. Delseny, and E. Guiderdoni, Highly Efficient Production and Characterization of T-DNA Plants for Rice (*Oryza sativa* L.) Functional Genomics, *Theor. Appl. Genet. 106*:1396–1408 (2003).
88. Buhr, T., S. Sato, F. Ebrahim, A. Xing, Y. Zhou, M. Mathiesen, B. Schweiger, A.J. Kinney, P. Staswick, and T. Clemente, Ribozyme Termination of RNA Transcripts Down Regulate Seed Fatty Acid Genes in Transgenic Soybean, *Plant J. 30*:155–163 (2002).
89. Padgette, S., K. Kolacz, X. Delannay, D.B. Re, B.J. LaVallee, C.N. Tinius, W.K. Rhodes, Y.I. Otero, G.F. Barry, D. Eichholtz, V.M. Peschke, D.L. Nida, R.B. Taylor, and G.M. Kishore, Development, Identification, and Characterization of a Glyphosate-Tolerant Soybean Line, *Crop Sci. 35*:1451–1461 (1995).
90. Armstrong, C.L., G.B. Parker, J.C. Pershing, S.M. Brown, P.R. Sanders, D.R. Duncan, T. Stone, D.A. Dean, D.L. DeBoer, J. Hart, A.R. Howe, F.M. Morrish, M.E. Pajeau, W.L. Peterson, B.J. Reich, R. Rodriguez, C.G. Santino, S.J. Sato, W. Schuler, S.R. Sims, S. Stehling, L.J. Tarochione, and M.E. Fromm, Field Evaluation of European Corn Borer Control in Progeny of 173 Transgenic Corn Events Expressing an Insecticidal Protein from *Bacillus thuringiensis*, *Crop Sci. 35*:550–557 (1995).
91. Fu, X., L.T. Duc, S. Fontana, B.B. Bong, P. Tinjuangjun, D. Sudhakar, R.M. Twyman, P. Christou, and A. Kohli, Linear Transgene Constructs Lacking Vector Backbone Sequences Generate Low-Copy-Number Transgenic Plants with Simple Integration Patterns, *Transgenic Res. 9*:11–19 (2000).

92. Hansen, G., Evidence for *Agrobacterium*-Induced Apoptosis in Maize Cells, *Mol. Plant Microbe Interact. 13*:649–657 (2000).
93. Olhoft, M., K. Lin, J. Galbraith, C. Nielsen, and A. Somers, The Role of Thiol Compounds in Increasing Agrobacterium-Mediated Transformation of Soybean Cotyledonary-Node Cells, *Plant Cell Rep. 20*:731–737 (2001).
94. Olhoft, M., and A. Somers, L-Cysteine Increases *Agrobacterium*-Mediated T-DNA Delivery into Soybean Cotyledonary-Node Cells, *Plant Cell Rep. 20*:706–711 (2001).
95. Clem, R.J., Baculoviruses and Apoptosis: The Good, the Bad, and the Ugly, *Cell Death Differ. 8*:137–143 (2001).
96. Dickman, M.B., Y.K. Park, T. Oltersdorf, W. Li, T. Clemente, and R. French, Abrogation of Disease Development in Plants Expressing Animal Antiapoptotic Genes, *Proc. Natl. Acad. Sci. USA 98*:6957–6962 (2001).
97. Gordon-Kamm, W., B.P. Dilkes, K. Lowe, G. Hoerster, X. Sun, M. Ross, L. Church, C. Bunde, J. Farrell, P. Hill, S. Maddock, J. Snyder, L. Sykes, Z. Li, Y.M. Woo, D. Bidney, and B.A. Larkins, Stimulation of the Cell Cycle and Maize Transformation by Disruption of the Plant Retinoblastoma Pathway, *Proc. Natl. Acad. Sci. USA 99*:11975–11980 (2002).
98. Mauro, A.O., T.W. Pfeiffer, and G.B. Collins, Inheritance of Soybean Susceptibility to *Agrobacterium tumefaciens* and Its Relationship to Transformation, *Crop Sci. 35*:1152–1156 (1995).
99. Yi, H., K.S. Mysore, and S.B. Gelvin, Expression of the *Arabidopsis* Histone H2a-1 Gene Correlates with Susceptibility to *Agrobacterium* Transformation, *Plant J. 32*:285–298 (2002).
100. Tzfira, T., M. Vaidya, and V. Citovsky, Increasing Plant Susceptibility to Agrobacterium Infection by Overexpression of the Arabidopsis Nuclear Protein Vip1, *Proc. Natl. Acad. Sci. USA 99*:10435–10440 (2002).
101. Tzfira, T., M. Vaidya, and V. Citovsky, Vip1, an Arabidopsis Protein That Interacts with *Agrobacterium* Vire2, Is Involved in Vire2 Nuclear Import and Agrobacterium Infectivity, *EMBO J. 20*:3596–3607 (2001).
102. Kovacs, L.G., and S.G. Pueppke, The Chromosomal Background of *Agrobacterium tumefaciens* Chry5 Conditions High Virulence on Soybean, *Mol. Plant Microbe Interact. 6*:601–608 (1993).
103. Hood, E.E., G.L. Helmer, R.T. Fraley, and M.D. Chilton, The Hypervirulence of *Agrobacterium tumefaciens* A281 Is Encoded in a Region of pTiBo542 Outside of T-DNA, *J. Bacteriol. 168*:1291–1301 (1986).
104. Torisky, R.S., L. Kovacs, S. Avdiushko, J.D. Newman, A.G. Hunt, and G.B. Collins, Development of a Binary Vector System for Plant Transformation Based on the Supervirulent *Agrobacterium tumefaciens* Strain Chry5, *Plant Cell Rep. 17*:102–108 (1997).
105. Narasimhulu, S.B., Y.B. Deng, R. Sarria, and S.B. Gelvin, Early Transcription of *Agrobacterium* T-DNA Genes in Tobacco and Maize, *Plant Cell Rep. 8*:873–886 (1996).
106. Hood, E.E., S.B. Gelvin, L.S. Melchers, and A. Hoekema, New Agrobacterium Helper Plasmids for Gene Transfer to Plants, *Transgenic Res. 2*:208–218 (1993).
107. Zambryski, P., H. Joos, C. Genetello, J. Leemans, M. Van Montagu, and J. Schell, Ti Plasmid Vector for the Introduction of DNA into Plant Cells without Alteration of Their Normal Regeneration Capacity, *EMBO J. 2*:2143–2150 (1983).

108. Palanichelvam, K., P. Oger, S.J. Clough, C. Cha, A.F. Bent, and S.K. Farrand, A Second T-Region of the Soybean-Supervirulent Chrysopine-Type Ti Plasmid pTiChry5, and Construction of a Fully Disarmed *Vir* Helper Plasmid, *Mol. Plant Microbe Interact. 13*:1081–1091 (2000).
109. Van der Fits, L., E.A. Deakin, J.H. Hoge, and J. Memelink, The Ternary Transformation System: Constitutive *VirG* on a Compatible Plasmid Dramatically Increases *Agrobacterium*-Mediated Plant Transformation, *Plant Mol. Biol. 43*:495–502 (2000).
110. Liu, C.N., X.Q. Li, and S.B. Gelvin, Multiple Copies of *Virg* Enhance the Transient Transformation of Celery, Carrot and Rice Tissues by *Agrobacterium tumefaciens*, *Plant Mol. Biol. 20*:1071–1087 (1992).
111. Hansen, G., A. Das, and M.D. Chilton, Constitutive Expression of the Virulence Genes Improves the Efficiency of Plant Transformation by *Agrobacterium*, *Proc. Natl. Acad. Sci. USA 91*:7603–7607 (1994).
112. Jin, S.G., T. Komari, M.P. Gordon, and E.W. Nester, Genes Responsible for the Supervirulence Phenotype of *Agrobacterium tumefaciens* A281, *J. Bacteriol. 169*:4417–4425 (1987).
113. Hiei, Y., S. Ohta, T. Komari, and T. Kumashiro, Efficient Transformation of Rice (*Oryza sativa* L.) Mediated by *Agrobacterium* and Sequence Analysis of the Boundaries of the T-DNA, *Plant J. 6*:271–282 (1994).
114. Ishida, Y., H. Saito, S. Ohta, Y. Hiei, T. Komari, and T. Kumashiro, High Efficiency Transformation of Maize (*Zea mays* L.) Mediated by *Agrobacterium tumefaciens*, *Nat. Biotechnol. 14*:745–750 (1996).
115. Nakazawa, M., T. Ichikawa, A. Ishikawa, H. Kobayashi, Y. Tsuhara, M. Kawashima, K. Suzuki, S. Muto, and M. Matsui, Activation Tagging, a Novel Tool to Dissect the Functions of a Gene Family, *Plant J. 34*:741–750 (2003).
116. Weigel, D., J.H. Ahn, M.A. Blazquez, J.O. Borevitz, S.K. Christensen, C. Fankhauser, C. Ferrandiz, I. Kardailsky, E.J. Malancharuvil, M.M. Neff, J.T. Nguyen, S. Sato, Z.Y. Wang, Y. Xia, R.A. Dixon, M.J. Harrison, C.J. Lamb, M.F. Yanofsky, and J. Chory, Activation Tagging in *Arabidopsis*, *Plant Physiol. 122*:1003–1013 (2000).
117. d'Erfurth, I., V. Cosson, A. Eschstruth, H. Lucas, A. Kondorosi, and P. Ratet, Efficient Transposition of the *Tnt1* Tobacco Retrotransposon in the Model Legume *Medicago truncatula*, *Plant J. 34*:95–106 (2003).
118. Izawa, T., T. Ohnishi, T. Nakano, N. Ishida, H. Enoki, H. Hashimoto, K. Itoh, R. Terada, C. Wu, C. Miyazaki, T. Endo, S. Iida, and K. Shimamoto, Transposon Tagging in Rice, *Plant Mol. Biol. 35*:219–229 (1997).
119. Martienssen, R.A., Functional Genomics: Probing Plant Gene Function and Expression with Transposons, *Proc. Natl. Acad. Sci. USA 95*:2021–2026 (1998).
120. Osborne, B.I., and B. Baker, Movers and Shakers: Maize Transposons as Tools for Analyzing Other Plant Genomes, *Curr. Opin. Cell Biol. 7*:406–413 (1995).
121. McCallum, C.M., L. Comai, E.A. Greene, and S. Henikoff, Targeted Screening for Induced Mutations, *Nat. Biotechnol. 18*:455–457 (2000).
122. Staswick, P.E., Z. Zhang, T.E. Clemente, and J.E. Specht, Efficient Down-Regulation of the Major Vegetative Storage Protein Genes in Transgenic Soybean Does Not Compromise Plant Productivity, *Plant Physiol. 127*:1819–1826 (2001).
123. Mazur, B., E. Krebbers, and S. Tingey, Gene Discovery and Product Development for Grain Quality Traits, *Science 285*:372–375 (1999).

124. Chuck, G., T. Robbins, C. Nijjar, E. Ralston, N. Courtney-Gutterson, and H.K. Dooner, Tagging and Cloning of a Petunia Flower Color Gene with the Maize Transposable Element *Activator, Plant Cell 5*:371–378 (1993).
125. Springer, P.S., W.R. McCombie, V. Sundaresan, and R.A. Martienssen, Gene Trap Tagging of PROLIFERA, an Essential MCM2-3-5-Like Gene in *Arabidopsis*, *Science 268*:877–880 (1995).
126. Speulman, E., P.L. Metz, G. van Arkel, B. te Lintel Hekkert, W.J. Stiekema, and A. Pereira, A Two-Component *Enhancer–Inhibitor* Transposon Mutagenesis System for Functional Analysis of the Arabidopsis Genome, *Plant Cell 11*:1853–1866 (1999).
127. Wellesen, K., F. Durst, F. Pinot, I. Beneniste, K. Nettesheim, E. Wisman, S. Steiner-Lange, H. Saedler, and A. Yephremov, Functional Analysis of the *Lacerata* Gene of *Arabidopsis* Provides Evidence for Different Roles of Fatty Acid W-Hydroxylation in Development, *Proc. Natl. Acad. Sci. USA 98*:9694–9699 (2001).
128. Sundaresan, V., P. Springer, T. Volpe, S. Haward, J.D.G. Jones, C. Dean, H. Ma, and R.A. Martienssen, Patterns of Gene Action in Plant Development Revealed by Enhancer Trap and Gene Trap Transposable Elements, *Genes Dev. 9*:1797–1810 (1995).
129. Babiychuk, E., M. Fuangthong, M. Van Montagu, D. Inze, and S. Kushnir, Efficient Gene Tagging in *Arabidopsis thaliana* Using a Gene Trap Approach, *Proc. Natl. Acad. Sci. USA 94*:12722–12727 (1997).
130. Kakimoto, T., Cki1, a Histidine Kinase Homolog Implicated in Cytokinin Signal Transduction, *Science 274*:982–985 (1996).
131. Wilson, K., D. Long, J. Swinburne, and G. Coupland, A Dissociation Insertion Causes a Semidominant Mutation That Increases Expression of *Tiny*, an Arabidopsis Gene Related to *APETALA2*, *Plant Cell 8*:659–671 (1996).
132. Wrather, J.A., T.R. Anderson, D.M. Arsyad, J. Gai, L.D. Ploper, A. Porta-Puglia, H.H. Ram, and J.T. Yorinoru, Soybean Disease Loss Estimates for the Top 10 Soybean Producing Countries in 1994, *Plant Dis. 81*:107–110 (1997).
133. Kim, H.S., C.H. Sneller, and B.W. Diers, Evaluation of Soybean Cultivars for Resistance to *Sclerotinia* Stem Rot in Field Environments, *Crop Sci. 39*:64–68 (1999).
134. Wegulo, S.N., X.B. Yang, and C.A. Martinson, Soybean Cultivar Responses to *Sclerotinia sclerotiorum* in Field and Controlled Environment Studies, *Plant Dis. 82*:1264–1270 (1998).
135. Kim, H.S., G.L. Hartman, J.B. Manandhar, G.L. Graef, J.R. Steadman, and B.W. Diers, Reaction of Soybean Cultivars to *Sclerotinia* Stem Rot in Field, Greenhouse, and Laboratory Evaluations, *Crop Sci. 40*:665–669 (2000).
136. Dickman, M.B., and A. Mitra, *Arabidopsis thaliana* as a Model for Studying *Sclerotinia sclerotiorum* Pathogenesis, *Physiol. Mol. Plant Pathol. 41*:255–263 (1992).
137. Donaldson, P.A., T.R. Anderson, B.G. Lane, A.L. Davidson, and D.H. Simmonds, Soybean Plants Expressing an Active Oligomeric Oxalate Oxidase from the Wheat *Gf-2.8* (Germin) Gene Are Resistant to the Oxalate-Secreting Pathogen *Sclerotinia sclerotiorum*, *Mol. Plant Pathol. 59*:297–307 (2001).
138. Tenhaken, R., A.J. Levine, L.F. Brisson, R.A. Dixon, and C. Lamb, Function of the Oxidative Burst in Hypersensitive Disease Resistance, *Proc. Natl. Acad. Sci. USA 92*:4158–4163 (1995).
139. Cober, E.R., S. Rioux, I. Rajcan, P.A. Donsaldson, and D.H. Simmonds, Partial Resistance to White Mold in a Transgenic Soybean Line, *Crop Sci. 43*:92–95 (2003).

140. Walker, D.R., J.N. All, R.M. McPherson, H.R. Boerma, and W.A. Parrott, Field Evaluation of Soybean Engineered with a Synthetic *Cry1ac* Transgene for Resistance to Corn Earworm, Soybean Looper, Velvetbean Caterpillar (Lepidoptera: Nocuidae), and Lesser Cornstalk Borer (Lepidoptera: Pyralidae), *J. Econ. Entomol. 93*:613–622 (2000).
141. Mascarenhas, R.N., and D.J. Boethel, Responses of Field-Collected Strains of Soybean Looper (Lepidoptera: Noctuidae) to Selected Insecticides Using an Artificial Diet Overlay Bioassay, *J. Econ. Entomol. 90*:1117–1124 (1997).
142. All, J.N., H.R. Boerma, and J.W. Todd, Screening Soybean Genotypes in the Greenhouse for Resistance to Insects, *Crop Sci. 29*:1156–1159 (1989).
143. Rowan, G.B., H.R. Boerma, J.N. All, and J.W. Todd, Soybean Maturity Effects on Expression of Resistance to Lepidopterous Insects, *Crop Sci. 33*:433–436 (1993).
144. Walker, D., H.R. Boerma, J.N. All, and W.A. Parrott, Combining cry1Ac and QTL Alleles from PI 229358 to Improve Soybean Resistance to Lepidopteran Pests, *Mol. Breed. 9*:43–51 (2002).
145. Abel, P.P., R.S. Nelson, B. De, N. Hoffmann, S.G. Rogers, R.T. Fraley, and R.N. Beachy, Delay of Disease Development in Transgenic Plants That Express the Tobacco Mosaic Virus Coat Protein Gene, *Science 232*:738–743 (1986).
146. Prins, M., and R. Goldbach, RNA-Mediated Virus Resistance in Transgenic Plants, *Arch. Virol. 141*:2259–2276 (1996).
147. Sanford, J.C., and S.A. Johnston, The Concept of Parasite-Derived Resistance-Deriving Resistance Genes from the Parasite's Own Genome, *J. Theor. Biol. 113*:395–405 (1985).
148. Mueller, E., J. Gilbert, G. Davenport, G. Brigneti, and D.C. Baulcombe, Homology-Dependent Resistance: Transgenic Virus Resistance in Plants Related to Homology-Dependent Gene Silencing, *Plant J. 7*:1001–1013 (1995).
149. Wang, X., A.L. Eggenberger, F.W. Nutter Jr., and J.H. Hill, Pathogen-Derived Transgenic Resistance to Soybean Mosaic Virus in Soybean, *Mol. Breed. 8*:119–127 (2001).
150. Tu, J.C., Effect of Different Strains of Soybean Mosaic Virus on Growth, Maturity, Yield, Seed Mottling, and Seed Transmission in Several Soybean Cultivars, *J. Phytopathol. 126*:231–236 (1989).
151. Goodman, R.M., and J.H. Oard, Seed Transmission and Yield Losses in Tropical Soybeans Infected by Soybean Mosaic Virus, *Plant Dis. 64*:913–914 (1980).
152. Eggenberger, A.L., D.M. Stark, and R.N. Beachy, The Nucleotide Sequence of a Soybean Mosaic Virus Coat Protein-Coding Region and Its Expression in *Escherichia coli*, *Agrobacterium tumefaciens*, and Tobacco Callus, *J. Gen. Virol. 70*:1853–1860 (1989).
153. Ross, J.P., Response of Early- and Late-Planted Soybeans to Natural Infection of Bean Pod Mottle Virus, *Plant Dis. 70*:222–224 (1986).
154. Windham, M.T., and J.P. Ross, Phenotypic Responses of Six Soybean Cultivars to Bean Pod Mottle Virus Infection, *Phytopathology 75*:305–309 (1985).
155. Quiniones, S.S., J.M. Dunleavy, and J.W. Fisher, Performance of Three Soybean Varieties Inoculated with Soybean Mosaic Virus and Bean Pod Mottle Virus, *Crop Sci. 11*:662–664 (1971).
156. Thanh, V.H., and K. Shibasaki, Major Proteins of Soybean Seeds: A Straightforward Fractionation and Their Characterization, *J. Agric. Food Chem. 24*:1117–1121 (1976).
157. Koshiyama, I., Chemical and Physical Properties of a 7S Protein in Soybean Globulins, *Cereal Chem. 45*:394–404 (1968).

158. Frisch, D.A., A.H.M. van der Geest, K. Dias, and T.C. Hall, Chromosomal Integration Is Required for Spatial Regulation of Expression from the ß-Phaseolin Promoter, *Plant J. 7*:503–512 (1995).
159. Slater, S., T.A. Mitsky, K.L. Houmiel, M. Hao, S.E. Reiser, N.B. Taylor, M. Tran, H.E. Valentin, D.J. Rodriguez, D.A. Stone, S.R. Padgette, G. Kishore, and K.J. Gruys, Metabolic Engineering of *Arabidopsis* and *Brassica* for Poly(3-Hydroxybutyrate-*Co*-3-Hydroxyvalerate) Copolymer Production, *Nature Biotechnol. 17*:1011–1016 (1999).
160. Koritala, S., and H.J. Dutton, Selective Hydrogenation of Soybean Oil (IV) Fatty-Acid Isomers Formed with Copper Catalysts, *J. Am. Oil Chem. Soc. 46*:245–248 (1969).
161. Carr, R., Processing of Oilseed Crops, in *Oil Crops of the World*, edited by G. Robbelen, and R.K. Downey, McGraw-Hill, New York, 1989, pp. 226–259.
162. Popescu, O., S. Koritala, and H.J. Dutton, High Oleic Oils by Selective Hydrogenation of Soybean Oil, *J. Am. Oil Chem. Soc. 46*:97–99 (1969).
163. Wolfram, G., Dietary Fatty Acids and Coronary Heart Disease, *Eur. J. Med. Res. 8*:321–324 (2003).
164. Ascherio, A., M.B. Katan, P.L. Zock, M.J. Stampfer, and W.C. Willett, Trans Fatty Acids and Coronary Heart Disease, *N. Engl. J. Med. 340*:1994–1998 (1999).
165. Allison, D.B., M.A. Denke, J.M. Dietschy, E.A. Emken, and R.J. Nicolosi, Trans Fatty Acids and Coronary Heart Disease Risk, *Am. J. Clin. Nutr. 62*:655–707 (1995).
166. Kinney, A.J., E.B. Cahoon, and W.B. Hitz, Manipulating Desaturase Activities in Transgenic Crop Plants, *Biochem. Soc. Trans. 30*:1099–1103 (2002).
167. Kinney, A.J., Production of Specialized Oils for Industry, in *Plant Lipid Biosynthesis: Fundamentals and Agricultural Applications*, edited by J.L. Harwood, Cambridge University Press, Cambridge, UK, 1998, pp. 273–285.
168. Kinney, A.J., and S. Knowlton, Designer Oils: The High Oleic Soybean, in *Genetic Engineering for Food Industry: A Strategy for Food Quality Improvement*, edited by S. Harander and S. Roller, Blackie Academic, London, 1997, pp. 193–213.
169. Heppard, E.P., A.J. Kinney, K.L. Stecca, and G.H. Miao, Developmental and Growth Temperature Regulation of Two Different Microsomal W-6 Desaturase Genes in Soybeans, *Plant Physiol. 110*:311–319 (1996).
170. Kinney, A.J., Genetic Engineering of Oilseeds for Desired Traits, in *Genetic Engineering*, edited by J.K. Setlow, Plenum Press, New York, 1997, pp. 149–166.
171. Kinney, A.J., Development of Genetically Engineered Soybean Oils for Food Application, *J. Food Lipids 3*:273–292 (1996).
172. Duffield, J., H. Shapouri, M. Graboski, R. McCormick, and R. Wilson, U.S. Biodiesel Development: New Markets for Conventional and Genetically Modificed Agricultural Products, Economic Research Service/USDA, Washington, D.C., 1998.
173. Gill, I., and R. Valivety, Polyunsaturated Fatty Acids, Part 1: Occurrence, Biological Activities and Applications, *Trends Biotechnol. 15*:401–409 (1997).
174. Griffiths, G., E.Y. Brechany, F.M. Jackson, W.W. Christie, and S. Stymne, Distribution and Biosynthesis of Stearidonic Acid in Leaves of *Borago officinalis*, *Phytochemistry 43*:381–386 (1996).
175. Sayanova, O., M.A. Smith, P. Lapinskas, A.K. Stobart, G. Dobson, W.W. Christie, P.R. Shewry, and J.A. Napier, Expression of a Borage Desaturase cDNA Containing an N-Terminal Cytochrome B_5 Domain Results in the Accumulation of High Levels of Δ^6-Desaturated Fatty Acids in Transgenic Tobacco, *Proc. Natl. Acad. Sci. USA 94*:4211–4216 (1997).

176. Sato, S., A. Xing, X. Ye, B. Schweiger, A.J. Kinney, G. Graef, and T. Clemente, Production of G-Linolenic Acid and Stearidonic Acid in Seeds of Marker-Free Transgenic Soybean, *Crop Sci. 46*:646–652 (2004).
177. Komari, T., Y. Hiei, Y. Saito, N. Murai, and T. Kumashiro, Vectors Carrying Two Separate T-DNAs for Cotransformation of Higher Plants Mediated by *Agrobacterium tumefaciens* and Segregation of Transformants Free from Selection Markers, *Plant J. 10*:165-174 (1996).

Chapter 14

Mapping and Sequencing the Soybean Genome

Gary Stacey[a,c] and Henry Nguyen[b,c]

[a]Department of Plant Microbiology and Pathology, [b]Department of Agronomy, and [c]National Center for Soybean Biotechnology, University of Missouri, Columbia, MO 65211

Introduction

Genomic research is revolutionizing plant biology. The recent completion of the Arabidopsis and rice genomes is a major milestone in plant science that is already yielding many exciting discoveries about genetic pathways, developmental networks, and genome evolution (1,2). This work, along with the completion of other eukaryotic genomes, clearly shows that there is no substitute for knowledge of complete genome sequence. For example, it is now clear that even with substantial investment, expressed sequence tag (EST) sequencing projects will probably not discover all of the genes in an organism. Many sequences are over-represented in cDNA libraries, and highly regulated genes may be absent or under-represented. As recently shown for chromosome 10 of rice, even draft sequencing may reveal only 50% of the genes that can be identified by complete sequencing of the chromosome (2). Phenotypic comparisons between phylogenetically close relatives (e.g., monkey and man; maize and teosinte) reveal that knowledge of gene regulation is critical to understanding genome function. This kind of information can come only from full genome sequencing. In addition, sequence information alone has limited utility unless fully integrated into the genetic and physical maps of the genome. Thus, mapping and sequencing are both essential elements for understanding the physical genomics of any organism.

Importance of Soybean Genomics

Soybean (*Glycine max* (L.) Merr.) is the second most valuable crop species in the United States, accounting for $12 to $17 billion in annual revenue (USDA-NASS Agricultural Statistics 2002, www.usda.gov/nass/). Although genomics is having a profound effect on plant biology, its direct impact on major crop species such as soybean remains limited. A primary difficulty is that the genomes of major crop species are 4 to 48 times larger than that of the model plant *Arabidopsis thaliana* (L.) Heyn. Nonetheless, the scientific community clearly understands the need for sequencing of crop genomes. A recent white paper arising from a U.S. Legume Crops Genomics Workshop (http://129.186.26.94/Legume_Initiative/LegGenomicsPaper10Oct01.html) strongly recommended the immediate sequencing of gene-rich regions of soybean, common bean, and peanut. This recommendation was echoed in a recent report from the National Research Council (http://books.nap.edu/books/0309085292/html/R1.html), which stated, "We recognize that soybean and wheat are key crop species and thus anticipate that draft sequencing of the gene rich regions of these genomes should commence.". Furthermore, this report also states that "tools for comparative genomics need to be developed, including large-insert libraries, physical maps, and mapping tools for agronomically relevant relatives within the . . . Fabaceae."

The Fabaceae (legumes) constitute the third-largest family of angiosperms, with over 650 genera (3). Legumes are a major source of human food (e.g., soybean, common bean, pea, peanut, lentil, chickpea), edible oils (peanut, soybean), and animal fodder and forage (alfalfa, clover). Legumes are grown on approximately 15% of the world's arable land (270 to 300 million ha) and provide one-third of humankind's protein requirements. The ability of legumes to obtain their nitrogen through symbiosis with rhizobia also has profound environmental benefits, in that it reduces fertilizer use and the associated environmental impact. Biological nitrogen fixation also makes legumes attractive in areas of subsistence farming throughout the developing world. Studies indicate that rotating a legume with corn can provide 12–15% of the corn's nitrogen needs. Legumes have also drawn attention because of the potential human health effects of soy protein and associated isoflavones.

Recently, considerable attention has been focused on two model legumes: *Medicago truncatula* Gaertner and *Lotus japonicus* (Regel) K. Larsen (4,5). Sequencing efforts are currently underway in both of these systems utilizing a bacterial artificial chromosome (BAC)-by-BAC approach. Indeed, the above-mentioned NRC report (http://books.nap.edu/books/0309085217/html) called for determining the complete genomic sequence of *M. truncatula*. However, it remains to be seen how relevant the information on these two temperate legumes will be to genomics efforts in tropical legumes (e.g., soybean). The intense selection practiced on cultivated crops exceeds natural selection by orders of magnitude and has resulted in different genome ploidies, sizes, and organizations. Given the importance of soybean, current knowledge, consistent with current recommendations, argues for a concerted effort in soybean genomics, including the mapping and sequencing of the gene-rich regions of the soybean genome.

Current Resources for Soybean Genomics

Because of soybean's tremendous agronomic importance, development of genetic, molecular, and genomic tools in soybean (like maize) has flourished. Soybean now has many of the experimental tools typically associated with model plant systems.

Genetic Map

Soybean is thought to be the product of a diploid ancestor ($n = 11$), which underwent aneuploid loss ($n = 10$), polyploidization ($n = 20$), and diploidization. Genes and quantitative trait loci (QTLs) in soybean segregate in simple Mendelian fashion, facilitating genetic analysis. The soybean composite genetic map is well developed. The classical map contains 63 loci on 19 linkage groups (6), while the molecular map encompasses 20 linkage groups and more than 2300 cM based on 600 SSRs mapped in six populations (Soybase, http://129.186.26.94/). In addition, the composite map contains greater than 800 restriction fragment length polymorphism (RFLP) markers, greater than 600 amplified restriction fragment length polymorphism (AFLP) markers, and hundreds of random amplified polymorphic DNA (RAPDs) (7).

BAC Libraries

A number of public-sector BAC libraries are available providing greater than 35-fold genome coverage. For example, a 4× to 6× Forrest BAC library (8) has been used to construct a physical map of this cultivar (see under next heading). Similarly, a 5× (about 130 kb, *Hind*III inserts) library exists for cultivar Williams 82 (9).

Physical Map

A double-digest-based map for the soybean cultivar Forrest was constructed (http://soybeangenome.siu.edu/). This map is under further development to reduce the number of contigs (presently greater than 3,000) and to add additional DNA markers (e.g., through BAC end sequencing).

EST Sequences

More than 300,000 soybean EST sequences are in Genbank (http://129.186.26.94/soybeanest.html).

Functional Genomics

Dr. Lila Vodkin's laboratory (University of Illinois) has utilized these ESTs to develop DNA microarrays for functional genomic analysis (10). In addition, proteomic efforts are under way that make use of the translated EST sequences for protein identification (e.g., Wan *et al.*, University of Missouri, personal communication).

Soybean Transformation and Mutagenesis

Contrary to the previously cited NRC report (http://books.nap.edu/books/0309085217/html), soybean transformation efficiencies are consistently greater than 5%; in some labs, efficiencies greater than 12% are now common (11). These improvements in transformation efficiencies have led to development of transposon tagging projects for soybean (T.E. Clemente, University of Nebraska, personal communication). Additionally, viral-induced gene silencing systems and TILLING populations for soybean are under development (A. Bent, University of Wisconsin, personal communication; www.anawah.com/). Soybean has more than 300 phenotypic mutants reported, some of which have already led to marketable advances (e.g., modified oils, low phytate soybean mutants).

Phenotype Analysis

Soybean biochemistry and physiology are well studied (12), and the agronomic knowledge of soybean far exceeds that of any model legume. The biochemical and developmental bases of pod and seed development, flowering, maturity, nodulation, nitrogen fixation, disease resistance, and abiotic stress responses are well studied in soybean. These traits are integrated with the genetic and physical maps by QTL analysis.

Soybean Research Community

Soybean has a large, diverse, and established research community to expeditiously apply new knowledge to crop improvement. This community includes molecular and functional genomicists, geneticists, breeders, physiologists, and agronomists. International collaboration in soybean genomics has long existed among the United States, Japan, China, Korea, and various countries of South America and Europe. Nearly 300 researchers routinely attend the biennial Cellular and Molecular Biology of Soybean Conferences (http://muconf.missouri.edu/soy2004), which have been held over the past two decades. Such an interdisciplinary community, together with a vibrant soybean industry, can transfer genome science to practical application effectively and efficiently.

DNA Methylation and Plant Genome Structure

Methylation of cytosine at CpG or CpNpG residues in plant genomes varies in frequency along the chromosome. Methylation at these sites is known to modify DNA structure and regulate gene expression (13). This regulation is mediated by changes in chromatin structure (14), prevention of regulatory proteins binding to DNA (15) or attraction of transcriptional repressors (13). On a larger scale, hypermethylation is thought to produce regions of transcriptionally inactive heterochromatin, since the level of methylation is much higher in heterochromatic than in euchromatic regions of the genome (16).

While several lines of evidence suggest that all seed-bearing plant species (angiosperms and gymnosperms) utilize roughly the same number of genes (17,18), seed plant genomes can vary in size by as much as 2,000-fold (19,20). Moreover, relationships between evolutionary lineage and genome size are tenuous, with different members of some plant families exhibiting up to 100-fold differences in 1C DNA content (21,22). Polyploidy accounts for some of the variation in plant genome sizes (generally less than 10-fold) (23), but the vast majority of variation is due to preferential amplification of nongenic repetitive elements. Many of the repetitive elements in plant genomes appear to have originated from intergenic proliferation of retroviruses (24), while others have uncertain origins (25). Regardless of their origin, repetitive DNA sequences are the principal complicating factor in plant genome sequencing.

Simple eukaryotic genomes (e.g., *Caenorhabditis elegans*, Arabidopsis, yeast) have genomes comprising relatively closely spaced genes interspersed with rare transposon-like, repetitive elements. Drosophila euchromatin is repeat poor, while the heterochromatin is rich in tandemly repeated sequences (26). In contrast, larger eukaryotic genomes, including soybean and most cultivated crops, have much higher levels of repetitive DNA. For example, approximately 60–80% of the maize genome is estimated to be repetitive (27,28). Early experiments using reassociation kinetics estimated that at least 60% of the soybean genome is repetitive (29). Recent studies suggest that genome size variation is correlated with the level of retrotransposons present and the number of different retrotransposon families present in the genome. In the relatively compact and simple genome of Arabidopsis, for example, long terminal repeat (LTR) and non-LTR retrotransposons account for only 4–10% of the nuclear DNA, while maize has literally thousands of different families of LTR and non-LTR retrotransposons (30). SanMiguel *et al.* (31) identified a region of nested retrotransposons in an intergenic region of the maize genome, where sequencing of a 280-kb YAC clone identified 10 retrotransposon families, each with reiteration frequencies in the total genome of 10,000 to 30,000 copies. Within the 280-kb region, these repeats represented 60% of the total DNA. Similar clusters of retroelements were dispersed throughout the genome. However, the relative under-representation of such sequences in the maize EST database suggests that such clusters of repeats are largely devoid of genes (excluding integrases, etc.). These results have been confirmed and extended through comparisons between methylation patterns and gene distributions of density-fractionated genomic DNA (32,33). In maize, most genes were found in a tight GC peak that represented only 10 to 20% of the genome. The remainder of the genome was found to be largely free of genes except for a few zein and ribosomal sequences. Similar observations were made in barley and rice. Arabidopsis, in contrast, had genes spread over a much larger portion of its fractionated DNA, and the genes seem to be fairly evenly distributed (34), a result largely confirmed by complete genome sequencing (35). Therefore, the emerging picture of complex plant genomes is a mosaic of highly repetitive, transcriptionally silent heterochromatin interspersed with regions of transcriptionally active genes that are often separated by islands of retroelement and transposon-related repeats. These repetitive sequences, both in heterochromatin and euchromatin, are hypermethylated relative to genes (36). In maize, the hypermethylated islands interspersed between genes may vary in size from 20 to 200 kb (36).

In a more recent study, Meyers *et al.* (37) found that retroelements were significantly underrepresented in a maize genomic BAC library constructed with a methylation-sensitive restriction enzyme. Retroelements found in hypomethylated sequences were distinct from those most abundant in the total genome. The placement of genes between retroelement islands is not restricted to plants, but appears to be a general feature of large-eukaryotic genomes. For example, hypermethylated, retroelement-repeat islands are a common feature of the slime mold, *Physarum polychephalum*, genome (38).

The study of gene distribution and evidence for gene islands has recently been extended to legumes. Kulikova *et al.* (39) reported a cytogenetic analysis of the *M. truncatula* genome using fluorescence in situ hybridization (FISH) of pachytene chromosomes. This analysis revealed a relatively simple arrangement of heterochromatic and euchromatic regions. Similar to the situation in Arabidopsis, heterochromatin was primarily pericentromeric. FISH analysis in predominantly euchromatic regions suggested that these areas were generally lacking in repetitive elements and relatively gene rich. Approximately 85% of the genome (about 395 Mb) was heterochromatic and 15% (about 105 Mb) euchromatic. These data suggest that the coding region of the *M. truncatula* genome is similar in size to that of Arabidopsis.

The limited data available for crop legumes suggest that their genomes are organized like those of other large-genome plants. Retroelements have been isolated from several legumes, including chickpea (40), *Vicia* spp. (41), pea (42), and soybean (43). The copy number of individual retroelement families varies from 10^2 to 10^6 in *Vicia* spp., 10^3 to 10^4 in pea, and 5,000 to 10,000 for a single family in soybean. However, the distribution of these repeats within legume genomes has not been as well studied as in maize. Gurley *et al.* (44) utilized DNA reassociation kinetics to examine the soybean genome. They found that at least 36% of the genome was highly repetitive DNA, though their analysis also suggested that single copy sequences were interspersed among regions of repetitive DNA. This analysis suggested that the soybean genome resembles other large-genome crop plants (e.g., maize) and may not have the simple genome arrangement suggested for *M. truncatula* (39). Morgante *et al.* (43) found that the STR120 repeat of soybean was arranged in tandem arrays on the genome and FISH to metaphase chromosomes revealed this element to be clustered in just two to eight locations (depending on stringency) throughout the genome. These data are consistent with the presence of retroelement islands in soybean.

Research based on RFLPs also has provided insight into the organization of the soybean genome and the distribution of methylated regions. For many years the methylation-sensitive enzyme *Pst*I was used to clone single-copy DNA for use as RFLP probes (45). Zhu *et al.* (46) isolated hypomethylated soybean genomic DNA by restriction with *Pst*I. Southern hybridizations to genomic DNA using these probes revealed that 15% of the hypomethylated sequences had only one copy in the soybean genome and the other 85% comprised either duplicated regions or middle-repetitive sequences. Further analysis of the duplicated regions indicated that both copies were hypomethylated in all cases. Young *et al.* (47) analyzed the distribution of hypo- and hypermethylated sequences on the soybean genome by mapping AFLP markers derived using either a methylation-sensitive (PstI) or methylation-insensitive enzyme (*Eco*RI). The two sets of AFLPs did not co-localize, and the *Eco*RI-AFLP markers showed a distinct, non-random distribution. These latter markers were clustered (e.g., some regions had up to 25 markers) and were particularly common in heterochromatic regions surrounding centromeres. These data, although limited, also suggest that the soybean genome resembles that of other large-genome plants. These data also suggest that targeting hypomethylated, euchromatic soybean DNA could be a useful way to isolate single-copy, gene-rich regions of the genome.

The Soybean Genome

Size and Sequence Complexity

The soybean genome has a 1C genome size of about 1.1×10^9 bp based upon flow cytometry (48), with similar values predicted by DNA reassociation kinetics (Cot analysis; 44,49). Both soybean Cot studies suggest that about 60% of the genome is repetitive; reassociation of different size fragments indicates that the majority of the repetitive sequences are physically linked while a smaller fraction is interspersed with single copy DNA. This latter conclusion is supported by sequence sampling of nearly 1,000 loci across all 20 soybean linkage groups (50). The predictions from this research are that the gene rich and repetitive regions occasionally are interspersed, but gene rich islands are present as well. Cytological studies show that euchromatin represents about 65% of the genome, with heterochromatin mainly localized to pericentromeric regions and the short arms of four chromosomes (51), one of which contains the single nucleolus organizer region. Collectively these studies suggest that gene space accounts for about one third of the soybean genome (i.e., about 3.7×10^8 bp).

Euchromatic Gene Density

Extensive sequence exists for only two euchromatic regions of the soybean genome. One region, encompassing 463 kb (in two segments separated by about 2 cM) on linkage group G (chromosome 18) contains a major gene (*rhg1*) for cyst-nematode resistance (Genbank AX196301-42; 53; GenBank AF506516 to AF506518). Gene density in these regions is about 5–7 kb/ORF, and repeat content is low (3–8%) (52,53), which is slightly higher than Arabidopsis (35) but similar to estimates for tomato (54). Similar figures are found for a second 512-kb sequence (53) that lies in the middle of the euchromatic arm of linkage group A2 (chromosome 8).

Synteny

Data clearly indicate that syntenic relationships exist among legumes and other angiosperms. However, genome duplications, insertions, deletions, and sequence divergence call into question how predictive the knowledge of one genome will be to the exploitation of another. For example, *M. truncatula* and soybean are estimated to have diverged approximately 40 million years ago, with duplications taking place in soybean 38 to 40 million years ago and the most recent only 14 to 15 million years ago (55–57).

Foster-Hartnett *et al.* (52) focused on the sequenced region (around *rhg1*) on linkage group G of soybean to examine gene distribution and microsynteny. The region was expanded beyond the sequenced segment by isolation of overlapping BAC clones. Consistent with earlier reports (58), there was considerable sequence homology between the linkage group G and Arabidopsis genome sequences. However, the linkage group G sequence exhibited homology to five different Arabidopsis chromosomes, suggesting that extensive chromosomal rearrangements differentiate soybean and Arabidopsis. Comparisons between the soybean regions homoeologous to the *rhg1* locus indicated extensive microsynteny, consistent with the recent duplications in the soybean genome. While BAC end sequences from the expanded *rhg1* region contained relatively few repetitive sequences (i.e., 3–8% vs. 18% from a whole-genome survey), BAC end sequences from homoeologous BAC clones were highly enriched for repetitive sequences (i.e., 33%). These latter data raise the question of whether the sequenced *rhg1* region is truly representative of the soybean genome. Yan *et al.* (57) utilized BAC hybridizations to estimate microsynteny between soybean, *M. truncatula*, and Arabidopsis. These data gave an estimate of only 6% microsynteny between soybean and Arabidopsis (3 of 50 BACs) and approximately 50% between *M. truncatula* and soybean (27 of 50 BACs).

In a more focused study, Cannon *et al.* (59) sequenced homoeologous BAC clones from *M. truncatula*, *L. japonicus*, and soybean. These clones were selected based on the presence of genes encoding apyrases (nucleotide phosphohydrolases, or NTPases). These legume genomic sequences were compared to one another and to corresponding regions in *A. thaliana*, revealing significant microsynteny. Genomic organization of the apyrase-containing BACs showed evidence of gene duplication, genomic rearrangement, and gene conversion among the apyrase homologs present. Detailed comparison of the sequenced regions showed only a low level of conservation in non-coding regions between the three legume species. These comparisons identified no systematic conservation outside of coding regions. Although *Medicago*, *Glycine*, and *Lotus* are all members of the Papilionoideae, they all fall within separate tribes (Viceae, Phaseoleae, and Loteae, respectively). If these three regions studied are related through speciation events, then low levels of conservation outside of coding sequences may be common among them. This would have implications about the relevance of noncoding sequences from the model species for understanding genome structure, regulation, and function in crop legumes (e.g., soybean).

Mapping and Sequencing

Physical Mapping

Two approaches are commonly used to sequence large genomes (60,61), a BAC-by-BAC approach and random sequencing approaches. Although the BAC-by-BAC approach provides sequences that are easier to assemble, the random, whole-genome shotgun sequencing approach is more amenable to high-throughput automation and also covers regions unclonable in BACs. A recent hybrid approach involves construction of a physical map, BAC end sequencing, and sequencing of gene-rich regions. The resulting sequenced contigs are positioned onto the map by comparison to the BAC end sequences. This approach is being used by the maize community (62). Although this strategy does not produce a complete sequence, it provides much information in a cost-effective way (as opposed to only a BAC-by-BAC approach). In the longer term, this strategy provides a framework for future whole genome sequencing. In the shorter term, if a laboratory is interested in a particular region, the BAC clones can be obtained and sequenced to completion and the information integrated into the larger resource.

Genome-wide physical mapping strategy based on BAC restriction fingerprinting has revolutionized genomics of large, complex genomes. The global fingerprinting approach employs a restriction fragment profile of a clone as its fingerprint and assumes that clones sharing a large number of, but not all, fragments are overlapping clones. Comparison of the fingerprints in a large population of clones is used to identify overlapping clones and is the basis for the construction of contigs. Two methods have been developed to generate fingerprints of DNA fragments cloned in BACs. One is polyacrylamide gel-based fingerprinting (PAGFP). The other is an agarose gel-based fingerprinting (AGFP). In the PAGFP, DNA is digested with one restriction enzyme, end-labeled, and either directly fractionated on a DNA sequencing gel or re-cut with a second restriction enzyme. The fingerprints are visualized by autoradiography. In the AGFP method, BAC clone DNA is digested with a restriction enzyme such as *Hind*III, and the fragments are resolved on agarose gels. The fingerprints are visualized by gel staining and subjected to extensive manual editing. The AGFP method was successfully used to construct a high-quality physical map of the rice genome (63). In theory, the PAGFP method has a much higher resolution (1 bp) than the AGFP method (>20 bp). Thus, the number of bands in each BAC fingerprint generated with the PAGFP method is much higher than that generated with the AGFP method. The PAGFP method can be done either manually, by using ^{33}P end labeling of the fragments, autoradiography, and scanning; or be automated, using an automated PAGE-based DNA sequencer for fragment sizing. The manual PAGFP approach has been used to construct physical maps of rice (64), *A. thaliana* (65), and soybean genomes (66).

As mentioned above, previous methods for physical mapping relied on restriction fragment analysis of large BAC clones after agarose or polyacrylamide gel electrophoresis (63,65,67). However, with the advent of automated sequencing machines, especially the capillary DNA sequencer, this methodology has been updated to provide increased accuracy of fingerprint assembly and reduce the number of contigs. High information-content fingerprinting (HICF) (68), is a capillary-sequencer-based fingerprinting method that generates a more accurate and detailed clone fingerprint, thus producing better assemblies with less redundancy in the clone coverage. This new methodology is currently being evaluated by applying it to the maize genome (www.genome.arizona.edu/fpc/maize) and was recommended as the method of choice in the recent National Research Council report (http://books.nap.edu/books/0309085217/html). Furthermore, this method has been successfully used to construct physical map contigs of the wheat D genome (69). In brief, the digestion of five restriction enzymes (four 6-bp-recognizing restriction endonucleases that generate 3' recessed ends and one 4-bp-recognizing restriction endonuclease that generates a blunt end) is performed simultaneously. Each of the four recessed 3' ends is labeled with a different fluorescent dye, and restriction fragments are sized on a capillary DNA analyzer. The resulting fingerprints are edited with the fingerprint-editing computer program Genoprofiler. Subsequent analysis indicated that the contigs constituting the map are accurate and reliable. Particularly important to proficiency are (i) a high-resolution, high-throughput DNA sizing capillary-based electrophoretic method for BAC fingerprinting; (ii) the use of several complementary large-insert BAC libraries; and (iii) computer-aided contig assembly. Use of several complementary libraries developed with different restriction enzymes minimized the gaps in the physical map (69). This study demonstrated that genome-wide BAC-based physical maps can be developed quickly and economically for complex genomes by restriction fingerprint analysis via DNA sizing capillary-based electrophoresis.

The HICF data can be directly uploaded into an adapted version of Fingerprint Contig (FPC) (70–72) software. Essential for successful physical mapping are powerful computational algorithms. This is especially relevant to soybean, with its relatively complex and recently duplicated genome organization. FPC is a program that assembles contigs based on both restriction digest fingerprints and markers, and the contigs are ordered and assigned to chromosomes based on anchored markers. The ability to use automatic data entry with the HICF method avoids tedious and error prone manual image editing. HICF has a five- to sevenfold increase in efficiency relative to gel based fingerprinting (68). The average number of fragments per clone ranged from 90 to 100 using the HICF system separated into four groups by color, while the public maps being assembled with gel-based *Hind*III fingerprints ranged from 25 to 75 fragments. The resulting increase in information and reduction in band number variability for each clone allows a more stringent FPC probability cutoff, in comparison to agarose (73,74) and acrylamide sequencing gels (64,65,75). The high cutoffs used for HICF provide a commensurate gain to allow building maps with increased accuracies.

The physical maps generated by HICF can be displayed over the web using WebFPC. For example, the current markers in the rice WebFPC link to Gramene (76,77), and the markers in maize WebFPC link to maizeDB (78). This provides a valuable resource for the plant community.

The integration of genetic, physical, and transcriptional maps is useful for anchoring gene sequences onto the physical map, cloning genes based on their map position (79,80) and comparing genomic structure among species (76,77,81). Moreover, it provides an important resource of BAC clones for genome sequencing. In principle, a high-resolution genetic map provides a large number of molecular markers (RFLP, SSR, STS and SNP) to colinearily locate recombination breakpoints of a mapping population to segments on the physical map. Transcripts (EST) also can be mapped on the genetic or physical maps with PCR-based or hybridization-based methods, which can provide a valuable resource for understanding the genome structure and function.

A variety of markers can be used to screen fingerprinted BAC libraries. The 250 core RFLP markers and 1000 SSR markers that serve as bin landmarks on the composite genetic map of soybean (7) would be a valuable starting source for anchoring BAC contigs to the genetic map. PCR-based screening requires pooling the BAC library into sets of clones, such as the six-dimensional pooling strategy (75), such that each clone can be uniquely identified by its detection in the sets of clones. Hybridization-based screening can be performed against high-density filters of an entire library or core BAC clones using radioactively labeled probes either singly or as a pool.

The combined strategy of two-dimensional or multidimensional pooling of overgo probes and hybridization to the high-density filters of BAC clones can be used to anchor EST unigenes to the physical map of BAC contigs (82). These overgo probes are made by annealing two overlapping oligonucleotides and filling in the overhanging bases with Klenow enzyme and radiolabeled nucleotides. More than 300,000 soybean EST sequences and STSs in the Genbank are the main source of overgos. As gene-rich sequences become available from a soybean genome sequencing project, they can be used to generate additional overgo probes, SNP, and STS markers.

For floating contigs that have no markers anchoring on them, the BAC end-sequencing approach can be applied to confirm the contig assembly and enhance the accuracy for integration of physical map and genetic map. BAC end sequences can produce a large number of locus-specific STS and/or SNP markers that can effectively eliminate the ambiguous locations of floating contigs on the genetic map.

Another emerging physical mapping technique, **HAP**loid equivalents of DNA and the **P**ol**Y**merase chain reaction (HAPPY), has been used for constructing a map of human chromosome 14 (83) and the *Dictyostelium* genome (84,85). HAPPY mapping is an *in vitro* mapping technique that involves randomly breaking genomic DNA by shearing or gamma-irradiation, segregating the fragments into aliquots known as a mapping panel, and subsequently detecting the frequency of breaks between markers. The principle is that if any two markers are close together on a chromosome, they will co-segregate more frequently (86). HAPPY mapping does not require construction of BAC libraries and provides a tool for comparative physical mapping. The method could be applied to soybean to distinguish duplicated regions in the physical map.

Capture of Gene-Rich Regions

Given the large size of the soybean genome and the greater interest in coding and associated regulatory sequences, focusing sequencing efforts on gene-rich regions is an obvious strategy. Two distinguished panels of plant scientists have recommended the use of gene enrichment approaches to obtain soybean gene sequences.

BAC Sequencing

When a plant genome is conveniently arranged into largely gene-rich, euchromatic, and gene-poor heterochromatic regions, then gene-rich BAC clones can be selected by hybridization to known genes of interest. In this way, selected BAC clones act as seed points from which contigous, presumably gene-rich BAC clones, can be selected. This is the approach currently being applied to the *M. truncatula* and *L. japonicus* genomes (http://www.genome.ou.edu/medicago.html; http://www.kazusa.or.jp/lotus/).

Methylfiltration (MF)

In some species, repetitive sequences in both heterochromatin and euchromatin are hypermethylated relative to genes (87). The data available for crop legumes suggest that their genomes possess hypermethylated repeats and hypomethylated genes (43,46,47). MF is a very simple and robust method for enriching for hypomethylated, gene-rich sequences in complex genomes (68). The technology is exclusively licensed commercially to Orion Genomics (http://www.oriongenomics.com/index.html) and marketed under the name *GeneThresher*. Briefly, genomic DNA is used to transform an *E. coli* strain that preferentially cleaves methylated DNA sequences. Consequently, only hypomethylated DNA inserts survive the cloning process. MF libraries can be shotgun sequenced. The selection of hypomethylated fragments by MF is a complementary approach to Cot-based enrichment (see under next heading).The MF strategy greatly reduces the cost of determining the sequences of many of the genes in some plant genomes, and has been successfully applied to genomes of 16 plant species (e.g., white clover, soybean, rape, tobacco, tomato, wheat, barley, corn, sorghum, rice, pine; N. Lakey, Orion Genomics, St. Louis, Missouri, personal communication).

An initial pilot study of MF in soybean was conducted at Orion Genomics (http://www.oriongenomics.com/; N. Lakey, personal communication). A total of 1,626 high-quality reads from a MF library (an McrABC$^+$ *E. coli* host strain) and 596 reads from unfiltered libraries (an McrABC$^-$ *E. coli* host strain) were generated. Methyl-filtered and unfiltered sequences were analyzed for gene content by comparison to a repeat-masked version of the Arabidopsis protein set using BLASTX (E value cutoffs from $E < 9.9 \times 10^{-5}$ to $E < 9.9 \times 10^{-20}$). A filter power value (an estimate of the increased gene representation found in filtered versus unfiltered libraries) was calculated at each cutoff by dividing the percentage of Arabidopsis matches in the filtered library by the percentage of matches for the unfiltered library. The soybean data indicated a filter power of 2.3 to 3.0, with a midpoint of 2.7. At a filter power of 2.7, 62% or about 686 megabases of the soybean genome, is removed from whole genome libraries. Comparison of filtered soybean shotgun sequences with public databases suggests that about 43% of the single copy sequence matches exons. This is consistent with the known exon density of sequenced soybean genes.

Thus, this MF study estimated the gene space of soybean at about 423 Mb (2.7-fold enrichment). This value is roughly in line with the gene-rich space estimated from earlier solution hybridization studies (about 370 Mb).

Cot-Based Cloning and Sequencing (CBCS)

CBCS is a powerful means by which single/low-copy DNA sequences can be preferentially captured (isolated and sequenced) from large, repetitive genomes in a manner independent of sequence methylation (89,90). CBCS is based on the observation that when heat-denatured genomic DNA is allowed to renature under controlled conditions, the rate at which a particular DNA sequence reassociates is proportional to the number of times it is found within the genome. Genomic DNA is denatured and reannealed so as to allow only reassociation and fractionation of highly repetitive (HR), moderately repetitive (MR), and single/low-copy (SL) sequences using hydroxyapatite chromatography. Fractionated HR, MR, and SL components are cloned to produce HRCot, MRCot, and SLCot genomic libraries. Like the *GeneThresher* technique, sequencing of the SLCot libraries elucidates gene-rich regions of the genome in a manner independent of sequence expression. Additionally, cloning Cot components in an *E. coli* strain, which does not restrict methylated DNA, provides libraries that are independent of sequence methylation patterns (89,90). This is an important feature since MF cloning strategies may result in the loss of important/interesting genes because the pattern and significance of DNA methylation can differ markedly between species (91), genes within an organism (92,93), developmental stages (94), and different regions of the same gene (95,96). Thus, CBCS is complementary to MF technology and should prevent the exclusion of genes that may fall in hypermethylated regions.

A Strategy for the Soybean Genome

Sequencing and mapping of the soybean genome represents a significant technical challenge with unique problems. The duplicated nature of the soybean genome, coupled with its large size, will present significant problems for BAC fingerprint analysis, as well as for sequence assembly. The presence of large paralogous gene families showing little or no sequence variation represents a major challenge for gene mapping. Given these constraints, model strategies previously applied to other plant genomes may not be applicable to soybean. Thus, a hybrid approach, using extant technologies in combination, would appear to be the best approach to any near term efforts to physically map and sequence soybean. Given the large size of the genome, it is unlikely that sufficient funds can be found to sequence the entire genome. Therefore, as suggested in the recent NRC report (http://books.nap.edu/books/0309085217/html), a practical approach would be to first focus on the gene rich regions of soybean.

Figure 14.1 presents a possible strategy for sequencing the gene-rich regions of soybean, as well as anchoring these sequences to the physical and genetic map. It is clear that gene sequences alone, without some means to relate them to phenotype,

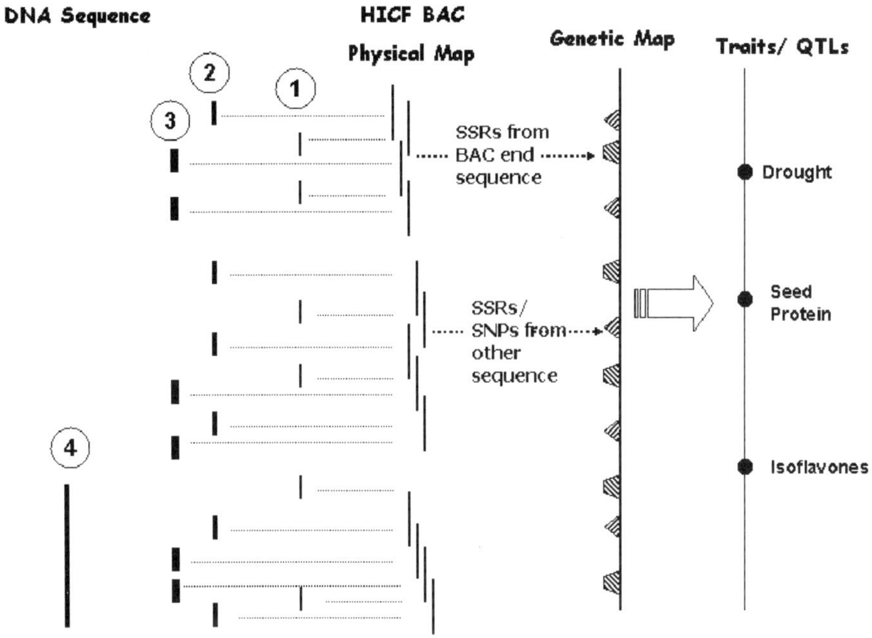

Figure 14.1. A possible scheme for sequencing the gene-rich regions of the soybean genome. A high information-content fingerprint (HICF) BAC physical map can be anchored to the genetic map using SSRs (simple sequence repeats) or SNPs (single nucleotide polymorphisms) identified by genomic sequencing. Sequence data can be generated by (1) BAC end sequencing, (2) EST sequencing, (3) Methylfiltration/Cot-based sequencing, and/or (4) full BAC sequencing (e.g., short contigs from gene-rich regions). These sequences can be anchored to the physical map via BLAST comparison to anchored BAC end sequence, overgo hybridizations to BAC filters, and/or PCR-based screening of BAC pools. Additionally, BAC clones can be anchored to chromosomes via fluorescent *in situ* hybridization. Ultimately, this resource will aid efforts to identify and map important soybean traits.

are of limited utility. Hence, placing the gene sequences onto the physical map and subsequently anchoring this physical map to the genetic map allows the correlation of gene sequence to known soybean traits (e.g., to important QTLs).

The best strategy for obtaining the gene rich regions of the genome depends on the structure of the genome and funds available. At present, a clear view of the soybean genome structure is not available. However, the data suggest that the genome, at least for some chromosomes, may be conveniently divided into gene-rich euchromatin and gene-poor heterochromatin. If this is indeed the case, then targeted sequencing of BAC clones selected to encode for genes may be the best approach. Indeed, this is the approach being applied to efforts to sequence the gene-rich segments of *M. truncatula* and *L. japonicus* (http://books.nap.edu/books/0309085292/html/R1.html). The attractiveness of this approach is that it yields contiguous sequence, which is easily placed onto the physical map by fingerprinting each of the sequenced BAC clones. However, this approach has additional costs due to the need for library construction and the sequence redundancy needed for full assembly.

Alternatively, the MF or CBCS approaches described above could be applied to soybean. These methods are currently being evaluated for the maize genome where genic and non-genic regions are clearly interspersed (http://maize.danforthcenter.org/). These methods provide a cost effective and relatively rapid means to capture the gene rich regions of the genome. When used in combination, they should provide data that are unbiased due to the variations in DNA methylation. However, the net result of MF and CBCS sequencing is short contigs (about 2 to 4 kb). Some of these can be placed onto the physical map by comparison to BAC end sequences (if available). However, many will have to be placed by other means and, given the duplicated nature of the soybean genome, could be inappropriately placed unless sufficient care is taken.

Regardless of the means by which the gene-rich regions are sequenced, any soybean genomic effort should begin with the development of a good physical map. This recommendation echoes that of previous Soybean Genomic workshops (see "Importance of Soybean Genomics" above). Given the increased resolution of the newer HICF method, it seems reasonable to apply this method to develop a soybean physical map that would improve on that already available for the cultivar Forrest. However, the duplicated nature of the soybean genome could present problems with appropriate assembly of a physical map even using the HICF method. Therefore, as described in Fig. 14.1, physical mapping should proceed in conjunction with BAC end sequencing and mapping of sequenced segments (sequence tagged sites, STS) to the physical map. As described above, BAC pooling methods coupled with overgo hybridizations or PCR-based methods provide high-throughput solutions for placing soybean sequences onto the physical map. BAC clones can also be anchored to chromosomes by fluorescent *in situ* hybridization (FISH). Finally, development of simple sequence repeat (SSR) or single nucleotide polymorphic (SNP) markers from the sequenced regions would allow the physical map to be anchored to the well-developed soybean genetic map.

The soybean genomic resource described would be an excellent intermediate step to full-genome sequence, which will undoubtedly be attempted when the cost of DNA sequencing is not prohibitive. However, this soybean genomic resource could be immediately used for gene discovery, development of molecular markers for important agronomic traits, comparative genomics, and other basic and applied uses.

Acknowledgments

The authors acknowledge the contributions of E. Butler, N. Lakey, T. Landewe, D. Lightfoot, K. Meksem, D. Peterson, R. Shoemaker, C. Soderlund, and C. Town. Cited work from Gary Stacey's laboratory was supported by a grant (DE-FG02-97ER20260) from the United States Department of Energy, Basic Energy Biosciences Program.

References

1. Goff, S. A., D. Ricke, T.H. Lan, G. Presting, R. Wang, M. Dunn, J. Glazebrook, A. Sessions, P. Oeller, H. Varma, D. Hadley, D. Hutchison, C. Martin, F. Katagiri, B.M. Lange, T. Moughamer, Y. Xia, P. Budworth, J. Zhong, T. Miguel, U. Paszkowski, S. Zhang, M. Colbert, W.L. Sun, L. Chen, B. Cooper, S. Park, T.C. Wood, L. Mao, P. Quail, R.Wing, R. Dean, Y. Yu, A. Zharkikh, R. Shen, S. Sahasrabudhe, A. Thomas, R. Cannings, A. Gutin, D. Pruss, J. Reid, S. Tavtigian, J. Mitchell, G. Eldredge, T. Scholl, R.M. Miller, S. Bhatnagar, N. Adey, T. Rubano, N. Tusneem, R. Robinson, J. Feldhaus, T. Macalma, A. Oliphant, and S. Briggs, A Draft Sequence of the Rice Genome (*Oryza sativa* L. ssp. *japonica*), *Science 296*:92–100 (2002).
2. Yu, J., S. Hu, J. Wang, G.K. Wong, S. Li, B. Liu, Y. Deng, L. Dai, Y. Zhou, X. Zhang, M. Cao, J. Liu, J. Sun, J. Tang, Y. Chen, X. Huang, W. Lin, C. Ye, W. Tong, L. Cong, J. Geng, Y. Han, L. Li, W. Li, G. Hu, X. Huang, W. Li, J. Li, Z. Liu, L. Li, J. Liu, Q. Qi, J. Liu, L. Li, T. Li, X. Wang, H. T. Lu, Wu, M. Zhu, P. Ni, H. Han, W. Dong, X. Ren, X. Feng, P. Cui, X. Li, H. Wang, X. Xu, W. Zhai, Z. Xu, J. Zhang, S. He, J. Zhang, J. Xu, K. Zhang, X. Zheng, J. Dong, W. Zeng, L. Tao, J. Ye, J. Tan, X. Ren, X. Chen, J. He, D. Liu, W. Tian, C. Tian, H. Xia, Q. Bao, G. Li, H. Gao, T. Cao, J. Wang, W. Zhao, P. Li, W. Chen, X. Wang, Y. Zhang, J. Hu, J. Wang, S. Liu, J. Yang, G. Zhang, Y. Xiong, Z. Li, L. Mao, C. Zhou, Z. Zhu, R. Chen, B. Hao, W. Zheng, S. Chen, W. Guo, G. Li, S. Liu, M. Tao, J. Wang, L. Zhu, L. Yuan, and H. Yang, A Draft Sequence of the Rice Genome (*Oryza sativa* L. ssp. *indica*), *Science 296*:79–92 (2002).
3. Doyle, J.J., Phylogeny of the Legume Family: An Approach to Understanding the Origins of Nodulation, *Ann. Rev. Ecol. Syst. 25*:325–349 (1994).
4. Bell, C.J., R.A. Dixon, A.D. Farmer, R. Flores, J. Inman, R.A. Gonzales, M.J. Harrison, N.L. Paiva, A.D. Scott, J.W. Weller, and G.D. May, The *Medicago* Genome Initiative: A Model Legume Database, *Nucleic Acids Res. 29*:114–117 (2001).
5. Handberg, K., J. Stougaard, *Lotus japonicus*, an Autogamous, Diploid Legume Species for Classical and Molecular Genetics, *Plant J. 2*:487–496 (1992).
6. Hedges, B.R., and R.G. Palmer, Mapping the w4 Locus in Soybean, *Soybean Genet. News 20*:20 26(1993).
7. Cregan, P.B., T. Jarvik, A.L. Bush, R.C. Shoemaker, K.G. Lark, A.L. Kahler, N. Kaya, T.T. VanTori, D.G. Lohnes, J. Chung, and J.E. Specht, An Integrated Genetic Linkage Map of the Soybean, *Crop Sci. 39*:1464–1490 (1999).
8. Meksem, K., H.B. Zhang, and D.A. Lightfoot, Two Transformation Ready Large Insert Clone Libraries for Soybean: Physical Mapping of Resistance to Soybean Cyst Nematode and Sudden Death Syndrome, *Theor. Appl. Genet. 101*:747–755 (2000).
9. Marek, L.F., and R.C. Shoemaker, BAC Contig Development by Fingerprint Analysis in Soybean, *Genome 40*:420–427 (1997).
10. Clough, S., R. Philip, R. Shealy, and L. Vodkin,. *NSF Soybean Microarray Workshop Manual*, University of Illinois, Champaign, IL, 2000.
11. Somers, D.A., D.A. Samac, and P.M. Olhoft. Recent Advances in Legume Transformation, *Plant Physiol. 131*:892–899 (2003).
12. World Soybean Research Conference VI: August 4–7, 1999, Chicago, IL, *Proceedings, Invited and Contributed Papers and Posters/Compiled and Arranged by Harold E. Kauffman*; National Soybean Research Laboratory, University of Illinois, Urbana, IL, 1999.
13. Kass, S.U., D. Pruss, and A.P. Wolffe, How Does DNA Methylation Repress Transcription, *Trends Genet. 13*:444–449 (1997).

14. Davey, C., S. Pennings, and J. Allen, CpG Methylation Re-models Chromatin Structure In Vitro, *J. Mol. Biol. 267*:276–288 (1997).
15. Inamdar, N.M., K.C. Ehrlich, and M. Ehrlich, CpG Methylation Inhibits Binding of Several Sequence-Specific DNA-Binding Proteins from Pea, Wheat, Soybean and Cauliflower, *Plant Mol. Biol. 17*:111–123 (1991).
16. Bird, A.P., CpG-Rich Islands and the Function of DNA Methylation, *Nature 321*:209–213 (1986).
17. Fu, H., Z. Zheng, and H.K. Dooner, Recombination Rates between Adjacent Genic and Retrotransposon Regions in Maize Vary by 2 Orders of Magnitude, *Proc. Natl. Acad. Sci. USA 99*:1082–1087 (2002).
18. Loomis, W.F., Similarities in Eukaryotic Genomes, *Comp. Biochem. Physiol. B 95*:21–27 (1990).
19. Bennett, M.D., and J.B. Smith, Nuclear DNA Amounts in Angiosperms, *Phil. Trans. R. Soc. Lond. B 274*:227–274 (1976).
20. Bennett, M.D., and J.B. Smith, Nuclear DNA Amounts in Angiosperms, *Phil. Trans. R. Soc. Lond. B 334*:309–345 (1991).
21. Bennett, M.D., and I.J. Leitch, Nuclear DNA Amounts in Angiosperms, *Ann. Bot. 76*:113–176 (1995).
22. Bennett, M.D., P. Bhandol, and I.J. Leitch, Nuclear DNA Amounts in Angiosperms and their Modern Uses—807 New Estimates, *Ann. Bot. 86*:859–909 (2000).
23. Bennett, M.D., and I.J. Leitch, Plant DNA C-values database (release 1.0, Sept. 2001), www.rbgkew.org.uk/cval/homepage.html, 2001.
24. SanMiguel, P., and J.L. Bennetzen, Evidence That a Recent Increase in Maize Genome Size Was Caused by the Massive Amplification of Intergene Retrotransposons, *Ann. Bot. 82*:37–44 (2000).
25. Lapitan, N.L.V., Organization and Evolution of Higher Plant Nuclear Genomes, *Genome 35*:171–181 (1992).
26. Gatti, M., and S. Pimpinelli, Functional Elements in *Drosophila melanogaster* Heterochromatin, *Ann. Rev. Genet. 26*:239–275 (1992).
27. Favell, R.B., M.D. Bennett, J.B. Smith, and D.B. Smith, Genome Size and Proportion of Repeated Nucleotide-Sequence in Plants, *Biochem. Genet. 12*:257–269 (1974).
28. Hake, S., and V. Walbot, The Genome of *Zea mays*, Its Organization and Homology to Related Grasses, *Chromosoma 79*:251–270 (1980).
29. Hepburn, A.G., W.B. Gurley, and J.L. Key, Organization of Soybean DNA. 1. Kinetic-Analysis of Total and Cs_2SO_4/Hg^{2+} Fractionated DNA, *Plant Physiol. 57*:13 (1976).
30. Kumar, A., and J.L. Bennetzen, Plant Retrotransposons, *Ann. Rev. Genet. 33*:479–532 (1999).
31. SanMiguel, P., T. Alexander, J. Young-Kwan, N. Motchoulskaia, D. Zakharov, A. Melake-Berhan, P.S. Springer, K.J. Edwards, M. Lee, Z. Avramova, and J.L. Bennetzen, Nested Retrotransposons in Intergenic Regions of the Maize Genome. *Science 274*:765–768 (1996).
32. Barakat, A., N. Carrels, and G. Bernardi, The Distribution of Genes in the Genomes of Gramineae, *Proc. Natl. Acad. Sci. USA 94*:6857–6861 (1997).
33. Carels, N., A. Barakat, and G. Bernardi, The Gene Distribution of the Maize Genome, *Proc. Natl. Acad. Sci. USA. 92*:11057–11060 (1995).
34. Barakat, A., G. Matassi, and G. Bernardi, Distribution of Genes in the Genome of *Arabidopsis thaliana* and Its Implication for the Genome Organization of Plants, *Proc. Natl. Acad. Sci. USA 95*:10044–10049 (1998).
35. *Arabidopsis* Genome Initiative Analysis of the Genome Sequence of the Flowering Plant *Arabidopsis thaliana*, *Nature 408*:796 (2000).

36. Bennetzen, J., K. Schrick, P. Springer, W. Brown, and P. SanMiguel, Active Maize Genes are Unmodified and Flanked by Diverse Classes of Modified, Highly Repetitive DNA, *Genome 37*:565–576 (1994).
37. Meyers, B.C., S.V. Tingley, and M. Morgante, Abundance, Distribution, and Transcriptional Activity of Repetitive Elements in the Maize Genome, *Genome Res. 11*:1660–1676 (2001).
38. Rothnie, H.M., K.J. Mccurrach, L.A. Glover, and N. Hardman, Retrotransposon-like Nature of tp1 Elements—Implications for the Organization of Highly Repetitive, Hypermethylated DNA in the Genome of *Physarum polycephalum*, *Nucleic Acids Res. 19*:279–286 (1991).
39. Kulikova, O., G. Gualtieri, R. Geurts, D.J. Kim, D. Cook, T. Huguet, J.H. de Jong, P.F. Fransz, and T. Bisseling, Integration of the FISH Pachytenet and Genetic Maps of *Medicago truncatula*, *Plant J. 27*:49–58 (2001).
40. Sant, V.J., M.N. Sainani, R. Sami-Subbu, P.K. Ranjekar, and V.S. Gupta, Ty1-copia Retrotransposon-like Elements in Chickpea Genome: Their Identification, Distribution and Use for Diversity Analysis, *Gene 257*:157–166 (2000).
41. Nouzova, M., P. Neumann, A. Navratilova, D.W. Galbraith, and J. Macas, Microarray-Based Survey of Repetitive Genomic Sequences in *Vicia* spp., *Plant Mol. Biol. 45*:229–244 (2001).
42. Neumann, P., M. Nouzova, and J. Macas, Molecular and Cytogenetic Analysis of Repetitive DNA in Pea (*Pisum sativum* L.), *Genome 44*:716–728 (2001).
43. Morgante, M., I. Jurman, L. Shi, T. Zhu, P. Keim, and J.A. Rafalski, The STR120 Satellite DNA of Soybean: Organization, Evolution and Chromosomal Specificity, *Chromosome Res. 5*:363–373 (1997).
44. Gurley, W.B., A.G. Hepburn, and J.L. Key, Sequence Organization of the Soybean Genome, *Biochim. Biophys. Acta 561*:167–183 (1979).
45. Burr, B., F.A. Burr, K.H. Thompson, M.C. Albertson, and C.W. Stuber, Gene Mapping with Recombinant Inbreds in Maize, *Genetics 118*:519–526 (1988).
46. Zhu, T., J.M. Schupp, A. Oliphant, and P. Keim, Hypomethylated Sequences—Characterization of the Duplicate Soybean Genome, *Mol. Gen. Genet. 244*:638–645 (1994).
47. Young, W.P., J.M. Schupp, and P. Keim, DNA Methylation and AFLP Marker Distribution in the Soybean Genome, *Theor. Appl. Genet. 99*:785–792 (1999).
48. Arumuganathan, K., and E.D. Earle, Estimation of Nuclear DNA Content of Plants by Flow Cytometry, *Plant Mol. Biol. Rep. 9*:229–241 (1991).
49. Goldberg, R.B., DNA Sequence Organization in the Soybean Plant, *Biochem. Genet. 16*:45–51 (1978).
50. Marek, L.F., J. Mudge, L. Darnielle, D. Grant, N. Hanson, M. Paz, Y. Huihuang, R. Denny, K. Larson, D. Foster-Hartnett, A. Cooper, D. Danesh, D. Larsen, T. Schmidt, R. Staggs, J.A. Crow, E. Retzel, N.D. Young, and R.C. Shoemaker, Soybean Genomic Survey: BAC-End Sequences near RFLP and SSR Markers, *Genome 44*:572–581 (2001).
51. Singh, R.J., and T. Hymowitz, The Genomic Relationship Between *Glycine max* (L.) Merr. and *G. soja* Sieb. and Zucc. As Revealed by Pachytene Chromosome Analysis, *Theor. Appl. Genet. 76*:705–711 (1988).
52. Foster-Hartnett, D., J. Mudge, D. Larsen, D. Danesh, H. Yan, R. Denny, S. Penuela, and N.D. Young, Comparative Genomic Analysis of Sequences Sampled from a Small Region on Soybean (*Glycine max*) Molecular Linkage Group G, *Genome 45*:634–645 (2002).

53. Triwitayakorn, K., Positional Cloning of the *Rfs* Loci. Ph.D. Thesis, Southern Illinois University-Carbondale, 2002, pp. 232.
54. Ku, H.M., T. Vision, J. Liu, and S.D. Tanksley, Comparing Sequenced Segments of the Tomato and Arabidopsis Genomes: Large-Scale Duplication Followed by Selective Gene Loss Creates a Network of Synteny, *Proc. Natl. Acad. Sci. USA 97*:9121–9126 (2000).
55. Shoemaker, R.C., K. Polzin, J. Labate, J. Specht, E.C. Brummer, T. Olson, N. D. Young, V. Concibido, J. Wilcox, J.P. Tamulonis, G. Kochert, and H.R. Boerma, Genome Duplication in Soybean (*Glycine* subgenus *soja*), *Genetics 144*:329–338 (1996).
56. Schlueter, J.A., P. Dixon, C. Granger, and R.C. Shoemaker, Mining the EST Databases to Determine Evolutionary Events in the Legumes and Grasses, in *Proceedings of the Stadler Genetics Symposium. 23rd Stadler Genetics Symposium at University of Missouri*, Columbia, Missouri, 2003.
57. Yan, H.H., J. Mudge, D.J. Kim, R.C. Shoemaker, D.R. Cook, and N.D. Young, Estimates of Conserved Microsynteny Among the Genomes of *Glycine max*, *Medicago truncatula*, and *Arabidopsis thaliana*, *Theor. Appl. Genet. 106*:1256–1265 (2003).
58. Grant, D., P. Cregan, and R.C. Shoemaker, Genome Organization in Dicots: Genome Duplication in *Arabidopsis* and Synteny between Soybean and *Arabidopsis*, *Proc. Natl. Acad. Sci. USA 97*:4168–4173 (2000).
59. Cannon, S.B., W.R. McCombie, S. Sato, S. Tabata, R. Denny, L. Palmer, M. Katari, N.D. Young, and G. Stacey, Evolution and Microsynteny of the Apyrase Gene Family in Three Legume Genomes, *Mol. Gen. Genomics 270*:347–361 (2003).
60. Green, P., Against a Whole-Genome Shotgun, *Genome Res. 7*:410–417 (1997).
61. Weber, L., and E. Myers, Human Whole-Genome Shotgun Sequencing, *Genome Res. 7*:410–419 (1997).
62. Chandler, V., and V. Brendal, The Maize Genome Sequencing Project, *Plant Physiol. 130*:1594–1597 (2003).
63. Chen, M., G. Presting, W.B. Barbazuk, J.L. Goicoechea, B. Blackmon, G. Fang, H. Kim, D. Frisch, Y. Yu, S. Sun, S. Higingbottom, J. Phimphilai, D. Phimphilai, S. Thurmond, B. Gaudette, P. Li, J. Liu, J. Hatfield, D. Main, K. Farrar, C. Henderson, L. Barnett, R. Costa, B. Williams, S. Walser, M. Atkins, C. Hall, M.A. Budiman, J.P. Tomkins, M. Luo, I. Bancroft, J. Salse, F. Regad, T. Mohapatra, N.K. Singh, A.K. Tyagi, C. Soderlund, R.A. Dean, and R.A. Wing, An Integrated Physical and Genetic Map of the Rice Genome, *Plant Cell 14*:537–545 (2002).
64. Tao, Q.Z., Y.L. Chang, J. Wang, H. Chen, C. Schuering, M.N. Islam-Faridi, B. Wang, D.M. Stelly, and H.B. Zhang, Bacterial Artificial Chromosome-Based Physical Map of the Rice Genome Constructed by Restriction Fingerprint Analysis, *Genetics 158*:1711–1724 (2001).
65. Chang, Y.L., Q. Tao, C. Schuering, K. Meksem, and H.B. Zhang, An Integrated Map of *Arabidopsis thaliana* for Functional Analysis of Genome Sequence, *Genetics 159*:1231–1241 (2001).
66. Meksem, K., J. Shultz, A. Jamai, K. Zobrist, K. Triwitakorn, H.B. Zhang, C.C. Wu, M.J. Iqbal, and D.A. Lightfoot, Development of Physical Maps Integrated with Genetic Markers and EST: Prelude to Genome Sequencing and Functional Analysis, Plant, Animal & Microbe Genomes X Conference, www.intl-pag.org/pag/10/abstracts/PAGX_W146.html (2002).
67. Shultz, J., C. Wu, F.A. Santos, P. Nimmakayala, R. Springman, C. LaMontague, K. Zobrist, K. Meksem, H.B. Zhang, and D.A. Lightfoot, A Physical Gene Map for the Soybean Genome, *Soybean Genetics Newsl. 28*:5–10 (2001).

68. Ding, Y., M.D. Johnson, W.Q. Chen, D. Wong, Y.J. Chen, S.C. Benson, J.Y. Lam, Y.M. Kim, and H. Shizuya, Five-Color-Based High-Information-Content Fingerprinting of Bacterial Artificial Chromosome Clones Using Type IIS Restriction Endonucleases, *Genomics 74*:142–154 (2001).
69. Luo, M.C., C. Thomas, F.M. You, J. Hsiao, S. Ouyang, C.R. Buell, M. Malandro, P.E. McGuire, O.D. Anderson, and J. Dvorak, High-Throughput Fingerprinting of Bacterial Artificial Chromosomes using the Snapshot Labeling Kit and Sizing of Restriction Fragments by Capillary Electrophoresis, *Genomics 82*:378–389 (2003).
70. Soderlund, C., S. Humphrey, A. Dunhum, and L. French, Contigs Built with Fingerprints, Markers and FPC V.4.7, *Genome Res. 10*:1772–1787 (2000).
71. Soderlund, C., I. Longden, and R. Mott, FPC: A System for Building Contigs from Restriction Fingerprinted Clones, *Comput. Appl. Biosciences 13*:523–535 (1997).
72. Soderlund C., F. Engler, J. Hatfield, S. Blundy, M. Chen, Y. Yeisoo, and R. Wing, Mapping sequence to Rice FPC, in *Computational Biology and Genome Informatics*, edited by P. Wang, J. Wang, and C. Wu, World Scientific Publ., 2002, pp. 59–79.
73. Marra, M.A., T.A. Kucaba, N.L. Dietrich, E.D. Green, B. Brownstein, R.K. Wilson, K.M. McDonald, L.W. Hillier, J.D. McPherson, and R.H. Waterston,. High Throughput Fingerprint Analysis of Large-Insert Clones, *Genome Res. 7*:1072–1084 (1997).
74. Marra, M., T. Kucaba, M. Sakhon, L. Hillier, R. Martienssen, A. Chinwalla, J. Crockett, J. Fedele, H. Grover, C. Gund, W. McCombie, K. McDonald, J. McPherson, N. Mudd, L. Parnell, J. Schein, R. Seim, P. Shelby, R. Waterston, and R. Wilson, A Map for Sequence Analysis of the *Arabidopsis thaliana* Genome, *Nature Genetics 22*:265–275 (1999).
75. Klein, P.E., R.R. Klein, S.W. Cartinhour, P.E. Ulanch, J. Dong, J.A. Obert, D.T. Morishige, S.D. Schlueter, K.L. Childs, M. Ale, and J.E. Mullet, A High-Throughput AFLP-Based Method for Constructing Integrated Genetic and Physical Maps: Progress toward a Sorghum Genome Map, *Genome Res. 10*:789–807 (2000).
76. Ware, D., P. Jaiswal, J. Ni, X. Pan, K. Chang, K. Clark, L. Teytelman, S. Schmidt, W. Zhao, S. Cartinhour, S. McCouch, and L. Stein, Gramene: A Resource for Comparative Grass Genomics, *Nucleic Acids Res. 30*:103–105 (2002).
77. Ware, D.H., P. Jaiswal, J. Ni, I.V. Yap, X. Pan, K.Y. Clark, L. Teytelman, S.C. Schmidt, W. Zhao, K. Chang, S. Cartinhour, L.D. Stein, and S.R. McCouch, Gramene, a Tool for Grass Genomics, *Plant Physiol. 130*:1606–1613 (2002).
78. Coe, E., K. Cone, M. McMullen, S. Chen, G. Davis, J. Gardiner, E. Liscum, M. Polacco, A. Paterson, H. Sanchez-Villeda, C. Soderlund, and R. Wing, Access to the Maize Genome: An Integrated Physical and Genetic Map, *Plant Physiol.* 128:9–12 (2002).
79. Li, L., S. Lu, D.M. O'Halloran, D.F. Garvin, and J. Vrebalov, High-Resolution Genetic and Physical Mapping of the Cauliflower High-Beta-Carotene Gene Or (Orange), *Mol. Gen. Genomics 270*:132–138 (2003).
80. Huang, L., S.A. Brooks, W. Li, J.P. Fellers, H.N. Trick, and B.S. Gill, Map-Based Cloning of Leaf Rust Resistance Gene Lr21 from the Large and Polyploid Genome of Bread Wheat, *Genetics 164*:655–664 (2003).
81. Ramakrishna, W., and J.L. Bennetzen, Genomic Colinearity as a Tool for Plant Gene Isolation, *Methods Mol. Biol. 236*:109–22 (2003).
82. Gardiner, J., S. Schroeder, M.L. Polacco, H. Sanchez-Villeda, M. Morgante, T. Landewe, K. Fengler, F. Useche, M. Hanafey, S. Tingey, H. Chou, S. Brinner, R. Wing, C. Soderlund, and E.H. Coe Jr., Anchoring 9371 Maize EST Unigenes to the BAC Contig Map by Two-Dimensional Overgo Hybridization, *Plant Physiol.* (in press) (2003).

83. Dear, P.H., A.T. Bankier, M.B. Piper, A High-Resolution Metric HAPPY Map of Human Chromosome 14, *Genomics 48*:232–241 (1998).
84. Konfortov, B.A., H.M. Cohen, A.T. Bankier, and P.H. Dear, A High-Resolution HAPPY Map of Dictyostelium Discoideum Chromosome 6, *Genome Res. 10*:1737–1742 (2000).
85. Williams, J.G., and R.A. Firtel, HAPPY Days for the Dictyostelium Genome Project, *Genome Res.10*:1658–1659 (2000).
86. Waugh, R., P.H. Dear, W. Powell, and G.C. Machray, Physical Education—New Technologies for Mapping Plant Genomes, *Trends Plant Sci. 7*:521–523 (2002).
87. Bennetzen, J.L., The Contributions of Retroelements to Plant Genome Organization, Function and Evolution, *Trends Microbiol. 4*:347353 (1996).
88. Rabinowicz, P.D., K. Schutz, N. Dedhia, C. Jordan, L.D. Parnell, L. Stein, W.R. McCombie, and R.A. Martienssen, Differential Methylation of Genes and Retrotransposons Facilitates Shotgun Sequencing of the Maize Genome, *Nature Genet. 23*:305–308 (1999).
89. Peterson, D.G., S.R. Schulze, E.B. Sciara, S.A. Lee, J.E. Bowers, A. Nagel, N. Jiang, D.C. Tibbitts, S. Wessler, and A.H. Paterson, Integration of Cot Analysis, DNA Cloning, and High-throughput Sequencing Facilitates Genome Characterization and Gene Discovery, *Genome Res. 12*:795–807 (2002).
90. Peterson, D.G., S.R. Wessler, and A.H. Paterson, Efficient Capture of Unique Sequences from Eukaryotic Genomes, *Trends Genet. 18*:547–550 (2002).
91. Simmen, M.W., S. Leitgeb, J. Charlton, S.J.M. Jones, B.R. Harris, V.H. Clark, and A. Bird, Nonmethylated Transposable Elements and Methylated Genes in a Chordate Genome, *Science 283*:1164–1167 (1999).
92. Lois, R., L. Freeman, B. Villeponteau, and H.G. Martinson, Active Beta-Globin Gene Transcription Occurs in Methylated, DNase I-Resistant Chromatin of Nonerythroid Chicken Cells, *Mol.Cell Biol. 10*:16–27 (1990).
93. Wölfl, S., M. Schräder, and B. Wittig, Lack of Correlation between DNA Methylation and Transcriptional Inactivation: The Chicken Lysozyme Gene, *Proc. Natl. Acad. Sci. USA 88*:271–27 (1991).
94. Heslop-Harrison, J.S., Comparative Genome Organization in Plants: From Sequence and Markers to Chromatin and Chromosomes, *Plant Cell 12*:617–635 (2000).
95. Li, E., C. Beard, and R. Jaenisch, Role for DNA Methylation in Genomic Imprinting, *Nature 366*:362–365 (1993).
96. Riesewijk, A.M., M.T. Schepens, T.R. Welch, E.M. Van Den Berg-Loonen, E.M. Mariman, H.H. Ropers, and V.M. Kalscheuer, Maternal-Specific Methylation of the Human *IGF2R* Gene Is Not Accompanied by Allele-Specific Transcription, *Genomics 31*:158–166 (1996).

Chapter 15

Bioinformatics: The Interpretation of Genomic Information

Volker Brendel,[a,b] Xiaokang Pan,[a] and Michael E. Sparks[a]

[a]Department of Genetics, Development and Cell Biology; and

[b]Department of Statistics, Iowa State University, 2112 Molecular Biology Building, Ames, IA 50011-3260, USA

Introduction

As noted elsewhere in this volume, legume crops are responsible for producing the majority of protein in the human diet, either directly through crops such as soybean, or indirectly via crops such as alfalfa, a staple feed for livestock. Therefore, legume crop improvement is of paramount importance to sustaining the world's burgeoning human population in the years to come. Biotechnological approaches to this end will produce abundant genome sequence data as a foundation for forward genetics and a starting point for reverse genetics. *Bioinformatics* is a term that encompasses the efforts of biologists, computer scientists, and statisticians to organize and analyze genomic sequences and associated data. These efforts are vast in scope. Each crop has traits in terms of morphology, physiology, and genome organization that are unique, so research efforts and resources must be devoted to each crop to do it justice. However, progress toward this goal will wisely seek to leverage knowledge gained from and resources developed for certain species that may serve as a model for others. Thus, current bioinformatics efforts for legumes concentrate on sequencing, display, and ultimately accurate genomic and functional annotation of standard legume model genomes, including *Medicago truncatula* and *Lotus japonicus*. The purpose of this chapter is to state the prospects and limitations of utilizing existing genetic data, software tools, and genome databases to help annotate these model systems.

The complete sequencing of the *Arabidopsis thaliana* genome (1) and extensive (though still incomplete) characterization of its gene space (2) provide an initial resource for understanding other plant genomes, including legumes. An intriguing question is to what extent *Arabidopsis* and legume species have similar genome organization. This issue is still being debated, with early research (3) supporting the notion that there is extensive conservation of gene order (synteny), but other studies suggesting much more limited synteny (4). Based on our survey of sequenced *M. truncatula* genome segments (retained as bacterial artificial chromosomes, or BACs), regions of extensive micro-colinearity between *Arabidopsis* and *Medicago* appear rare: Few syntenic blocks containing three or more genes in conserved order were found. Thus, it seems likely that blocks of colinearity between *Arabidopsis* and other legume species will also be rarely found. Interestingly, a recent study comparing the *bz* loci of different maize cultivars demonstrated that colinearity may not even hold within a species (5).

Despite the high variability in gene order that is often observed between plant species (5,6), it is estimated that >90% of a given plant's gene repertoire is highly similar to all other plants (7). Therefore, the information gleaned from work in *Arabidopsis* annotation can still be effectively leveraged for functional annotation of a significant fraction of the gene space of other plant species. Indeed, this approach is being used to characterize the gene space of the model monocot, rice (8), the annotations for which will in turn be used to infer gene function for a number of other commercially important cereal crops, such as maize and sorghum (9). The cases where micro-colinearity is evident will no doubt lead investigators to probe the functional significance of the evolutionarily conserved unit (9,10).

This approach to plant genomics underscores the critical necessity for highly accurate annotation of the model species, starting with *Arabidopsis* (8). Efforts to this effect are ongoing (2), with strong emphasis on incorporating evidence from expressed sequence tags (ESTs) and cDNAs at present (11–13). A less direct way, which is highly powerful in practice, is to use putative homologous gene products from a reference species to infer gene structure in a novel genome (reviewed in 14). Because of the conservation of plant gene space, genome annotators will likely be able to determine the structure of a large number of legume genes in this way. There are some serious limitations to this type of approach: Gene loss/absence in the reference species relative to the test species—e.g., *Arabidopsis* relative to soybean—can confound the annotation effort. This is a problem for legume species in particular, as they posses a complement of genes required to establish symbiosis with nitrogen-fixing bacteria and mycorrhizal fungi as well as to initiate the development of root nodules, structures that are not present in *Arabidopsis*. It is unlikely that a species outside of the *Fabaceae* family could be used to model these peculiarities. The focus of the subsequent sections of this chapter, however, will be on methods to identify and display common blocks of genes and common gene structure between a query and a model species.

Genome Informatics

Although DNA sequencing is routine nowadays, genome sequencing still involves assembly of many relatively small fragments into larger contigs, eventually entire chromosomes. This bioinformatics task has been reviewed elsewhere (15) and is not discussed here. The tasks after sequencing is complete (or, in practice, during approximations of completeness) concern genome annotation: identification of gene locations and gene structure in terms of exon/intron boundaries (including possible alternative splicing events), promoter elements, and other transcription signals, and characterization of the translation products of protein-coding genes. Typically, annotation proceeds in stages. Initial annotation is large-scale, automated, and progressive for rapid dissemination to the public along with the genome sequence. A large fraction of gene structures annotated in this way will remain tentative (and are commonly referred to as hypothetical, or putative, genes and gene products). Subsequent annotation involves successive re-evaluation, extension, and correction of the annotation, removing many tentative assignments on the basis of novel experimental evidence; it therefore becomes necessary to implement databases that integrate sequence data, analysis tools, and the community in a synergistic manner (16). Sequencing technology is advancing so rapidly that we can expect to acquire dozens of completed plant genome sequences in the next couple of decades. Indeed, such foundation has already become reality for bacterial research, with more than a hundred prokaryotic genomes publicly available (17). Thus, it is clear that genome informatics must rise to the challenge to quickly and accurately annotate genomes and to make the inferred knowledge easily accessible. We will briefly review the two main strategies for gene structure annotation, as well as for dedicated databases to facilitate annotation and a novel tool being developed by one of the authors (X.Pan) for syntenic display and analysis.

Ab Initio *Algorithms for Gene Finding*

During the in- or post-sequencing phase of a given genome project, an error-prone, first-pass analysis of the raw sequence data using *ab initio* gene-prediction software is standard. For bacterial genomes, GeneMark.hmm (18) and Glimmer (19), both based on variants of Hidden Markov Models, are quite effective at predicting prokaryotic gene structures. In large part, the task of developing gene-annotation algorithms for prokaryotic genomes is done. However, algorithms to accurately predict the more complex gene structures of eukaryotes lag behind.

Two of the more popular modern eukaryotic gene-finding programs are GENSCAN (20) and Fgenesh (21). The former uses a fixed-order, three-periodic Hidden Markov Model, as well as transcriptional and translational signal scoring schemes to predict gene structures. Fgenesh has a similar design. Unfortunately, even when ample, unambiguous data are available to parameterize the algorithms, it is estimated that the majority of eukaryotic gene structures predicted by *ab initio* software is incorrect (22). Given the current state of the art of computational gene structure prediction, gene structures must be confirmed with experimental evidence to generate reliable annotations.

Spliced Alignment

Spliced alignment is an important step in genome annotation that facilitates both the identification of errors made during the *ab initio* annotation process and the detection of spliced isoforms that might surface in the organism's transcript population.

In general, spliced alignment involves the following: A query sequence (EST/cDNA or polypeptide) is threaded onto a genomic template such that long gaps are introduced into the query, corresponding to introns, and short gaps and substitution pairs are tolerated in the exon-exon alignment, provided these do not cause a similarity score to fall below some threshold value. In principle, this technique will identify the cognate locus that produced the transcript/protein and resolve the gene structure found at that locus. In addition, aligning a query from one species against the genome of another will help to identify potentially homologous genes common to the test and reference species; the function of the gene in the reference species, if known, can be used to infer the function of its homologous partner(s) in the test species. This presents a nice framework in which to use transcript data from leguminous plants and the *Arabidopsis* genome annotation to gain better insight into legume biology, despite the current lack of a complete legume genome sequence.

A variety of good spliced-alignment programs exist and are freely available for research purposes. Some programs, including sim4 (23) and Spidey (24) follow a strategy of post-processing output from a heuristic, word-based alignment tool, such as BLAST, for spliced alignment. The GeneSeqer program implements a full dynamic programming–based algorithm, optimized for plants, that produces an optimally scoring spliced alignment of an EST/cDNA or protein sequence against a genomic template (25).

Databases

There are a growing number of increasingly comprehensive databases available to the modern legume researcher. The Legume Information System has been established by the USDA and the NCGR to serve as a current, public repository for legume transcript and genomic sequence data, genetic maps, and metabolic pathway schematics (26). Another useful resource is SoyBase, which can be found at http://soybase.agron.iastate.edu/. Specialized resources of this ilk, coupled with other differentiated databases such as TAIR (27), MaizeGDB (28), and SGN (29), along with more general plant databases such as PlantGDB (16), all operating under a unified, controlled vocabulary as being developed by the Plant Ontology Consortium (30), constitute a potent information network within which to advance plant biology.

A Prototype Syntenic Browser

Informative and visually appealing displays of genome information have become indispensable tools for genome research, allowing users to quickly link and analyze data concerning particular genomic loci, genes, or proteins. As model genomes are becoming available, there is a pressing need for both identification and display of synteny between a query genomic sequence and regions of the model genome. More generally, biologists want to assess similarity on various scales, ranging from gene order to individual gene structure and to the sequence level. We have developed prototype software for web-based display of genomic sequence similarity, with zooming capabilities that allow assessment of similarity on all scales. The tool consists of two components: a web-based graphical interface and a MySQL-based relational database. The graphical, user-friendly interface lets users search and browse detailed plant cross-genome comparisons for syntenic regions, highlighting conserved coding sequences between a reference sequence and a query sequence. The tool helps to identify uncharacterized genes, putative regulatory elements, and novel structural features of crop species (e.g., soybean) by comparison with an annotated reference sequence (e.g., *A. thaliana*); and it facilitates classifying a similar pair of genes as orthologous or paralogous (7). It also helps plant genome curators to refine and edit annotations of plant species such as *M. truncatula* and *Lotus japonicus*. The browser is currently able to visualize sequence comparisons between *A. thaliana* and *M. truncatula*. The code is heavily borrowed from Lincoln Stein's implementation of a syntenic browser specific to WormBase (31). We have developed the basic graphical interface and are enhancing it in several aspects, including comparison tracks, portability, and robustness.

The displayed data are for the most part results of concluded analysis of available genomic sequences, using a variety of the tools reviewed above. All results are stored in the underlying MySQL database. Currently, *ab initio* gene predictions are derived with Fgenesh (21) to annotate *Medicago* and *Lotus* BAC sequences available from GenBank. Nucleotide level alignments are based on the BLASTZ program (32). Spliced alignments are made with the GeneSeqer program (25). The browser has a flexible application interface that can accommodate essentially all other analysis tools, provided the tools generate output in some standard format such as GFF (33). We are continuing to develop the syntenic browser by enriching the MySQL-based database and improving the graphical interface. We plan to incorporate *Lotus japonicus* and *Glycine max* data into the syntenic browser to make it a more comprehensive tool for legume comparative genomics.

Applications

A MySQL-based database has been developed for the synteny browser by collecting, annotating, and comparing genomic sequences of *A. thaliana* and *M. truncatula* with respect to genes, repeats, ESTs, and other features. For *Arabidopsis*, we downloaded the *Arabidopsis* genome, release 4, and its annotation data from the FTP site of the TIGR *Arabidopsis* genome annotation database (2). We mined the annotation for information such as the location of exons, introns, 5′- and 3′-UTRs, protein sequences, and the listed functions of individual genes. For *Medicago*, we retrieved 187,759 ESTs and 197 complete BAC sequences from GenBank. GeneSeqer (25) was used to align *Arabidopsis* proteins and *Medicago* ESTs to the *Medicago* BACs (Fig. 15.1). BLASTZ was used to align *Medicago* BACs to the *Arabidopsis* genome (Fig. 15.2). Since *Medicago* genome annotation is not currently available from GenBank, we annotated the *Medicago* BACs using Fgenesh (21). These putative gene structures were manually scrutinized for accuracy by comparison with EST

```
Alignment of AC123572:92516..94771 vs At1g67730 (Generated by GeneSeqer2.
Version of March 25, 2002):

 Exon  1  92516  93147 (632 n); Protein 1 211 (211 aa); score: 0.713
  Intron  1  93148  93287 ( 140 n); Pd: 0.995  Pa: 0.992
 Exon  2  93288  93357 (70 n);  Protein 212 234 (23 aa); score: 0.894
  Intron  2  93358  94516 (1159 n); Pd: 0.957  Pa: 0.999
 Exon  3  94517  94771 (255 n); Protein 235 318 (84 aa); score: 0.616

MATCH   AC123572+     At1g67730      0.702   957    1.000   P


TGTTGCATAA TCAGTAAACT CAAAACCCAA CCATTCTGGT TCCTCATTCT CTTCTCCCTC    92575
 C  C  I   I  S  K  L   K  T  Q   P  F  W   F  L  I  L   F  S  L
       |      +     .   |  +  |   |     |   .  |  |  |   |     |
 M  E  I   C  T  Y  F   K  S  Q   P  T  W   L  L  I  L   F  V  L       20

///

CCTCTTTATG CTGTTTATGC TGCTACAAAA GCGTGAGCTT TTTTTTTCTT CTTTTTTTGT    93175
 P  L  Y   A  V  Y  A   A  T  K   A
 |  .  |   +  |  |  |   .  .  |   .
 P  F  Y   S  V  Y  A   G  A  K   T....... ......... .........      211

///

TATGGATGTT ATGGTGTTAA TGAGGGTTTT TGTGTTTGTT TTGTTGTTTC AGGTACATTG    93295
                                                        Y  I
                                                        |  +
......... ......... ......... ......... ......... ..    Y  V         213

///

AGGTATTGTG CTGCTGCTTT GAGTCACTCA ATTTACATAC TTCTATTTCT AGCTTAGTAA    93415
Q
|
Q ....... ......... ......... ......... ......... .........          234

///

GATATTTTTT CTTGATTTCA GGTTCCATTA TATGTGGCAA CAAAGATGGC ATCAATCAGG    94555
                       V  P  L   Y  V  A   T  K  M  A   S  I  R
                       |  |  |   |  |  |   |  |  |  .   .  |  |
......... ......... .  V  P  L   Y  V  A   T  K  M  T   K  I  R       247

///

AGGGGGCAGC TCAAAGATTC GAAGAAGCAG GAATAA     94771
 R  G  Q   L  K  D  S   K  K  Q   E  *
 +  |         |  |  |      |  +   |
 K  G  L   Q  K  D  S   M  K  K   E  *        319
```

Figure 15.1. GeneSeqer alignment of *Arabidopsis* gene At1g67730 on *Arabidopsis* chromosome 1 and gene AC123572.18 onto *Medicago* BAC 123572.

```
Alignment of AC123572:92516..94771 vs At1g67730 & its promoter (Generated
by BLASTZ)

Blue (80-90% identity)
       1 25457049 tagttaattc cggtgg
AC123572     92350 tacttaattt ctgttg

White (40-50% identity)
       1 25457116 aattctctct tcctttctct ctctttctcc ttaaaaactc ///
AC123572     92437 aaatccattg ctgctactct catttttttc ataaataaat ///

Blue (80-90% identity)
       1 25457212 cacttacttc aaatcacaac ccacttggct cctcattctc ///
AC123572     92527 cagtaaactc aaaacccaac cattctggtt cctcattctc ///

white (40-50% identity)
       1 25457868 gagatgatgc ttttgagtgg tttaaagtag tgac
AC123572     93209 gatttgttct tttttgaggc tttgaaatat ggat

Gray (50-60% identity)
       1 25458630 ttatagGTTC CCTTGTATGT TGCAACAAAG ATGACAAAAA ///
AC123572     94511 tttcagGTTC CATTATATGT GGCAACAAAG ATGGCATCAA ///
```

Figure 15.2. BLASTZ alignments of *Arabidopsis* gene At1g67730 onto *Arabidopsis* chromosome 1 and gene AC123572.18, as well as their corresponding promoter sequences on BAC123572 of *Medicago*.

alignments. *Arabidopsis* repeat data were obtained using the RepeatMasker program (34) and the most recent version of RepBase, 8.12 (35).

Identification of Syntenic Regions

Using the synteny browser, we have identified more than 20 syntenic regions containing three or more orthologous gene pairs in 197 *Medicago* BACs compared to the *Arabidopsis* genome. Figure 15.3 shows a region of microsynteny between *M. truncatula* BAC AC123572 and *A. thaliana* chromosome 1. In this syntenic region, there are five orthologous gene pairs. For each gene pair, the relative transcriptional orientation and gene structure are conserved, with some slight variations: Two gene structures, *Arabidopsis* gene At1g67710 and *Medicago* gene AC123572.10, display different splicing patterns, as highlighted by the vertical gray lines. The DNA level alignments have been depicted by rectangles with different colors, of which black represents the highest percent identity (100%), lighter colors represent lower percent identity, and white represents the lowest percent identity (40%). The graph shows that noncoding regions, such as introns and the At1g67730 promoter, are also conserved.

Gene Annotation

Single, putatively paralogous, gene pairs are often found between *Medicago* and *Arabidopsis*. Consider, for example, the comparison, depicted in Fig. 15.4, of the

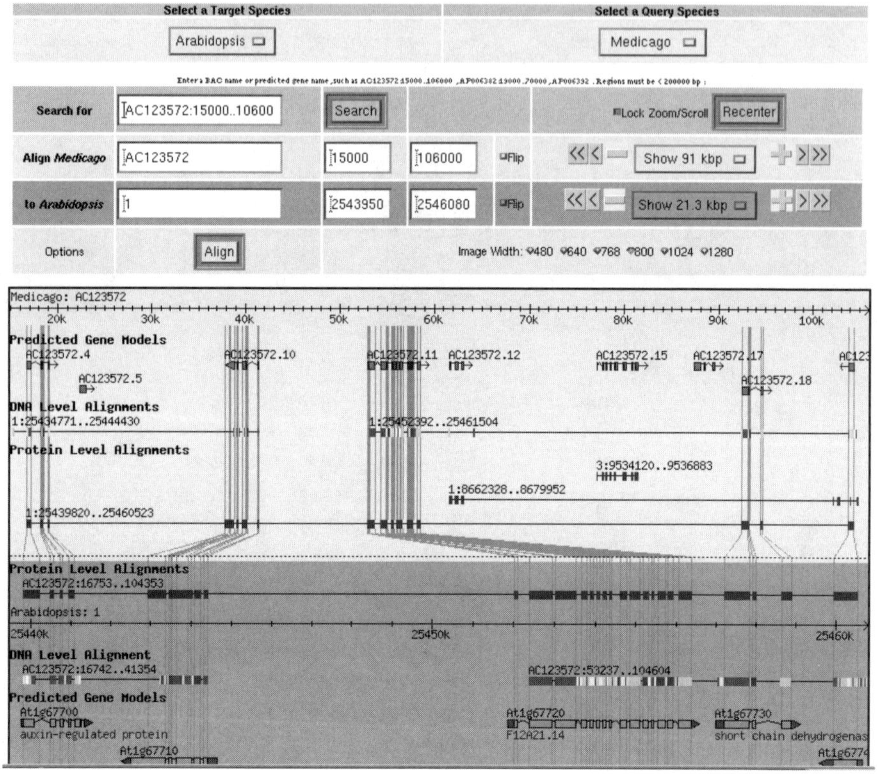

Figure 15.3. Microsynteny between *Arabidopsis* and *Medicago*: a syntenic region between *M. truncatula* BAC AC123572 and *Arabidopsis thaliana* chromosome 1. When a user selects *Arabidopsis* as a target species and *Medicago* as a query species and then searches *Medicago* BAC AC123572 within the region from 15,000 to 10,600 bp, it displays the longest syntenic region we have found between *Medicago* and *Arabidopsis*. The user can adjust the window size by changing the data in the text fields and clicking the "Align" button to get a high-resolution figure as shown here. The predicted *Arabidopsis* gene models link out to the gene annotation page—just click the graphical representation of a gene model. The predicted *Medicago* gene models were named based on the BAC name and order of the genes beginning from the 5′-end of the BAC. Clicking on a rectangle in the bottom window produces the text representation of the alignments for the user's perusal.

second gene on the *Medicago* BAC AC119419, predicted by Fgenesh, and the *Arabidopsis* gene At1g71820. This second gene of the *Medicago* BAC has 21 exons, but the confirmed Arabidopsis gene model has 25. The *Medicago* gene's 21 exons exactly match 21 exons of At1g71820. However, *Medicago* EST alignments indicate that the four additional exons found in At1g71820 should have been included in the

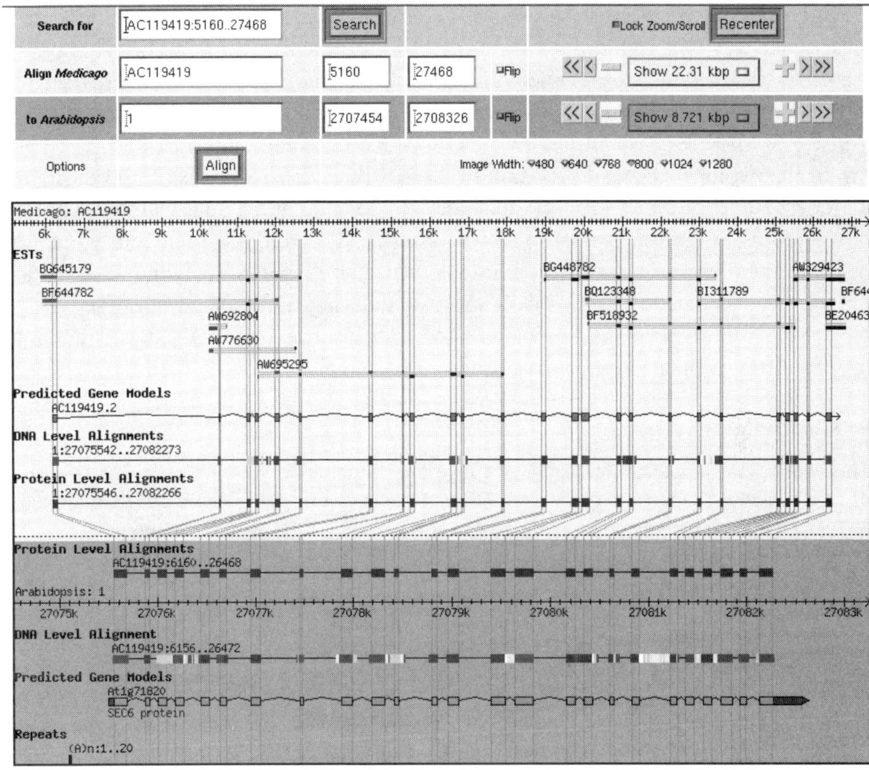

Figure 15.4. Annotation of homologous genes: a putatively paralogous gene pair between *Arabidopsis thaliana* and *Medicago truncatula*. The results shown here correspond to selecting *Arabidopsis* and *Medicago* as a target and query species, respectively, and then searching for a *Medicago* BAC AC119419 within the region from 5,160 to 27,468 bp. The second gene of *Medicago* BAC AC119419 and the *Arabidopsis* gene At1g71820 both have 25 exons and are highly conserved in order and content. EST alignments confirm the Fgenesh-predicted *Medicago* gene model.

Fgenesh-suggested *Medicago* gene annotation. The At1g71820 has weak similarity to the yeast SEC6 protein, involved in exocytosis, which was the ascribed putative function for the gene we identified in *Medicago*. This comparison makes a good case for the importance of experimentally confirming putative gene structures, and it also underscores the potential of using our syntenic browser to refine genome annotations.

Summary

Advances in biotechnology are bringing unprecedented opportunities for understanding the genetic basis of important crop plants. Bioinformatics is challenged to keep pace with the technological innovations and the bonanza of genomic data they produce. Great strides have already been made, and better computational tools, databases, and web-accessible displays are facilitating genomic research. We have reviewed various tools for genome annotation and introduced a syntenic browser that allows cross-species genome comparisons. Such tools will greatly facilitate the transfer of knowledge from one (model) species to another and aid our understanding of the evolutionary and breeding processes that have given rise to modern crop varieties. Because much progress has been made and can be anticipated toward the development of computational tools for genomic research, complementary technology must also be developed to enable easy sharing of expert knowledge via feedback to crop species databases (36). We anticipate exciting times ahead that will witness the integration of tools and knowledge and provide a revolutionary new foundation for crop improvement in the next decades.

Acknowledgments

This work was supported in part by NSF Plant Genome Research Projects grants DBI-0110254 and by a subcontract from the National Center for Genome Research (NCGR), Santa Fe, New Mexico, as part of NCGR's USDA-ARS Specific Cooperative Agreement No. 58-3625-2-109. We are grateful to our colleague Carolyn Lawrence for critical reading and editing of an earlier version of the manuscript.

References

1. The Arabidopsis Genome Initiative. Analysis of the genome sequence of the flowering plant *Arabidopsis thaliana*, *Nature 408*:796–815 (2000).
2. Wortman, J., B.J. Haas, L.I. Hannick, R.K. Smith, Jr., R. Maiti, C.M. Ronning, A.P. Chan, C. Yu, M. Ayele, C.A. Whitelaw, O.R. White, and C.D. Town, Annotation of the *Arabidopsis* Genome, *Plant Physiol. 132*:461–468 (2003).
3. Grant, D., P. Cregan, and R. Shoemaker, Genome Organization in Dicots: Genome Duplication in *Arabidopsis* and Synteny between Soybean and *Arabidopsis*, *Proc. Natl. Acad. Sci. USA 97*:4168–4173 (2000).
4. Yan, H.H., J. Mudge, D.J. Kim, D. Larsen, R.C. Shoemaker, D.R. Cook, and N.D. Young, Estimates of Conserved Microsynteny among the Genomes of *Glycine max*, *Medicago truncatula* and *Arabidopsis thaliana*, *Theor. Appl. Genet. 106*:1256–1265 (2003).
5. Fu, H., and H.K. Dooner, Intraspecific Violation of Genetic Colinearity and Its Implications in Maize, *Proc. Natl. Acad. Sci. USA 99*:9573–9578 (2002).
6. Tarchini, R., P. Biddle, R. Wineland, S. Tingey, and A. Rafalski, The Complete Sequence of 340 kb of DNA around the Rice *Adh1-Adh2* Region Reveals Interrupted Colinearity with Maize Chromosome 4, *Plant Cell 12*:381–391 (2000).
7. Bennetzen, J.L., Comparative Sequence Analysis of Plant Nuclear Genomes: Microcolinearity and Its Many Exceptions, *Plant Cell 12*:1021–1029 (2000).
8. Schoof, H., and W.M. Karlowski, Comparison of Rice and *Arabidopsis* Annotation, *Curr. Opin. Plant Biol. 6*:106–112 (2003).
9. Morishige, D.T., K.L. Childs, D. Moore, and J.E. Mullet, Targeted Analysis of Orthologous *Phytochrome A* Regions of the Sorghum, Maize, and Rice Genomes Using Comparative Gene-Island Sequencing, *Plant Physiol. 130*:1615–1625 (2002).

10. Oh, K., K. Hardeman, M.G. Ivanchenko, M. Ellard-Ivey, A. Nebenführ, T.J. White, and T.L. Lomax, Fine Mapping in Tomato Using Microsynteny with the *Arabidopsis* Genome: The *Diageotropica (Dgt)* Locus, *Genome Biol. 3*:research0049.1-0049.11 (2002).
11. Haas, B., A.L. Delcher, S.M. Mount, J.R. Wortman, R.K. Smith, Jr., L.I. Hannick, R. Maiti, C.M. Ronning, D.B. Rusch, C.D. Town, S.L. Salzberg, and O. White, Improving the *Arabidopsis* Genome Annotation Using Maximal Transcript Alignment Assemblies, *Nucleic Acids Res. 31*:5654–5666 (2003).
12. Haas, B.J., N. Volfovsky, C.D. Town, M. Troukhan, N. Alexandrov, K.A. Feldman, R.B. Flavell, O. White, and S.L. Salzberg, Full-Length Messenger RNA Sequences Greatly Improve Genome Annotation, *Genome Biol. 3*:research0029.1-0029.12 (2002).
13. Zhu, W., S.D. Schlueter, and V. Brendel, Refined Annotation of the *Arabidopsis thaliana* Genome by Complete EST Mapping, *Plant Physiol. 132*:469–484 (2003).
14. Brendel, V., and W. Zhu, Computational Modeling of Gene Structure in *Arabidopsis thaliana*, *Plant Mol. Biol. 48*:49–58 (2002).
15. Huang, X., J. Wang, S. Aluru, S.P. Yang, and L. Hillier, PCAP: A Whole-Genome Assembly Program, *Genome Res. 13*:2164–2170 (2003).
16. Dong, Q., S.D. Schlueter, and V. Brendel, PlantGDB, Plant Genome Database and Analysis Tools, *Nucleic Acids Res. 32*:D354–D359 (2004).
17. Bernal, A., U. Ear, and N. Kyrpides, Genomes OnLine Database (GOLD): A Monitor of Genome Projects World-Wide, *Nucleic Acids Res. 29*:126–127 (2001).
18. Lukashin, A.V., and M. Borodovsky, GeneMark.hmm: New Solutions for Gene Finding, *Nucleic Acids Res. 26*:1107–1115 (1998).
19. Delcher, A.L., D. Harmon, S. Kasif, O. White, and S.L. Salzberg, Improved Microbial Gene Identification with GLIMMER, *Nucleic Acids Res. 27*:4636–4641 (1999).
20. Burge, C., and S. Karlin, Prediction of Complete Gene Structures in Human Genomic DNA, *J. Mol. Biol. 268*:78–94 (1997).
21. Salamov, A.A., and V.V. Solovyev, *Ab Initio* Gene Finding in *Drosophila* Genomic DNA, *Genome Res. 10*:516–522 (2000).
22. Pavy, N., S. Rombauts, P. Déhais, C. Mathé, D.V. Ramana, P. Leroy, and P. Rouzé, Evaluation of Gene Prediction Software Using a Genomic Data Set: Application to *Arabidopsis thaliana* Sequences, *Bioinformatics 15*:887–899 (1999).
23. Florea, L., G. Hartzell, Z. Zhang, G.M. Rubin, and W. Miller, A Computer Program for Aligning a cDNA Sequence with a Genomic DNA Sequence, *Genome Res. 8*:967–974 (1998).
24. Wheelan, S.J., D.M. Church, and J.M. Ostell, Spidey: A Tool for mRNA-to-Genomic Alignments, *Genome Res. 11*:1952–1957 (2001).
25. Brendel, V., L. Xing, and W. Zhu, Gene Structure Prediction from Consensus Spliced Alignment of Multiple ESTs Matching the Same Genomic Locus, *Bioinformatics*, Advance Access Feb. 5, 2004, 10.1093/bioinformatics/bth058 (2004).
26. Legume Information System (LIS), www.comparative-legumes.org/.
27. Rhee, S.Y., W. Beavis, T.Z. Berardini, G. Chen, D. Dixon, A. Doyle, M. Garcia-Hernandez, E. Huala, G. Lander, M. Montoya, N. Miller, L.A. Mueller, S. Mundodi, L. Reiser, J. Tacklind, D.C. Weems, Y. Wu, I. Xu, D. Yoo, J. Yoon, and P. Zhang, The Arabidopsis Information Resource (TAIR): A Model Organism Database Providing a Centralized, Curated Gateway to Arabidopsis Biology, Research Materials and Community, *Nucleic Acids Res. 31*:224–228 (2003).

28. Lawrence, C.J., Q. Dong, M.L. Polacco, T.E. Seigfried, and V. Brendel, MaizeGDB, the Community Database for Maize Genetics and Genomics, *Nucleic Acids Res. 32*:D393–D397 (2004).
29. Solanaceae Genomics Network (SGN), www.sgn.cornell.edu/.
30. The Plant Ontology Consortium, www.plantontology.org/.
31. Harris, T.W., N. Chen, F. Cunningham, M. Tello-Ruiz, I. Antoshechkin, C. Bastiani, T. Bieri, D. Blasiar, K. Bradnam, J. Chan *et al.* WormBase: A Multi-Species Resource for Nematode Biology and Genomics, *Nucleic Acids Res. 32*:D411–D417 (2004).
32. Schwartz, S., W.J. Kent, A. Smit, Z. Zhang, R. Baertsch, R.C. Hardison, D. Haussler, and W. Miller, Human-Mouse Alignments with BLASTZ, *Genome Res. 13*:103–107 (2003).
33. General Feature Format (GFF), www.sanger.ac.uk/Software/formats/GFF/GFF_Spec.shtml.
34. Smit, A.F.A., and P. Green, http://ftp.genome.washington.edu/cgi-bin/RepeatMasker.
35. Jurka, J., Repbase Update: A Database and an Electronic Journal of Repetitive Elements, *Trends Genet. 16*:418–420 (2000).
36. Brendel, V., Novel Tools for Plant Genome Annotation and Applications to *Arabidopsis* and Rice, In J.P. Gustafson, R. Shoemaker, and J.W. Snape (eds.), *Genome Exploitation: Data Mining*, Stadler Genetics Symposia Series, 23rd Symposium (to appear 2004).

Chapter 16

Soybean Microarrays: A Genomic Tool for Crop Improvement

Steven J. Clough[a] and Lila O. Vodkin[b]

[a]USDA-ARS and the Department of Crop Sciences, University of Illinois, Urbana IL; and

[b]Department of Crop Sciences, University of Illinois, Urbana IL

Introduction

DNA, the genetic material common to all life forms, has the inherent property of self-adhesion. The famous double helix consists of two strands of DNA that adhere to each other because of the attractive forces of hydrogen bonds that occur between DNA's four chemical residues: the bases A, T, C, and G. This adhesion occurs in a very specific manner such that the A bases bind to T's and the C's to G's (Fig. 16.1). Stretches of sequence bind only to complementary stretches that have a high percentage of the bases matching up correctly. For example, the sequence ATTCGAGTTA will bind more tightly to its exact complement, TAAGCTCAAT (where every A aligns with a T and every C with a G), than it will bind to TAACCTCATT, which has only 8 out of 10 bases lining up with its complement. Geneticists and molecular biologists have been taking advantage of the nature of one strand to bind (or hybridize) to another strand and especially how the strength of this binding is related to the number of bases that are complementary. For decades researchers have been analyzing nucleic acids (DNA and RNA) that have been size separated through gel matrixes, as in the use of Southern blots (1) to analyze gene structure/organization, and in the use of northern blots (1) to examine gene expression (mRNA abundance). In these techniques, single-stranded DNA or RNA is stabilized on a membrane and is then hybridized with radioactively labeled nucleic acid to identify or quantify fragments of interest.

Researchers can also take advantage of this specific binding between complementary strands of nucleic acid to simultaneously analyze hundreds of different genes by spotting cloned DNA gene fragments of interest and hybridizing them with labeled nucleic acid probes. The combined use of fluorescently labeled nucleic acids, precise spotting robots, and microscopy has allowed researchers to miniaturize the platform of hybridization analysis such that thousands of DNA spots representing thousands of different genes may be done on the surface of a glass microscope slide. Such spotted slides are referred to as microarrays (2–8). It is physically possible to fit over 80,000 spots on a standard 1-inch by 3-inch microscope slide, but in practice most labs spot

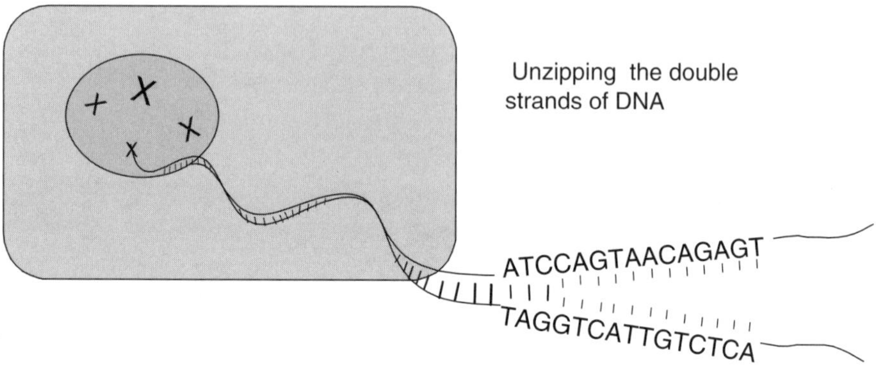

Figure 16.1. DNA consists of two complementary strands that adhere because of specific base pairing.

about 10,000 to 20,000 genes per slide (9). Because it is estimated that most organisms contain fewer than 80,000 genes, one could theoretically spot the entire genome of an organism on a single microscope slide.

Microarrays are being developed for organisms across the spectrum of life, for example, bacteria (10–12), yeast (13–15), nematodes (16), fly (17–19), honeybee (20), fish (21, 22), and many mammals (5, 8, 23, 24). Many array projects for the study of plant gene expression are being developed, including Arabidopsis (8, 16, 25–34), barley (35,36), cacao (37), maize (38,39), pine (40), poplar (41), rice (42–44), soybean (L.O. Vodkin, unpublished data), strawberry (45), and tomato (46,47). A review of plant microarrays was published recently (4).

Soybean Microarray Construction

The development of soybean cDNA microarrays was facilitated greatly by two recent publicly funded genomics projects. The Public EST Project for Soybean (48), funded by U.S. soybean grower associations, led to the creation of over 80 cDNA libraries (collections of copy DNA [cDNA], which is DNA enzymatically produced from messenger RNA [mRNA] that have been cloned into *E. coli*). The cDNA clones in each library came from a different source tissue, thereby generating a collection that represents genes expressed in many different soybean tissues, organ systems, and developmental stages, as well as in pathogen- and stress-challenged plant tissues. This grower-funded project resulted in over 280,000 partial sequences (known as expressed sequence tags, or ESTs). Bioinformatic analysis directly compared the relatedness of each sequence of individual cDNA clones and allowed the identification of possible duplicative clones, thus reducing this list to 60,000 potentially unique cDNA clones. One of the major goals of the second project, the Functional Genomics Program for Soybean funded by the National Science Foundation, was to physically assemble these unique cDNA clones into unique gene subsets, called unigene sets, and to use these unigene sets to develop soybean cDNA microarrays.

The complex process of preparing cDNA microarrays involves robotic picking of selected soybean cDNA clones from plates of clones that are maintained as individual recombinant *E. coli* cultures into new sets of plates containing 384 wells per plate. From these unigene plates the plasmids carrying the cDNA are extracted, purified, and sequenced from the end of the cDNA opposite the end originally sequenced from each clone. The sequence information is used to verify accurate clone picking and to obtain additional sequence information about each of the unique genes. After the sequence data are analyzed to verify that the clones were correctly picked, the soybean cDNA inserts are amplified by polymerase chain reaction (PCR). Approximately one nanoliter of each PCRed cDNA insert is spotted onto specially treated glass microscope slides. This process (Fig. 16.2) has been completed for three unique sets of 9,216 each, or a total of 27,648 potentially unique genes.

Currently (as of February 2004), each soybean microarray slide set consists of 9,216 theoretically different genes represented as spotted PCR products derived from 9,216 cloned cDNAs (Fig. 16.2). Set 1 (sequence-driven unigene library Gm-r1070, consisting of 9,216 cDNA clones) is highly representative of genes expressed in the developing flowers and buds, young pods, developing seed coats, and immature cotyledons. Set 2 (unigene libraries Gm-r1021 + Gm-r1083, 9,216 cDNA clones) is highly representative of genes in the roots of seedlings and adult plants, including roots infected with the nodulating bacterium *Bradyrhizobium japonicum*. Set 3 (unigene library Gm-r1088, 9,216 cDNA clones) is highly representative of clones selected from libraries derived from germinating cotyledons, germinating seedlings subjected to various stresses, and leaves of 2-week-old plants, including some challenged by pathogens. Set 4, currently under construction, is being assembled from libraries derived from soybean-tissue-culture embryos and from stressed or pathogen-challenged tissues. Completion of all four sets will bring the total number of genes represented to 36,864. Because the selection process is not foolproof, however, duplications will most likely occur and the actual number of unique gene representatives may be nearer to 30,000.

Microarray Use

The microarray slides of spotted nucleic acid are first washed in boiling water to weaken the hydrogen bonding between the two strands of the double helix. This weakening leads to the separation and removal of one of the complementary strands. This step leaves behind single strands of DNA that have been chemically fixed to the slide surface and are now available to hybridize to fluorescently labeled fragments (Fig. 16.3). Genetic material from a sample under study is fluorescently labeled as single-stranded DNA and is applied as a solution onto the slide of DNA spots. Time is allowed for fragments to find the spots containing closely complementary partners

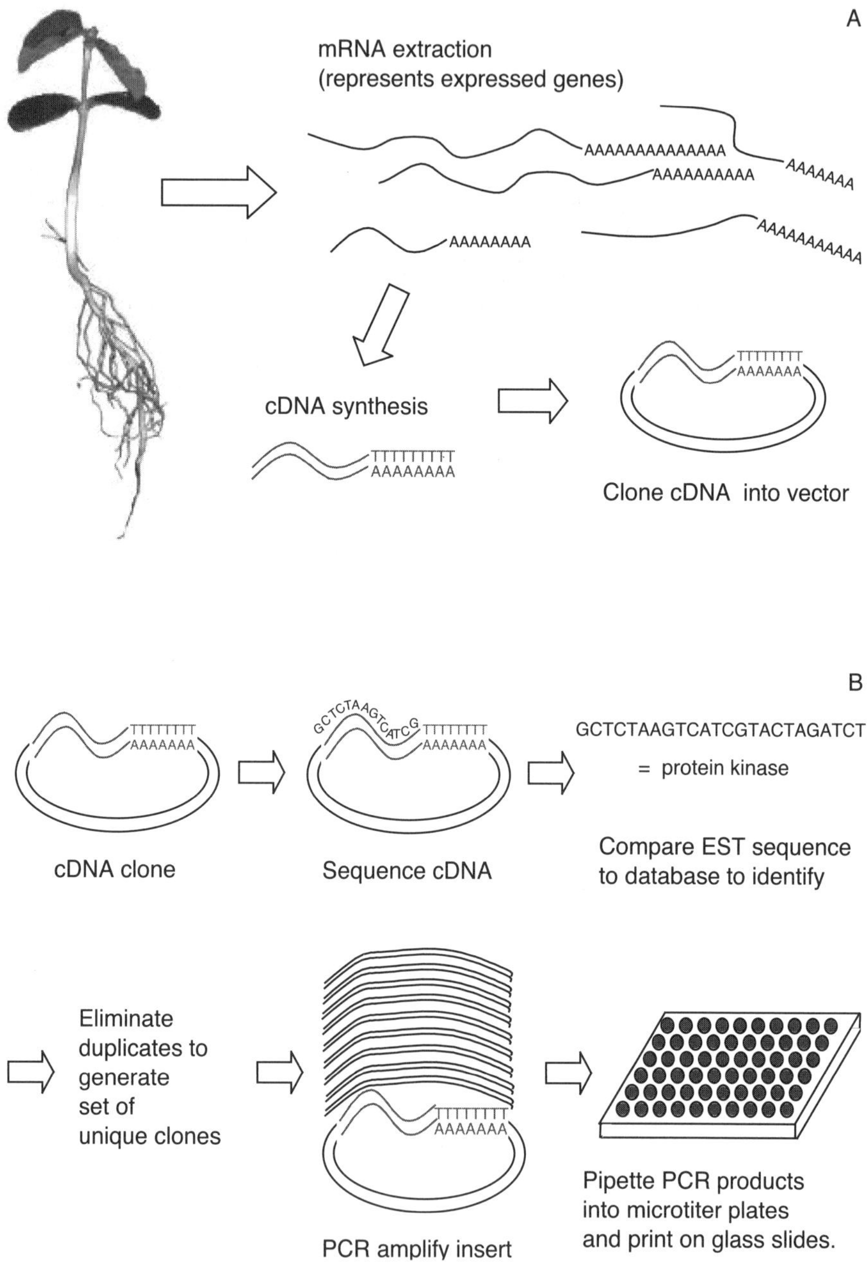

Figure 16.2. Soybean microarray construction: (A) Process begins with cDNA libraries; (B) continues with identification of unique sets of clones, PCR amplification of the soybean cDNA, pipetting PCR products into 384-well microtiter plates, and printing of PCR products onto specially coated glass slides.

and to bind to them. Unbound labeled DNA is gently washed away, leaving behind fluorescently glowing spots that are detectable by special microscopic fluorescent scanners and whose intensities are proportional to the number of fluorescent strands bound to that particular spot (Fig. 16.4).

A powerful use of cDNA microarrays is the direct comparison of samples from two different sources (control and treatment) to determine gene expression ratios. mRNA, which is proportional to the level of gene expression, is isolated from tissues of interest and enzymatically converted into single-stranded cDNA. The mRNA prep from the treatment sample is converted into cDNA labeled with fluorescent dye A (ex. Cy3), and the mRNA of the control is converted to cDNA labeled with dye B (ex. Cy5). Because fluorescent dye A fluoresces at a wavelength that is distinguishable from dye B, one may mix the two labeled cDNA samples onto the same microarray slide, where the fluorescent probes compete for the available spotted strands of DNA and bind. Since the labels fluoresce at different wavelengths, one can determine which genes were more abundant in which sample. Computer analysis of the fluorescent images allows one to quickly obtain fluorescence ratios (and therefore relative expression ratios) of tens of thousands of genes on a single slide and thus to determine how the two samples differed in gene activity at the particular time point when the samples were collected (Fig. 16.5).

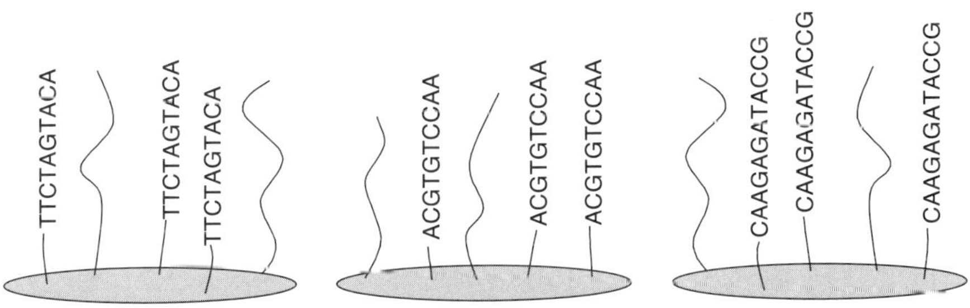

Figure 16.3. Spotted PCR products are chemically bonded to the slide surface. The complementary strand is then removed, leaving single-stranded DNA in each spot.

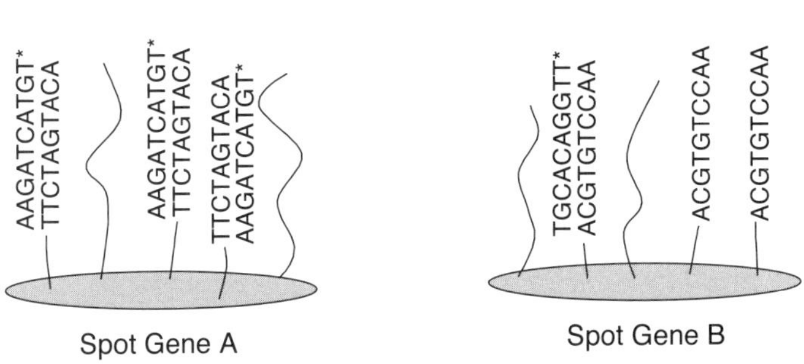

Figure 16.4. Hybridization of fluorescently labeled DNA (*) to spotted DNA. In this example, Gene A has three times more fluorescent strands attached than that for Gene B; therefore the Gene A spot will be three times brighter. Note: drawing is simplified for clarity. Actual DNA strands are hundreds of bases long.

Soybean Microarray Applications

Genes are responsible for producing the macromolecular raw materials that make up a cell, as well as for the synthesis of the enzymes that conduct and direct the chemistry of a cell. As specific materials or functions are needed, specific genes are turned on or off to meet the needs of the organism. Since microarrays allow for the simultaneous analysis of the expression of thousands of genes at any given time, one may use them to analyze the genetic responses to any stimulus or growth condition, or simply to compare any two samples. Below we discuss several uses that may serve as examples of soybean microarray applications.

Mutation and Transgene Analysis

Microarrays can be used to examine closely related cultivars that contain single-gene differences leading to variant or mutant phenotypes (traits). This approach has the

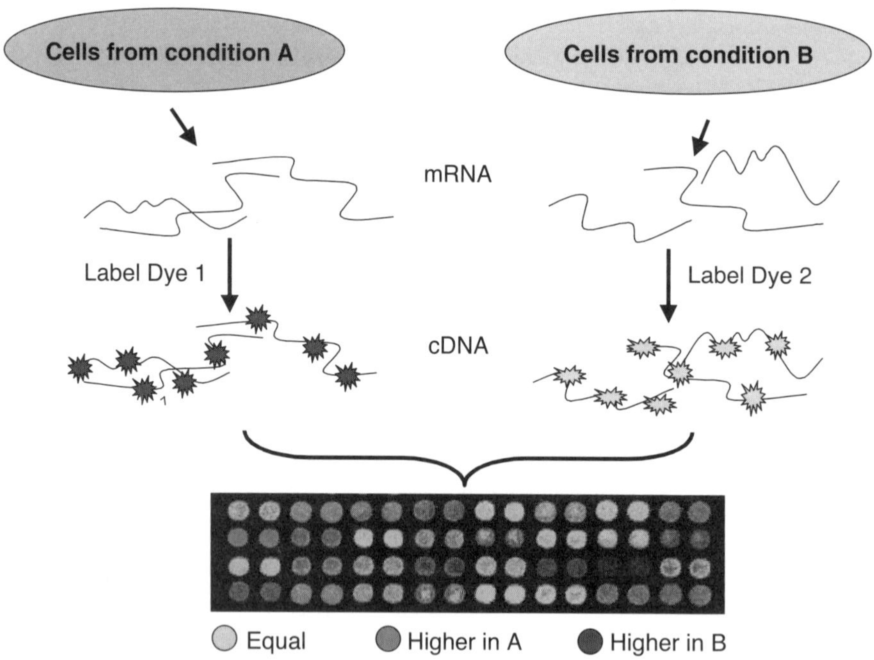

Figure 16.5. Obtaining the gene expression ratio between two samples. One sample is labeled with Dye 1 (ex: Cy5) and the other with Dye 2 (ex: Cy3), and samples compete for DNA spotted on the slide surface. The more abundantly a gene is present in a sample, the stronger the fluorescent signal from that gene's corresponding spot. Computer software color-codes the relative expression levels between samples. Many software programs label higher expression from the treatment as red, and higher in control as green.

potential to aid in the discovery of the molecular basis of a phenotype if the mutant gene affects the levels of its RNA product, and to identify the genetic basis of a mutation. For example, the RNA from a mutant plant can be compared to the RNA of the parent. If the mutated gene is no longer expressed, it may be identified by microarray analysis. As an illustration of the potential for this approach, near-isogenic lines that differed in the production of a specific seed protein, soybean lectin, were examined. The lectin null mutation had previously been shown to block production of the lectin mRNA, leading to absence of the protein (49). For the microarray analysis, mRNA from the lectin positive cultivar was fluorescently labeled with the Cy5 dye, and the mRNA from the lectin negative cultivar was labeled with the Cy3 dye. After hybridization to the Gm-r1070 soybean microarray, a clear difference was observed between the two different mRNA populations. The fluorescent intensities of the spots representing the lectin gene were much more intense in the parent than in the mutant (Fig. 16.6). While most genes are represented only once on the microarrays, a series of control clones is spotted repetitively throughout the array. In this case, there were eight separate spots of the lectin cDNA on the microarray slide, all of which were significantly weaker in the lectin mutant. As one can see, if this mutation had been uncharacterized and the mutagenized gene unknown, this microarray analysis would quickly have narrowed the number of gene candidates to a relatively few possible genes instead of thousands.

In another example, the Vodkin lab compared mRNA from seed coats of two near-isogenic lines differing at the T (tawny) locus, which encodes for the flavonoid 3' hydroxylase gene and is responsible for the tawny or gray color of the soybean trichome hairs on the plant. The T locus is also the first genetic locus to have been defined in soybean by crossing and segregation analysis (50). Microarray analysis suggests that the two lines in question differ only with regard to the expression of the flavonol 3' hydroxylase gene, as it was the only spot (out of 9,126 cDNAs on the array) that varied consistently by more than twofold. The microarray data also agreed with RNA blots using the isogenic lines at the T locus (51).

Like mutational analyses that examine the effect of the loss of a single gene, microarray analyses could be performed to examine the effect of adding a single gene on the expression of the rest of a plant genome. Such data could be useful to support or refute the proposition that a specific foreign gene does not affect metabolic processes in the plant other than the one it is intended to change in the construction of a transgenic plant. Therefore, microarray analysis is a valuable diagnostic for the biotech industry and government regulatory agencies alike.

Tissue-Specific Expression Profiles

Plants contain tens of thousands of genes, but many are expressed only in specific tissues or at specific times during growth and development. Microarrays are excellent tools to assay for tissue-specific genes. Maguire *et al.* (52) used microarrays of library Gm-r1021 (described above) to examine tissue expression differences in soybean. They easily identified the most prominently expressed genes in these tissues and, of

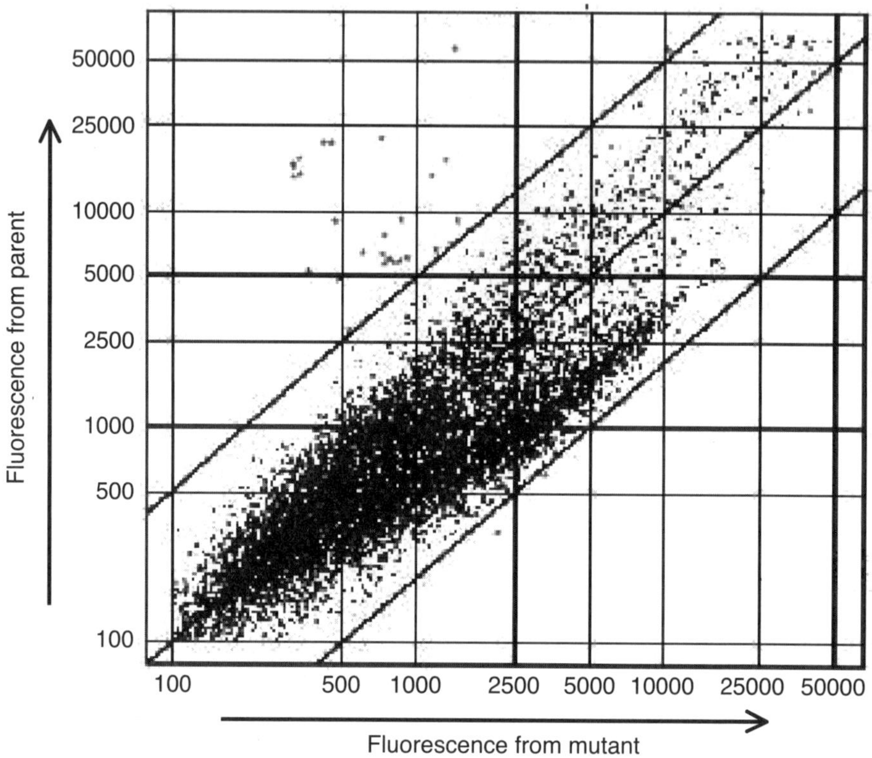

Figure 16.6. Scatterplot of fluorescent levels for each spot of one microarray slide that was hybridized with fluorescent cDNA from a lectin mutant (*x*-axis) and its parent (*y*-axis). Because most genes were expressed at a similar level in each sample, most spots fit along the slope line of 1.0. However, a handful of spots, most of which were lectin genes that were spotted multiple times on the slide, were more fluorescent in the parent than mutant.

the 4,089 gene representatives analyzed, they identified 11 putative shoot-specific genes and 4 putative root-specific genes. These tissue-specific genes may be utilized to isolate tissue-specific promoters that could be used in the strategic design of transgenic soybeans (i.e., to control expression of an introduced gene such that it is expressed only in that specific tissue). This study also presented a careful comparison of microarray data to that of the more traditional techniques of northern analysis and RT-PCR. The high correlation reported between the techniques indicates that microarrays offer a reliable platform for the study of gene expression.

Gene Expression Changes During Tissue and Organ Differentiation

Some methods of generating transgenic plants involve the generation of somatic embryos that are propagated in Petri dishes on various media containing plant growth regulators, such as auxin. Thibaud-Nissen *et al.* (53) conducted a detailed analysis of the induction of soybean somatic embryos during culture of cotyledon tissue on auxin-containing media using the Gm-r1070 array. Of the 9,216 cDNAs represented, they reported a set of 495 genes whose expression varied during the course of the experiments. The authors subsequently subjected the expression data to cluster analysis, which revealed that certain transcripts had similar patterns of expression and that many of the genes of the same pattern also had similar functions. These data revealed many aspects of the events in the genetic reprogramming of the cotyledon cells during the embryo-induction process. For example, the data suggest that auxin induces dedifferentiation of the cotyledon and provokes a surge of genes involved in cell division and oxidative burst. The data also indicate that the formation of somatic globular embryos is accompanied by the accumulation of storage protein transcripts as well as transcripts for the synthesis of gibberellic acid. These data will assist researchers in future design of media for culturing plant tissue, and the study provides further understanding of organ differentiation. The raw data from the 22 hybridizations used in this study have been entered in the publicly accessible Gene Expression Omnibus (GEO) database at: www.ncbi.nlm.nih.gov/geo.

Plant-Microbe Interactions

In addition to having tissue- and development-specific genes, plants have many genes that are rarely expressed except during interactions with microbes. Microarrays allow one to identify genes that are involved with a plant's response to a pathogenic or symbiotic organism (27,47,54,55). In the Clough lab, we have demonstrated that very few genes (approximately 20 out of 9,216) are dramatically altered in roots (at the whole root level) during early interactions with *Bradyrhizobium japonicum*. On the other hand, hundreds of genes were identified as being differentially expressed (both up and down) during interactions with pathogens. Figure 16.7 depicts how the number of genes showing differential expression increased over time as the disease "white mold" progressed into stem tissue. By examining how genes are changing during the development of disease versus the development of a defense, it is possible to identify candidate defense-related genes that could then be introduced into elite soybean lines. This approach is especially promising for cases where defense-related genes have been difficult to map precisely, such as when resistance is governed by many genes (QTLs, or quantitative trait loci) and each gene has a minor contribution to the defense response.

Identifying Genes Governing Quantitative Traits

Many agronomically important traits of soybean (such as yield; levels of oil, protein, and isoflavones; drought resistance; and defense against pathogens) are determined by multiple genes and are referred to as quantitative traits. Identifying the specific genes that constitute a quantitative trait is desirable because breeders can follow gene introgression into new seed lines. However, identifying these genes is difficult

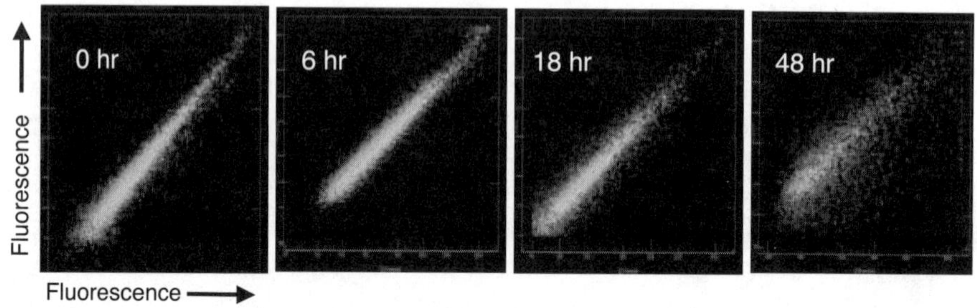

Figure 16.7. Scatterplots of *Sclerotinia*-inoculated plants (*y*-axis) versus mock inoculated controls (*x*-axis). The figure shows how the number of genes that were differentially expressed increased (scattering increases) as the disease progressed from time 0 to 48 hours post inoculation.

because each gene in a QTL is contributing only a fraction of the trait. Therefore, although breeders have been mapping QTLs for many important quantitative traits, precise mapping leading to gene identification has been elusive. Combining microarray expression data with QTL maps can greatly reduce the number of gene candidates associated with quantitative traits. For example, Wayne and McIntyre (56) were able to reduce the gene candidate pool for identifying QTLs for a drosophila quantitative trait from the approximately 550 possible genes identified by mapping down to 34 by combining map data with microarray expression data. Because many traits of interest in soybean are governed by QTLs, being able to reduce the gene candidate pool by 10- to 15-fold is very promising. The more completely the soybean genome is sequenced, the more powerful these studies will become; for if the area of a QTL is sequenced, then one will know immediately if a gene identified from a microarray analysis is present within the QTL.

Other Stresses

Plants are subjected to a wide variety of assaults by the environment. As plants cannot get up and look for a better homeland when times get tough, they must adjust their genetic expression to deal with both rapid and gradual environmental changes. Microarrays may be used to identify genes that are providing better adaptation to any given environmental stress. For example, microarray studies have been reported on plant gene expression response to drought (36,57), extreme temperatures (57,58), and high salt (36,43,57). Similar studies can now be performed with soybean microarrays to identify soybean's genetic response to environmental stresses. Comparing how two different cultivars respond may be useful for identifying key genes governing soybean's ability to effectively cope with specific stresses.

Genotyping

Instead of labeling RNA, one may use DNA as the starting material to label and hybridize to microarrays. Hybridizing with DNA permits the identification of absence or presence of the spotted genes in the samples being genotyped (59). Under ideal conditions, the arrays will also distinguish copy number as determined by relative strengths of signal for each spotted gene (60). Array genotyping performs almost like a sophisticated, biological bar-coding system in being able to characterize a cultivar based on the absence or presence of thousands of genes.

Oligonucleotide Microarrays

Although this chapter has focused on the use of cDNA microarrays, another platform—oligonucleotide microarrays, or oligo arrays—is also readily available for many species and will soon be available for soybean research as well. The sequencing information generated on the set of 36,000 soybean cDNAs described above will be critical for developing soybean oligo arrays because these cDNAs have been sequenced from both ends of their cloned fragments. These sequence data are available in the public databases of the National Center for Biotechnology Information (www.ncbi.nih.gov).

Oligo arrays involve the chemical attachment of short nucleotide sequences that are designed to be highly specific for just one gene (i.e., no cross-hybridization from highly similar genes). Because oligo arrays are so highly specific, a single base difference may lead to the inability of a labeled fragment to bind to a given spotted oligo. Therefore, in addition to gene expression analysis studies and mapping, oligo arrays are also ideally suited for specific applications such as single nucleotide polymorphism (SNP) analysis (61). With oligo arrays, one can genotype a given cultivar or offspring for several thousand SNPs with a single overnight hybridization. Another clever application of microarrays to mapping is the use of a heteroduplex pre-screen to identify the genetic alleles that have descended to an offspring (identical by descent). Cheung *et al.* (62,63) refer to this technique as genomic mismatch scanning (GMS), and they have demonstrated that one may analyze thousands of alleles simultaneously.

Summary

Microarrays, whether cDNA or oligo based, are expensive and time-consuming to construct. However, once they are produced, their uses are many and varied, because they can be used to obtain a gene expression report from virtually any tissue, at any time, after any treatment. The limits are dependent only on the investigator's curiosity. The greater the number of diverse experiments that are performed, the greater will be the researcher's ability to determine more completely what it is that a given gene or stimulus does, as well as when and why the gene is expressed. The data will also identify new genes that are critical for specific functions. Mutant analysis is highly valuable, as it will allow scientists to decipher pathways and identify how one gene may block a particular behavior or response. Of additional interest is the fact that microarrays should be able to be used across species as long as there is a high degree of sequence identity (at least 85%) among the genomes. For example, the soybean microarrays described in this chapter were constructed based on the *Glycine max* genome; however, one can use them to analyze gene expression in other *Glycine* species. As a greater number of gene sequences are identified, microarrays will become an even more powerful genomic tool. As costs decrease, the number of labs conducting microarray experiments will increase. In the future, it will be common to check the genomic response of a mutation or treatment by hybridizing to microarray slides prior to publication. Collecting the effects on gene expression that all these treatments are provoking will require massive databases and new tools to simplify analyses in order to compare microarray experiments. Such databases will allow one to compare how a given set of genes responded under multiple treatments, without having to repeat all the treatments previously completed in another lab. The future of biology is looking bright, for as the years go on, our understanding of many biological responses is becoming rapidly more detailed and more complete.

The following Web sites contain more information on the soybean sequence analysis and soybean microarray production: http://web.ahc.umn.edu/biodata and http://soybeangenomics.cropsci.uiuc.edu/.

References

1. Sambrook, J., and D.W. Russell, *Molecular Cloning: A Laboratory Manual*, Cold Spring Harbor Laboratory Press, Cold Spring Harbor, New York, 2001.
2. Shalon, D., S.J. Smith, and P.O. Brown, A DNA Microarray System for Analyzing Complex DNA Samples Using Two-Color Fluorescent Probe Hybridization, *Genome Res. 6*:639–645 (1996).
3. Lemieux, B., A. Aharoni, and M. Schena, Overview of DNA Chip Technology, *Mol. Breed. 4*:277–289 (1998).
4. Aharoni, A., and O. Vorst, DNA Mcroarrays for Functional Plant Genomics, *Plant Mol. Biol. 48*:99–118 (2001).
5. Brown, P.O., and D. Botstein, Exploring the New World of the Genome with DNA Microarrays, *Nat. Genet. 21*:33–37 (1999).
6. Heller, M.J., DNA Microarray Technology: Devices, Systems, and Applications. *Annu. Rev. Biomed. Eng. 4*:129–153 (2002).
7. Schena, M., R.A. Heller, T.P. Theriault, K. Konrad, E. Lachenmeier, and R.W. Davis, Microarrays: Biotechnology's Discovery Platform for Functional Genomes, *Trends in Biotechnol. 16*:301–306 (1998).
8. Schena, M., D. Shalon, R.W. Davis, and P.O. Brown, Quantitative Monitoring of Gene Expression Patterns with a Complementary DNA Microarray, *Science 270*:467–470 (1995).
9. Schena, M., and R.W. Davis, Technology Standards for Microarray Research, in *Microarray Biochip Technology*, edited by M. Schena, Eaton Publishing, Natick, Massachusetts, 2000, pp. 1–18.
10. Fouts, D.E., R.B. Abramovitch, J.R. Alfano, A.M. Baldo, C.R. Buell, S. Cartinhour, A.K. Chatterjee, M. D'Ascenzo, M.L. Gwinn, S.G. Lazarowitz, N.C. Lin, G.B. Martin, A.H. Rehm, D.J. Schneider, K. van Dijk, X. Tang, and A. Collmer, A Genome-Wide Identification of *Pseudomonas syringae* pv. Tomato DC3000 Promoters Controlled by the HrpL Alternative Sigma Factor, *Proc. Natl. Acad. Sci. USA. 99*:2275–2280 (2002).

11. Wei, Y., J.-M. Lee, C. Richmond, F.R. Blattner, J.A. Rafalski, and R.A. LaRossa, High-Density Microarray-Mediated Gene Expression Profiling of *Escherichia coli, J. Bacteriol. 183*:545–556 (2001).
12. Khodursky, A.B., B.J. Peter, N.R. Cozzarelli, D. Botstein, P.O. Brown, and C. Yanofsky, DNA Microarray Analysis of Gene Expression in Response to Physiological and Genetic Changes That Affect Tryptophan Metabolism in *Escherichia coli, Proc. Natl. Acad. Sci. USA. 97*:12170–12175 (2000).
13. Lashkari, D.A., J.L. DeRisi, J.H. McCusker, A.F. Namath, C. Gentile, S.Y. Hwang, P.O. Brown, and R.W. Davis, Yeast Microarrays for Genome Wide Parallel Genetic and Gene Expression Analysis, *Proc. Natl. Acad. Sci. USA. 94*:13057–13062 (1997).
14. Wodicka, L., H. Dong, M. Mittmann, M.H. Ho, and D.J. Lockhart, Genome-Wide Expression Monitoring in *Saccharomyces cerevisiae, Nat. Biotechnol. 15*:1359–1367 (1997).
15. Nantel, A., D. Dignard, C. Bachewich, D. Harcus, A. Marcil, A.P. Bouin, C.W. Sensen, H. Hogues, M. van het Hoog, P. Gordon, T. Rigby, F. Benoit, D.C. Tessier, D.Y. Thomas, and M. Whiteway, Transcription Profiling of *Candida albicans* Cells Undergoing the Yeast-to-Hyphal Transition, *Mol. Bio. Cell 13*:3452–3465 (2002).
16. Hill, A.A., C.P. Hunter, B.T. Tsung, G. Tucker-Kellogg, and E.L. Brown, Genomic Analysis of Gene Expression in *C. elegans, Science 290*:809–812 (2000).
17. Andrews, J., G.G. Bouffard, C. Cheadle, J. Lu, K.G. Becker, and B. Oliver, Gene Discovery Using Computational and Microarray Analysis of Transcription in the *Drosophila melanogaster* Testis, *Genome Res. 10*:2030–2043 (2000).
18. Stathopoulos, A., and M. Levine, Whole-Genome Expression Profiles Identify Gene Batteries in *Drosophila, Dev. Cell 3*:464–465 (2002).
19. Nuwaysir, E.F., W. Huang, T.J. Albert, J. Singh, K. Nuwaysir, A. Pitas, T. Richmond, T. Gorski, J.P. Berg, J. Ballin, M. McCormick, J. Norton, T. Pollock, T., Sumwalt, L. Butcher, D. Porter, M. Molla, C. Hall, F. Blattner, M.R. Sussman, R.L. Wallace, F. Cerrina, and R.D. Green, Gene Expression Analysis Using Oligonucleotide Arrays Produced by Maskless Photolithography, *Genome Res. 12*:1749–1755 (2002).
20. Whitfield, C.W., M.R. Band, M.F. Bonaldo, C.G. Kumar, L. Liu, J. Pardinas, H.M. Robertson, M.B. Soares, and C.E. Robinson, Annotated Expressed Sequence Tags and cDNA Microarrays for Studies of Brain and Behavior in the Honey Bee, *Genome Res. 12*:555–566 (2002).
21. Clark, M.D., S. Hennig, R. Herwig, S.W. Clifton, M.A. Marra, H. Lehrach, S.L. Johnson, and W. Groupt, An Oligonucleotide Fingerprint Normalized and Expressed Sequence Tag Characterized Zebrafish cDNA Library, *Genome Res. 11*:1594–1602 (2001).
22. Ju, Z., R.A. Dunham, and Z. Liu, Differential Gene Expression in the Brain of Channel Catfish (*Ictalurus punctatus*) in Response to Cold Acclimation, *Mol. Genet. and Genomics 268*:87–95 (2002).
23. DeRisi, J., L. Penland, P.O. Brown, M.L. Bittner, P.S. Meltzer, M. Ray, Y. Chen, Y.A. Su, and J.M. Trent, Use of a cDNA Microarray to Analyze Gene Expression Patterns in Human Cancer, *Nat. Genet. 14*:457–460 (1996).
24. Schena, M., D. Shalon, R. Heller, A. Chai, P.O. Brown, and R.W. Davis, Parallel Human Genome Analysis: Microarray-Based Expression Monitoring of 1000 Genes, *Proc. Natl. Acad. Sci. USA. 93*:10614–10619 (1996).
25. Zhu, T., P. Budworth, B. Han, D. Brown, H. Chang, G. Zou, and X. Wang, Toward Elucidating the Global Gene Expression Patterns of Developing Arabidopsis: Parallel

Analysis of 8300 Genes by a High-Density Oligonucleotide Probe Array, *Plant Physiol. Biochem. 39*:221–242 (2001).
26. Ruan, Y., J. Gilmore, and T. Conner, Towards Arabidopsis Genome Analysis: Monitoring Expression Profiles of 1400 Genes Using cDNA Microarrays, *Plant J. 15*:821–833 (1998).
27. Schenk, P.M., K. Kazan, I. Wilson, J.P. Anderson, T. Richmond, S.C. Somerville, and J.M. Manners, Coordinated Plant Defense Responses in Arabidopsis Revealed by Microarray Analysis, *Proc. Natl. Acad. Sci. USA. 97*:11655–11660 (2000).
28. Girke, T., J. Todd, S. Ruuska, J. White, C. Benning, and J. Ohlrogge, Microarray Analysis of Developing Arabidopsis Seeds, *Plant Physiol. 124*:1570–1581 (2000).
29. Maleck, K., A. Levine, T. Eulgem, A. Morgan, J. Schmid, K.A. Lawton, J.L. Dangl, and R.A. Dietrich, The Transcriptome of *Arabidopsis thaliana* During Systemic Acquired Resistance, *Nat. Genet. 26*:403–410 (2000).
30. Schaffer, R., J. Landgraf, M. Accerbi, V. Simon, M. Larson, and E. Wisman, Microarray Analysis of Diurnal and Circadian-Regulated Genes in Arabidopsis, *Plant Cell 13*:113–123 (2001).
31. Rowland, O., and J.D. Jones, Unraveling Regulatory Networks in Plant Defense Using Microarrays, *Genome Biol. 2*:1001 (2001).
32. Finkelstein, D., R. Ewing, J. Gollub, F. Sterky, J.M. Cherry, and S. Somerville, Microarray Data Quality Analysis: Lessons From the AFGC Project, Arabidopsis Functional Genomics Consortium, *Plant Mol. Biol. 48*:119–131 (2002).
33. Donson, J., Y. Fang, G. Espiritu-Santo, W. Xing, A. Salazar, S. Miyamoto, V. Armendarez, and W. Volkmuth, Comprehensive Gene Expression Analysis by Transcript Profiling, *Plant Mol. Biol. 48*:75–97 (2002).
34. Chen, W.Q., N.J. Provart, J. Glazebrook, G. Katagiri, H. Chang, T. Eulgem, F. Mauch, S. Luan, G. Zou, S. Whitham, P.R. Budworth, Y. Tao, Z. Xie, X. Chen, S. Lam, J.A. Kreps, J.F. Harper, A. Si-Ammour, B. Mauch-Mani, M. Heinlein, K. Kobayashi, T. Hohn, J.L. Dangl, X. Wang, and T. Zhu, Expression Profile Matrix of Arabidopsis Transcription Factor Genes Suggests Their Putative Functions in Response to Environmental Stresses, *Plant Cell 14*:559–574 (2002).
35. Negishi, T., H. Nakanishi, J. Yazaki, N. Kishimoto, F. Fujii, K. Shimbo, K. Yamamoto, K. Sakata, T. Sasaki, S. Kikuchi, S., Mori, and N.K. Nishizawa, cDNA Microarray Analysis of Gene Expression During Fe-Deficiency Stress in Barley Suggests That Polar Transport of Vesicles Is Implicated in Phytosiderophore Secretion in Fe-Deficient Barley Roots, *Plant J. 30*:83–94 (2002).
36. Oztur, Z.N., V. Talame, M. Deyholos, C.B. Michalowski, D.W. Galbraith, N. Gozukirmizi, R. Tuberosa, and H.J. Bohnert, Monitoring Large-Scale Changes in Transcript Abundance in Drought- and Salt-Stressed Barley, *Plant Mol. Biol. 48*:551–573 (2002).
37. Jones, P.G., D. Allaway, D.M. Gilmour, C. Harris, D. Rankin, and E.R. Retzel, Gene Discovery and Microarray Analysis of Cacao (*Theobroma cacao L.*) Varieties, *Planta 216*:255–264 (2002).
38. Nakazono, M., F. Qiu, L.A. Borsuk, and P.S. Schnable, Laser-Capture Microdissection, a Tool for the Global Analysis of Gene Expression in Specific Plant Cell Types: Identification of Genes Expressed Differentially in Epidermal Cells or Vascular Tissues of Maize [Erratum appears in *Plant Cell 2003* Apr; 15(4):1049], *Plant Cell 15*:583–596 (2003).
39. Fernandes, J., V. Brendel, X. Gai, S. Lal, V.L. Chandler, R.P. Elumalai, D.W. Galbraith, E.A. Pierson, and V. Walbot, Comparison of RNA Expression Profiles Based on Maize Expressed Sequence Tag Frequency Analysis and Microarray Hybridization, *Plant Physiol. 128*:896-910 (2002).

40. Whetten, R., Y.H. Sun, Y. Zhang, and R. Sederoff, Functional Genomics and Cell Wall Biosynthesis in Loblolly Pine, *Plant Mol. Biol. 47*:275–291 (2001).
41. Hertzberg, M., H. Aspeborg, J. Schrader, A. Andersson, R. Erlandsson, K. Blomqvist, R. Bhalerao, M. Uhlen, T.T. Teeri, J. Lundeberg, B. Sundberg, P. Nilsson, and G. Sandberg, A Transcriptional Roadmap to Wood Formation, *Proc. Natl. Acad. Sci. USA. 98*:14732–14737 (2001).
42. Leach, J.E., M. Ryba-White, Q. Sun, C.J. Wu, E. Hilaire, C. Gartner, O. Nedukha, E. Kordyum, M. Keck, H. Leung, and J.A. Guikema, Plants, Plant Pathogens, and Microgravity—A Deadly Trio, *Gravit. Space Biol. Bull. 14*:15–23 (2001).
43. Kawasaki, S., C. Borchert, M. Deyholos, H. Wang, S. Brazille, K. Kawai, D. Galbraith, and H.J. Bohnert, Gene Expression Profiles During the Initial Phase of Salt Stress in Rice, *Plant Cell 13*:889–905 (2001).
44. Babu, P.R., A.C. Sekhar, N. Ithal, G. Markandeya, and A.R. Reddy, Annotation and BAC/PAC Localization of Nonredundant ESTs from Drought-Stressed Seedlings of an Indica Rice, *J. Genet. 81*:25–44 (2002).
45. Aharoni, A., L.C.P. Keizer, H.J. Bouwmeester, Z. Sun, M. Alvarez-Huerta, H.A. Verhoeven, J. Blaas, A.M.L. van Houwelingen, R.C.H. De Vos, H. van der Voet, R.C. Jansen, M. Guis, J. Mol, R.W. Davis, M. Schena, A.J. van Tunen, and A.P. O'Connell, Identification of the SAAT Gene Involved in Strawberry Flavor Biogenesis by Use of DNA Microarrays, *Plant Cell 12*:647–661 (2000).
46. Frick, U.B., and A. Schaller, cDNA Microarray Analysis of Fusicoccin-Induced Changes in Gene Expression in Tomato Plants, *Planta 216*:83–94 (2002).
47. Zwiesler-Vollick, J., A.E. Plovanich-Jones, K. Nomura, S. Bandyopadhyay, V. Joardar, B.N. Kunkel, and S.Y. He, Identification of Novel *Hrp*-Regulated Genes through Functional Genomic Analysis of the *Pseudomonas syringae* pv. Tomato DC3000 Genome, *Mol. Microbiol. 45*:1207–1218 (2002).
48. Shoemaker, R., P. Keim, L. Vodkin, E. Retzel, S.W. Clifton, R. Waterston, D. Smoller, V. Coryell, A. Khanna, J. Erpelding, X. Gai, V. Brendel, C. Raph-Schmidt, E.G. Shoop, C.J. Vielweber, M. Schmatz, D. Pape, Y. Bowers, B. Theising, J. Martin, M. Dante, T. Wylie, and C. Granger, A Compilation of Soybean ESTs: Generation and Analysis, *Genome 45*:693–701(2002).
49. Vodkin, L.O., P.R. Rhodes, and R.B. Goldberg, A Lectin Gene Insertion Has the Structural Features of a Transposable Element, *Cell 34*:1023–1031 (1983).
50. Woodworth, C.M., Inheritance of Cotyledon, Seed-Coat, Hilum, and Pubescence Colors in Soy-beans, *Genetics 6*:487-553 (1921).
51. Zabala, G.C., and L.O. Vodkin, Cloning of the Pleiotropic *T* Locus in Soybean and Two Recessive Alleles That Differentially Affect Structure and Expression of the Encoded Flavonoid 3' Hydroxylase, *Genetics 163*:295–309 (2003).
52. Maguire, T., S. Grimmond, A. Forrest, I. Iturbe-Ormaetxe, K. Meksem, and P.M. Gresshoff, Tissue-Specific Gene Expression in Soybean (*Glycine max*) Detected by cDNA Microarray Analysis, *J. Plant Physiol. 159*:1361–1374 (2002).
53. Thibaud-Nissen, F., R.T. Shealy, A. Khanna, and L.O. Vodkin, Clustering of Microarray Data Reveals Transcript Patterns Associated with Somatic Embryogenesis in Soybean, *Plant Phys. 132*:118–136 (2003).
54. Rao, Z.M., H.T. Dong, J.Y. Zhuang, R.Y. Chai, Y.Y. Fan, D.B. Li, and K.L. Zheng, Analysis of Gene Expression Profiles During Host-Magnaporthe Grisea Interactions in a Pair of Near Isogenic Lines of Rice, *I. Chuan Hsueh Pao—Acta Genetica Sinica 29*:887–893 (2002).

55. Tao, Y., Z. Xie, W. Chen, J. Glazebrook, H.-S. Chang, B. Han, T. Zhu, G. Zou, and F. Katagiri, Quantitative Nature of Arabidopsis Responses During Compatible and Incompatible Interactions with the Bacterial Pathogen *Pseudomonas syringae, Plant Cell 15*:317–330 (2003).
56. Wayne, M.L., and L.M. McIntyre, Combining Mapping and Arraying: An Approach to Candidate Gene Identification, *Proc. Nat. Acad. Sci. USA 99*:14903–14906 (2002).
57. Seki, M., M. Narusaka, J. Ishida, T. Nanjo, M. Fujita, Y. Oono, S. Kamiya, M. Nakajima, A. Enju, T. Sakurai, M. Satou, K. Akiyama, T. Taji, K. Yamaguchi-Shinozaki, P. Carninci, J. Kawai, Y. Hayashizaki, and K. Shinozaki, Monitoring the Expression Profiles of 7000 *Arabidopsis* Genes under Drought, Cold and High-Salinity Stresses Using a Full-Length cDNA Microarray, *Plant J. 31*:279–292 (2002).
58. Fowler, S., and M.F. Thomashow, Arabidopsis Transcriptome Profiling Indicates That Multiple Regulatory Pathways Are Activated During Cold Acclimation in Addition to the CBF Cold Response Pathway, *Plant Cell 14*:675–1690 (2002).
59. Jaccoud, D., K. Peng, D. Feinstein, and A. Kilian, Diversity Arrays: A Solid State Technology for Sequence Information Independent Genotyping, *Nucleic Acids Res. 29*:E25 (2001).
60. Pollack, J.R., C.M. Perou, A.A. Alizadeh, M.B. Eisen, A. Pergamenschikov, C.F. Williams, S.S. Jeffrey, D. Botstein, and P.O. Brown, Genome-Wide Analysis of DNA Copy-Number Changes Using cDNA Microarrays, *Nat. Genet. 23*:41–46 (1999).
61. Cho, R.J., M. Mindrinos, D.R. Richards, R.J. Sapolsky, M. Anderson, E. Drenkard, J. Dewdney, T.L. Reuber, M. Stammers, N. Federspiel, A. Theologis, W.H. Yang, E. Hubbell, M. Au, E.Y. Chung, D. Lashkari, B. Lemieux, C. Dean, R.J. Lipshutz, F.M. Ausubel, R.W. Davis, and P.J. Oefner, Genome-Wide Mapping with Biallelic Markers in *Arabidopsis thaliana, Nat. Genet. 23*:203–207 (1999).
62. Cheung, V.G., J.P. Gregg, K.J. Gogolin-Ewens, J. Bandong, C.A. Stanley, L. Baker, M.J. Higgins, N.J. Nowak, T.B. Shows, W.J. Ewens, S.F. Nelson, and R.S. Spielman, Linkage-Disequilibrium Mapping without Genotyping, *Nat. Genet. 18*:225–230 (1998).
63. Cheung, V.G., and S.F. Nelson, Genomic Mismatch Scanning Identifies Human Genomic DNA Shared Identical by Descent, *Genomics 47*:1–6 (1998).

Chapter 17

DNA Marker-Assisted Selection for Improvement of Soybean Oil Concentration and Quality

Vincent R. Pantalone,[a] David R. Walker,[b] Ralph E. Dewey,[c] and Istvan Rajcan[d]

[a]Department of Plant Sciences, 2431 Joe Johnson Dr., University of Tennessee, Knoxville, TN 37996-4561;

[b]Center for Applied Genetic Technologies, 111 Riverbend Rd., University of Georgia, Athens, GA 30605-6810;

[c]Department of Crop Science, 1244 Williams Hall, Box 7620, North Carolina State University, Raleigh, NC 27695-7620; and

[d]Department of Plant Agriculture, University of Guelph, Ontario, Canada, N1G 2W1

Introduction

Soybean (*Glycine max* (L.) Merr.) is currently the most prominent of the world's major oilseed crops. The versatility of soy oil enables its leading role in edible vegetable oils, baking and frying fats, and a diverse array of industrial oil products, including pharmaceuticals, inks, dyes, lubricants, and bio-based fuels. Advances in the knowledge of soybean lipid metabolism and biological mechanisms have stimulated great interest in genetic improvement of soybean oil quality. United States oilseed processors and food manufacturers currently face a pressing challenge to produce quality products with lower levels of *trans* isomers of unsaturated fatty acids and lower levels of saturated fats. These two challenges affect approximately 96% of the domestic soybean oil market—44% of domestic consumption is in liquid soy oil applications such as cooking oils, and 52% of domestic applications include margarine, shortening, and frying fats (1).

The U.S. Food and Drug Administration (FDA) has issued guidelines related to health claims that may be included on food product labels. For example, a product must contain less than 1 gram of total saturated fat per serving to qualify as a low–saturated fat food (2). Genetic improvements in soybean will be necessary to meet this target because soy oil currently contains approximately 15% total saturated fat. Reduction of soy oil to less than 7% total saturated fat would provide opportunities to meet the low–saturated fat health claim. The recent FDA requirement (3) for food processors to list the amount of *trans* isomers (a product of soy oil hydrogenation) and an estimate of their amount of the total saturated fat on all food product labels is having an immediate impact on the global vegetable oil industry. Technological advances in genetics, oil processing, and food product formulation are needed to balance the demands of health-conscious consumers with the needs for functionality, stability, and other quality aspects of soy oil. Fundamental genetic changes in soy oil composition are now necessary to have a domestic crop that will meet the demands of consumers and industry.

There is currently a lack of standards for evaluating soy oil content at the grain elevator and throughout the oil-processing industry. High analytical error rates overlook true potential genetic variability for soy oil concentration. Thus, leaders in the soybean industry recently highlighted the necessity of promoting standards and accepted measurement techniques for soybean traits, including soy oil, which are compliant to the International Organization for Standardization (ISO). A U.S. committee, Soybean Quality Traits Analytical Standards Program (SQT), which is composed of representatives from the American Oil Chemists' Society (AOCS), industries, universities, the USDA-ARS, and other agencies, is currently engaged in developing standards for analytical assessment of soybean quality (www.SoybeanQualityTraits.org, verified February 11, 2004). These standards will build a high level of confidence in identity-preserved products and pave the way for the development, transport, storage, and processing of superior quality soybeans for trade in the global soy oil industry.

Molecular Genetic Markers for Oil Concentration and Quality

Oil Concentration

Current soybean production is founded on approximately five decades of genetic improvement, coupled with technological developments in refining and processing of vegetable proteins and oils. The domestication of the wild progenitor (*G. soja* Sieb. & Zucc.) by Asian farmers in ancient times altered gene frequencies and changed aspects of the plant's morphology, growth, development, and seed metabolism. *Glycine soja* accessions are typically wild, weedy, trailing vines that produce hard, small, black seeds. The cultivated soybean (*G. max*) is an erect species that produces a high yield of round, yellow seeds with relatively high (about 20%) oil content.

Additional genetic gains for soybean oil concentration should be possible as understanding of genomic regions regulating seed oil accumulation increases. The soybean genome is estimated to comprise between 30,000 and 100,000 genes that are distributed among 20 homologous pairs of chromosomes. The current integrated molecular genetic map of soybean comprises 20 linkage groups with approximately 1,845 markers consisting of amplified fragment length polymporphism (AFLP), classical genetic, isozyme, random amplification of polymorphic DNA (RAPD), restriction fragment length polymorphism (RFLP), and simple sequence repeat (SSR) markers (4). SOYBASE (5) indicates that genomic regions or quantitative trait loci (QTLs) that govern soybean oil accumulation are located on 17 different linkage groups (Table 17.1).

Distribution of oil QTLs over various linkage groups provides researchers with opportunities to breed for genetic gains by combining favorable oil-producing genes from several chromosomes to develop new, higher oil soybean lines. Moreover, some of the chromosomes have multiple regions controlling seed oil. Molecular breeding can be used to target the specific accumulation of oil-producing genes and build new sources of germplasm which stack favorable alleles among multiple chromosomes to develop a superior line. This process does not need to involve foreign DNA and can readily achieve gains in either non–genetically modified or genetically modified backgrounds by employing a marker-assisted selection (MAS) strategy to enhance soy oil synthesis.

This chapter will explore the concept of molecular breeding by outlining, on a linkage group by linkage group basis, the location of currently reported oil QTLs. Readers should note, however, that many oil QTLs have been detected, but only a few have been confirmed. The lack of confirmed QTLs for all soybean traits represents a gap in our knowledge base. Although most oil QTLs remain unconfirmed, Table 17.1 shows that several oil QTLs have been verified as being present in multiple populations, and that some novel oil QTLs may exist in specific germplasm. The opportunities for positively impacting genetic gains in soybean are substantial.

Detection of QTLs is based on a statistical association between DNA band classes and a measured trait (such as oil concentration). When the detection of an oil QTL is inferred via single-factor analysis of variance, an R^2 value indicates the proportion of the total observed variation for oil concentration among all individuals in the population that is due to genetic effects at the specific locus. The remaining variation is due to genetic effects at other loci plus environmental factors and experimental error. By convention, if the association (R^2) is 10% or higher, the locus is considered to be a major QTL; if it is less than 10%, it is considered to be a minor QTL. Oil concentration is a quantitative trait where many genes, each with small or large effects, contribute to the production of oil. Significant increases in oil concentration would thus require the accumulation of multiple QTLs across the genome.

Many oil QTLs are environmentally stable because they are detected across multiple environments. However, some oil QTLs may be environmentally sensitive, meaning that they were detected only in a single environment. In addition to genetic effects, oil biosynthesis can be affected by temperature, rainfall, and other environmental factors. If the basis of environmentally sensitive QTLs can be characterized, their incorporation into soybean lines may lead to genetic gains for oil in specifically targeted areas of seed production. In lieu of that genomic information, molecular breeders would benefit from utilizing environmentally stable QTLs in their selection strategies for cultivars deployed over broad geographical areas. These concepts can guide geneticists toward designing a better soybean with superior oil concentration. The currently reported oil QTLs are outlined in the following paragraphs.

LG A1. Linkage Group A1 has four regions relevant to seed oil production: RFLP marker A329_2 at map position 30.3 for Oil 3-3 (6) and SSR marker Satt591 at map position 31.1 (David Hyten, Soybean Genomics and Improvement Laboratory, USDA,

TABLE 17.1
QTL, Linkage Group, and Current Map Position[a] for Molecular Markers Reported in SOYBASE[b] Associated with Soybean Oil Concentration

QTL	LG	Map position	Marker	R^2	LOD
Oil 3-3	A1	30.3	A329_2	0.05	N/A
Oil 4-1	A1	53.4	K400_1	0.14	N/A
Oil 4-2	A1	75.4	A975_1	0.11	N/A
Oil 8-1	A1	88.6	Satt174	0.10	4.9
Oil 4-3	A1	92.3	A104_1	0.19	N/A
Oil 3-2	A1	93.6	T155_1	0.07	N/A
Oil 10-1	A1	93.6	T155_1	0.13	3.4
Oil 13-1	A1	94.9	B170_1	0.04	4.6
Oil 1-1	A2	50.4 to 67.3	T153_1 to A111_1	0.36	5.5
Oil 4-4	A2	132.3	A505_1	0.09	N/A
Oil 4-5	B1	29.2	A109_1	0.31	N/A
Oil 2-6	B2	33.1	A242_1	0.39	N/A
Oil 14-1	B2	72.1	Satt020	0.03	N/A
Oil 8-2	C1	10.3	SOYGPATR	0.11	3.3
Oil 9-1	C1	10.3	SOYGPATR	0.07	3.4
Oil 6-1	C1	90.7	A063_1	0.13	N/A
Oil 9-2	C2	38.0	Satt432	0.11	3.3
Oil 4-6	C2	97.2	L148_1	0.08	N/A
Oil 13-2	D1a	69.9	Satt468	0.09	4.7
Oil 5-5	D2	73.5	K258_2	0.09	N/A
Oil 5-4	D2	N/A	CR142_1	0.13	N/A
Oil 5-6	D2	N/A	CR326_1	0.09	N/A
Oil 2-5	E	6.3	SAC7_1	0.43	N/A
Oil 2-3	E	13.6	Pb	0.27	N/A
Oil 2-8	E	28.3	K229_1	0.22	N/A
Oil 2-4	E	30.9	A454_1	0.23	N/A
Oil 2-9	E	34.6	A203_1	0.18	N/A
Oil 5-1	E	N/A	A069_2	0.07	N/A
Oil 13-3	F	71.4	Satt510	0.06	3.3
Oil 4-7	G	65.6	A584_1	0.11	N/A
		65.6	A584_1	0.18	N/A
Oil 4-8	G	67.5	A816_1	0.11	N/A
Oil 4-9	G	67.7	A890_1	0.15	N/A
Oil 6-4	G	97.7	L002_2	0.14	N/A
Oil 6-2	G	99.3	L154_1	0.17	N/A
Oil 6-3	G	N/A	A235_4	0.15	N/A
Oil 4-10	H	33.2	A069_1	0.18	N/A
Oil 7-1	H	124.0	B072_1	0.21	N/A
Oil 6-5	H	N/A	A566_2	0.10	N/A
Oil 14-3	I	22.8	Satt562	0.06	N/A
Oil 11-1	I	32.4	A144_1	0.39	4.0
Oil 12-1	I	32.4	A144_1	0.15	N/A
Oil 13-4	I	35.2	BLT002_1	0.10	5.1
Oil 2-1	I	38.1	K011_1	0.27	N/A

QTL	Marker	P-value	Parent 1[c]	Parent 2[c]
Oil 3-3	A329_2	0.001	Minsoy	**Noir 1**
Oil 4-1	K400_1	0.02	A87-296011	CX1039-99
Oil 4-2	A975_1	0.009	CX1763	CX1159-49-1
Oil 8-1	Satt174	N/A	Minsoy	**Archer**
Oil 4-3	A104_1	0.003	M82-806	HHP
Oil 3-2	T155_1	0.001	**Minsoy**	Noir 1
Oil 10-1	T155_1	N/A	**Minsoy**	Noir 1
Oil 13-1	B170_1	N/A	Minsoy	**Noir 1**
Oil 1-1	T153_1 to A111_1	N/A	Minsoy	**Noir 1**
Oil 4-4	A505_1	0.03	C1763	CX1039-99
Oil 4-5	A109_1	0.0001	C1763	CX1159-49-1
Oil 2-6	A242_1	0.0001	**A81356022**	PI468916
Oil 14-1	Satt020	0.001	**Ma. Belle**	Proto
Oil 8-2	SOYGPATR	N/A	**Minsoy**	Archer
Oil 9-1	SOYGPATR	N/A	Minsoy	**Archer**
Oil 6-1	A063_1	0.05	**PI97100**	Coker237
Oil 9-2	Satt432	N/A	Noir 1	**Archer**
Oil 4-6	L148_1	0.03	A87-296011	CX1039-99
Oil 13-2	Satt468	N/A	Minsoy	**Noir 1**
Oil 5-5	K258_2	0.05	**Young**	PI416937
Oil 5-4	CR142_1	0.05	**Young**	PI416937
Oil 5-6	CR326_1	0.05	**Young**	PI416937
Oil 2-5	SAC7_1	0.0001	**A81356022**	PI468916
Oil 2-3	Pb	0.0001	**A81356022**	PI468916
Oil 2-8	K229_1	0.001	**A81356022**	PI468916
Oil 2-4	A454_1	0.0008	**A81356022**	PI468916
Oil 2-9	A203_1	0.006	**A81356022**	PI468916
Oil 5-1	A069_2	0.05	**Young**	PI416937
Oil 13-3	Satt510	N/A	Minsoy	**Noir 1**
Oil 4-7	A584_1	0.01	M84-492	Sturdy
	A584_1	0.009	C1763	CX1039-99
Oil 4-8	A816_1	0.007	A87-296011	CX1039-99
Oil 4-9	A890_1	0.004	M81-382	PI423949
Oil 6-4	L002_2	0.05	PI97100	**Coker237**
Oil 6-2	L154_1	0.05	PI97100	Coker237
Oil 6-3	A235_4	0.05	PI97100	**Coker237**
Oil 4-10	A069_1	0.0003	C1763	CX1159-49-1
Oil 7-1	B072_1	0.002	Peking	**Essex**
Oil 6-5	A566_2	0.05	PI97100	**Coker237**
Oil 14-3	Satt562	0.01	Ma. Belle	**Proto**
Oil 11-1	A144_1	0.0001	**A81356022**	PI468916
Oil 12-1	A144_1	0.0008	**Parker**	PI468916
Oil 13-4	BLT002_1	N/A	Minsoy	**Noir 1**
Oil 2-1	K011_1	0.0002	**A81356022**	PI468916

continues

TABLE 17.1 (*cont'd*)
QTL, Linkage Group, and Current Map Position[a] for Molecular Markers Reported in SOYBASE[b] Associated with Soybean Oil Concentration

QTL	LG	Map position	Marker	R^2	LOD
Oil 2-2	I	39.4	A407_1	0.28	N/A
Oil 13-5	I	119.1	L026_2	0.05	2.6
Oil 5-2	J	54.8	B122_1	0.07	N/A
Oil 1-2	K	28.7	A315_1	0.24	2.9
Oil 4-11	K	98.9	K387_1	0.16	N/A
Oil 14-2	K	104.8	Satt196	0.06	N/A
Oil 2-7	L	36.7	A023_1	0.32	N/A
Oil 5-3	L	36.7	A023_1	0.07	N/A
Oil 3-1	L	92.0	Satt006	0.08	N/A
Oil 9-3	L	95.4	A489_1	0.19	6.1

[a]Map position (cM), within each linkage group, based on the July 2003 integrated soybean map, http://bldg6.arsusda.gov/~pooley/soy/cregan/soymap.htm (verified February 11, 2004).
[b]SOYBASE, http://soybase.org (verified February 11, 2004).

ARS, Beltsville, MD, personal communication) likely represent a single underlying minor oil locus. Marker K400_1 at map position 53.4 is a major oil locus that is environmentally stable (7). RFLP marker A975_1 may be a distinct oil locus from another QTL reported upstream or downstream in this region. This marker maps to position 75.4 and is a major oil locus that is also environmentally stable (7). A 6.3-cM interval between SSR marker Satt174 (map position 88.6) and RFLP marker B170_1 (map position 94.9) was found to govern oil concentration by multiple researchers who utilized different mapping populations at Oil 8-1, Oil 4-3, Oil 3-2, Oil 10-1, and Oil 13-1 (6–9).

LG A2. Linkage Group A2 has two regions relevant to seed oil production. Oil 1-1 was detected within the 16.9-cM map interval between RFLP markers T153_1 and A111_1. This major oil locus increased oil by 7.0 g kg^{-1} (10). Oil 4-4 is an independent minor locus on LG A2, located by RFLP marker A505_1, toward the bottom of the linkage group at map position 132.3. This locus is environmentally stable (7).

LG B1. Only a single oil QTL has been detected on LG B1. This suggests that opportunities may be available to screen populations from unique germplasm sources to discover novel oil loci on that chromosome. RFLP marker A109_1 at map position 29.2 on LG B1 is a major oil locus, and it was the most significant of all the environmentally stable QTLs detected by Brummer et al. (7).

LG B2. Linkage Group B2 has two independent regions governing oil production: Oil 2-6 was detected by RFLP marker A242_1 at map position 33.1. This major oil locus increased oil concentration by 14 g kg^{-1} (11). Oil 14-1, detected by SSR

QTL	Marker	P-value	Parent 1[c]	Parent 2[c]
Oil 2-2	A407_1	0.0005	**A81356022**	PI468916
Oil 13-5	L026_2	N/A	Minsoy	**Noir 1**
Oil 5-2	B122_1	0.05	Young	PI416937
Oil 1-2	A315_1	N/A	Minsoy	**Noir 1**
Oil 4-11	K387_1	0.002	CX1763	CX1039-99
Oil 14-2	Satt196	0.01	Ma. Belle	**Proto**
Oil 2-7	A023_1	0.0001	**A81356022**	PI468916
Oil 5-3	A023_1	0.05	Young	**PI416937**
Oil 3-1	Satt006	0.001	Minsoy	**Noir 1**
Oil 9-3	A489_1	N/A	**Noir 1**	Archer

[c]The favorable form of the allele (increased oil concentration) is indicated in bold type for those QTLs for which genetic effect data were available.

marker Satt020 at map position 72.1, represents a minor oil locus with a genetic effect that increases oil by only 2 g kg^{-1} (12).

LG C1. The QTLs Oil 8-2 and Oil 9-1 are found by SSR marker SOYGPATR at map position 10.3 and have been detected as either major or minor oil loci, depending upon the population (8). Indeed, in the Minsoy × Archer population, the Minsoy allele increased oil, whereas in a related Noir 1 × Archer population, the Archer allele increased oil at the same locus. This oil locus needs to be further verified before it can be considered effective for MAS. In contrast, RFLP marker A063_1 at map position 90.7 is an oil locus that was verified in an independent mapping population. This confirmed QTL, designated *cqOil-003*, is one of only three currently confirmed oil loci (13).

LG C2. Oil 9-2 was detected by SSR marker Satt432 at map position 38.0 and is a major oil locus (8). Further downstream, Oil 4-6 was detected by RFLP marker L148_1 at map position 97.2 (7) and Satt557 at map position 112.2 (David Hyten, personal communication). Although it may influence oil concentration, the QTL may actually be a maturity locus because these genomic regions are strongly associated with soybean maturity (8,14).

LG D1a. SSR marker Satt184 at map position 26.0 cM has recently been identified as a minor oil locus (David Hyten, personal communication). Oil 13-2, detected by SSR marker Satt468 at map position 69.9 cM, is a major oil locus. SSR marker Satt147 at map position 108.0 is a third locus that contributes to oil concentration on this linkage group (David Hyten, personal communication).

LG D1b. No oil loci have been detected on LG D1b.

LG D2. Satt458 at map position 24.5 is a minor locus that increases oil by 2.3 g kg^{-1} (David Hyten, personal communication). Oil 5-5 was detected with RFLP marker K258_2 at map position 73.5 and is an independent minor locus that increases oil by 3.0 g kg^{-1} (15). Satt082 at map position 87.0 is a third oil locus on LG D2 (David Hyten, personal communication). Oil 5-4 is a major locus that increased oil by 3.0 g kg^{-1} (15). Oil 5-6 is a minor locus that slightly increases oil, by 2.0 g kg^{-1} (15).

LG E. Seven oil QTLs are reported on LG E; however, it appears that there may be only two to three distinct regions on this chromosome that control oil synthesis. Oil 2-5, detected by RFLP marker SAC7_1 at map position 6.3, and Oil 2-3, detected by morphological marker Pb (blunt-sharp pubescence tip) at nearby map position 13.6, are major oil QTLs. The allele from the *G. max* accession A81-356022 increased oil by 17.0 and 13.0 g kg^{-1}, respectively, compared with the *G. soja* allele at these loci (11). The high association between the *G. max* allele class means and seed oil (R^2 = 0.43 and 0.27 for Oil 2-5 and Oil 2-3, respectively) and their close map proximity suggest that a single major oil locus may reside in this region. This also may be one of the major loci involved in the domestication of cultivated soybean. A second region between map positions 28.3 and 34.6—Oil 2-8, Oil 2-4, and Oil 2-9 (11)—may represent a single underlying major oil locus. All three reported QTLs have similar association values (R^2 = 0.22, 0.23, and 0.18, respectively) and similar genetic effects (11, 10, and 10 g kg^{-1} increased oil, respectively). A potentially different oil locus, further downstream at Satt268 (map position 44.0), has been recently identified (David Hyten, personal communication). Researchers should note that the previously reported Oil 5-1 at RFLP marker A069_2 (15) on this linkage group was not validated in a recent study by these authors; hence, Oil 5-1 should not be considered effective for oil MAS.

LG F. A single locus between map position 71.4 (Satt510; Oil 13-3 [9] and 77.7 (Satt335 [David Hyten, personal communication]) is a major oil locus.

LG G. Although five oil QTLs are reported on LG G, it appears that this LG contains only one region that may be useful for increasing oil quantities. This narrow region, between RFLP markers A584_1 (map position 65.6) and A890_1 (map position 67.7), is where Brummer *et al.* (7) detected Oil 4-7, Oil 4-8, and Oil 4-9. Perhaps all three QTLs represent the same major environmentally stable oil locus. The other two QTLs reported by SOYBASE in a second region between map positions 97.7 and 99.3 were not detected through a recent validation study (13); hence, that region likely will have little or no impact on oil improvement.

LG H. LG H contains at least three regions that may provide molecular breeders with opportunities for MAS to increase seed oil concentration. Oil 4-10 was detected with RFLP marker A069_1 at map position 33.2 and is a major locus, environmentally stable, and highly significant in each of three study populations (7). A putatively independent oil locus was recently reported greater than 10 cM downstream, at Satt192 (14). Oil 7-1, detected with RFLP marker B072_1 at map position 124.0, is a third and independent major oil locus. Oil 6-5 is an important minor oil locus, originally mapped by Lee *et al.* (15) by RFLP marker A566_2 and has recently been validated as *cqOil-002* (13).

LG I. Seven QTLs have been assigned on this linkage group, yet it appears to contain only three discrete regions, two minor and one major region controlling oil. Oil 14-3 was detected with SSR marker Satt562 at map position 22.8 and is a minor locus where the "Proto" allele increased oil by 5.5 g kg^{-1}. This locus is environmentally stable (12). A 7-cM region between map positions 32.4 and 39.4 contains five oil QTLs (Oil 11-1, Oil 12-1, Oil 13-4, Oil 2-1, and Oil 2-2). Each of these was detected as a major oil QTL (9,11,16). This region may represent a single major oil locus. Oil 13-5, detected by RFLP marker L026_2 at map position 119.1, is an independent minor oil locus.

LG J. Currently there are no known oil loci on LG J. The previously reported Oil 5-2 (15) has failed to be validated (13). If new oil QTLs are confirmed on this LG, they may represent novel genes for introgression programs to improve oil concentration.

LG K. Oil 1-2, detected with RFLP marker A315_1 at map position 28.7, is a major oil locus where the Noir 1 allele increased oil by 5.0 g kg^{-1} (10). Oil 4-11 was detected as a major oil locus with RFLP marker K387_1 at map position 98.9, and Oil 14-2 was detected as a minor oil locus with SSR marker Satt196 at map position 104.8. These QTLs were found independently in the same region of this linkage group by two different teams of researchers (7,12). Whether this represents a single major or minor oil locus, and whether it is environmentally stable, remains to be confirmed.

LG L. Marker Satt523 at map position 27.9 was recently detected as an oil locus (David Hyten, personal communication). Oil 2-7 and Oil 5-3, both detected with RFLP marker A023_1 at map position 36.7, were found by two different groups who worked with different populations (11,15). This locus was recently confirmed as *cqOil-003* (13) using SSR marker Satt398 (map position 30.6) and Satt313 (map position 34.5), where the Japanese plant introduction PI 416937 increased seed oil by about 5 g kg^{-1}. Marker Satt166 at map position 66.5 was recently detected as an oil locus (14). Oil 3-1 was detected with SSR marker Satt006 at map position 92.0 (6), Satt229 at map position 93.9 (14), and Oil 9-3 detected with RFLP marker A489_1 at map position 95.4 (8). This represents another oil governing locus on this linkage group, but it remains to be verified whether this is a major or minor oil locus.

LG M. Only one oil QTL has been reported on this LG by SSR marker Satt540 at map position 39.9. The QTL is either a minor or a major oil locus, depending upon the environment (14).

LG N. Currently, no known oil QTLs have been detected on LG N.

LG O. Currently, no known oil QTLs have been detected on LG O.

Molecular breeders interested in increasing soybean oil concentration should consider prioritizing oil loci targets for MAS. An initial target to consider would be the accumulation of the environmentally stable major oil loci (such as those reported on LG A1, B1, G, and H). A second priority would be the accumulation of those major oil loci for which environmental stability is unknown (such as those on LG A2, B2, C1, C2, D1a, D2, E, F, H, I, K, and M). After successful completion of these initial priorities, modest additional gains in seed oil concentration should be achieved through the accumulation of minor oil loci (such as those reported on LG A1, A2, B2, D1a, D2, H, I, and K). It will be important to confirm that recombinant progeny actually express enhanced levels of oil concentration. In addition, certified laboratory analyses of the oil phenotype per se will be important for MAS programs to enhance genetic progress.

Fatty Acids

Palmitic Acid. Palmitic acid (16:0) is the predominant saturated fatty acid in oil produced from soybean seeds (17). Oil from the seeds of typical cultivars contains about 120 g kg^{-1} palmitic acid (18–20). Because of health risks associated with the cholesterogenic properties of saturated fatty acids, the development of cultivars with reduced levels of palmitic acid is a breeding objective in many soybean improvement programs (21). There are other health concerns about the *trans* isomers of unsaturated fatty acids that are produced when soybean oil is hydrogenated to increase the oxidative stability of the oil and to increase its saturated fatty acid content for the manufacture of semisolid fat products. Soybean seeds with elevated palmitic acid contents may therefore be useful for the production of margarines and shortenings, although seeds from germplasm high in stearic acid contents would be preferable (22).

Modified soybean lines with reduced levels of palmitic acid will help achieve the FDA target of low saturated fats (<7% total saturates) to meet current demands of health-conscious consumers. Low–palmitic acid lines have been developed through chemical mutagenesis, recurrent selection, and hybridization (18,20,23,24). Even greater reductions are possible through genetic engineering (25,26), but consumer suspicions and the high cost of obtaining regulatory approval for transgenic crops hinder commercialization of germplasm developed using this approach. Combining mutant alleles developed through chemical mutagenesis has produced a range of genotypes with seed oil palmitic acid contents ranging from <40 to >400 g kg^{-1} (19,27).

Classical genetics studies show that at least six independent loci are involved in determining the palmitic acid content of soybean seed (28). Mutant and wild-type alleles at palmitic acid QTLs act in either an additive or partially dominant manner (18,28,29). Reduced palmitic acid content in soybean seed is conditioned by at least three major alleles which are all recessive (1). The fap_{nc} allele from the N79-2077 line (about 60 g kg^{-1}) is the only allele that originated from a natural mutation (20). Two other loci, fap_1 occurring in line C1726 and fap_3 occurring in line A22, were created using chemical mutagenesis (17–19). There is no evidence for maternal genetic effects (30), and the alleles reside at independent loci (19). Segregation ratios indicate that fap_1 is independent of fap_{nc}, but it is not yet known whether fap_3, fap_{nc}, and a gene in ELLP2 that is temporarily designated as fap^*, are independent or allelic to one another (1). Modifying genes also have an effect on palmitate levels (19,31,32). Reduced palmitate levels are often accompanied with a reduction in oil content (31) and/or seed yield (33).

Four recessive alleles have been identified that increase palmitic acid levels to 150% or more of the typical level (34). Inheritance studies of the fap_2 (C1727), fap_{2b} (A21), fap_4 (A24), and fap_5 (A27) alleles suggest that they reside at independent loci (30). Various combinations of these alleles can increase palmitic acid concentration to 35% of crude oil (34). The proteins encoded by these genes are unknown, but the possibilities include at least eight of the enzymes in the fatty acid biosynthesis pathway. Some of these enzymes may be associated with stearic acid levels as well, since selection for low palmitic acid content in some populations also leads to a reduction in stearic acid (32).

Reduced yields have been observed in lines with either reduced levels of palmitic acid (33,35) or elevated palmitic acid (36). In the case of the low–palmitic acid lines, this appears to result from a close linkage between the locus with the major palmitic acid gene and either an important seed yield gene (35) or a gene conditioning growth habit (33). In comparison to related lines with normal palmitic acid levels, high-palmitic acid $BC_1F_{2:4}$ lines from three different populations typically had a shorter height, smaller seed size, higher protein percentages, and lower levels of oil and oleic and linoleic acids averaged across the populations (36). These associations suggest that pleiotropy and/or linkage drag could hinder efforts to develop competitive cultivars with extreme palmitic acid phenotypes.

Molecular markers have been used to obtain more accurate estimates of the number, locations, and relative contributions of loci that affect the levels of palmitic acid. In addition, they have provided information about which loci are associated with more than one type of fatty acid, epistatic interactions between loci, and the effect of a locus in different genetic backgrounds. Markers linked to confirmed QTLs can be used for MAS to transfer alleles conditioning modified fatty acid composition from donor germplasm to agronomically competitive breeding lines adapted to broad geographic regions or specific environments.

Diers and Shoemaker (11) used RFLPs to map QTLs associated with variation in the levels of several fatty acids in a population of 62 *G. max* × *G. soja* F_2-derived lines. Three markers on different linkage groups were associated with variation in palmitic acid levels. Markers pA-343a ($R^2 = 0.24$) and pA-18 ($R^2 = 0.19$) are on LG B2 (formerly M), and pK-375 ($R^2 = 0.18$) is on LG J (formerly Q) (5). The low-palmitate alleles at the LG B2 QTL came from *G. soja*, whereas the one at the LG J QTL was from the *G. max* line (Table 17.2). The fap_2 allele which conditions elevated palmitic acid in C1727 was mapped to LG D2 (37).

The development of simple sequence repeat (SSR) markers for soybean has permitted the construction of linkage maps using a type of co-dominant marker that is both highly polymorphic and more practical than RFLPs for MAS as well as mapping. Li *et al.* (38) used SSR markers to map genes associated with reduced palmitic acid levels in F_2 and $F_{2:3}$ populations from a cross of Cook × N87-2122-4. N87-2122-4 carries a major low–palmitic acid gene and modifiers inherited from N79-2077-12. A major gene ($R^2 = 0.31$ to 0.38) near Satt684 on LG A1 and a minor gene ($R^2 = 0.08$ to 0.12) on LG M were identified (Table 17.2), as well as a significant interaction between the markers in the F_2 generation. Since the data suggested that the gene on LG A1 might be located distal to Satt684, the terminal polymorphic marker on that linkage group, a simulated marker was constructed to better estimate the position of the gene. This simulated marker mapped 14.5 cM distal to Satt684 and accounted for 87% of the variation in the two generations. Although this location and effect are imprecise, they are nevertheless useful as estimates of the importance of the locus and in deciding which markers to use for MAS.

QTLs associated with reduced palmitic acid in C1726 are being mapped in a BC_1F_2 population from a cross of Cook × C1726. SSR markers on LGs B2 and H have been associated with variations in palmitic acid, and there is an epistatic interaction between these loci (David Hulbert, Department of Crop and Soil Sciences, University of Georgia, personal communication). The lowest palmitic acid concentrations are observed when the QTL on LG B2 is homozygous for the allele from Cook (the parent with normal palmitic acid levels), and the QTL on LG H is homozygous for the C1726 allele. The QTL detected on LG B2 is unlikely to be the QTL reported by Diers and Shoemaker (11) because of the 30-cM distance between their estimated locations.

Identification of markers linked to major palmitic acid QTLs should facilitate the transfer of mutant alleles to elite breeding lines, especially where backcrossing is employed. Selection for the recurrent parent genome in the regions flanking the marker(s) nearest a QTL might allow identification of segregants in which linkages between desirable fatty acid alleles and undesirable yield alleles have been broken. Unfortunately, many of the QTLs associated with palmitic acid levels have only minor to moderate effects on the phenotype (i.e., have an $R^2 < 0.10$) and may be involved in epistatic and/or genotype × environment interactions. These interactions and possible pleiotropic effects on the levels of other fatty acids will make MAS less efficient than it would be for a trait determined by the genotype with few QTLs with major effects by delaying recovery of the recurrent parent genome in the region around a QTL. However, even the ability to eliminate plants that lack the major alleles will permit more time and resources to be devoted to plants or lines most likely to have the desired phenotype.

Stearic Acid. Stearic acid (18:0) concentration in soybean may be genetically altered by mutations at gene loci, designated *Fas*, which typically result in elevated 18:0 concentration (22). Nearly all of these variants have been induced by chemical or X-ray mutagenesis. An exception is the newly developed line, FAM94-41 (9% 18:0), which carries a natural mutation, temporarily designated as the recessive fas_{nc} allele (39). Five additional germplasm lines were reported to carry homozygous recessive alleles that influence 18:0 concentration in soybean oil: fas^a (A6, [40]), fas^b (FA41545, [41]), *fas* (A81-606085, [41]), st_1 (KK-2 [42]), and st_2 (M25, [42]). As reported (41), fas^a (30% 18:0), fas^b (15% 18:0), and *fas* (19% 18:0) are allelic to each other, and represent different mutations in the same gene. In an F_3 population segregating for st_1 and st_2, the observation of progeny with 18:0 concentrations above and below the respective parental means provides evidence for two independently inherited genes governing higher 18:0 concentration in soybean. The proposed $st_1\,st_1\,st_2\,st_2$ genotype may elevate 18:0 concentration beyond 35% of crude oil (42). However, it remains unknown whether st_1 or st_2 are allelic to fas^a, fas^b, or *fas*. Other evidence supports the hypothesis that genes at different *Fas* loci may contribute to a very high 18:0 phenotype. In the mating of A6 and ST2 (another line derived from chemical mutagenesis), the F_1 progeny mean was greater than either parent's (23). Based on this information, it is probable that mutations in at least two independent genes determine the activity of one or more enzymes involved in 18:0 synthesis in soybean.

Two additive alleles (designated fas_2 and fas_x) determine lower 18:0 concentration in N-nitroso-N methylurea (NMU)-treated sunflower (43). Also, two alleles in ethylmethane-sulfonate (EMS)-treated Arabidopsis elevate 18:0 concentrations in both seed and leaf tissue (44). These alleles are identified as $fab_{2\text{-}1}$ (20% 18:0) and $fab_{2\text{-}2}$ (6% 18:0) and represent mutations at loci designated *Fab* in Arabidopsis. A notable observation is that both $fab_{2\text{-}1}$ and $fab_{2\text{-}2}$ alleles caused up to a fivefold reduction in plant size. This finding is interesting in view of reports that the yielding ability of soybean is severely depressed by fas^a, fas^b, or *fas* alleles (45,46). Poor yielding ability apparently is a reason for slow progress in the development of acceptable commercial soybean cultivars with higher 18:0 content. Our experience with FAM94-41 suggests that this problem may be overcome. The interaction of *fas* and other alleles that influence fatty acid composition in a segregating population often provides important clues to help guide the identification of the probable metabolic site(s) within the fatty acid synthetic pathway that is affected by these alleles, thus providing information which can be applied toward genetic gains in stearic acid.

Two soybean germplasm lines, A6 and FAM94-41, that exhibit higher levels of 18:0 were mated to evaluate the inheritance of the alleles responsible for this trait (39). A6 typically contains about six times the 18:0 concentration that is commonly found in soybean oil. FAM94-41 germplasm contains a two-fold increase in 18:0

TABLE 17.2
QTL, Linkage Group, and Current Map Position[a] for Molecular Markers Reported in SOYBASE[b] Associated with Soybean Fatty Acid Composition

Fatty Acid	QTL	LG	cM	Marker
16:0	Palm 2-1	A1	3.5	Satt684
16:0	Palm 1-1	B2	38.0	A343_1
16:0	Palm 1-2	B2	53.5	A018_1
16:0	Palm 1-3	J	68.3	K375_1
16:0	Palm 2-2	M	67.0	Satt175
18:0	Stear 1-1	J	83.3	A233_1
18:1	Ole 1-2	A1	92.3	A104_1
18:1	Ole 1-3	A1	92.6	A170_1
18:1	Ole 1-1	A1	102.3	A082_1
18:1	Ole 1-5	B2	N/A	A619_1
18:1	Ole 1-4	E	11.0	A242_2
18:1	Ole 1-6	E	13.6	Pb
18:2	Linole 1-2	A1	92.3	A104_1
18:2	Linole 1-3	A1	92.6	A170_1
18:2	Linole 1-1	A1	102.3	A082_1
18:2	Linole 1-5	B1	58.9	A118_1
18:2	Linole 1-4	E	11.0	A242_2
18:2	Linole 1-6	E	13.6	Pb
18:3	Linolen 1-1	E	6.3	SAC7_1
18:3	Linolen 1-4	E	11.0	A242_2
18:3	Linolen 1-7	E	28.3	K229_1
18:3	Linolen 1-5	E	30.9	A454_1
18:3	Linolen 1-3	E	34.6	A203_1
18:3	Linolen 1-6	K	N/A	A065_3
18:3	Linolen 1-2	L	36.7	A023_1

[a]Map position (cM), within each linkage group, based on the July 2003 integrated soybean map, http://bldg6.arsusda.gov/~pooley/soy/cregan/soymap.htm (verified February 11, 2004).
[b]SOYBASE, http://soybase.org (verified February 11, 2004).

concentration. Comparing the composition of these lines, the gain in 18:0 concentration appears to come at the expense of the sum of all unsaturated C18 fatty acids (total-18:1), more so than 16:0. As a result, the high-18:0 trait contributes to significantly elevated total saturates in the oil. In A6, 18:0 accounts for about 76% of total saturates.

The frequency distribution for 18:0 concentration among F_2 progeny of A6 × FAM94-41 segregating for fas_{nc} and fas^a showed a wide range of values from 6 to 26% 18:0. The pattern was indicative of allelic segregation among progeny, whereby the elevated 18:0 phenotype in both FAM94-41 and A6 is governed by the same gene. Chi-square analysis of these data supported a 3:1 progeny ratio ($\chi^2 = 1.33$) fitting a single gene model for differing alleles between FAM94-41 and A6 at the same gene locus (39). A similar conclusion was reported from the inheritance pattern for

QTL	Marker	R^2	P-value	Parent 1[c]	Parent 2[c]
Palm 2-1	Satt684	0.33[d]	0.001[d]	**Cook**	N87-2122-4
Palm 1-1	A343_1	0.24	0.001	**A81356022**	PI 468916
Palm 1-2	A018_1	0.19	0.005	**A81356022**	PI 468916
Palm 1-3	K375_1	0.18	0.005	A81356022	**PI 468916**
Palm 2-2	Satt175	0.09[d]	0.01[d]	**Cook**	N87-2122-4
Stear 1-1	A233_1	0.19	0.005	A81356022	**PI 468916**
Ole 1-2	A104_1	0.26	0.0003	A81356022	**PI 468916**
Ole 1-3	A170_1	0.23	0.001	A81356022	**PI 468916**
Ole 1-1	A082_1	0.28	0.0001	A81356022	**PI 468916**
Ole 1-5	A619_1	0.19	0.004	A81356022	**PI 468916**
Ole 1-4	A242_2	0.20	0.002	**A81356022**	PI 468916
Ole 1-6	Pb	0.21	0.001	**A81356022**	PI 468916
Linole 1-2	A104_1	0.33	0.0001	**A81356022**	PI 468916
Linole 1-3	A170_1	0.30	0.0001	**A81356022**	PI 468916
Linole 1-1	A082_1	0.38	0.0001	**A81356022**	PI 468916
Linole 1-5	A118_1	0.20	0.006	A81356022	**PI 468916**
Linole 1-4	A242_2	0.21	0.002	A81356022	**PI 468916**
Linole 1-6	Pb	0.20	0.006	A81356022	**PI 468916**
Linolen 1-1	SAC7_1	0.31	0.0005	A81356022	**PI 468916**
Linolen 1-4	A242_2	0.23	0.008	A81356022	**PI 468916**
Linolen 1-7	K229_1	0.20	0.003	A81356022	**PI 468916**
Linolen 1-5	A454_1	0.22	0.001	A81356022	**PI 468916**
Linolen 1-3	A203_1	0.22	0.002	A81356022	**PI 468916**
Linolen 1-6	A065_3	0.20	0.003	**A81356022**	PI 468916
Linolen 1-2	A023_1	0.26	0.0005	A81356022	**PI 468916**

[c]The form of the allele exhibiting the higher concentration of that fatty acid is indicated in bold type.
[d]Palm 2-1 and Palm 2-2 data for the $F_{2:3}$ generation; additional data are available for the F_2 generation.

fas^a, *fas*^{b}, and *fas* alleles (41). Thus, all five of these *fas* alleles probably contribute major genetic effects upon the activity of the same enzyme in the fatty acid or glycerolipid synthetic pathway.

Although the gene product (enzyme) of these *fas* alleles is unknown, further interpretation of the A6 × FAM 94-41 population segregating for *fas*$_{nc}$ and *fas*^{a} provides useful information regarding the nature of these genetic effects. A strong negative correlation (R^2 = 0.96) was found between total 18:1 and 18:0 concentration, while only a weak negative correlation (R^2 = 0.28) is seen between 16:0 and 18:0. This suggests that increases in 18:0 concentration are achieved primarily at the expense of all C18-unsaturated fatty acids (total 18:1). In fact, the decline in total 18:1 accounts for 89% of the increase in 18:0 within these F_2 progeny. Among all enzymes in the fatty acid and glycerolipid pathways, it is generally believed that the enzyme activities of 3-keto-acyl-ACP synthetase II (KAS-II), 18:0-ACP desaturase (Δ9DES), or 18:1-ACP thioesterase (18:1-ACP TE) have the highest probability of effecting change in 18:0 concentration in oilseeds and other plant tissues. KAS-II catalyzes the elongation of 16:0-ACP to 18:0-ACP, Δ9DES desaturates 18:0-ACP to 18:1-ACP, and 18:1-ACP TE initiates conversion of 18:0-ACP or 18:1-ACP (with lesser activity on 18:0-ACP) to their respective acyl-CoA derivatives (47). In view of the apparently strong negative relation between 18:0 and total 18:1 concentration, it is likely that the high-18:0 trait arises from *fas* gene action that either reduces Δ9DES or perhaps 18:1-ACP TE activity. However, it is conceivable that the mutations resident in fas_{nc} and fas^{a} might effect increased conversion of 16:0 to 18:0, as may be envisioned by greater KAS-II activity.

A new population (FAM94-41 × C1727) was created to test the theory that *fas* alleles influence KAS-II activity. C1727 germplasm carries a homozygous recessive mutation (fap_2) at gene locus designated as *Fap* (18). The fap_2 mediates an oil phenotype with about 18% 16:0 and 3% 18:0, and appears to effect reduced KAS-II activity (34). Given that assumption, if fas_{nc} encodes greater KAS-II activity, the segregating population should produce no progeny with 16:0 plus 18:0 concentration greater than the original parents' because of opposing genetic effects on KAS-II activity. Conversely, if fas_{nc} mediates either reduced Δ9DES or perhaps reduced 18:1-ACP TE activity, a small proportion of the progeny from this population should represent transgressive segregants with a high-16:0-plus-high-18:0 phenotype.

Analysis of seed from FAM94-41 × C1727 showed a wide range of values, from 13.8 to 23.2% total saturates, where transgressive segregants occurring above the high-parent value and below the low-parent value were found in this population. This type of phenotypic variation may be expected with the combination of independently inherited genes. Furthermore, Chi-square analyses of these data fit a 9:3:3:1 ratio, which supports a two-gene model. Based on these data, the elevated total saturates phenotype in this population is governed by two different genes, and the genetic interaction of these alleles appears to be additive in nature (Fig. 17.1). Therefore, we may reject the hypothesis that *fas* alleles encode or regulate the KAS-II enzyme. Given this conclusion, and based on what is known about the enzymes for this pathway from the literature, reduced Δ9DES or perhaps reduced 18:1-ACP TE activity becomes the leading candidate for the causal mechanism for elevated 18:0 concentration in soybean. Researchers may now focus attention upon characterization of Δ9DES and 18:1-ACP TE genes, and evaluation of their respective enzyme activities in soybean. Geneticists may also consider hybridizing germplasm carrying appropriate *fas* and *fap* alleles which would enable TS to be enhanced to levels which may be desirable for low-*trans*-fat margarine and shortening applications (1).

Marker-assisted selection can increase the efficiency of realizing targeted objectives. We recently identified and mapped SSR markers linked to the *Fas* locus using bulk segregant analysis (22). The genetic material used consisted of F_2 and $F_{2:3}$

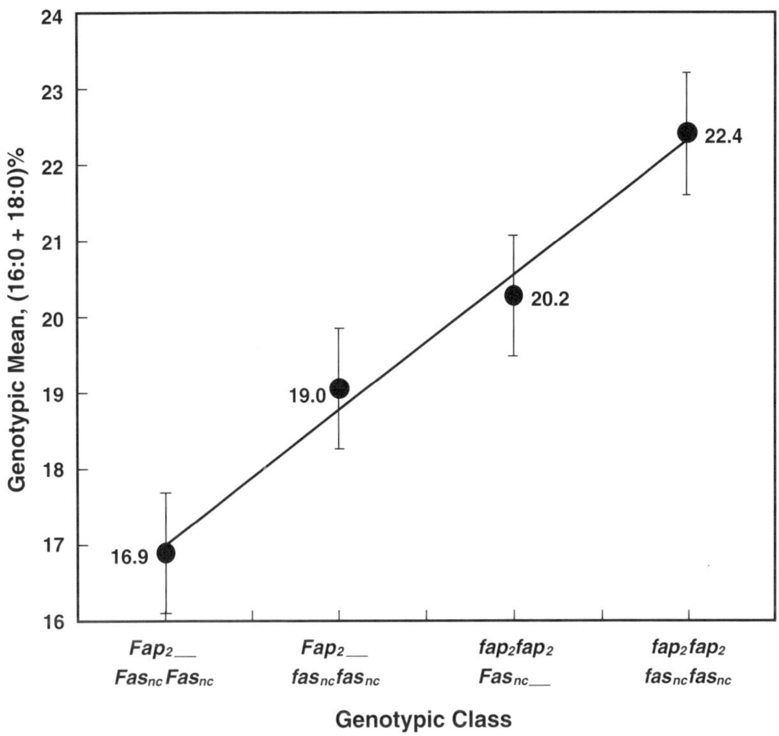

Figure 17.1. Gene combinations governing 18:0 and 16:0 concentration that influence total saturates in oil of mature seed from an F_2 population of FAM94-41 × C1727. The linear relation among mean estimates of total saturated fatty acid concentration within a genotypic class indicates additive genetic effects; $R^2 = 0.962$.

individuals from the cross between the parental lines Dare (normal stearic acid content) and FAM94-41 (high stearic acid content). From the total of 101 SSR primers tested, three (satt070, satt474, and satt556, all of which are located on LG B2) were highly significant ($P < 0.0001$), were major QTLs ($R^2 > 0.61$), and showed consistent polymorphisms between the two extreme DNA bulks. Rennie and Tanner (48) previously showed that *Fas* and *Fan* loci are linked and also that *Fan* and *Idh2* are linked in the suggested order *Fas*, *Fan*, and *Idh2*. Brummer *et al.* (49) placed the *Fan* locus close to RFLP markers pB194-1 and pB124 on LG B2. The three SSR markers identified by Spencer *et al.* (22) mapped on LG B2 in the most likely order as *Fas*, Satt070, Satt474, and Satt556. These markers would be useful for recognizing high stearic acid lines in an MAS breeding program for modified oil quality.

Oleic Acid: Soybean seed oil typically contains about 20% oleic acid (18:1), and increasing this concentration is an important breeding objective (1). Diets containing higher ratios of monounsaturated to polyunsaturated fatty acids can be beneficial in reducing the level of undesirable low density lipoprotein cholesterol in blood plasma (50,51). Monounsaturated fatty acids are also more effective at preventing arteriosclerosis than are polyunsaturated fatty acids (52). Vegetable oils with low levels of polyunsaturated fatty acids and increased concentrations of oleic acid have higher oxidative stability and fewer oxidative products than oil with the fatty acid composition of typical soybean cultivars. In addition, oil with high oleic acid content can be heated to a higher temperature, which reduces cooking time and absorption of oil into food (53).

The line N78-2245 contains 42% oleic acid (1). FAD_2 is the predominant ω-6 desaturase in soybean seed. Increases of up to 70% have been achieved by selection in a segregating population, and one line selected from this population, N97-3363-4 (60% 18:1) appears to carry mutations in both the *fad* and fad_1 alleles, which reside at different loci (54). Seed oil from the mutant line M23 contains twice the level of oleic acid as oil of Bay, the cultivar from which it was developed (55). Oleic acid content of segregating progeny indicated that the trait was controlled by two alleles at a single locus, and the mutant allele was designated *ol*. Segregation at the locus was inversely correlated with linoleic acid content. No maternal or cytoplasmic effects were observed, but gene action was partially dominant.

In studies of the inheritance of alternative desaturase genes in populations derived from *G. max* × *G. soja* crosses, Pantalone *et al.* (56) found both transgressive segregants and a frequency distribution that was skewed toward higher oleic acid concentrations. The 13:3 phenotypic ratio observed in the F_2 (with $Fad_2\_Fad_3\_$ as the minority class) suggested that epistatic gene interaction was involved.

Phenotypic selection for modified oleic acid levels in some populations is complicated by maternal effects (18,57). When these effects on hybrid fatty acid levels are significant, neither single F_2 seeds nor F_3 means from individual F_3 plants provide an accurate indication of F_2 segregation. Environmental effects on fatty acid profiles can also interfere with the efficiency of phenotypic selection for high oleic acid (58). Variation in temperature plays a large role in fatty acid synthesis in soybean and has a particularly large effect on the synthesis of unsaturated fatty acids. In such cases, molecular markers linked to genes affecting oleic acid levels could be used to select segregants that possess the desired alleles.

Diers and Shoemaker (11) detected three QTLs associated with variation for oleic acid levels in a population of 60 F_2-derived lines from a *G. max* × *G. soja* cross. One QTL was close to RFLP marker pA-82 ($R^2 = 0.28$) on LG A1 (formerly LG B) and another close to pA-619 ($R^2 = 0.19$) on LG B2 (formerly LG J) (Table 17.2). A QTL linked to the *pb* locus on LG E (formerly LG A) that determines blunt-sharp pubescence tips is notable because it had a similar effect on variation in both oleic ($R^2 = 0.21$) and linoleic acid levels ($R^2 = 0.20$).

A study is currently underway at the University of Georgia to map QTLs associated with oleic acid concentration using SSR markers and a population of 316 $F_{2:3}$ lines from a cross between the breeding lines G99-G725 (about 17% 18:1) and

N00-3350 (about 51% 18:1). N00-3350 is descended from N78-2245, and it also has the low palmitic acid line C1726 in its pedigree. Markers significantly associated with oleic acid levels ($P \leq 0.01$) have thus far been found on LG A1 (R^2 = 0.04), G (R^2 = 0.06 to 0.08), L (R^2 = 0.06 to 0.11), and M (R^2 = 0.10), with evidence that more than one QTL may be present on LGs G and L (Maria Monteros, Department of Crop & Soil Sciences, University of Georgia, personal communication). Analysis of a subset of 94 random lines suggests that some of these QTLs also affect the levels of other fatty acids, including palmitic (LGs A1 and G), stearic (LG A1), and linoleic (LGs G and L). Evidence for epistatic interactions between some markers linked to the QTLs has also been found, which is in accord with conclusions from other studies (56). LG A1 was the only linkage group with a putative oleic acid QTL that was also identified by Diers and Shoemaker (11) in their *G. max* × *G. soja* population. It will therefore be important to confirm these putative QTLs in independent populations and in different environments. It will also be important to determine whether some linkage groups have multiple QTLs, and the extent to which linkage drag could hinder the development of high-oleic lines adapted to particular regions.

Identification of SSR markers closely linked to QTLs affecting oleic acid levels will facilitate transfer of the moderate– to high–oleic acid trait to elite breeding lines with other desirable characters. The quantitative nature of this trait and the relatively small effect of individual QTLs complicate MAS. The apparent pleiotropic effects of some QTLs on levels of two or more fatty acids may be either helpful or problematic, depending on the specific fatty acid profile desired.

Polyunsaturated Fatty Acids. Soybean oil has been recognized as the most prominently utilized vegetable oil because its fatty acid composition consists of relatively low levels of saturated fat, primarily 16:0, and relatively high levels of the healthy monounsaturated 18:1 and essential polyunsaturated fatty acids. Linoleic (18:2) and linolenic (18:3) acids are the predominant polyunsaturated fatty acids in soybean oil. As essential fatty acids, they are considered desirable components of soybean oil. Because of the relatively high level of naturally occurring 18:2, there has been little effort by soybean researchers to reduce its levels through breeding or mutagenesis. On the other hand, although considered nutritious and an essential fatty acid, 18:3 is particularly prone to oxidation because of its three unsaturated double bonds. Once oxidized, oil containing high levels of 18:3 turns rancid, resulting in the shortening of its shelf life. The level of 18:3 and polyunsaturated fatty acids can be reduced by partial hydrogenation of the oil; however, this process results in the formation of *trans* fatty acids. *Trans* fatty acids have been associated with coronary heart disease in some studies (59) and currently are an important concern to food product manufactures. Efforts to modify the concentration of polyunsaturated fatty acids in soybean oil thus have focused primarily on the reduction of 18:3 through plant breeding as a healthier alternative to hydrogenation.

Linoleic acid in soybean seeds is produced primarily by desaturation of oleic acid (18:1) in phosphatidylcholine (PC) on the endoplasmatic reticulum (ER) (60). This is the first unsaturated fatty acid that is thought to be synthesized on the ER in the cytosol once 18:1 has been transported out of the plastids following its production through *de novo* fatty acid biosynthesis. Therefore, an ω-6 desaturase has been implicated as the enzyme responsible for the production of 18:2 in soybean seeds (61).

The genetic control of 18:2 has scarcely been studied, perhaps due to the lack of need for its modification. No gene symbols exist for 18:2. It has been determined that the level of 18:2 tends to be skewed toward lower values in material that has been treated with ethyl-methane sulfonate (EMS) (62). Plant transformation efforts have managed to reduce the level of 18:2 in soybean oil from 65 to 3% using a co-suppressed oleate desaturase gene (63).

Molecular markers have been used to map the genes for 18:2. Most markers and QTLs associated with 18:2 (11) were located on linkage groups A1, B1, and one on E of the composite soybean linkage map (5). A total of five RFLP markers (pA-82, pA-104, pA-170, pA-242b, and pA-118) and one morphological marker (pb^c) were found in that study (11). All but one (pA-118 on LG B) were also associated with 18:1, making it difficult to discern if these markers were associated with the production of 18:1 or the desaturation step from 18:1 to 18:2. Since the latter involves markers that are significantly associated with 18:1 and 18:2 (and not 18:0), it might be possible that the mapped loci were involved in the desaturation step. However, the authors did not speculate on this concept. In a recent study conducted using SSR markers with interspecific populations generated from crossing *G. max* and *G. soja*, Weir (64) found QTLs that were associated with 18:2 content located on several previously unreported LGs (N, C1, C2, and I). The amount of phenotypic variation (R^2) associated with each marker ranged from 6.4 to 14.8%. The finding of novel regions responsible for 18:2 (Weir, A.D., and I. Rajcan, unpublished data) is consistent with the notion that alternative desaturases may be involved in the regulation of the enhanced polyunsaturated fatty acids levels in wild soybeans (56). This finding is further supported by recent reports on the common and genome-wide duplication found in soybean (65,66).

Linolenic acid has been identified as an unstable component of soybean oil (67). In order to increase oil stability through breeding, several research groups have developed soybean mutants that have reduced levels of 18:3, with four mutant alleles reported: *fan*, *fan2*, *fan3*, and *fanx* (68–71). It is believed that mutations in genes coding ω-3 desaturases which catalyze the final desaturation step of 18-carbon fatty acids incorporating a third double bond into 18:3 may be responsible for the reduction in 18:3 levels.

The *Fan* locus has been mapped to LG B2 of soybean (49). Four RFLP markers (pB153, pB124, pB194_1, and pA_183) were found using a population from the cross C1640 × PI 479750 and were successfully anchored to markers mapping the *Linolen 1-2* QTL on LG B2 of the USDA/ISU soybean map (11,49). Therefore, there is strong evidence for the location of the *Fan* locus within a region on LG B2. In another study, a DNA fragment missing from the ω-3 desaturase gene in soybean line A5 was mapped to the same region of LG B2 as the *Fan* locus, suggesting that a deletion in this gene was responsible for the reduction in 18:3 in A5 (72).

In addition to the *Fan* locus, seven other QTLs were found for 18:3 concentration, five of which were located on LG E and one each on LGs K and L of the consensus map (5). Using simple sequence repeat markers, a major new QTL was found on LG C2 (64) of the consensus map (4), which contributed 27% of the phenotypic variation in an interspecific soybean population.

A study conducted to try to determine the number of ω-3 desaturase genes present in the soybean genome has revealed the presence of three distinct genetic sequences named Gm*FAD3A*, Gm*FAD3B*, and Gm*FAD3C* (73). A deletion in the Gm*FAD3A* copy of the ω-3 desaturase was associated with the *Fan* locus and is responsible for the reduced 18:3 in line A5 (73). These results concurred with those previously reported on the association between the *Fan* locus and the microsomal ω-3 desaturase (72) and identified the seed-specific expression of Gm*FAD3A* in developing soybean seeds. It is postulated further that Gm*FAD3B* and Gm*FAD3C*, which do not have seed-specific expression, may be candidate sites for the mutations with minor effects on 18:3 found in other mutant soybean lines (*fan2*, *fan3*, and *fanx*) (73). Even if the ω-3 desaturases are associated with three of the QTLs for linolenic acid, a total of five independent QTLs remain unaccounted for by these genes. Further research using map-based cloning may reveal the true nature of the genes harbored in the latter group of QTLs as, for example, transcription factors or independent genes with pleiotropic effects on 18:3 levels. It may also be possible that future QTL mapping studies could reveal novel and thus far unreported 18:3 QTLs as a reflection of soybean genome duplication and germplasm diversity.

Function of Genes Governing Lower Linolenic and Palmitic Acid Concentration

For the past three decades, soybean breeders have been very successful in amassing an impressive collection of lines whose storage oil compositions vary significantly from those found in conventional soybean cultivars. As described in the preceding sections, molecular markers have been identified that are closely linked to the loci responsible for many of these modified oil traits. This has all been accomplished in the absence of any direct knowledge concerning the nature of the genetic polymorphism responsible for a given novel fatty acid phenotype. However, advances in the biochemistry and molecular genetics of plant lipid biosynthesis have led to the elucidation of the specific molecular mutations that give rise to a particular oil phenotype.

By deciphering the precise gene mutations responsible for altering soybean oil composition, one should have sufficient information to generate specific molecular markers that can be used to unambiguously distinguish individuals possessing the mutations of interest. Molecular markers that are based on the actual genetic polymorphism that underlies a specific phenotype are particularly ideal for large-scale breeding efforts because the marker will never become separated from the trait of interest through recombination. Furthermore, the characterization of the genes and specific mutations responsible for desirable oil phenotypes will enhance overall understanding of the mechanisms by which soybeans synthesize and regulate the production of storage lipids during seed development. A comprehensive understanding of such mechanisms will enable researchers to predict more accurately the outcomes (and consequences) of pyramiding multiple oil traits or combining specific oil traits with other seed traits that compete for the same pool of available photosynthate. To date, researchers have been successful in defining the specific gene mutations found in soybean lines displaying a low–linolenic acid phenotype and a low–palmitic acid phenotype.

Molecular basis of low linolenic acid. The synthesis of α-linolenic acid is catalyzed by ω-3 desaturase enzymes that introduce a third double bond into linoleic acid. Although unique ω-3 desaturase activities are found within both the plastids and microsomes of plant cells, it is the microsomal enzymes (encoded by genes designated FAD_3) that are primarily responsible for the production of the linolenic acid found in seed storage oils (47). Because deficiencies in microsomal ω-3 desaturase activity should result in lower linolenic acid concentration of the seed, FAD_3 genes represent good targets for analysis in soybean lines possessing reduced levels of this fatty acid. Soybean genotype A5 represents a reduced linolenic acid line that was selected using a chemical mutagenesis strategy (40). Inheritance studies established the trait to be transmitted as a single recessive gene at a locus designated *fan* (68). When Byrum *et al.* (72) conducted a DNA gel blot analysis of genomic DNAs from a select number of normal and low–linolenic acid lines using a soybean FAD_3 cDNA as a hybridization probe, it became apparent that soybeans possess more than one copy of the FAD_3 gene and that line A5 was missing all or part of one copy of the gene. Segregation analysis confirmed the association of the low–linolenic acid trait with the FAD_3 deletion.

Although the above study clearly established the A5 line as a FAD_3 deletion mutant, the specific number of FAD_3 genes that exist in the soybean genome and the identity of the specific FAD_3 isoform that is deleted in A5 remained unknown. A recent study by Beuselinck and colleagues, however, has resolved these issues. Computer analysis of the gene databases generated by the publicly funded soybean genome initiative led the authors to conclude that three closely related, but distinct, FAD_3 isoforms are expressed in soybean (73). Full-length sequence analysis of the three genes (designated Gm*FAD3A*, Gm*FAD3B*, and Gm*FAD3C*) facilitated the generation of specific PCR primers that could uniquely distinguish each isoform. PCR analysis revealed Gm*FAD3A* to be the specific FAD_3 gene that was deleted in genotype A5. Allele-specific primers based on the Gm*FAD3A* sequence can now serve as effective molecular markers to identify individuals in segregating populations that are homozygous for the low–linolenic acid trait originating from A5.

Molecular basis of low palmitic acid. Given our current understanding of lipid biosynthesis in plants, several mechanisms can be envisioned that could result in an overall reduction in the palmitic acid content of the storage oil in a mutant soybean line. For example, reduced affinity for palmitic acid by the acyltransferase enzymes that attach acyl groups to the glycerol backbone could decrease the overall 16:0 content of the storage oil. Alternatively, enhancement of the KAS-II condensing enzyme that is responsible for elongating the 16:0-acyl carrier protein (16:0-ACP) thioester to 18:0-ACP could, in theory, reduce the pool of 16:0 available for introduction into storage triacylglycerols. Finally, alterations in the enzymes that release free palmitic acid from 16:0-ACP could also influence the amount of palmitic acid that is available for incorporation into the final oil product. Although each of these possibilities must be considered when evaluating germplasm with altered palmitic acid content, transgenic plant experiments suggest that this last option (i.e., the 16:0-ACP thioesterase-mediated release of palmitic acid) should be given particular consideration given the strong control that this step of the pathway has been shown to exert in determining the ultimate palmitic acid concentration of the seed (74,75).

Soybean genotypes N79-2077-12 and C1726 represent two low–palmitic acid lines that carry nonallelic mutations at loci designated fap_{nc} and fap_1, respectively (17,18,20). N79-2077-12 carries a serendipitous natural mutation, whereas the mutation that defines C1726 was induced by chemical mutagenesis. To assist in pinpointing the specific step in triacylglycerol metabolism that is compromised in N79-2077-12 and C1726, Wilson *et al.* (76) investigated the saturation kinetics of immature soybean seeds treated with labeled acetate and 16:0-CoA. These studies ruled out the involvement of acyltransferases in mediating the low–palmitic acid phenotypes and focused attention on the KAS II and 16:0-ACP thioesterase steps that precede the exportation of free palmitic acid from the plastid to the cytosol. RNA blot analysis of genotype N79-2077-12 revealed a significant decrease in transcripts corresponding to a soybean *FATB* gene that encodes 16:0-ACP thioesterase activity (34); DNA blot assays demonstrated that this line possesses a *FATB* deletion mutation (77).

Similar to the observation that the soybean genome contains multiple FAD_3 genes as described in the previous section, soybeans also possess more than one genomic copy of *FATB*. In order to derive a facile molecular marker for the low–palmitic trait found in N79-2077-12, it was first necessary to have sequence information for each *FATB* isoform to enable the design of specific PCR primers that could unambiguously amplify each unique *FATB* gene. Using computer database analysis in combination with gene cloning and sequencing, four distinct *FATB* genes have been identified and designated Gm*FATB1*, Gm*FATB2*, Gm*FATB3*, and Gm*FATB4* (R.E. Dewey, unpublished data). PCR analysis using allele-specific primers revealed that N79-2077-12 and lines derived from it do not possess the Gm*FATB1* gene. This information has enabled the design of highly specific molecular markers that can readily identify individuals homozygous for the low–palmitic acid trait originating from N79-2077-12 (R.E. Dewey and A.J. Cardinal, Department of Crop Science, North Carolina State University, unpublished data).

Although the successful characterizations of the molecular mutations found in modified oil lines A5 and N79-2007-12 both involved major gene deletions, the above described strategies should work equally well for deciphering more subtle lesions, such as point mutations. Because nearly all of the genes of the triacylglycerol biosynthetic pathway have been identified in plants (78), the elucidation of any modified oil soybean genotype that results from a mutation in one of these genes should be tractable. However, it is likely that some of the modified oil breeding lines are the consequence of mutations in genes that regulate certain steps of the pathway (such as transcription factors or protein kinases), as opposed to mutations in genes directly involved in lipid biosynthesis. Until more is known concerning the molecular and cellular processes involved in the regulation of lipid metabolism during seed development, identifying mutations of this nature will be a much greater challenge.

Summary

A current challenge to the food-processing industry involves meeting the demands of health-conscious consumers for reduced *trans* fats and saturated fats. Genetic improvements in soy oil quality will be required for a domestic crop to fill this need. Currently there are a total of 16 linkage groups that report loci likely to control total oil biosynthesis. Linkage groups A1, B1, G, and H each contain one or more major, environmentally stable oil QTL loci. Additional major and minor oil loci can be accumulated through a MAS strategy to enhance soybean oil concentration.

Molecular markers can be used for MAS of individuals carrying donor parent alleles at marker loci linked to a gene or QTL associated with a desired fatty acid phenotype. In addition, markers permit confirmation that plants are true F_1 hybrids or backcrosses, rather than the result of self-pollination. This is especially important in selecting for a trait that cannot be determined prior to plant reproductive maturity, such as a modified fatty acid profile. Molecular markers have also proven useful for confirming the parentage of a particular line or cultivar, and as a tool that can be used to trace the inheritance of a certain region of the genome through a pedigree. The latter application can provide convincing evidence that a major QTL is real and located in the conserved region of DNA.

A major low–palmitic acid allele is located near Satt684 on LG A1, and a minor locus conditioning this trait is on LG M. These markers could help molecular breeders assemble germplasm which targets achieving less than 7% total saturates for a low–saturated fat food label. A major locus conditioning elevated stearic acid is located near markers Satt070, Satt474, and Satt556. Any of these three markers would support a MAS strategy for increasing a natural saturate in order to reduce the amount of *trans* fat in semi-solid fat products such as margarines. Markers significantly associated with increased oleic acid levels ($P \leq 0.01$) have thus far been found on LG A1, G, L, and M, with evidence that more than one QTL may be present on LG G and L. Increased oleic acid will be an important achievement to improved soybean oxidative stability and other frying oil attributes. Markers near the *fan* locus on LG B2 will help reduce the level of linolenic acid for improved oxidative stability and reduced need for hydrogenation, for further reductions in *trans* fatty acids.

Our understanding of the biochemistry and molecular genetics of plant lipid biosynthesis has enabled the possible elucidation of the specific molecular mutations that give rise to a particular oil phenotype. Identification of the precise gene mutations responsible for modified soybean oil composition could generate specific molecular markers that can be used to unambiguously distinguish individuals possessing the mutations of interest. Molecular markers that are based on the actual genetic polymorphism that underlies a specific phenotype are ideal for large-scale molecular breeding efforts. The characterization of the genes and specific mutations responsible for desirable oil phenotypes will enhance our overall understanding of the mechanisms by which soybeans synthesize and regulate the production of storage lipids. A comprehensive understanding of such mechanisms will enable researchers to predict more accurately the outcomes of pyramiding multiple oil traits or combining specific oil traits with other desirable seed traits.

Acknowledgments

Support by the United Soybean Board (Projects 9208, 1239, and 3222) and the Tennessee Soybean Promotion Board is greatly appreciated in making portions of this work possible. We wish to thank H. Roger Boerma, Maria Monteros, and Andrea Cardinal for their insightful contributions to this chapter. A. D. Weir and D. L. Hyten, Jr. are gratefully acknowledged for providing contributions from their M.S. thesis research. We wish to thank Warren Rayford and Donna Thomas of the USDA National Center for Agricultural Research (Peoria, IL) for conducting analyses for seed oil concentration.

References

1. Wilson, R.F., J.W. Burton, V.R. Pantalone, and R.E. Dewey, New Gene Combinations Governing Saturated and Unsaturated Fatty Acid Composition in Soybean, in *Lipid Biotechnology*, edited by T.M. Kuo and H.W. Gardner, Marcel Dekker, New York, 2002, pp. 95–114.
2. U.S. Food and Drug Administration, Center for Food Safety and Applied Nutrition, *A Food Labeling Guide—Appendix A: Definitions of Nutrient Content Claims* (1999), www.cfsan.fda.gov/~dms/flg-toc.html, verified February 11, 2004.
3. U.S. Food and Drug Administration, FDA Proposes New Rules for Trans Fatty Acids in Nutrition Labeling, Nutrient Content Claims and Health Claims, *Federal Register*, November 12, 1999.
4. Cregan, P.B., T. Jarvik, A.L. Bush, R.C. Shoemaker, K.G. Lark, A.L. Kahler, N. Kaya, T.T. VanToai, D.G. Lohnes, and J. Chung, An Integrated Genetic Linkage Map of the Soybean Genome, *Crop Sci. 39*:1464–1490 (1999).

5. Grant, D., M.I. Imsande, and R.C. Shoemaker, SoyBase, The USDA-ARS Soybean Genome Database, http://soybase.org (verified February 11, 2004).
6. Mansur, L.M., J.H. Orf, K. Chase, T. Jarvik, P.B. Cregan, and K.G. Lark, Genetic Mapping of Agronomic Traits Using Recombinant Inbred Lines of Soybean, *Crop Sci. 36*:1327–1336 (1996).
7. Brummer, E.C., G.L. Graef, J. Orf, J.R. Wilcox, and R.C. Shoemaker, Mapping QTL for Seed Protein and Oil Content in Eight Soybean Populations, *Crop Sci. 37*:370–378 (1997).
8. Orf, J.H., K. Chase, T. Jarvik, L.M. Mansur, P.B. Cregan, F.R. Adler, and K.G. Lark, Genetics of Soybean Agronomic Traits I: Comparison of Three Related Recombinant Inbred Populations, *Crop Sci. 39*:1642–1651 (1999).
9. Specht, J.E., K. Chase, M. Macrander, G.L. Graef, J. Chung, J.P. Markwell, M. Germann, J.H. Orf, and K.G. Lark, Soybean Response to Water: A QTL Analysis of Drought Tolerance, *Crop Sci. 41*:493–509 (2001).
10. Mansur, L.M., J. Orf, and K.G. Lark, Determining the Linkage of Quantitative Trait Loci to RFLP Markers Using Extreme Phenotypes of Recombinant Inbreds of Soybean (*Glycine max* L. Merr.), *Theor. Appl. Genet. 86*:914–918 (1993).
11. Diers, B.W., and R.C. Shoemaker, Restriction Fragment Length Polymorphism of Soybean Fatty Acid Content, *J. Am. Oil Chem. Soc. 69*:1242–1244 (1992).
12. Csanadi, G., J. Vollmann, G. Stift, and T. Lelley, Seed Quality QTLs Identified in a Molecular Map of Early Maturing Soybean, *Theor. Appl. Genet. 103*:912–919 (2001).
13. Fasoula, V.A., D.K. Harris, and H.R. Boerma, Validation and Designation of Quantitative Trait Loci for Seed Protein, Seed Oil, and Seed Weight from Two Soybean Populations, *Crop Sci. 44* (2004, in press).
14. Hyten, D.L., Jr., QTL Mapping and Identification of G×E Interactions of Agronomic and Seed Quality Traits in Soybean, M.S. thesis, University of Tennessee, Knoxville, TN, 2002, pp. 24–39.
15. Lee, S.H., M.A. Bailey, M.A.R. Mian, T.E. Carter, Jr., E.R. Shipe, D.A. Ashley, W.A. Parrott, R.S. Hussey, and H.R. Boerma, RFLP Loci Associated with Soybean Seed Protein and Oil Content across Populations and Locations, *Theor. Appl. Genet. 93*:649–657 (1996).
16. Sebolt, A.M., R.C. Shoemaker, and B.W. Diers, Analysis of a Quantitative Trait Locus Allele from Wild Soybean that Increases Seed Protein Concentration in Soybean, *Crop Sci. 40*:1438–1444 (2000).
17. Wilcox, J.R., J.W. Burton, G.J. Rebetzke, and R.F. Wilson, Transgressive Segregation for Palmitic Acid in Seed Oil of Soybean, *Crop Sci. 34*:1248–1250 (1994).
18. Erickson, E.A., J.R. Wilcox, and J.F. Cavins, Inheritance of Altered Palmitic Acid Percentage in Two Soybean Mutants, *J. Heredity 79*:465–468 (1988).
19. Fehr, W.R., G.A. Welke, E.G. Hammond, D.N. Duvick, and S.R. Cianzano, Inheritance of Reduced Palmitic Acid Content in Seed Oil of Soybean, *Crop Sci. 31*:88–89 (1991).
20. Burton, J.W., R.F. Wilson, and C.A. Brim, Registration of N79-2077-12 and N87-2122-4, Two Soybean Germplasm Lines with Reduced Palmitic Acid in Seed Oil, *Crop Sci. 34*:313 (1994).
21. Wilson, R.F., J.W. Burton, and C.A. Brim, Progress in the Selection for Altered Fatty Acid Composition in Soybean, *Crop Sci. 21*:788–791 (1981).
22. Spencer, M.M., V.R. Pantalone, E.J. Meyer, D. Landau-Ellis, and D.J. Hyten, Jr., Mapping the *Fas* Locus Controlling Stearic Acid Content in Soybean, *Theor. Appl. Genet. 106*:615–619 (2003).

23. Bubeck, D.M., W.R. Fehr, and E.G. Hammond, Inheritance of Palmitic and Stearic Acid Mutants of Soybean, *Crop Sci. 29*:652–656 (1989).
24. Wilcox, J.R., and J.F. Cavins, Registration of C1726 and C1727 Soybean Germplasm with Altered Levels of Palmitic Acid, *Crop Sci. 30*:240 (1990).
25. Töpfer, R., N. Martini, and J. Schell, Modification of Plant Lipid Synthesis, *Science 268*:681–686 (1995).
26. Thelen, J.J., and J.B. Ohlrogge, Metabolic Engineering of Fatty Acid Biosynthesis in Plants, *Metab. Eng. 4*:12–21 (2002).
27. Stolzfus, D.L., W.R. Fehr, G.A. Welke, E.G. Hammond, and S.R. Cianzio, A fap_5 Allele for Elevated Palmitate in Soybean, *Crop Sci. 40*:647–650 (2000).
28. Narvel, J.M., W.R. Fehr, J. Ininda, G.A. Welke, E.G. Hammond, D.N. Duvick, and S.R. Cianzio, Inheritance of Elevated Palmitate in Soybean Seed Oil, *Crop Sci. 40*:635–639 (2000).
29. Primomo, V.S., D.E. Falk, G.R. Ablett, J.W. Tanner, and I. Rajcan, Inheritance and Interaction of Low Palmitic and Low Linolenic Soybean, *Crop Sci. 42*:31–36 (2002).
30. Schnebly, S.R., W.R. Fehr, G.A. Welke, E.G. Hammond, and D.N. Duvick, Inheritance of Reduced and Elevated Palmitate in Mutant Lines of Soybean, *Crop Sci. 34*:929–833 (1994).
31. Horjesi, T.F., W.R Fehr, G.A. Welke, D.N. Duvick, E.G. Hammond, and S.R. Cianzio, Genetic Control of Reduced Palmitate Content in Soybean, *Crop Sci. 34*:331–334 (1994).
32. Rebetzke, G.J., J.W. Burton, T.E. Carter, Jr., and R.F. Wilson, Genetic Variation for Modifiers Controlling Reduced Saturated Fatty Acid Content in Soybean, *Crop Sci. 38*:303–308 (1998).
33. Rebetzke, G.J., J.W. Burton, T.E. Carter, Jr., and R.F. Wilson, Changes in Agronomic and Seed Characteristics with Selection for Reduced Palmitic Acid Content in Soybean, *Crop Sci. 38*:297–302 (1998).
34. Wilson, R.F., T.C. Marquardt, W.P. Novitzky, J.W. Burton, J.R. Wilcox, A.J. Kinney, and R.E. Dewey, Metabolic Mechanisms Associated with Alleles Governing the 16:0 Concentration of Soybean Oil, *J. Am. Oil Chem. Soc. 78*:335–340 (2001).
35. Ndzana, X., W.R. Fehr, G.A. Welke, E.G. Hammond, D.N. Duvick, and S.R. Cianzio, Influence of Reduced Palmitate Content on Agronomic and Seed Traits of Soybean, *Crop Sci. 34*:646–649 (1994).
36. Hayes, M.F., W.R. Fehr, and G.A. Welke, Association of Elevated Palmitate with Agronomic and Seed Traits of Soybean, *Crop Sci. 42*:1117–1120 (2002).
37. Nickell, A.D., J.R. Wilcox, L.L. Lorenzen, J.F. Cavins, R.G. Guffy, and R.C. Shoemaker, The Fap_2 Locus in Soybean Maps to Linkage Group D, *J. Hered. 85*:160–162 (1994).
38. Li, Z., R.F. Wilson, W.E. Rayford, and H.R. Boerma, Molecular Mapping Genes Conditioning Reduced Palmitic Acid Content in N86-2122-4 Soybean, *Crop Sci. 42*:373–378 (2002).
39. Pantalone, V.R., R.F. Wilson, W.P. Novitzky, and J.W. Burton, Genetic Regulation of Elevated Stearic Acid Concentration in Soybean Oil, *J. Am. Oil Chem. Soc. 79*:549–553 (2002).
40. Hammond, E.G., and W.R. Fehr, Registration of A5 Germplasm Line of Soybean, *Crop Sci. 23*:192 (1983).
41. Graef, G.L., W.R. Fehr, and E.G. Hammond, Inheritance of Three Stearic Acid Mutants of Soybean, *Crop Sci. 25*:1076–1079 (1985).

42. Rahman, S.M., Y. Takagi, and T. Kinoshita, Genetic Control of High Stearic Acid Content in Seed Oil of Two Soybean Mutants, *Theor. Appl. Genet. 95*:772–776 (1997).
43. Miller, J.F., and B.A. Vick, Inheritance of Reduced Stearic and Palmitic Acid Content in Sunflower Seed Oil, *Crop Sci. 39*:364–367 (1999).
44. Lightner, J., E. Lark, D. James, and J. Browse, Novel Mutations Affecting Leaf Stearate Content and Plant Size in Arabidopsis, *Theor. Appl. Genet. 94*:975–981 (1997).
45. Lundeen, P.O., W.R. Fehr, E.G. Hammond, and S.R. Cianzio, Association of Alleles for High Stearic Acid with Agronomic Characters of Soybean, *Crop Sci. 27*:1102–1105 (1987).
46. Hartmann, R.B., W.R. Fehr, G.A. Welke, E.G. Hammond, D.N. Duvick, and S.R. Cianzio, Association of Elevated Stearate with Agronomic and Seed Traits of Soybean, *Crop Sci. 37*:124–127 (1997).
47. Ohlrogge, J., and J. Browse, Lipid Biosynthesis, *Plant Cell 7*:957–970 (1995).
48. Rennie, B.D., and J.W. Tanner, Mapping a Second Fatty Acid Locus to Soybean Linkage Group 17, *Crop Sci. 29*:1081–1083 (1989).
49. Brummer, E.C., A.D. Nickell, J.R. Wilcox, and R.C. Shoemaker, Mapping the *Fan* Locus Controlling Linolenic Acid Content in Soybean Oil, *J. Hered. 86*:245–247 (1995).
50. Grundy, S.M., Composition of Monounsaturated Fatty Acids and Carbohydrates for Lowering Plasma Cholesterol, *New Eng. J. Med. 314*:745-748 (1986).
51. Wardlaw, G.M., and J.T. Snook, Effect of Diets High in Butter, Corn Oil, or High-Oleic Acid Sunflower Oil on Serum Lipids and Apolipoproteins in Men, *Am. J. Clin. Nutr. 51*:815–821 (1990).
52. Chang, N.W., and P.C. Huang, Effects of the Ratio of Polyunsaturated and Monounsaturated Fatty Acid on Rat Plasma and Liver Lipid Concentration, *Lipids 33*:481–487 (1998).
53. Miller, J.F., D.C. Zimmermann, and B.A. Vick, Genetic Control of High Oleic Acid Content in Sunflower Oil, *Crop Sci. 27*:923–926 (1987).
54. Wilson, R.F., *Seed Composition in Soybeans: Improvement, Production, and Uses*, 3rd. ed., edited by H.R. Boerma and J.E. Specht, Am. Soc. Agronomy–Crop Sci. Soc. Am.–Soil Sci. Soc. Am., Madison, WI, 2003 (in press).
55. Takagi, Y., and S.M. Rahman, Inheritance of High Oleic Content in the Seed Oil of Soybean Mutant M23, *Theor. Appl. Genet. 92*:179–182 (1996).
56. Pantalone, V.R., G.J. Rebetzke, J.W. Burton, and R.F. Wilson, Genetic Regulation of Linolenic Acid Concentration in Wild Soybean *Glycine soja* Accessions, *J. Am. Oil Chem. Soc. 74*:159–163 (1997).
57. Brim, C.A., W.M. Schutz, and F.I. Collins, Maternal Effect on Fatty Acid Composition and Oil Content of Soybean, *Glycine max* (L.) Merrill, *Crop Sci. 8*:517–518 (1968).
58. Primono, V.S., D.E. Falk, G.R. Ablett, J.W. Tanner, and I. Rajcan, Genotype × Environment Interactions, Stability, and Agronomic Performance of Soybean with Altered Fatty Acid Profiles, *Crop Sci. 42*:37–44 (2002).
59. Hu, F.B., M.J. Stampfer, J.E. Manson, E. Rimm, G.A. Colditz, B.A. Rosner, C.H. Hennekens, and W.C. Willett, Dietary Fat Intake and the Risk of Coronary Heart Disease in Women, *New Eng. J. Med. 337*:1491–1499 (1997).
60. Ohlrogge, J., and J. Browse, Lipid biosynthesis, *Plant Cell 7*:957–970 (1995).
61. Hitz, W.D., T.J. Carlson, J.R. Booth, Jr., A.J. Kinney, K.L. Stecca, and N.S. Yadav, Cloning of a Higher Plant Plastid T-6 Fatty Acid Desaturase cDNA and its Expression in a Cyanobacterium, *Plant Physiol. 105*:635–641 (1994).

62. Brossman, G.D., and J.R. Wilcox, Induction of Genetic Variation for Oil Properties and Agronomic Characteristics of Soybean, *Crop Sci. 24*:783–787 (1984).
63. Yadav, N.S., Genetic Modification of Soybean Oil Quality, in *Soybean Biotechnology*, edited by D.P.S. Verma and R.C. Shoemaker, CAB International, London, 1996, pp. 165–188.
64. Weir, A.D., Genetic Mapping of Quantitative Trait Loci for Seed Quality Traits in an Interspecific Soybean Population, M.S. Thesis, University of Guelph, Ontario, Canada, 2003, pp. 86–90.
65. Shoemaker, R., P. Keim, L. Vodkin, E. Retzel, S.W. Clifton, R. Waterston, D. Smoller, V. Coryell, A. Khanna, J. Edpelding, X.W. Gai, V. Brendel, C. Raph-schmidt, E.G. Shoop, C.J. Vielweber, M. Scmatz, D. Pape, Y. Bowers, B. Theising, J. Martin, M. Dante, T. Wylie, and C. Granger, A Compilation of Soybean ESTs: Generation and Analysis, *Genome 45*:329–338 (2002).
66. Shoemaker, R.C., K. Polzin, J. Labate, J. Specht, E.C. Brummer, T. Olson, N. Young, V. Concibido, J. Wilcox, J.P. Tamulonis, G. Kochert, and H.R. Boerma, Genome Duplication in Soybean (*Glycine* subgenus *soja*), *Genetics 144*:329–338 (1996).
67. Liu, H.R., and P.J. White, Oxidative Stability of Soybean Oils with Altered Fatty Acid Compositions, *J. Am. Oil Chem. Soc. 69*:528–532 (1992).
68. Fehr, W.R., G.A. Welke, E.G. Hammond, D.N. Duvick, and S.R. Cianzio, Inheritance of Reduced Linolenic Acid Content in Soybean Genotypes A16 and A17, *Crop Sci. 32*:903–906 (1992).
69. Rahman, S.M., and Y. Takagi, Inheritance of Reduced Linolenic Acid Content in Soybean Seed Oil, *Theor. Appl. Genet. 94*:299–302 (1997).
70. Ross, A.J., W.R. Fehr, G.A. Welke, and S.R. Cianzio, Agronomic and Seed Traits of 1%-Linoleate Soybean Genotypes, *Crop Sci. 40*:383–386 (2000).
71. Wilcox, J.R., and J.F. Cavins, Gene Symbol Assigned for Linolenic Acid Content in the Soybean, *J. Heredity 78*:410 (1987).
72. Byrum, J.R., A.J. Kinney, K.L. Stecca, D.J. Grace, and B.W. Diers, Alteration of Omega-3 Fatty-Acid Desaturase Gene is Associated with Reduced Linolenic Acid in the A5 Soybean Genotype, *Theor. Appl. Genet. 94*:356–359 (1997).
73. Bilyeu, K.D., L. Palavalli, D.A. Sleper, and P.R. Beuselinck, Three Microsomal Omega-3 Fatty-Acid Desaturase Genes Contribute to Soybean Linolenic Acid Levels, *Crop Sci. 43*:1833–1838 (2003).
74. Jones, A., H.M. Davies, and T.A. Voelker, Palmitoyl-Acyl Carrier Protein (ACP) Thioesterase and the Evolutionary Origin of Plant Acyl-ACP Thioesterases, *Plant Cell 7*:359–371 (1995).
75. Kinney, A.J., Improving Soybean Seed Quality, in *FAO/IAEA International Symposium on the Use of Induced Mutations and Molecular Techniques from Crop Improvement*, Vienna, Austria, 1995, pp. 101–113.
76. Wilson, R.F., T.C. Marquardt, W.P. Novitzky, J.W. Burton, J.R. Wilcox, and R.E. Dewey, Effect of Alleles Governing 16:0 Concentration on Glycerolipid Composition in Developing Soybeans, *J. Am. Oil Chem. Soc. 78*:329–334 (2001).
77. Wilson, R.F., J.W. Burton, W.P. Novitzky, and R.E. Dewey, Current and Future Innovations in Soybean (*Glycine max* (L.) Merr.) Oil Composition, *J. Oleo Sci. 50*:353–358 (2001).
78. Mekhedov, S., O.M. de Ilarduya, and J. Ohlrogge, Toward a Functional Catalog of the Plant Genome. A Survey of Genes for Lipid Biosynthesis, *Plant Physiol. 122*:389–401 (2000).

Chapter 18

Genomic Approaches for Developing Soybeans with Resistance to Pests (Soybean Cyst Nematode)

Benjamin F. Matthews

USDA, ARS, Soybean Genomics and Improvement Laboratory, Beltsville, MD 20705-2350

Introduction

A fundamental goal of many plant scientists is to broaden resistance of plants to pests and pathogens. Plant genomics promises to provide many new tools for plant breeders to develop elite germplasm with broader, stronger resistance more quickly and more economically. Plant genomics will improve methods for traditional breeding of resistance, and it will provide new tools and knowledge for devising strategies to broaden resistance to a multitude of stresses. This chapter will review recent advances in genomics in relation to developing resistance in soybean to the soybean cyst nematode, the most damaging pest of soybean.

Plant breeding has been responsible for remarkable gains in productivity, and it promises to provide even more gains with the knowledge and tools provided through genomics research. Traditionally, plant breeders have used available genetic and phenotypic data to identify parents most suitable for breeding; however, the breeder was often forced to make decisions based on incomplete data. Furthermore, the breeder had to screen plants laboriously to identify progeny possessing a particular advantageous phenotype, and the genotype of these individuals was often unknown, confounding the breeder's efforts and requiring additional generations to determine genotypes. Thus, the traditional method of plant breeding is expensive in terms of time, labor, and infrastructure. Screening for resistance to pests and pathogens can take several months and may require specialized equipment and greenhouses. Genomics can help provide the plant breeder with information to make more informed and better decisions and to more quickly identify useful progeny. The complete, annotated DNA sequence of a plant genome combined with locations for important traits provides the breeder with information necessary to develop breeding strategies and molecular screening techniques in order to identify useful progeny faster and more easily, and it helps the biologist to determine gene function (Fig. 18.1).

Knowledge of the full DNA sequence of a plant genome combined with mapping of important agronomic traits will provide molecular markers for marker-assisted selection (MAS) in plant breeding schemes and provide a framework for identifying candidate genes responsible for the trait through map position–based cloning. The plant DNA sequence can be used to design and develop oligomer primers for polymerase chain reaction (PCR) assays and identify simple sequence

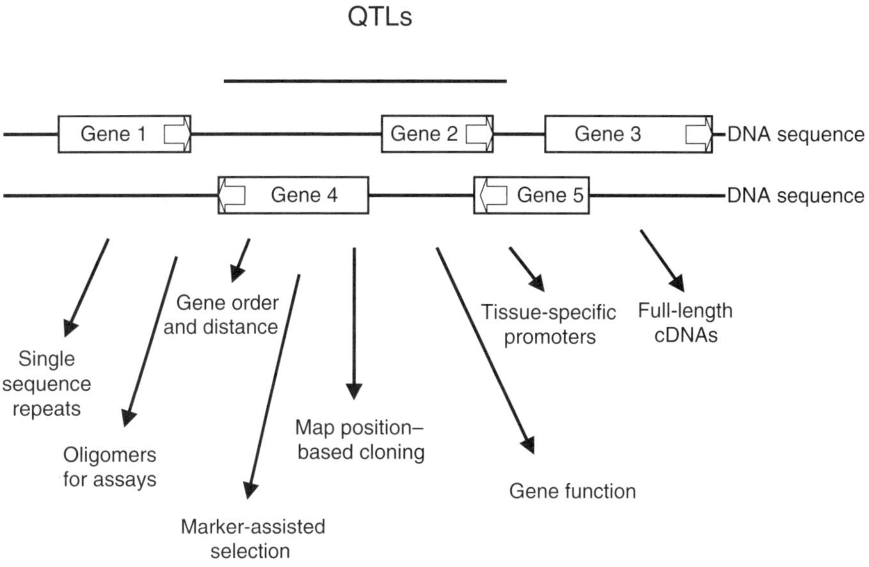

Figure 18.1. Complete annotation of the genome, including the location of important quantitative trait loci (QTLs). Open block arrows indicate the direction of gene transcription; genes are located on solid lines indicating DNA sequences.

repeats (SSRs) for rapid genotyping of a plant population, to identify desirable progeny. By providing accurate linkage and physical distance data, the complete sequence of a plant genome and the correlation of DNA sequences with important agronomic traits will allow the plant breeder to devise more efficient strategies for choosing the best parental material and for identifying the most useful progeny possessing the correct genotype.

Other data developed through genomic studies will open doors for developing new methods and approaches for broadening resistance. For example, gene expression data will help to identify target genes for transgenic approaches to pest and pathogen resistance and to promoters controlling precise expression of genes in specific tissues and cells.

A Case Study of Traditional Strategies: SCN Plant Resistance

The soybean cyst nematode (SCN), *Heterodera glycines* Ichinohe, is the most devastating worldwide pest of soybean, *Glycine max* (L.) Merr. (see Fig. 18.2). In recent years the estimated yield losses in the United States from SCN ranged between 142 million and 279 million bushels, reflecting $0.7 to $1.5 billion in lost revenue (1). SCN infection of soybean causes various symptoms that may include chlorosis, root necrosis, loss

Figure 18.2. Three female soybean cyst nematodes attached to a susceptible soybean root.

in seed yield, and suppression of root and shoot growth; it is the major yield-limiting factor in the midwestern United States (2). Chemical fumigation is used to control field infestations by nematodes for some crops with a high cash value, but it is not economically or environmentally feasible to control SCN populations using fumigation. Crop rotation and planting resistant cultivars constitute the norm, yet farmers report yield losses despite planting SCN-resistant cultivars. Levels of resistance of soybean to SCN are not as high as necessary, and diversity of sources of resistance is lacking. Most SCN resistance in currently grown soybean cultivars is derived from PI 88788 and Peking. Approximately two-thirds of the varieties entered in trials have resistance to SCN, and 90% of this resistance is from PI 88788, which provides resistance to races 3 and 14, while Peking offers resistance to races 1 and 3. Thus, most SCN resistance is derived from a very narrow germplasm base. The SCN race most often encountered is race 3, which is found in 70% of fields testing positive for SCN, while race 1, found in 20% of the fields, is the second most encountered race. PI 88788 and Peking do therefore possess resistance genes to counter the most common races; however, it is difficult to ensure that all of the genes contributing to resistance are retained in released elite cultivars. Also, field populations of SCN are composed of a mixture of SCN races. Thus, races other than race 1 and race 3 can proliferate and predominate in fields grown with soybeans resistant to only one or two races. Fourteen races of SCN have been defined by their interaction with selected soybean cultivars and with introductions that differ in their response to different races of SCN (3). In the soil, SCN populations are diverse and are composed of numerous phenotypes, or races. This variability in the SCN populations complicates the effective use of cultivars resistant to SCN by farmers and confounds rotation schemes. The 14 races of SCN reflect the multitude of SCN genotypes and the complexity breeders must face to breed broad resistance in soybean to SCN. A newer classification system for SCN recently was proposed (4), and it provides more detailed information on the type of SCN being classified. However, it further emphasizes the challenges facing the soybean breeder that are due to the range of soybean interactions with the numerous nematode population variations.

Resistance to SCN is a multigenic trait, making breeding for resistance even more complicated (5–8). Quantitative trait loci (QTLs) associated with resistance to SCN have been mapped to linkage groups A2 (*Rhg4*), D, G (*Rhg1*), the opposite end of G, J, K, L, and N (7–9). Two major loci (*Rhg1* and *Rhg4*) have been identified in cv. Peking and PI 437654, and both are required to confer resistance to race 3. With all these different resistance genes needed to confer resistance to the numerous races of SCN in soybean, the soybean breeder must screen thousands of progeny to identify individuals that possess the appropriate SCN resistance genes to provide protection to several races. Because SCN assays are usually conducted in the greenhouse, where mature females are counted 30 days after plants are inoculated with juvenile nematodes, the phenotypic screening for SCN resistance is a long and tedious task. A faster and more economical approach would be to use a marker-assisted selection scheme by conducting rapid assays for molecular markers tightly associated with the different SCN resistance loci.

Application of Genomics to Broaden Resistance Using Plant Breeding

Development of a soybean map that contains gene locations for agronomically important phenotypes and that is also saturated with easily assayed molecular markers will allow soybean breeders to identify useful siblings from a segregating breeding population without extensive biological screens. Certainly for the model system *Arabidopsis*, this process is made easier because the genome is sequenced. For soybean, whose genome is not sequenced, a large amount of effort is focused on developing a map saturated with molecular markers and important phenotypes. Cregan and colleagues (10) developed an integrated genetic linkage map containing the locations of more than 600 simple sequence repeat (SSR) markers, numerous restriction fragment length polymorphisms (RFLPs), and many phenotypes. The molecular markers developed for this map can be used for marker-assisted selection in breeding programs and for map-based cloning efforts. Information supporting the soybean genetic map is becoming vast and complex, and it will need to be integrated with databases providing gene function information.

Identifying molecular markers near SCN-resistance loci would allow breeders to use marker-assisted selection to track the numerous resistance genes in segregating progeny. Thus, selection of resistant lines will be faster, and marker-assisted selection will help confirm homozygous, resistant genotypes. The first molecular markers developed for screening for SCN resistance were RFLP markers (pBLT65 and pBLT24) that were located on soybean Linkage Group A2. The markers are close to the *Rhg4* locus and provided soybean breeders "with an alternative method of selecting for the *Rhg4* gene by allowing selection for the two RFLP markers rather than selection by phytopathological assay for disease resistance" (11). This report set a precedent for the use of molecular markers for monitoring cyst nematode resistance in soybean. From these molecular markers a rapid PCR assay was developed from which the closer marker (pBLT65) was identified for breeders to screen for *Rhg4* (Fig. 18.3) (12). The marker pBLT65 was used for bacterial artificial chromosome (BAC) mapping to the region containing *Rhg4*, and these were subsequently used to identify several SSR markers that are tightly linked to the *Rhg4* locus (13). These markers should be useful to soybean breeders for marker-assisted selection of progeny carrying the *Rhg4* locus. Also, these markers have been invaluable in efforts to clone the *Rhg4* gene. BAC mappings to the region containing *Rhg4* have been analyzed from both cv. Williams, susceptible to SCN race 3 (14), and PI 437654, resistant to SCN race 3 (15).

Similarly, markers mapping near the *Rhg1* locus have been developed for molecular-assisted selection. Mudge *et al.* (16) identified two SSR markers, BARC-Satt038 and BARC-Satt130, flaking the *Rhg1* locus on linkage group G, which is an important gene conditioning resistance to SCN. Sequence-tagged site markers, recently developed, are also useful for monitoring inheritance at this locus (17). Markers closely associated with SCN resistance genes, such as those associated with *Rhg4* and *Rhg1*, should help traditional soybean breeders select soybean progeny with the appropriate resistance genes faster, more accurately, and more economically by making time-consuming greenhouse SCN assays unnecessary. By identifying homozygous, resistant genotypes the breeder will be able to accumulate numerous different SCN-resistance genes into elite cultivars more easily.

This technology is being adapted by the soybean seed industry, as indicated by a series of patents disclosing methods to use molecular markers to monitor SCN re-

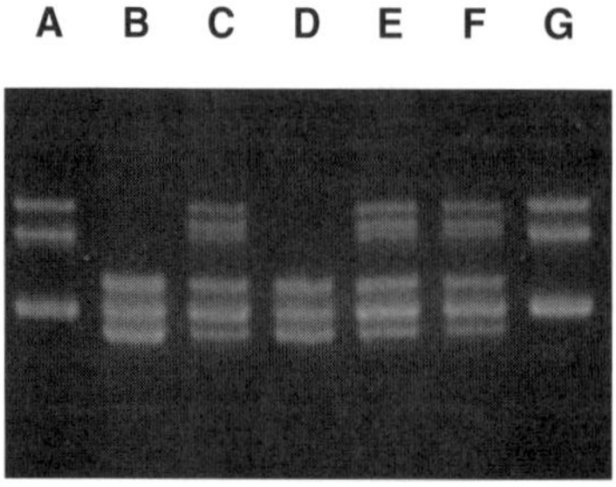

Figure 18.3. A rapid molecular assay of molecular markers mapping close to the *Rhg4* locus. PCR amplifications of DNA fragments are separated on an agarose gel: lane (A) cv. Peking (SCN resistant); (B) cv. Kent (SCN susceptible); (C, E, F) heterozygous; (D) homozygous susceptible; (G) homozygous resistant.

sistance in segregating progeny. It has also been used in map position–based cloning of DNA sequences encompassing SCN-resistance loci. For example, Pioneer Hi-Bred International, Inc. (now a Dupont company) disclosed a method for using marker-assisted selection of SCN resistance using pBLT24 and pBLT65 (18–20). Recently, Monsanto sequenced BACs and other DNA fragments encompassing the region around *Rhg4* and *Rhg1* and was issued a patent related to producing soybean plants with these genes by screening a segregating population with DNA hybridizing within 500 kb of the region (21). This sequencing effort also revealed a gene that encodes a leucine-rich repeat similar to known resistance genes in other species. As yet, there are no reports confirming that this candidate gene confers SCN resistance by complementation or by gene silencing.

As more of the soybean genome is sequenced, molecular-marker mapping will be associated with QTLs, and candidate genes conferring agronomically important phenotypes, such as SCN resistance, will be identified. Validation of the function of the candidate genes through transformation and other methods will allow QTLs to be associated with specific DNA sequences. These genes and molecular assays can be integrated into breeding programs to broaden resistance of soybean to SNC and other pests and pathogens.

New Strategies for Broadening Resistance

The complete genome sequence of Arabidopsis has provided scientists access to whole genes, open reading frames, promoters of genes, and other essential information for using gene technology to understand the biology of plants. The research should lead to modifying and improving plant productivity. However, improving and modifying soybean will require knowledge of specific soybean genes and DNA sequences. For instance, genes involved in resistance to pests and pathogens specific to soybean will often require studying at the whole-organism, cell, protein, and DNA levels. Using soybean-specific genes, DNA fragments, and promoters may be essential to developing strategies for controlling soybean pests and pathogens.

Numerous techniques are being used to identify genes and their functions in soybean. Currently, there are more than 300,000 expressed sequence tags (ESTs) in the TIGR Soybean Gene Index (www.tigr.org/tdb/tgi/gmgi/). The DNA sequences have been compared, aligned, organized, and annotated. However, many ESTs are singletons, many do not represent mature gene transcripts, and many have no known function. Also, EST sequencing in a plant genus does not replace sequencing of an individual genome, because many genes are not revealed by EST sequencing but remain undiscovered until the genome is sequenced. Thus, the job of obtaining the DNA sequences of all the reading frames in the soybean genome is incomplete, and annotating and determining the function of these genes has just begun.

Recently, we used EST and microarray analysis to identify genes involved in the resistance response of soybean roots to SCN invasion. We constructed three cDNA libraries using RNA from cv. Peking resistant to SCN population NL1-Rhp (similar to race 3). Specifically, EST analysis was used to identify genes involved in the resistance response of soybean to SCN and to obtain genes specifically induced by SCN for gene expression profiling using microarrays. One library used RNA isolated from soybean roots 12 hours after invasion by SCN population NL1-Rhp; the second was made from pooled RNAs extracted from 2 and 4 days post invasion; and the third was made from RNA pooled after extraction from roots 6 and 8 days post invasion (22). Approximately 1,000 clones from each library were individually placed in wells of microtiter dishes to give each clone a unique address. This allowed later recovery of clones, the ability to track clones in microarray experiments, and for storing additional information in our database about each clone. A replicate plate was archived at –80°C for long-term preservation of the clones, while another replicate was used in the laboratory for one-pass DNA sequencing at the 5' end of the clone. The one-pass sequence provided DNA sequence information for computerized searches to determine if this gene or a similar gene had already been sequenced and thus to provide clues to the identity and function of the gene. The abundance of specific ESTs in a library can reflect the relative abundance of the gene's transcript, hence, the relative amount of its expression compared to that of other genes. Specific ESTs in each of these libraries were compared to those found in root libraries from the soybean EST database. ESTs representing genes encoding glutamine S-transferase and SAM-22 were among those elevated in the three libraries made from SCN-infected roots as compared to uninoculated roots. ESTs representing ascorbate peroxidase and lipoxygenase were more abundant in the 12-hour library compared to those in the 6- to 8-day library.

These results could be confirmed and extended by monitoring gene expression over time after SCN invasion using microarrays (Fig. 18.4). Expression analysis of more than a few genes was cumbersome until microarray technology recently provided a rapid method for measuring the expression of thousands of genes at the same time in a high-density, high-throughput format (23). By this method, thousands of genes are printed as spots of DNA on chemically treated microscope slides (Fig. 18.5). These spots may be DNA inserts from cDNA, or as oligonucleotides. They are printed in grids with each spot correlated with the clone address and DNA sequence. In our laboratory, gene expression profiles are obtained by isolating mRNAs from two related sources and labeling each with a different fluorescent molecule during reverse transcription to cDNA. The fluorescently labeled cDNAs are mixed and then hybridized to the spotted DNAs printed on the microscope slide. The two fluorescence signals are independently measured using a laser beam and detector for each gene. The ratio of fluorescence of each gene is calculated to determine if the gene is induced, suppressed, or unchanged between the sample treatments (Fig. 18.6).

Microarray technology has several advantages over other gene profiling methods. Microarrays offer (i) the ability to quantify transcription of thousands of genes in parallel at high sensitivity; (ii) a much more systematic approach to gene analysis; and (iii) easy replication by multiple spotting of clones within an array as well as replication of arrays within a single experiment. These advantages allow confirmation of gene expression and reduce error variance of transcript expression evaluation. Microarrays using cDNA inserts provide fast retrieval of archived, full-

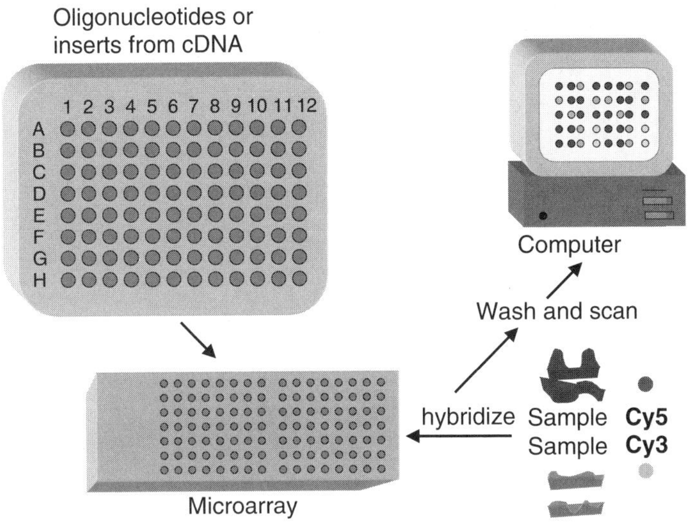

Figure 18.4. Experimental flow for a typical microarray experiment using cDNA inserts. The cDNA inserts are amplified by PCR in 96- or 384-well microtiter plates, then printed on chemically treated microscope slides. RNA from two biological sources is labeled during reverse transcription to cDNA with different fluorescent tags, mixed, then hybridized to the array. A scanner detects the amount of each fluorescence tag for each cDNA, and the data are stored in a computer and analyzed.

length or near full-length clones, which avoids their construction from PCR and RT-PCR or by library screening (i.e., colony hybridization) with short fragments. Microarrays using oligonucleotides offer well-defined probes that may distinguish among members of a gene family. Microarrays now make genome exploration more orderly because each clone has a unique address in a microtiter dish and is archived. A relational database can be constructed in an ordered and logical manner that allows other important information to augment gene expression data. New DNA sequences can be added to the microarray as required for further genome exploration. However, microarray technology is expensive, and it requires good database management, gene tracking, and bioinformatic support, as well as access to sequenced cDNA or oligonucleotides for array manufacture.

The first report describing the use of microarrays to monitor gene expression reported the expression of 45 Arabidopsis genes in parallel (23). Other early examples include monitoring genes in a transgenic line of Arabidopsis (24), monitoring over 1,000 human genes as they were subjected to heat stress (24), identifying genes uniquely expressed in human cancer cells (25), and surveying 864 yeast genomic DNA clones (26). In the plant world, Arabidopsis is at the forefront of gene expression profiling using microarrays.

Recently, microarrays constructed with soybean cDNA inserts were used to compare gene expression in soybean roots and shoots to demonstrate their utility

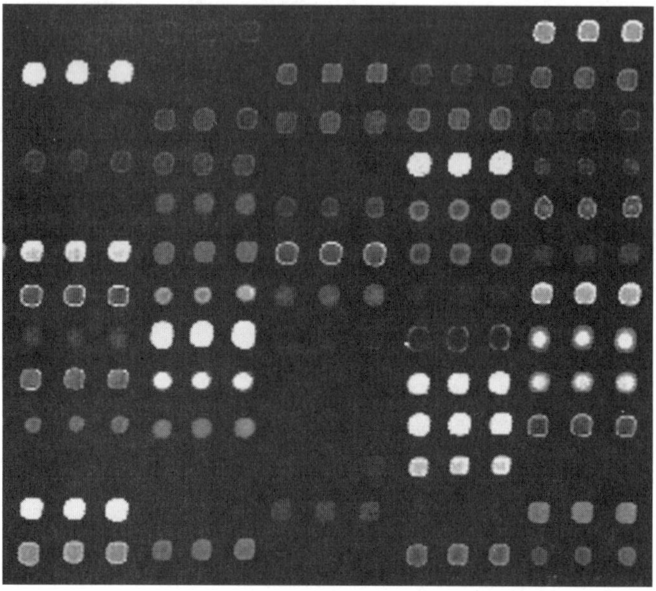

Figure 18.5. A grid of cDNA printed on a slide. Each cDNA insert was robotically printed in triplicate, then hybridized with fluorescently labeled cDNA representing the mRNA transcribed in the biological tissue.

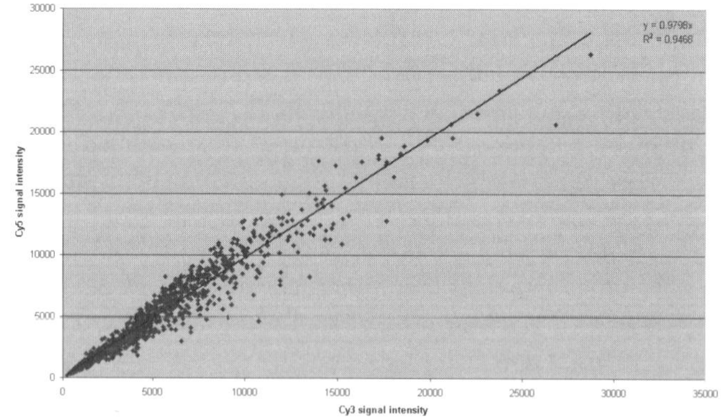

Figure 18.6. A plot of fluorescently labeled RNA from roots of cv. Kent two days after treatment with SNC compared to untreated control Kent roots. Genes with no change in transcript levels between the treated and control samples have a theoretical Cy5/Cy3 ratio of one. Statistically significant deviations from this ratio indicate an increase or decrease in transcript level.

using RNA extracted from these organs (27). Thibaud-Nissen *et al.* (28) examined gene expression patterns over time during somatic embryogenesis in soybean. Their results indicated that dedifferentiation of the cotyledon occurred before the appearance of somatic embryos on the adaxial side of the cotyledon after growth on auxin-containing medium. Preliminary experiments examining the expression of 468 soybean cDNAs in soybean roots two days after invasion by the SCN indicated possible increases in transcript levels of genes involved in phenylpropanoid synthesis and in a number of genes of unknown function (29).

To confirm and extend our EST and microarray analysis, my laboratory has initiated a time-course examination of gene expression in several cultivars of soybean under attack by different SCN genotypes in an attempt to identify genes expressed in both the susceptible and the resistance reaction by different races. The resistance reaction begins 2 to 4 days after SCN invasion; we therefore examined time points from 6 hours through 8 days post invasion. Inserts from over 3,000 cDNA clones from the aforementioned libraries were amplified by PCR and the products checked by electrophoresis for robustness of reaction and presence of only one amplification product. They were then printed in triplicate robotically on specially coated microscope slides for hybridization with fluorescently labeled cDNAs derived from RNA extracted from SCN-inoculated and -uninoculated roots of soybean. Preliminary interpretation of results from one time course experiment suggests that numerous genes involved in the defense response are increased in levels of expression in plants susceptible to SCN. Genes that were induced several days after invasion by SCN included those encoding enzymes involved in jasmonic acid production, cell wall rearrangement, signaling, plant defense, and transcription factors. For more information see http://bldg6.arsusda.gov/benlab/.

Our microarray data may identify genes important to the resistance response of soybean to SCN, thus providing targets and tools for broadening resistance. Over-expression of genes at or near the syncytium may decrease nematode survival and reproduction. Also, microarrays may identify genes important to syncytium formation and maintenance; these may be candidate target genes to silence, kill, or debilitate syncytial cells, thus starving the nematode. Also, genes identified that are expressed at and near the syncytium may have promoters useful for expressing DNA at and near the syncytium, allowing better control of genes expressed to control the nematode or alter syncytium cell vitality and function. Possibilities for controlling the nematode include killing the nematode, interfering with nematode feeding and digestion, and interfering with the nematode's life cycle. Tools developed through genomics will help attain these goals.

Using inserts from cDNA microarrays has both benefits and disadvantages. Use of cDNA library inserts identified by EST sequencing provides a fast and economical approach to obtaining DNA for microarray printing. However, the inserts often are only partially sequenced, with the result that some clones are chimeric, containing a portion of another cDNA. The chimera provides faulty data that will be unknown to the researcher. Custom oligomer arrays can substitute for cDNA microarrays, but they are expensive. Furthermore, little is known about the genome of most plant species. For laboratories working with plants less frequently studied than rice and Arabidopsis, using custom oligomer arrays may not be practical. For laboratories with access to oligomer microarrays, these arrays can often be designed to hybridize specifically to a gene family member, thus decreasing cross-hybridization problems that confound microarray analysis when cDNA inserts are used and thus providing high-quality data. Yet cDNA microarrays can be a useful and economical screening tool, providing good insights into the workings of a biological process.

Future Directions

The future promises to bring higher-throughput, faster, and more accurate analysis; use of smaller sample size; and use of fewer reagents. Integration of data obtained from multiple experimental tools through computer databases, analytical tools, and software packages that integrate multiple types of information will become standard. This means that genomics data will be integrated with metabolic data, protein data, and cellular and tissue information to develop a more complete biological picture at the organismal, cellular, protein, metabolite, and DNA levels. Techniques such as quantitative real-time PCR can be used to measure gene expression and to confirm microarray data by providing very accurate measurements. Protein profiling using two-dimensional polyacrylamide gel electrophoresis and other techniques provide information on translation of expressed genes, while metabolic profiles of cells and tissues can be integrated with gene and protein expression data to provide a better picture of what is happening. A large effort is being made in proteomics and metabolic profiling of *Medicago* (30–32), which may be useful to scientists studying other legumes such as soybean. Soybean transformation is still labor intensive, and procuring regenerated plants is a lengthy, tedious process; therefore, gene silencing and gene knockout technologies are not readily available in soybean. These technologies do, however, provide provocative opportunities for research in the future. Microscale analytical tools will allow the capture and processing of single cells and homogenous populations of cells. For example, laser capture microscopy, now in its infancy, will allow scientists to study the molecular contents of single cells (Fig. 18.7).

Data from genomic, proteomic and metabolomic research will provide information that will allow the development of strategies for controlling pests and pathogens and will identify likely targets for exerting control. Genes that are highly expressed at the nematode feeding sites will have promoters that may be useful to express transgenes or RNAi that can trigger the collapse and destruction of the syncytium or can effect starvation of the nematode by inhibiting nematode feeding and digestion, through the use of protease inhibitors, for example. Other targets may be genes involved in the nematode's reproduction and development of viable offspring, including cell division, development, movement, energy formation, and use; egg shell production; and larval hatching. Should these strategies be effective in control of the soybean cyst nematode in soybean, such strategies may be expanded to control other nematode pests of soybean, and of other crop plants, and perhaps even to control other pests and pathogens.

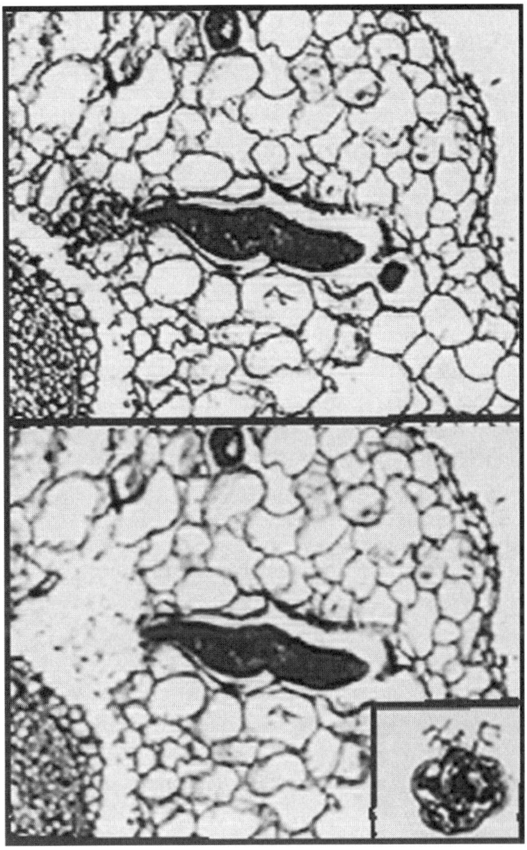

Figure 18.7. Laser capture microdissection of soybean root syncytial cells near the anterior portion of the SCN. The root samples were photographed (top panel); then cells were removed by LCM (bottom panel). Cells removed by LCM are shown in the insert.

Certainly, knowing the DNA sequence of the soybean genome would greatly promote advances in developing soybean with broader resistance to the soybean cyst nematode, to biotic and abiotic stresses in general, and to higher soybean productivity and value-added soybean for new markets. It would help traditional breeding by providing DNA sequences for marker-assisted selection and would provide information for faster development of promoters and gene candidates for transgenic modification of soybean.

Summary

Full application of genomics to soybean awaits full sequencing and annotation of the soybean genome. Bioinformatics tools, databases, ESTs, available portions of small sections of the genomes, and microarray analysis provide tools and a framework for application of the use of genomics to soybean breeding and to basic science. Ultimately, the soybean industry will benefit by having soybean with broader resistance to diseases and pests (such as SCN), by increased productivity, and by development of new soybean cultivars with value-added traits to increase the breadth of the soybean market.

Acknowledgments

This research was supported by USDA-CSREES project no.99-35302-8189 and United Soybean Board project no. 3214.

References

1. Wrather, J.A., Soybean Disease Loss Estimates for the United States 1996–2002, http://aes.missouri.edu/delta/research/soyloss.stm (2003; updated yearly).
2. Doupnik, B., Jr., Soybean Production and Disease Loss Estimates for North-Central United States from 1989 to 1991, *Plant Dis. 77*:1170–1171 (1993).
3. Riggs, R.D., and D.P. Schmitt, Optimization of the *Heterodera glycines* Race Test Procedure, *J. Nematol. 23*:149–154 (1991).
4. Niblack, T.L., R.D. Riggs, P.R. Arelli, G.R. Noel, C.H. Opperman, J.H. Orf, D.P. Schmitt, J.G. Shannon, and G.L. Tylka, A Revised Classification Scheme for Genetically Diverse Populations of *Heterodera glycines*, *J. Nematol. 34*:279–288 (2002).
5. Caldwell, B.E., C.A. Brim, and J.P. Ross, Inheritance of Resistance of Soybeans to Cyst Nematode, *Heterodera glycines*, *Agron. J. 52*:635–636 (1960).
6. Matson, A.L., and L.F. Williams, Evidence of a Fourth Gene for Resistance to Soybean Cyst Nematode, *Crop Sci. 5*:447 (1965).
7. Concibido, V.C., R.L. Denny, D.A. Lang, J.H. Orf, and N.D. Young, RFLP Mapping and Marker-Assisted Selection of Soybean Cyst Nematode Resistance in PI 209332, *Crop Sci. 36*:1643–1650 (1996).
8. Webb, D.M., B.M. Baltazar, A.P. Rao-Arelli, J. Schupp, K. Clayton, P. Keim, and W.D. Beavis, Genetic Mapping of Soybean Cyst Nematode Race-3 Resistance Loci in the Soybean PI 427.654, *Theor. Appl. Genet. 91*:574–581 (1995).
9. Concibido, V.C., D.A. Lange, R.L. Denny, J.H. Orf, and N.D. Young, Genome Mapping of Soybean Cyst Nematode Resistance Genes in 'Peking', PI 90763, and PI 88788 Using DNA Markers, *Crop Sci. 37*:258–264 (1997).
10. Cregan, P.B., T. Jarvik, A.L. Bush, R.C. Shoemaker, K.G. Lark, A.L. Kahler, N. Kaya, T.T. VanToai, D.G. Lohnes, J. Chung, and J.E. Specht, An Integrated Genetic Linkage Map of the Soybean Genome, *Crop Sci. 39*:1464–1490 (1999).
11. Weisemann, J.M., B.F. Matthews, and T.E. Devine, Molecular Markers Located Proximal to the Soybean Cyst Nematode Resistance Gene *Rhg4*, *Theor. Appl. Genet. 85*:136–138 (1992).
12. Matthews, B.F., M.H. MacDonald, J.S. Gebhardt, and T.E. Devine, PCR Markers Residing Close to the *Rhg4* Locus Conferring Resistance to Soybean Cyst Nematode Race 3 on Linkage Group A of Soybean, *Theor. Appl. Genet. 97*:1047–1052 (1998).
13. Cregan, P., J. Mudge, E.W. Fickus, L. Fredrick Marek, D. Danesh, R. Denny, R.C. Shoemaker, B.F. Matthews, K.G. Lark, and N.D. Young, Targeted Isolation of Simple Sequence Repeat Markers through the Use of Bacterial Artificial Chromosomes, *Theor. Appl. Genet. 98*:919–928 (1999).
14. Lewers, K., R. Heinz, H. Beard, L. Marek, and B. Matthews, A Physical Map of a Gene-Dense Region in Soybean Linkage Group A2 near the Black Seed Coat and *Rhg4* Loci, *Theor. Appl Genet. 104*:254–260 (2002).
15. Lewers, K.S, S.D. Nimalgoda, A.L. Warner, H.T. Knap, and B.F. Matthews, Physical Mapping of Resistant and Susceptible Soybean Genomes near the Soybean Cyst Nematode Resistance Gene *Rhg4*, *Genome 44*:1057–1064 (2001).

16. Mudge, J., P.B. Cregan, J.P. Kenworthy, W.J. Kenworthy, J.H. Orf, and N.D. Young, Two Microsatellite Markers That Flank the Major Soybean Cyst Nematode Resistance Locus, *Crop Sci. 37*:1611–1615 (1997).
17. Meksem, K., E. Ruben, D. Hyten, K. Triwitayakorn, and D.A. Lightfoot, Conversion of AFLP Bands into High-Throughput DNA Markers, *Mol. Genet. Genomics 265*:710–717 (2001).
18. Webb, D.M., U.S. Patent 5,491,081 (1996).
19. Webb, D.M., U.S. Patent 6,162,967 (2000).
20. Webb, D.M., U.S. Patent 6,538,175 (2003).
21. Hauge, B.M., M.L. Wang, J.D. Parsons, and L.D. Parnell, International Patent Number WO 01/51627 A2 (2001).
22. Alkharouf, N., R. Khan, and B. Matthews, Analysis of Expressed Sequence Tags from Roots of Resistant Soybean Infected by the Soybean Cyst Nematode, *Genome* (in press) (2004).
23. Shena, M., D. Shalon, R.W. Davis, and P.O. Brown, Quantitative Monitoring of Gene Expression Patterns with a Complementary DNA Microarray, *Science 270*:467–470 (1995).
24. Shena, M., D. Shalon, R. Heller, A. Chai, P.O. Brown, and R.W. Davis, Parallel Human Genome Analysis: Microarray-Based Expression Monitoring of 1000 Genes, *Proc. Natl. Acad. Sci. USA. 93*:10614–10619 (1996).
25. DeRisi, J., L. Penland, P.O. Brown, M.L. Bittner, P.S. Meltzer, M. Ray, Y. Chen, Y.A. Su, J.M. Su, and J.M. Trent, Use of a cDNA Microarray to Analyze Gene Expression Patterns in Human Cancer, *Nat. Genet. 14*:457–460 (1996).
26. Shalon, D., S.J. Smith, and P.O. Brown, A DNA Microarray System for Analyzing Complex DNA Samples Using Two-Color Fluorescent Probe Hybridization, *Genome Res. 6*:639–645 (1996).
27. Maguire, T.L., S. Grimmond, A. Forrest, I. Iturbe-Ormaetxe, K. Meksem, and P. Gresshoff, Tissue-Specific Expression in Soybean (*Glycine max*) Detected by cDNA Microarray Analysis, *J. Plant Physiol. 159*:1361–1374 (2002).
28. Thibaud-Nissen, F., R.T. Shealy, A. Khanna, and L.O. Vodkin, Clustering of Microarray Data Reveals Transcript Patterns Associated with Somatic Embryogenesis in Soybean, *Plant Physiol. 132*:118–136 (2003).
29. Matthews, B.F, Microarrays: Application to Soybean Gene Expression During SCN Invasion, in *Recent Advances in Phytochemistry*, edited by J. Romeo, J. Saunders, and B. Matthews, Elsevier Science, Oxford, 2001, Vol. 35, pp.15–30.
30. Watson, B.S., V.S. Asirvatham, L. Wang, and L.W. Sumner, Mapping the Proteome of Barrel Medic (*Medicago truncatula*), *Plant Physiol. 131*:1004–1123 (2003).
31. Asirvatham, V.S., B.S. Watson, and L.W. Sumner, Analytical and Biological Variances Associated with Proteomic Studies of *Medicago truncatula* by Two-Dimensional Polyacrylamide Gel Electrophoresis, *Proteomics 2*:960–968 (2002).
32. Huhman, D.H., and L.W. Sumner, Metabolic Profiling of Saponin Glycosides in *Medicago sativa* and *Medicago truncatula* Using HPLC Coupled to an Electrospray Ion-Trap Mass Spectrometer, *Phytochemistry 59*:347–360 (2002).

Chapter 19

Allergenic Reponses to Legume Proteins

Eliot M. Herman

Plant Genetics Research Unit, USDA/ARS, Donald Danforth Plant Science Center, 975 N. Warson Rd., St. Louis MO 63132

Introduction

Food allergies are recognized as a growing problem. Potentially allergenic compounds in milk, eggs, legumes, fruit, shellfish, fish, cereals, and tree nuts account for the vast majority of sensitivity to food. Overall, 5–8% of infants and children and 1–2% of adults (1,2) exhibit food allergies that range from mild rashes to severe, life-threatening anaphylaxis (3). The increased diversity of cuisines and food sources has created opportunities for much wider food choices, and with this there is a much greater potential to encounter a food that elicits a food allergy. An example of a new allergenic introduction is kiwi fruit. Kiwi fruit contains a potent allergen for about 5% of the population, who must practice avoidance because of sensitivity (4). However, avoidance of major foods such as wheat, dairy, eggs, and soybean that are most often cited as sources of food allergies is very difficult because these foods are a major component of most diets in industrialized countries. Other plant sources of allergens include peanuts, tree nuts, sesame, and sunflower. Although these foods are less prominent components of diets, they can provoke severe and sometimes life-threatening anaphylaxis (Fig. 19.1).

Legumes have a prominent role in the manifestation of food allergies, with significant allergies occurring from peanut, soybean, pea, chickpea, and lentil. Peanut sensitivity has a special prominence in the public perception of food allergies because of its potential to provoke fatal anaphylactic shock (5–7). In the United States each year there are about 30,000 emergency room visits and about 200 deaths from food allergies, with a large fraction from accidental peanut ingestion.

Manifestation of Clinical Allergies

The basic outline of an immunoglobulin E (IgE)-mediated allergy (8) is that the immune system's T cells direct B cells to produce antibodies with a specificity for the allergens that coat (arm) mast cells, basophils, and other cells with receptor sites for IgE. These cells are then primed to react to exposure to the allergen. When the sensitized individual encounters the allergen source upon subsequent exposures, the allergen cross-links cell-bound IgE, provoking the immediate release of histamine and other mediators that cause a cascade of physiological events leading to the overt response of a clinical allergy. Within minutes, respiratory distress, skin reactions, gastrointestinal distress, and changes

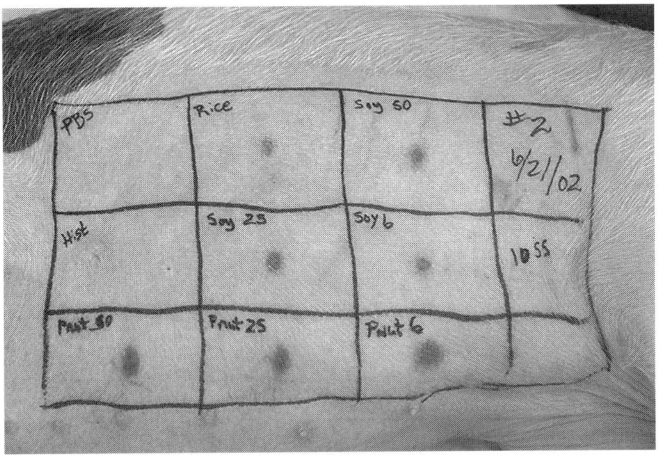

Figure 19.1. Demonstration of the atopic skin response of a swine previously sensitized with peanut and soybean proteins using subdermal injections into marked squares. The top row, from the left, is phosphate-buffered saline (PBS) control, rice protein, and 50 μg of soybean protein. The middle row, from the left, is histamine control, 25 μg of soybean, and 6 μg of soybean proteins. The bottom row, from the left, shows 50 μg, 25 μg, and 6 μg of peanut proteins. The red welts are the atopic response of the animal to the soybean and peanut proteins, which is an experimental model of hives that frequently results from ingestion of allergenic proteins.

in the cardiovascular system are manifested in this IgE-mediated allergic response. A systemic allergenic reaction or anaphylaxis can result that may be fatal if not appropriately treated (about three deaths per 10,000 people per year in the United States).

Food allergy is one of the several IgE-mediated abnormal immunoglobulin events that characterize hypersensitivity reactions, and it develops after exposure to an individual predisposed to develop sensitivity. There may be a significant genetic component, with allergies occurring in related people. Although there is much anecdotal evidence for a genetic predisposition for food allergies, there are relatively few data to support this. Among the best evidence for a genetic predisposition are twin studies showing that have shown siblings develop similar sensitivity to peanuts (9). For a predisposed individual, exposure that induces the allergy usually occurs at an early age. Studies have shown that peanut exposure during the first few years of life may induce life-long peanut sensitivity (3,7). Other studies have suggested that exposure may occur *in utero* as a consequence of the mother's consumption of peanuts (10). Considered together, the acquisition of peanut sensitivity is complex and involves both genetic and exposure parameters, some of which may occur prior to birth. Once sensitivity is acquired it tends to remain for peanuts but not necessarily for other legume allergens. Only about 20% of children with peanut sensitivity outgrow it by adulthood (6,11). In contrast, about 8% of U.S. children have soybean sensitivity but only 2% of adults remain sensitive. This indicates that 75% of soybean-sensitive children will outgrow this allergy. Other studies have indicated that perhaps 50% of soybean-sensitive babies and children will outgrow the sensitivity.

The sensitivity to food allergens also can be transferred from one person to another by tissue transplantation. Both bone marrow and liver transplants from peanut-sensitive donors have transferred peanut sensitivity to the recipient (12,13). While food, or ingestion, allergies appear to develop most often in children and persist into adulthood, there are also inhalation allergies of aeroallergens that appear to result from exposure at any age. The occupational allergies, such as to soybean, consist most often of the asthma that is associated with workers in agricultural transport and food production industries and that appears to develop later in life (14). The soybean inhalation allergy is directed at proteins different from those associated with the soybean ingestion allergy. Thus, this allergy develops only as the consequence of industrial exposure. The persistence of the inhalation allergies has not been studied.

Legume allergies have been documented for farm and companion animals. The allergies are primarily to soybean, which is extensively used as a component of animal feed. Soybean allergies are documented for neonatal pigs (15–21), salmon and trout in aquaculture (22), young cattle (23), and dogs (24–26). The allergy manifested in these animals mimics those of people with documented atopic skin reactions and gastrointestinal distress. The gastrointestinal distress induces histological changes in the digestive tract that affect growth rates. For farm animals, soybean allergies have economic consequences because of decreased weight gain of young animals and the increased feed-to-weight ratios. Soybean allergy in dogs causes similar distress resulting in veterinary visits and treatment. That animals exhibit responses to peanut and soybean allergens similar to those of people indicates that these animals could provide suitable models to investigate food allergies (27).

Preventing Allergic Reactions

The current primary treatment for food allergies is avoidance. Although this appears to be a simple solution, its implementation in the real world is far more complex. For example, soybean protein is widely used in processed foods destined for human and animal consumption. Add to this the wide distribution of milk (casein) and wheat in processed and prepared foods, and avoidance of these allergens by food-sensitive people is very difficult. Other allergens such as tree nuts and peanuts are more easily avoided because of their restricted use in most countries. However, accidental exposure to peanuts and tree nuts still results in many hospital visits and deaths. In the United States and Europe, a large percentage of processed foods contains soybean proteins that may or may not be labeled with its soybean origin (28). There is an increasing trend for content labeling, especially for products containing peanuts and tree nuts, because of the potential liability for compromised adverse reactions that may occur.

Peanut Allergens

The peanut ingestion allergy is one of the most prominent food allergies and causes thousands of emergency room visits and 100 or more deaths in the United States each year (6,7). Serum-specific IgE from peanut-sensitive individuals has been used to identify specific allergens and short linear amino acid sequences (epitopes) within these proteins as IgE-binding proteins. Peanut has seven distinct allergenic proteins (Ara h 1 to Ara h 7) that together include the vast majority of the proteins in seeds and include both storage protein families (29). Ara h 1 is a 614-kDa amino acid vicilin (7S) type storage protein that has 21 distinct epitopes that are recognized by IgEs from people with peanut sensitivity (30,31). The critical amino acids of each epitope have been determined, and substitution at one or more sites can abolish the IgE cross-reactivity. The 21 epitopes are distributed throughout the sequence from the amino-terminal to carboxy-terminal region. Ara h 1 has a single N-linked glycan on amino acid 516 (32). Ara h 2 is a 17.5-kDa glycoprotein that is a member of the conglutin family. Ara h 2 has 10 distinct epitopes that are recognized IgEs (33). Ara h 3 is a 507 amino acid member of the legumin family that has four distinct epitopes recognized by IgEs from peanut-sensitive people (34). Ara h 4 is another isoform of the legumin (35). Ara h 5 is a 131 amino acid profilin with an actin binding protein associated with the cytoskeleton (35). Ara h 6 and 7 are additional conglutin isoforms (35). The epitopes recognized by IgEs from peanut-sensitive people have not been determined for Ara h 4 to 7. In addition to Ara h 1 to 7, the 18-kDa oleosin oil body protein and the peanut agglutinin also may be minor allergens (36,37).

Soybean and Other Legume Allergens

Soybean ingestion allergies occur not only from obvious soybean foods such as tofu and soy milk, but also from a wide range of processed foods that contain soybean protein (38–42). Other exposures can occur from nonfood uses of soybean such as in cosmetics (43). The wide use of soybean-based baby formula exposes babies to soybeans and creates the potential for soybean sensitivity (44).

Soybeans possess as many as 15 different proteins that show some cross-reactivity to IgEs from sensitive people with ingestion allergy. Similar experiments identifying allergens in soybean-sensitive animals have not been reported. The major allergen is Gly m Bd 30K, which accounts for a large majority of the IgE cross-reactivity (45). Gly m Bd 28K (46) and Gly m Bd 60K (47) are secondary, but significant, allergens (Fig. 19.2). Gly m Bd 30K is an unusual member of the papain family of cysteine proteases that is accumulated to about 0.5% of the total protein of soybean seeds. This protein is also known as P34 and has the structural characteristics of a typical cysteine protease, except it lacks the cysteine in the catalytic site, and is therefore presumptively an inactive protease (48). Gly m Bd 60K is the beta subunit of conglycinin and is one of the major storage proteins of soybean (47). Gly m Bd 28K is a conglutin protein with a complex glycan side chain that also is recognized by IgEs from soybean-sensitive people (49). Soybean is unusual among seeds that provoke serious allergies because the storage proteins that account for the vast majority of the protein content are a minority of the soybean antigens recognized by IgEs from sensitive people.

P34 was recognized as a major soybean allergen by Ogawa *et al.* (40,45) who surveyed proteins that bound IgEs derived from soybean-sensitive people. When they determined the amino terminal sequence of an approximately 30-kDa IgE binding protein, they found it corresponded exactly with the P34 sequence previously determined by Kaliniski et al. (48,50). The human allergenicity of P34 is shown in SDS/PAGE and immunoblot of Fig. 19.2. The first panel shows a gel of the total protein from a mature soybean seed extract. The middle panel shows the cross-reactivity with an anti-P34 monoclonal antibody, resulting in the labeling of a single 34-kDa polypeptide. The third panel shows the consequences of probing a replicate blot with sera obtained from a soybean-sensitive person and labeling the bound IgE antibodies with a second antibody. Note that the immunodominant protein indicated by IgE binding is the 34-kDa P34 protein. The other, weaker IgE immunoreactive bands include the beta subunit of conglycinin and trypsin inhibitors.

P34/Gly m Bd 30K is not unique as a cysteine protease human allergen. The dust mite fecal allergen Der1p is also a cysteine protease that has been extensively characterized (51,52). The primary cysteine protease of kiwi fruit, actinidin, is a food allergen for sensitive people (4). Other plant cysteine proteases induce allergies through their use to tenderize food (53,54). An allergy to papain (55–57), the archetype cysteine protease, occurs in people from its uses in contact lens treatment solution where

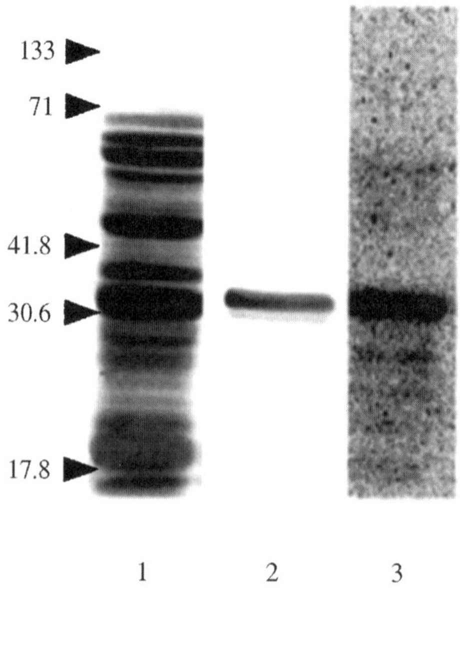

Figure 19.2. A SDS/PAGE immunoblot of total soybean proteins (lane 1) probed with a monoclonal antibody against P34/Gly m Bd 30K (lane 2) and sera from a pool of human baby with soybean sensitivity (lane 3). These blots show that P34/Gly m Bd 30K is the primary or immunodominant allergen.

it strips proteins from the lens (58–60), from exposure in fruit juices in throat lozenges (61), and from its use as a meat tenderizer (62). One model for allergenicity of the papain superfamily is that the proteolytic activity has a major role in inducing the immune response to these proteins (52,55,63). However, this would not apply to P34/Gly m Bd 30K because of its apparent lack of intrinsic protease activity.

The human immunology of P34/Gly m Bd 30K in soy-sensitive people has been investigated in great detail (64,65). The human IgE epitopes of P34/Gly m Bd 30K from soy-sensitive people has been determined to consist of 14 epitopes distributed across the mature P34/Gly m Bd 30K sequence. The individual epitopes vary in intensity, with some epitopes much "hotter" than others. The same procedure was used to examine immunological differences among unrelated soy-sensitive people. This study showed that although P34/Gly m Bd 30K is an immunodominant allergen, the individual IgE epitopes varied in intensity among these subjects, and some of the epitopes tended be stronger in most of the subjects (64). The immunodominant epitopes were further examined by producing glycine mutations within the epitope to map the critical amino acids (66). Among six epitopes examined, the antigenicity was entirely abolished by glycine substitution in any position in one epitope, whereas in the other eight epitopes there was only a single critical amino acid.

The human IgE epitopes of the 11S legumin storage protein glycinin have been examined. The acidic chain of glycinin shares some epitopes in common with the peanut allergen Ara h 3 (67). Eleven linear epitopes were mapped on the basic subunit of the 11S glycinin and modeled on a three-dimensional structure based on the phaseolin crystal structure (68). Of these 11 epitopes, only one could be converted to a form that did not bind IgE by sequential alanine substitution (66). The epitopes of the two other significant soybean allergens Gly m Bd 60K (a beta conglycinin) and Gly m Bd 28K (a conglutin) have not been reported. Gly m Bd 28K allergenicity includes IgEs directed at the complex glycan side chain (49).

Other legume allergens may also pose significant allergenic risks, especially in geographically discrete regions where these legumes can be a significant part of the diet. Lupine is used as an additive in wheat flour in parts of Europe. The IgEs of peanut-sensitive people were found to cross-react with lupine proteins, indicating that lupine and peanut allergens share epitopes (69). This indicates that individuals with peanut sensitivity also have the potential to have a severe reaction to wheat products containing lupine additive as a hidden allergen. Chickpea, cowpea, and lentil also are significant allergens in regions of the world where there is high consumption (70–72). Like peanut and lupine, it appears that storage proteins are among the major allergens. Unlike peanut and soybean, there are no detailed immunological studies on the allergenic proteins of these other legumes.

Allergens and Seed Oil

The question of whether seed oils, especially soybean and peanut oil, are allergenic has been investigated. This is particularly important for peanut allergy where there is a higher rate of anaphylaxis. In theory, vegetable oil should be protein-free and pose no risk. However, the situation is much more complex because of the availability of oil with varying levels of refinement. Tests with IgE samples from peanut- and soybean-sensitive people have shown no cross-reactivity with highly refined oil (73–77). However, there is evidence that less highly refined peanut oil contains sufficient levels of protein to pose a risk. One of the minor allergens of peanut is the 18-kDa peanut oleosin that is part of the seed's oil storage organelle and is likely to be present in less refined oil (37).

Food Processing Affects Allergenicity

Food processing has been shown to alter allergenic properties of legume proteins and the products derived from them. Food processing involves diverse treatments including heating and cooling; exposure to solvents, varying pH and salt concentrations; and fermentation, self-digestion, and enzyme treatments. The alterations to composition can be very complex and in most instances not well characterized. Each of these treatments can (and often does) alter the conformation of proteins and expose previously cryptic allergenic epitopes or, alternately, destroy previously existing allergenic epitopes.

For peanuts, the processing methods appear to be a significant factor in the development and risk of severe allergenic reactions. The incidence of peanut allergy is much greater in the United States than in China even though the consumption rate of peanuts is high in China (78). In the United States they are primarily dry roasted, while in China peanuts are primarily consumed after boiling or frying. Using sera samples from humans who are peanut sensitive, researchers showed that fried and boiled peanuts bind significantly less IgE for the Ara h 2 and Ara h 3 allergens compared to roasted peanuts (79,80). The fried and boiled peanuts also appeared to lose some of the Ara h 1 protein compared to the dry roasted peanuts (79,80). These results show that cooking methods have a major role in the overall allergenicity of peanuts, with dry roasting either exposing previously cryptic epitopes or creating significant new epitopes. The potential that new epitopes are created by dry roasting could be an important factor in the higher incidence of severe peanut sensitivity in the United States. Exposure of Chinese and American babies to boiled/fried or roasted peanuts, respectively, may have a major role in the different incidences in the two countries for development of severe peanut sensitivity. Important questions remain to be investigated. For example, are the stronger epitopes of roasted peanuts newly created by the dry roasting treatment; or are they the already characterized epitopes, but ones which are now either exposed by the roasting or hidden by boiling/frying? In other experiments, the IgE and IgG binding for both peanut and soybean was tested after thermal denaturation treatment and showed heat treatment did not abolish the antibody-binding capacity of these allergens (81).

A comparison between boiled and raw chickpeas and lentils showed there is no alteration of the IgE binding capacity (82,83). The chickpea allergens, like those of peanuts, include the major storage proteins. Boiled lentils maintain similar IgE binding patterns compared to raw lentils, but there was a decrease in the IgE binding capacity.

Food processing can also decrease allergenicity by either extracting the allergen during processing or modifying the protein so that the allergen is no longer recognized by the IgEs. The destruction of epitopes by protein fermentation digestion is one method by which food allergens can be inactivated. However, the use of fungi in the fermentation processes, especially *Aspergillus*, may introduce other allergens. Hence, people who can tolerate soybean products such as tofu cannot tolerate other fermented soybean products (84). The human immunodominant soybean ingestion allergen Gly m Bd 30K is differentially extracted in the preparation of textured vegetative protein when used as an extender/component of many processed foods (85). However, other studies have shown that Gly m Bd 30K is abundant in many soybean products prepared and processed by other techniques (41). This indicates that the removal of allergens by processing can be a narrow and specific event.

Occupational Legume Allergies

Inhalation allergies are among the most significant occupational disorders (14). The inhalation allergies are largely identified by either the occupation or the geographic location of individuals who suffer from the disorder. This type of allergy is referred to as baker's allergy or Barcelona syndrome for outbreaks of asthma that occurred during the 1980s among dockworkers and other exposed individuals. Flour and dust, especially from industrial quantities of soybeans and other legumes, raise aerosol particles that provoke these allergies (86). Inhalation allergies are not restricted to workers; at least one study showed that people living in the area adjacent to docks and loading areas who received similar exposure also experienced soybean dust allergies (87). Similar inhalation allergies may result from use of beanbag chairs (88). Although the baker's allergy and transportation-associated asthma are grouped together as inhalation allergies, there is evidence to indicate that these are two very different types of allergies and are caused by different components of the soybean seed. The Barcelona syndrome outbreak in the 1980s from soybean dust resulted in a large majority (over 80%) developing IgE antibodies directed at several low Mr (8 to 14 kDa) proteins (89–91). The 8-kDa hull aeroallergen has been purified (92). In a study in Argentina on people associated with the soybean production industry, additional high levels of Mr proteins were associated with occupational asthma. These proteins included a 50-kDa protein that is a member of the chlorophyll a/b binding protein family (93).

In contrast, baker's allergies appear to be directed at higher molecular weight Mr soybean proteins (94,95), including beta amylase and proteins similar to these that cause ingestion allergies. Crude soybean lecithin, a common additive to baked goods, is allergenic due to its protein content (73,96). However, differences have been noted. For example, the 20-kDa trypsin inhibitor (which is one of many secondary allergens in babies and children) is also a major allergen in the manifestation of baker's allergy (86). Because baker's allergy is induced primarily in adults as an occupational hazard, it is not known what role prior soybean/peanut exposure as a child may play in its development. Detailed immunological studies on the proteins and epitopes of the baker's allergy have not yet been conducted, and a comparison with the ingestion allergy of babies and children would be of interest.

The differences between the transportation industry soybean dust allergy and baker's allergy lie in the proteins that elicit the allergenic response. The dust from soybeans produced during shipping and handling may preferentially be sourced from the fragments of hulls. In particular, the low molecular weight Mr hydrophobic protein has been shown to be localized on the hull surface where it should be easily dislodged into aerosols (97). The lower molecular weight Mr proteins associated with the transportation-associated asthma should facilitate incorporation into the aerosols. Bakers and other food processors are exposed to flour, which is ground whole seeds. The hull constitutes only a small fraction of the total mass of seeds, and therefore, the primary exposure of bakers and food processors is to the entire complex of storage and other high Mr proteins. The difference in exposure agrees very well with the observation that these are two different inhalation allergies provoked by distinct allergens.

Hypoallergenic Seed Lines

Allergen null lines of allergenic crops would be desirable and could be used to produce specialty hypoallergenic products for sensitive people; for example, in soybean-based baby formula, in dog food, or in neonatal animal feed for swine and calves. Reduced or eliminated allergens in peanuts could provide an increased margin of safety for accidental ingestion that can cause death in highly sensitive people. The major varieties of peanut do not appear to differ in the content of the allergenic proteins (29); however, surveys of large collections have not been conducted. Because the major peanut allergens are caused by proteins that account for a large majority of the peanut's protein content, it is unlikely that a low or nonallergenic line will be found.

For soybeans, minor proteins account for the majority of the allergenicity. Hence, screening for naturally occurring variants in USDA soybean germplasm seed collection could lead to isolation of lines with reduced or modified allergenicity. Yaklich *et al.* (98) surveyed a small collection of diverse soybean cultivars and wild soybean relatives with immunoblots and a sera pool from soybean-sensitive people. The major soybean allergen Gly m Bd 30K was abundant in all domesticated soybeans, but the quantity of minor allergenic proteins showed some variability. A conglycinin null was identified (99), as well as a naturally occurring null for trypsin inhibitors (100). Among wild soybean relatives, the variability of allergenicity is far more extreme. Using sera from soybean-sensitive people to probe protein blots of

Glycine species, it was shown that the wild soybeans contain proteins that more intensely bind IgEs. This interesting and curious result hints that soybeans domesticated in Asia were selected for less allergenic properties. A large survey of the soybean collection for allergen content is in progress; this should indicate the range of variation of their allergenic content.

For other legumes with significant allergen problems, such as chickpea and lentil, there have not been surveys to examine the potential for finding naturally occurring variants of lower allergenicity. To the extent that these legumes (like peanuts) have the abundant storage proteins as the major allergens, there is a low probability of finding desirable variants.

Biotechnology and Allergens

The potential of any new food to induce allergic reactions has been seized upon by critics of genetically modified crops that contain foreign or engineered proteins. The potential for any introduced protein to be a new food allergen is difficult to predict. Furthermore, it is difficult to assay a new and/or engineered protein for its allergenic potential. In order to provide a basis to assess potential risks of introduced allergenicity, decision trees that outline protocols to guide the prediction of novel protein allergenicity have been proposed. However, these models still provide, at best, an educated guess of potential allergenicity. One of the primary methods for predicting an increased likelihood of allergenicity is to evaluate resistance of the protein to pepsin digestion by animal gut proteases representing simulated gastric fluid (SGC) or simulated intestinal fluid (SIF) as enzyme sources (101). However, resistance to digestion is only a single characteristic that increases the probability of a protein being an allergen, and results are not conclusive. Resistance to enzymatic degradation is not universal among known food allergens; for example, many of the allergens associated with oral allergy syndrome are labile and easily digested by enzyme treatment of unstable proteins.

To acquire sources of allergens for skin prick testing, a prick-prick test of fresh fruits to the skin of suspected allergic individuals is required. These tests are used by the World Health Organization to develop a widely adopted allergen risk assessment decision tree that involves asking questions about whether the protein exhibits various characteristics that might increase its probability of being an allergen. These include sequence-relationship to other known allergens as well the stability of the protein to gastric digestion (101). Such a decision tree, however, is only a guide and not a good rigorous predictor of allergenic potential or potency. Much has been learned about the protein sequence epitopes of food allergens by using epitope maps constructed with overlapping short peptides encompassing the entire protein. The linear epitopes recognized by IgEs from sensitive people have been mapped, and for most food allergens the linear epitopes consist of a few too many noncontiguous sequences that map to the surface of the allergenic proteins. A database of allergenic proteins and epitopes has been compiled, and public access is available on the Web (www.allergome.org/ and http://fermi.utmb.edu/SDAP/sdap_man.html). By using these data it is feasible to screen the known epitopes for similarity to the sequences of other proteins. The database is also useful in evaluating known proteins from related species. The problem with this approach is that it assesses only proteins or amino acid sequences previously identified as allergens or allergen epitopes and requires regular updating as new allergens are identified and characterized. However, much of the concern about allergens is with novel or engineered proteins, either those transferred from other organisms (many of which are not plants), or shifting a plant protein normally found in a nonfood plant or tissue to an edible organ, thus creating a new novel food not previously existing in nature. Developing tests of allergenic potential for all proteins presents difficulties. No existing characteristic or combination of characteristics has been identified that can be used to accurately predict the potential allergenicity of new or novel proteins because exceptions to each characteristic can be identified. Human testing of large numbers of proteins is not feasible because it would require identifying potentially sensitive people to be used as test subjects. Animal models would be useful, providing that controlled and standardized populations were used to conduct allergen tests, and there are efforts to develop animals as allergenic-potential test subjects. Rodents, dogs, and swine (25,27,102,103) are potential animal models, but much work remains before these subjects can accurately predict human responses to potential allergens. However, model systems are needed to guide development of prediction models to test the possible allergenicity of proteins that were not previously part of the food supply.

Allergy testing has become a critical part of the evaluation of new transgenic products. Current regulatory rules require allergenicity to be addressed as a broad part of the question of substantial equivalence. Soybeans possessing resistance to the herbicide Roundup® are now a large majority of the total planting in the United States. Analysis of Roundup Ready® soybeans for allergenic proteins has shown that transgenic soybeans do not differ from conventional soybeans. No additional allergenic proteins or alterations in the relative content of the intrinsic soybean allergens have been detected (104,105).

Introduction of an Allergen (Brazil Nut 2S Protein) into Soybean

There is only a single instance where a major allergen has been introduced into soybeans. Lore surrounding this experiment has become one of the most prominent examples for both the advocates and critics of biotechnology. In order to increase the methionine content of methionine-deficient soybeans, a gene encoding a high-methionine 2S storage protein that originated from Brazil nuts was inserted into soybeans. The resulting seeds contain increased amounts of this essential amino acid. Before commercialization occurred it was recognized that the Brazil nut 2S protein was a potent allergen that had the potential to induce anaphylaxis in sensitive people (106). The project was terminated prior to cultivar release because of the increased allergy risk, and thus no one was harmed or placed at risk. This project was undertaken in the early days of development of biotechnology and before allergy problems were routinely addressed. In the current regulatory climate it would not be possible to obtain approval to insert known serious allergens into soybeans or other grains to improve base composition. This and other examples of known allergens being scheduled to be inserted into food sources (but programs were discontinued) clearly demonstrates the usefulness of decision trees in greatly reducing the risk of introducing known allergens into our food supply. Individual biotechnologists, biotechnology companies, and regulatory agencies must be responsible for protecting our food from causing adverse allergic reactions.

Targeted Suppression of Food Allergens

It is feasible to dramatically reduce risk by eliminating or greatly reducing endogenous allergenicity by using genomics. There are two general types of targeted suppression of allergenicity that can be undertaken: epitope site-directed alterations to alter the allergenic portion of the molecule, and suppression of the entire allergenic molecule. Modification of each epitope with a substitution at each amino acid within an epitope has shown that abolishing IgE binding capability is feasible (30). Glycine or alanine substitutions have been shown to abolish the IgE immunoreactivity of the major peanut allergens (30) and the soybean Gly m Bd 30K (66). Using these data, it is feasible to engineer site-directed mutations to sequentially remove each allergenic epitope and gradually build a nonallergenic variant (107). Substituting a glycine or alanine places a neutral substitution into each site. Epitopes that lack bulky and/or charged side groups are much less likely to elicit an immunological response; however, exceptions do occur in which alanine or glycine substitution can increase IgE binding to short amino acid sequences. Using this approach has resulted in recombinant peanut allergens with reduced IgE binding (30). The resulting proteins have been produced in heterologous systems and shown not to cross-react with the IgEs from sensitive people. Such recombinant hypoallergenic variants have not been designed for substituting the disarmed allergen for the intrinsic protein. At present, swapping the engineered hypoallergenic protein for the intrinsic allergen has technical challenges that remain to be resolved. This would include the complete suppression of the intrinsic protein to eliminate the seed's allergenicity and then expressing the engineered protein at a level similar to that of the removed protein. Even with this in hand, there are other potential pitfalls for developing hypoallergenic plants. Prior attempts to alter seed proteins to improve amino acid composition for legumes have involved improving the sulfur deficiency of storage proteins (108,109). While it is feasible to engineer amino acid substitutions and to express this gene in transgenic plants, some of the engineered storage proteins are post-translationally unstable and degrade in the storage vacuoles shortly after synthesis (110). This is likely the result of the amino substitution altering the three-dimensional structure of the protein and thus rendering it susceptible to the action of proteases present in the storage vacuole (68).

Experiments designed to determine if epitope-disarmed storage proteins are stable in seeds of transgenic plants have not been accomplished.

Allergenic epitopes located on the protein's surface present a smaller risk of structural alteration than modification of interior amino acids that are critical to the protein's structure. Examples of the major storage proteins that have been crystallized to determine their structures have permitted the effects of amino acid substitution to be modeled. Knowing this information increases the probability that epitope destruction by amino acid substitution will maintain the native protein conformation. Even if substitution of the allergenic-disarmed protein for the intrinsic allergen were to be accomplished, assuring complete safety would require many tests. In particular, the risk associated with peanut exposure is high and caution in exposing highly sensitive people to low or non-allergenic peanuts would still be present. Hence, instead of non-allergenic peanuts, the engineered allergen-disarmed proteins were created to reduce the allergenicity of the proteins to states that will not cause anaphylactic episodes in individuals who may accidentally consume foods with hidden allergen sources or for hidden immunotherapy options.

Hypoallergenic variants of the peanut allergen could be effectively used for immunotherapy to desensitize highly sensitive individuals. Desensitizing individuals at high risk of anaphylaxis (111) would provide a larger margin of safety in cases of accidental ingestion. A mouse model has been tested and the disarmed peanut protein was shown to provide a measure of protection against anaphylaxis. Immunotherapy for people at risk for anaphylaxis could greatly reduce the recurring human costs of injury and death from accidental peanut exposure. Similar non-allergenic variants might be found in seed collections or result from induced mutations that could result in a safe peanut. Finding non-allergenic lines is unlikely because of the many different epitopes and a few different, but equally highly allergenic proteins. Hence, there is a small probability of finding a line or mutant that preserves the protein while eliminating its allergenic properties. Mutant lines may alter the levels of seed allergens, potentially increasing as well as decreasing levels. Takahashi *et al.* (112) hybridized two soybean mutants, one lacking the 7S storage family and the other lacking the 11S storage protein family. Results indicated an increase in the level of P34/Gly m Bd 30K, the major soybean allergen, in some progeny.

The alternative is complete removal of allergens by targeted suppression. There has been limited application of gene suppression technology to remove food allergens. A reduction of the level of a rice allergen was achieved by antisense technology. Although reduction of the wild-type level by two- to threefold was achieved, relatively high levels of the allergen still remained in the seeds (113).

The use of targeted allergen suppression has been successful in creating a soybean line that lacks the immunodominant human allergen Gly m Bd 30K (114). Sense co-suppression was used to produce a knockout of this seed allergen. P34/Gly m Bd 30K cDNA was cloned into a vector in the sense direction with the conglycinin promoter to control expression. Transformation was accomplished by particle bombardment of embryogenic suspension cells. After selection, embryos were regenerated and assayed for P34/Gly m Bd 30K content using a monoclonal antibody. Several lines with apparent suppression were identified as well as other lines with apparent over-accumulation. Soybean plants were regenerated and seeds of these plants produced lines with little or no apparent P34/Gly m Bd 30K content. Assays of T4 generation homozygous plants showed stable and virtually complete seed-specific suppression of P34/Gly m Bd 30K accumulation (Fig. 19.3). Soybean plants with suppressed seed-specific P34/Gly m Bd 30K have no apparent divergence from the normal phenotype other than the absence of the protein. P34/Gly m Bd 30K suppressed soybean seeds germinate, grow, mature, and complete reproduction without any apparent differences compared to the conventional type. The

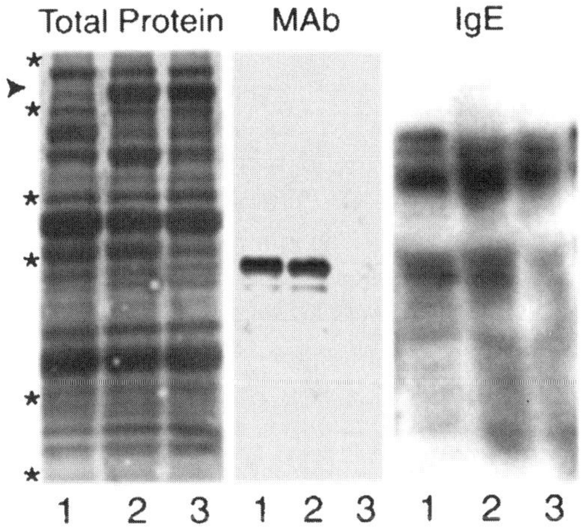

Figure 19.3. The suppression of P34/Gly m Bd 30K by sense co-suppression is shown in the adjacent blots. The three panels show replicate samples stained for total protein with amido black (left) and labeled with either anti-P34 monoclonal antibody (MAb) (middle) or with IgEs from a soybean-sensitive person (right). Within each panel Lane 1 is a transgenic control of another soybean suppressing the α/α' conglycinin (arrowhead) that does contain P34. Lane 2 is a soybean control for unaltered traits. Lane 3 is a third-generation homozygous transgenic soybean with the P34 construct inducing the complete suppression of P34 accumulation. Note the complete absence of P34 assayed by the monoclonal antibody and IgE cross-reactivity using sera from a soybean-sensitive person in Lane 3, showing that this transgenic lane has a complete suppression of the immunodominant soybean allergen.

subcellular structure of the protein storage vacuoles appears identical to that of the conventional type, except for the absence of P34/Gly m Bd 30K. The Gly m Bd 30K suppressed seeds were analyzed with two-dimensional electrophoresis and proteomic analysis for detection of collateral changes in protein content. Analysis of more than 1,400 different polypeptides in the wild-type and the Gly m Bd 30K suppressed seeds showed the only apparent change in the polypeptide composition was the targeted antigen. This indicates that the Gly m Bd 30K suppressed soybeans are substantially equivalent to the conventional soybeans and constitute a potential means to mitigate soybean food allergies in humans and animals. Research is in progress to test the effect of these soybeans on gastrointestinal problems that occur in neonatal pigs in order to produce a model system to study food allergies.

Summary

During the past 10 years there have been major developments in understanding and controlling legume allergies, which are among the most prominent and serious of the food and food-industry allergies. Allergens have been identified in the major commodities, and a great deal has been learned about the human immunology of the allergens. It is now recognized that farm and companion animals share similar, if not identical, allergies that have a large economic impact. The animal allergies may provide good models to study the human allergies, and by solving the problem of animal allergies their health and productivity will be improved. The solution to the problems associated with food allergies can, in large part, be found in utilizing biotechnology techniques to produce therapeutic reagents (such as epitope disarmed allergens) or to produce genetic knockouts of allergens. A recent report by the Pew Trust (115) pointed out how little of the total biomedical research resource is committed to the study of food allergens. However, considerable progress has been made with efforts from the few research projects. Hopefully, the support of new initiatives will have great success on preventing the life and economic impact of food allergies.

References

1. Sampson, H.A., and S.M. Scanlon, Natural History of Food Hypersensitivity in Children with Atopic Dermatitis, *J. Pediatr. 115*:23–27 (1989).
2. Sampson, H.A., and C.C. McCaskill, Food Hypersensitivity and Atopic Dermatitis: Evaluation of 113 Patients, *J. Pediatr. 107*:669–675 (1985).
3. Wood, R.A., The Natural History of Food Allergy, *Pediatrics 111*:1631–1637 (2003).
4. Pastorello, E.A., A. Conti, V. Pravettoni, L. Farioli, F. Rivolta, R. Ansaloni, M. Ispano, C. Incorvaia, M.G. Giuffrida, and C. Ortolani, Identification of Actinidin as the Major Allergen of Kiwi Fruit, *J. Allergy Clin. Immunol. 101*:531–537 (1998).
5. Skolnick, H.S., M.K. Conover-Walker, C.B. Koerner, H.A. Sampson, W. Burks, and R.A. Wood, The Natural History of Peanut Allergy, *J. Allergy Clin. Immunol. 107*:367–374 (2001).
6. Al-Muhsen, S., A.E. Clarke, and R.S. Kagan, Peanut Allergy: An Overview, *Can. Med. Assoc. J. 168*:1279–1285 (2003).

7. Burks, W., Peanut Allergy: A Growing Phenomenon, *J. Clin. Invest. 111*:950–952 (2003).
8. Kraft, D., F. Ferreira, S. Vrtala, H. Breiteneder, C. Ebner, R. Valenta, M. Susani, M. Breitenbach, and O. Scheiner, The Importance of Recombinant Allergens for Diagnosis and Therapy of IgE-Mediated Allergies, *Arch. Allergy Immunol. 118*:171–176 (1999).
9. Sicherer, S.H., T.J. Furlong, H.H. Maes, R.J. Desnick, H.A. Sampson, and B.D. Gelb, Genetics of Peanut Allergy: A Twin Study, *J. Allergy Clin. Immunol. 106*:53–56 (2000).
10. Frank, L., A. Marian, M. Visser, E. Weinberg, and P.C. Potter, Exposure to Peanuts in Utero and in Infancy and the Development of Sensitization to Peanut Allergens in Young Children, *Pediatr. Allergy Immunol. 10*:27–32 (1999).
11. Marshall, H., 20% of Sufferers Could Outgrow Peanut Allergy, *Trends Immunol. 22*:183 (2001).
12. Bellou, A., G. Kanny, S. Fremont, and D.A. Moneret-Vautrin, Transfer of Atopy Following Bone Marrow Transplantation, *Ann. Allergy Asthma Immunol. 78*:513–516 (1997).
13. Legendre, C., S. Caillat-Zucman, D. Samuel, S. Morelon, H. Bismuth, J.F. Bach, and H. Kreis, Transfer of Symptomatic Peanut Allergy to the Recipient of a Combined Liver-And-Kidney Transplant, *N. Engl. J. Med. 337*:822–825 (1997).
14. Baur, X., P.O. Degens, and I. Sander, Baker's Asthma: Still Among the Most Frequent Occupational Respiratory Disorders, *J. Allergy Clin. Immunol. 102*:984–997 (1998).
15. Bailey, M., B.G. Miller, E. Telemo, C.R. Stokes, and F.J. Bourne, Specific Immunological Unresponsiveness Following Active Primary Responses to Proteins in the Weaning Diet of Piglets, *Int. Arch. Allergy Immunol. 101*:266–271 (1993).
16. Barratt, M.E., P.J. Strachan, and P. Porter, Antibody Mechanisms Implicated in Digestive Disturbances Following Ingestion of Soya Protein in Calves and Piglets, *Clin. Exp. Immunol. 31*:305–312 (1978).
17. Dreau, D., J.P. Lalles, R. Toullec, and H.B. Salmon, T Lymphocytes are Enhanced in the Gut of Piglets Fed Heat-Treated Soyabean Proteins, *Vet. Immunol. Immunopathol. 47*:69–79 (1995).
18. Dreau, D., J.P. Lalles, C. LeJan, R. Toullec, and H. Salmon, Hypersensitivity to Soybean Proteins in Early Weaned Piglets: Humoral and Cellular Components, *Adv. Exp. Med. Biol. 37*:865–869 (1995).
19. Friesen, K.G., R.D. Goodband, J.L. Nelssen, F. Blecha, D.N. Reddy, P.G. Reddy, and L.J. Kats, The Effect of Pre- and Postweaning Exposure to Soybean Meal on Growth Performance and on the Immune Response in the Early-Weaned Pig, *J. Anim. Sci. 71*:2089–2098 (1993).
20. Li, D.F., J.L. Nelssen, P.G. Reddy, F. Blecha, J.D. Hancock, G.L. Allee, R.D. Goodband, and R.D. Klemm, Transient Hypersensitivity to Soybean Meal in the Early-Weaned Pig, *J. Anim. Sci. 68*:1790–1799 (1990).
21. Li, D.F., J.L. Nelssen, P.G. Reddy, F. Blecha, R. Klemm, and R.D. Goodband, Interrelationship Between Hypersensitivity to Soybean Proteins and Growth Performance in Early-Weaned Pigs, *J. Anim. Sci. 10*:4062–4069 (1991).
22. Nordrum, S., A.M. Bakke-McKellep, A. Krogdahl, and R.K. Buddington, Effects of Soybean Meal and Salinity on Intestinal Transport of Nutrients in Atlantic Salmon (*Salmo salar* L.) and Rainbow Trout (*Oncorhynchus mykiss*), *Comp. Biochem. Physiol. Biochem. Mol. Biol. 125*:317–335 (2000).
23. Gardner, R.W., M.G. Shupe, W. Brimhall, and D.J. Weber, Causes of Adverse Responses to Soybean Milk Replacers in Young Calves, *J. Dairy Sci. 73*:1312–1317 (1990).

24. Jeffers, J.G., E.K. Meyer, and E.J. Sosis, Responses of Dogs with Food Allergies to Single-Ingredient Dietary Provocation, *J. Am. Vet. Med. Assoc. 209*:608–611 (1996).
25. Buchanan, B.B., and O.L. Frick, The Dog as a Model for Food Allergy, *Ann. New York Acad. Sci. 964*:173–183 (2002).
26. Foster, A.P., T.G. Knowles, A.H. Moore, P.D. Cousins, M.J. Day, and E.J. Hall, Serum IgE and IgG Responses to Food Antigens in Normal and Atopic Dogs, and Dogs with Gastrointestinal Disease, *Vet. Immunol. Immunopathol. 92*:113–124 (2003).
27. Helm, R.M., R.W. Ermel, and O.L. Frick, Nonmurine Animal Models of Food Allergy. *Environ. Health Perspect. 111*:239–244 (2003).
28. Vidal, C., C. Perez-Carral, and B. Chomon, Unsuspected Sources of Soybean Exposure, *Ann. Allergy Asthma Immunol. 79*:350–352 (1997).
29. Koppelman, S.J., R.A. Vlooswijk, L.M. Knippels, M. Hessing, E.F. Knol, F.C. Reijsen, and C.A. Bruijnzeel-Koomen, Quantification of Major Peanut Allergens Ara h 1 and Ara h 2 in the Peanut Varieties Runner, Spanish, Virginia, and Valencia, Bred in Different Parts of the World, *Allergy 56*:132–137 (2001).
30. Burks, A.W., G. Cockrell, J.S. Stanley, R.M. Helm, and G.A. Bannon, Recombinant Peanut Allergen Ara h I Expression and IgE Binding in Patients with Peanut Hypersensitivity, *J. Clin. Invest. 96*:1715–1721 (1995).
31. Shin D.S., C.M. Compadre, S.J. Maleki, R.A. Kopper, H. Sampson, S.K. Huang, A.W. Burks, and G.A. Bannon, Biochemical and Structural Analysis of the IgE Binding Sites on Ara h 1, an Abundant and Highly Allergenic Peanut Protein, *J. Biol. Chem. 273*:13753–13759 (1998).
32. Kolarich, D., and F. Altmann, N-Glycan Analysis by Matrix-Assisted Laser Desorption/Ionization Mass Spectrometry of Electrophoretically Separated Nonmammalian Proteins: Application to Peanut Allergen Ara h 1 and Olive Pollen Allergen Ole e 1, *Anal. Biochem. 285*:64–75 (2000).
33. Stanley, J.S., N. King, A.W. Burks, S.K. Huang, H. Sampson, G. Cockrell, R.M. Helm, C.M. West, and G.A. Bannon, Identification and Mutational Analysis of the Immunodominant IgE Binding Epitopes of the Major Peanut Allergen Ara h 2, *Arch. Biochem. Biophys. 342*:244–253 (1997).
34. Rabjohn, P., R.M. Helm, J.S. Stanley, C.M. West, H.A. Sampson, A.W. Burks, and G.A. Bannon, Molecular Cloning and Epitope Analysis of the Peanut Allergen Ara h 3, *J. Clin. Invest. 103*:535–542 (1999).
35. Kleber-Janke, T., R. Crameri, U. Appenzeller, M. Schlaak, and W.M. Becker, Selective Cloning of Peanut Allergens, Including Profilin and 2S Albumins, by Phage Display Technology, *Int. Arch. Allergy Immunol. 119*:265–274 (1999).
36. Burks, A.W., G. Cockrell, C. Connaughton, J. Guin, W. Allen, and R.M. Helm, Identification of Peanut Agglutinin and Soybean Trypsin Inhibitor as Minor Legume Allergens, *Int. Arch. Allergy Immunol. 105*:143–149 (1994).
37. Pons, L., C. Chery, A. Romano, F. Namour, M.C. Artesani, and J.L. Gueant, The 18 kDa Peanut Oleosin is a Candidate Allergen for IgE-Mediated Reactions to Peanuts, *Allergy Suppl. 72*:88–93 (2002).
38. Bando, N., H. Tsuji, M. Hiemori, K. Yoshizumi, R. Yamanishi, M. Kimoto, and T. Ogawa, Quantitative Analysis of Gly M Bd 28K in Soybean Products by a Sandwich Enzyme-Linked Immunosorbent Assay, *J. Nutr. Sci. Vitaminol. 44*:655–664 (1998).
39. Herian, A.M., S.L. Taylor, and R.K. Bush, Allergenicity of Various Soybean Products as Determined by RAST Inhibition, *J. Food Sci. 58*:385 (1993).

40. Ogawa, T., N. Bando, H. Tsuji, H. Okajima, K. Nishikawa, and K. Sasaoka, Investigation of the IgE-Binding Proteins in Soybeans by Immunoblotting with the Sera of the Soybean-Sensitive Patients with Atopic Dermatitis, *J. Nutr. Sci. Vitaminol. 37*:555–565 (1991).
41. Ogawa, A., M. Samoto, and K. Takahashi, Soybean Allergens and Hypoallergenic Soybean Products, *J. Nutr. Sci. Vitaminol 46*:271–279 (2000).
42. Tsuji, H., N. Okada, R. Yamanishi, N. Bando, M. Kimoto, and T. Ogawa, Measurement of Gly m Bd 30K, a Major Soybean Allergen, in Soybean Products by a Sandwich Enzyme-Linked Immunosorbent Assay, *Biosci. Biotechnol. Biochem. 59*:150–151 (1995).
43. Shaffrali, F.C., and D.J. Gawkrodger, Contact Dermatitis from Soybean Extract in a Cosmetic Cream, *Contact Dermatitis 44*:51–52 (2001).
44. Heppell, L.M., J.W. Sissons, and H.E. Pedersen, A Comparison of the Antigenicity of Soya-Bean-Based Infant Formulas, *British J. Nutr. 58*:393–403 (1987).
45. Ogawa, T., H. Tsuji, N. Bando, K. Kitamura, Y.L. Zhu, H. Hirano, and K. Nishikawa, Identification of the Soybean Allergenic Protein, Gly m Bd 30K, with the Soybean Seed 34-kDa Oil-Body-Associated Protein, *Biosci. Biotechnol. Biochem. 57*:1030–1033 (1993).
46. Tsuji, H., M. Hiemori, M. Kimoto, H. Yamashita, R. Kobatake, M. Adachi, T. Fukuda, N. Bando, M. Okita, and S. Utsumi, Cloning of cDNA Encoding a Soybean Allergen, Gly m Bd 28K, *Biochim. Biophys. Acta 1518*:178–182 (2000).
47. Ogawa, T., N. Bando, H. Tsuji, K. Nishikawa, and K. Kitamura, Alpha-Subunit of Beta-Conglycinin, an Allergenic Protein Recognized by IgE Antibodies of Soybean-Sensitive Patients with Atopic Dermatitis, *Biosci. Biotechnol. Biochem. 59*:831–833 (1995).
48. Kalinski, A.J., D.L. Melroy, R.S. Dwivedi, and E.M. Herman, A Soybean Vacuolar Protein (P34) Related to Thiol Proteases which is Synthesized as a Glycoprotein Precursor During Seed Maturation, *J. Biol. Chem. 267*:12068–12076 (1992).
49. Hiemori, M., N. Bando, T. Ogawa, H. Shimada, H. Tsuji, R. Yamanishi, and J. Terao, Occurrence of IgE Antibody-Recognizing N-Linked Glycan Moiety of a Soybean Allergen, Gly m Bd 28K, *Int. Arch. Allergy Immunol. 122*:238–245 (2000).
50. Kalinski, A.J., J. Weisemann, B.F. Matthews, and E.M. Herman, Molecular Cloning of a Protein Associated with Soybean Oil Bodies which is Homologous to Thiol Proteases of the Papain Family, *J. Biol. Chem. 265*:13843–13848 (1990)
51. Topham, C.M., N. Srinivasan, C.J. Thorpe, J.P. Overington, and N.A. Kalsheker, Protein Comparative Modeling of Major House Dust Mite Allergen Der p I: Structure Validation Using an Extended Environmental Amino Acid Propensity Table, *Protein Eng. 7*:869–894 (1994).
52. Yasuhara, T., T. Takai, T. Yuuki, H. Okudaira, and Y. Okumura, Biologically Active Recombinant Forms of a Major House Dust Mite Group 1 Allergen Der f 1 with Full Activities of Both Cysteine Protease and IgE Binding, *Clin. Exp. Allergy 31*:116–124 (2001).
53. Nettis, E., G. Napoli, A. Ferrannini, and A. Tursi, IgE-Mediated Allergy to Bromelain, *Allergy 56*:257–258 (2001).
54. Soto-Mera, M.T., M.R. Lopez-Rico, J.F. Filgueira, E. Villamil, and R. Cidras, Occupational Allergy to Papain, *Allergy 55*:983–984 (2000).
55. Chambers, L., A. Brown, D.I. Pritchard, S. Sreedharan, K. Brocklehurst, and N.A. Kalsheker, Enzymatically Active Papain Preferentially Induces an Allergic Response in Mice, *Biochem. Biophys. Res. Commun. 253*:837–840 (1998).

56. Niinimaki, A., K. Reijula, T. Pirila, and A.M. Koistinen, Papain-Induced Allergic Rhino Conjunctivitis in a Cosmetologist, *J. Allergy Clin. Immunol. 92*:492–493 (1993).
57. Quarre, J.P., J. Lecomte, D. Lauwers, P. Gilbert, and J. Thiriaux, Allergy to Latex and Papain, *J. Allergy Clin. Immunol. 95*:922 (1995).
58. Bernstein, D.I., J.S. Gallagher, M. Grad, and I.L. Bernstein, Local Ocular Anaphylaxis to Papain Enzyme Contained in a Contact Lens Cleansing Solution, *J. Allergy Clin. Immunol. 74*:258–260 (1984).
59. Fisher, A.A., Allergic Reactions to Contact Lens Solutions, *Cutis 36*:209–211 (1985).
60. Santucci, B., A. Cristaudo, and M. Picardo, Contact Urticaria from Papain in a Soft Lens Solution, *Contact Dermatitis 12*:233 (1985).
61. Iliev, D., and P. Elsner, Generalized Drug Reaction Due to Papaya Juice in Throat Lozenges, *Dermatology 194*:364–366 (1997).
62. Mansfield, L.E., and C.H. Bowers, Systemic Reaction to Papain in a Nonoccupational Setting, *J. Allergy Clin. Immunol. 71*:371–374 (1983).
63. Deleuran, M., A.R. Ellingsen, K. Paludan, C. Schou, and K. Thestrup-Pedersen, Purified Der p1 and p2 Patch Tests in Patients with Atopic Dermatitis: Evidence for Both Allergenicity and Proteolytic Irritancy, *Acta Derm. Venereol. 78*:241–243 (1998).
64. Helm, R.M., G. Cockrell, E. Herman, A.W. Burks, H.A. Sampson, and G.A. Bannon, Cellular and Molecular Characterization of a Major Soybean Allergen, *Int. Arch. Allergy Immunol. 117*:29–37 (1998).
65. Helm, R.M., G. Cockrell, C. Connaughton, H.A. Sampson, G.A. Bannon, V.D. Beilinson Livingstone, N.C. Nielsen, and A.W. Burks, Soybean G2 Glycinin Allergen 1 Identification and Characterization, *Int. Arch. Allergy Immunol. 123*:205–212 (2000).
66. Helm, R.M., G. Cockrell, C.M. West, E.M. Herman, H.A. Sampson, G.A. Bannon, and A.W. Burks, Mutational Analysis of the IgE-Binding Epitopes of P34/Gly m1, *J. Allergy Clin. Immunol. 105*:378–384 (2000).
67. Beardslee, T.A., M.G. Zeece, G. Sarath, and J.P. Markwell, Soybean Glycinin G1 Acidic Chain Shares IgE Epitopes with Peanut Allergen Ara h 3, *Int. Arch. Allergy Immunol. 123*:299–307 (2000).
68. Lawrence, M.C., T. Izard, M. Beuchat, R.J. Blagrove, and P.M. Colman, Structure of Phaseolin at 2.2 A Resolution. Implications for a Common Vicilin/Legumin Structure and the Genetic Engineering of Seed Storage Proteins, *J. Mol. Biol. 238*:748–776 (1994).
69. Moneret-Vautrin, D.A., L. Guerin, G. Kanny, J. Flabbee, S. Fremont, and M. Morisset, Cross-Allergenicity of Peanut and Lupine: The Risk of Lupine Allergy in Patients Allergic to Peanuts, *J. Allergy Clin. Immunol. 104*:883–888 (1999).
70. Kalogeromitros, D., M. Armenaka, I. Galatas, O. Capellou, and A. Katsarou, Anaphylaxis Induced by Lentils, *Ann. Allergy Asthma Immunol. 77*:480–482 (1996).
71. Patil, S.P., P.V. Niphadkar, and M.M. Bapat, Chickpea: A Major Food Allergen in the Indian Subcontinent and its Clinical and Immunochemical Correlation, *Ann. Allergy Asthma Immunol. 87*:140–145 (2001).
72. Rao, T.R., D.N. Rao, K. Kotilingam, and R.R. Athota, Isolation and Characterization of Allergens from the Seeds of *Vigna sinensis*, *Asian Pac. J. Allergy Immunol. 18*:9–14 (2000).
73. Awazuhara, H., H. Kawai, M. Baba, T. Matsui, and A. Komiyama, Antigenicity of the Proteins in Soy Lecithin and Soy Oil in Soybean Allergy, *Clin. Exp. Allergy 28*:1559–1564 (1998).

74. Bush, R.K., S.L. Taylor, J.A. Nordlee, and W.W. Busse, Soybean Oil is not Allergenic to Soybean-Sensitive Individuals, *J. Allergy Clin. Immunol. 76*:242–245 (1985).
75. Errahali, Y., M. Morisset, D.A. Moneret-Vautrin, G. Kanny, M. Metche, J.P. Nicolas, and S. Fremont, Allergen in Soy Oils, *Allergy 57*:648–649 (2002).
76. Fremont, S., Y. Errahali, M. Bignol, and J.P. Metche, Allergenicity of Oils, *Allerg. Immunol. (Paris) 34*:91–94 (2002).
77. Hourihane, J.O., S.J. Bedwani, T.P. Dean, and J.O. Warner, Randomised, Double Blind, Crossover Challenge Study of Allergenicity of Peanut Oils in Subjects Allergic to Peanuts, *Brit. Med. J. 314*:1084–1088 (1997).
78. Beyer, K., E. Morrow, X.M. Li, L. Bardina, G.A. Bannon, A.W. Burks, and H.A. Sampson, Effects of Cooking Methods on Peanut Allergenicity, *J. Allergy Clin. Immunol. 107*:1077–1081 (2001).
79. Maleki, S.J., O. Viquez, T. Jacks, H. Dodo, E.T. Champagne, S.Y. Chung, and S.J. Landry, The Major Peanut Allergen, Ara h 2, Functions as a Trypsin Inhibitor, and Roasting Enhances this Function, *J. Allergy Clin. Immunol. 112*:190–195 (2003).
80. Maleki, S.J., S.Y. Chung, E.T. Champagne, and J.P. Raufman, The Effects of Roasting on the Allergenic Properties of Peanut Proteins, *J. Allergy Clin. Immunol. 106*:763–768 (2000).
81. Burks, A.W., L.W. Williams, W. Thresher, C. Connaughton, G. Cockrell, and R.M. Helm, Allergenicity of Peanut and Soybean Extracts Altered by Chemical or Thermal Denaturation in Patients with Atopic Dermatitis and Positive Food Challenges, *J. Allergy Clin. Immunol. 90*:889–897 (1992).
82. Martinez, S.I., M.I. Sandin, E. Fernandez-Caldas, F. Maranon, M.C. Munoz, M.C. Martinez, and M.T. Laso Borrego, The Diagnostic Value of Crude or Boiled Extracts to Identify Tolerant Versus Nontolerant Lentil-Sensitive Children, *Ann. Allergy Asthma Immunol. 86*:686–690 (2001).
83. Sanchez-Monge, R., C.Y. Pascual, A. Diaz-Perales, J. Fernandez-Crespo, M. Martin-Esteban, and G. Salcedo, Isolation and Characterization of Relevant Allergens from Boiled Lentils, *J. Allergy Clin. Immunol. 106*:955–961 (2000).
84. Tsuchiya, Y., K. Shimokata, H. Ohara, K. Nishiwaki, and T. Kino, Hypersensitivity Pneumonitis in a Soy Sauce Brewer Caused by *Aspergillus oryzae*, *J. Allergy Clin. Immunol. 91*:688–689 (1993).
85. Franck, P., D.A. Moneret Vautrin, B. Dousset, G. Kanny, P. Nabet, L. Guenard-Bilbaut, and L. Parisot, The Allergenicity of Soybean-Based Products is Modified by Food Technologies, *Int. Arch. Allergy Immunol. 128*:212–219 (2002).
86. Baur, X., M. Pau, A. Czuppon, and G. Fruhmann, Characterization of Soybean Allergens Causing Sensitization of Occupationally Exposed Bakers, *Allergy 51*:326–330 (1996).
87. Ballester, F., J.B. Soriano, I. Otero, M.L. Rivera, J. Sunyer, A. Merelles, H. Verea, J. Marin, and J.M. Anto, Asthma Visits to Emergency Rooms and Soybean Unloading in the Harbors of Valencia and A Coruna, Spain, *Am. J. Epidemiol. 149*:315–322 (1999).
88. Falleroni, A.E., and C.R. Zeiss, Bean Bag Allergy Revisited: A Case of Allergy to Inhaled Soybean Dust, *Ann. Allergy Asthma Immunol. 77*:298–302 (1996).
89. Codina, R., R.F. Lockey, E. Fernandez-Caldas, and R. Rama, Identification of the Soybean Hull Allergens Responsible for the Barcelona Asthma Outbreaks, *Int. Arch. Allergy Immunol. 119*:69–71 (1999).

90. Gonzalez, R., L. Zapatero, F. Caravaca, and J. Carreira, Identification of Soybean Proteins Responsible for Respiratory Allergies, *Int. Arch. Allergy Appl. Immunol. 95*:53–57 (1991).
91. Swanson, M.C., J.T. Li, P.E. Wentz-Murtha, W.L. Trudeau, E. Fernandez-Caldas, A. Greife, M.A. Rodrigo, F. Morell, and C.E. Reed, Source of the Aeroallergen of Soybean Dust: A Low Molecular Mass Glycopeptide From the Soybean Tela., *J. Allergy Clin. Immunol. 87*:783–788 (1991).
92. Codina, R., R.F. Lockey, E. Fernandez-Caldas, and R. Rama, Purification and Characterization of a Soybean Hull Allergen Responsible for the Barcelona Asthma Outbreaks. II. Purification and Sequencing of the Gly m 2 Allergen, *Clin. Exp. Allergy 27*:424–430 (1997).
93. Codina, R., L. Ardusso, R.F. Lockey, C.D. Crisci, C. Jaen, and N.H. Bertoya, Identification of the Soybean Hull Allergens Involved in Sensitization to Soybean Dust in a Rural Population from Argentina and N-terminal Sequence of a Major 50 KD Allergen, *Clin. Exp. Allergy 32*:1059–1063 (2002).
94. Quirce, S., F. Polo, E. Figueredo, R. Gonzalez, and J. Sastre, Occupational Asthma Caused by Soybean Flour in Bakers—Differences with Soybean-Induced Epidemic Asthma, *Clin. Exp. Allergy 30*:839–846 (2000).
95. Quirce, S., M. Fernandez-Nieto, F. Polo, and J. Sastre, Soybean Trypsin Inhibitor is an Occupational Inhalant Allergen, *J. Allergy Clin. Immunol. 109*:178 (2002).
96. Lavaud, F., D. Perdu, A. Prevost, H. Vallerand, C. Cossart, and F. Passemard, Baker's Asthma Related to Soybean Lecithin Exposure, *Allergy 49*:159–162 (1994).
97. Gijzen, M., S.S. Miller, K. Kuflu, R.I. Buzzell, and B.L.A. Miki, Hydrophobic Protein Synthesized in the Pod Endocarp Adheres to the Seed Surface, *Plant Physiol. 120*:951–960 (1999).
98. Yaklich, R., R. Helm, and E. Herman, Analysis of the Distribution of the Major Soybean Allergen in a Core Collection of *Glycine max* Accessions, *Crop Sci. 39*:1444–1447 (1999).
99. Hayashi, M., K. Harada, T. Fujiwara, and K. Kitamura, Characterization of a 7S Globulin Deficient Mutant of Soybean (*Glycine max* (L.) Merrill), *Mol. Gen. Genet. 258*:208–214 (1998).
100. Jofuku, K.D., R.D. Schipper, and R.B. Goldberg, A Frameshift Mutation Prevents Kunitz Trypsin Inhibitor mRNA Accumulation in Soybean Embryos, *Plant Cell 1*:427–435 (1989).
101. Astwood, J.D., J.N. Leach, and R.L. Fuchs, Stability of Food Allergens to Digestion in vitro, *Nat. Biotechnol. 14*:1269–1273 (1996).
102. Li, X.M., D. Serebrisky, S.Y. Lee, C.K. Huang, L. Bardina, B.H. Schofield, J.S. Stanley, A.W. Burks, G.A. Bannon, and H.A. Sampson, A Murine Model of Peanut Anaphylaxis: T- and B-cell Responses to a Major Peanut Allergen Mimic Human Responses, *J. Allergy Clin. Immunol. 106*:150–158 (2000).
103. Li, X.M., G. Kleiner, C.K. Huang, S.Y. Lee, B. Schofield, N.A. Soter, and H.A. Sampson, Murine Model of Atopic Dermatitis Associated with Food Hypersensitivity, *J. Allergy Clin. Immunol. 107*:693–702 (2001).
104. Burks, A.W., and R.L. Fuchs, Assessment of the Endogenous Allergens in Glyphosate-Tolerant and Commercial Soybean Varieties, *J. Allergy Clin. Immunol. 96*:1008–1010 (1995).
105. Padgette, S.R., N.B. Taylor, D.L. Nida, M.R. Bailey, J. MacDonald, L.R. Holden, and R.L. Fuchs, The Composition of Glyphosate-Tolerant Soybean Seeds is Equivalent to that of Conventional Soybeans, *J. Nutr. 126*:702–716 (1996).

106. Nordlee, J.A., S.L. Taylor, J.A. Townsend, L.A. Thomas, and R.K. Bush, Identification of a Brazil-Nut Allergen in Transgenic Soybeans, *N. Eng. J. Med. 334*:688–692 (1996).
107. Ferreira, F., C. Ebner, B. Kramer, G. Casari, P. Briza, A.J. Kungl, R. Grimm, B. Jahn-Schmid, H. Breiteneder, D. Kraft, M. Breitenbach, H.J. Rheinberger, and O. Scheiner, Modulation of IgE Reactivity of Allergens by Site-Directed Mutagenesis: Potential use of Hypoallergenic Variants for Immunotherapy, *FASEB J. 12*:231–242 (1998).
108. Hoffman, L.M., D.D. Donaldson, and E.M. Herman, A Modified Storage Protein is Synthesized, Processed, and Degraded in Seeds of Transgenic Plants, *Plant Mol. Biol. 11*:717–729 (1988).
109. Kjemtrup, S., E.M. Herman, and M.J. Chrispeels, Correct Posttranslational Modification and Stable Vacuolar Accumulation of Phytohemmagglutinin Engineered to Contain Multiple Methionine Residues, *Eur. J. Biochem. 226*:385–391 (1994).
110. Pueyo, J.J., M.J. Chrispeels, and E.M. Herman, Degradation of Transport-Competent Destabilized Phaseolin with a Signal for Retention in the Endoplasmic Reticulum Occurs in the Vacuole, *Planta 196*:586–596 (1995).
111. Nelson, H.S., J. Lahr, R. Rule, S.A. Bock, and D.Y.M. Leung, Treatment of Anaphylactic Sensitivity to Peanuts by Immunotherapy with Injections of Aqueous Peanut Extract, *J. Allergy Clin. Immunol. 99*:744–751 (1997).
112. Takahashi, M., Y. Uematsu, K. Kashiwaba, K. Yagasaki, M. Hajika, R. Matsunaga, K. Komatsu, and M. Ishimoto, Accumulation of High Levels of Free Amino Acids in Soybean Seeds Through Integration of Mutations Conferring Seed Protein Deficiency, *Planta 217*:577–586 (2003).
113. Tada, Y., M. Nakase, T. Adachi, R. Nakamura, H. Shimada, M. Takahashi, T. Fujimura, and T. Matsuda, Reduction of 14-16 kDa Allergenic Proteins in Transgenic Rice Plants by Antisense Gene, *FEBS Lett. 391*:341–345 (1996).
114. Herman, E.M., R. Helm, R. Jung, and A.C. Kinney, Targeted Gene Silencing Removes an Immunodominant Allergen from Soybean Seeds, *Plant Physiol. 132*:36–43 (2003).
115. Bucchini, L., and L.R. Goldman, A Snapshot of Federal Research on Food Allergy: Implications for Genetically Modified Food, Pew Initiative on Food and Biotechnology, http://pewagbiotech.org/research/allergy.pdf (2002).

Index

T